Aeronomy

PART A

AERONOMY

PART A

P. M. Banks
Department of Applied Physics
and Information Science
University of California
San Diego, California

G. Kockarts
Institut d'Aéronomie Spatiale
de Belgique
Bruxelles, Belgium

ACADEMIC PRESS New York and London 1973

A Subsidiary of Harcourt Brace Jovanovich, Publishers

ACADEMIC PRESS, INC.
111 Fifth Avenue, New York, New York 10003

United Kingdom Edition published by
ACADEMIC PRESS, INC. (LONDON) LTD.
24/28 Oval Road, London NW1

LIBRARY OF CONGRESS CATALOG CARD NUMBER: 72-88332

PRINTED IN THE UNITED STATES OF AMERICA

Contents

Chapter 4. **The Transition from the Homosphere to the Heterosphere**

Chapter 5. **Temperature Gradient in the Thermosphere**

Chapter 6. **Photoabsorption**

Chapter 7. **Solar Radiation and Photoionization**

Chapter 8. **Photodissociation**

Chapter 9. **Collision Processes**

Chapter 10. **Aeronomic Reactions**

Chapter 11. **Oxygen and Ozone**

Chapter 12. **Nitrogen and Nitrogen Oxides**

Chapter 13. Hydrogen Compounds and Their Effects in the Chemosphere

Preface

Aeronomy is the scientific discipline devoted to the study of the composition, movement, and thermal balance of planetary atmospheres. The need for such an area of specialization has developed during the past twenty years as sophisticated experiments have revealed the complexities of the upper atmospheres of the earth and other planets.

As a field of research, aeronomy demands an understanding of the basic concepts of both chemistry and physics as applied to a highly rarefied medium composed of neutral and charged particles. The first major accomplishments of aeronomy were associated with attempts to understand basic photochemical processes in the earth's upper atmosphere. Additional studies emphasized the importance of molecular diffusion and its effect in producing the transition from homogeneous molecular gases of the lower atmosphere to the hetrogeneous atomic and molecular gases of the upper atmosphere. As might be expected, progress in aeronomy has frequently been linked with advances in laboratory measurements or the development of new concepts in allied scientific specialities.

In this text we have attempted to make a comprehensive exposition of the basic processes involved in the aeronomy of the earth's upper atmosphere. Accordingly, it has been necessary to reach a balance between generalized descriptions of different phenomena in terms of various models and the actual observed behavior. It has been our goal to emphasize the fundamental character of the upper atmosphere and ionosphere without detailed descriptions of global morphology. Consequently, considerable discussion has been given to chemical and ionic reactions and to the different collisional interactions involving particles and radiation. The data presented have then been used to provide quantitative reaction rates, cross sections, and interaction coefficients of aeronomical interest which reflect current best values and possible ranges consistent with present knowledge.

Owing to the interrelatedness of many aeronomical problems the question of self-consistency of the different atmospheric parameters is always important; i.e., when certain processes are examined it is necessary to question their relationship to the overall atmospheric structure. To avoid as much as possible an inconsistent choice of densities, temperatures, and similar parameters, the calculations given in Part A of this text are largely based upon the sets of atmospheric models described in Chapter 3, while those in Part B are based on Chapter 14 and Appendix B. Such an approach is somewhat unique and reflects the large number of calculations which have gone into the preparation of the text material and figures. This approach is consistent with our desire to present descriptions of basic processes rather than predictions of absolute values.

The material of this text has been divided into two volumes. The first volume is principally concerned with the characteristics of the atmosphere as a whole and a presentation of the outstanding problems of chemical aeronomy. To assist later discussion, the topics of photoabsorption, solar radiation, photodissociation, collision processes and aeronomic reactions are introduced. Next, the chemistry of the atmospheric regions below 100 km is discussed with emphasis on oxygen, ozone, nitrogen compounds, and the effects of hydrogen compounds.

The second volume is concerned with phenomena which occur, for the most part, above 100 km in the heterosphere. The first topics are concerned with the neutral atmosphere and include the behavior of the thermosphere, the derivation of atmospheric models, and the formation of the exosphere. With regard to charged particles, subsequent chapters are devoted to the formation of the ionosphere, the basic ideas of plasma transport and the plasma thermal balance.

Acknowledgments

Portions of the present text are based on manuscripts prepared by Professor M. Nicolet between 1952 and 1965 and by the present authors and Professor Nicolet in the period 1967 to 1971. The development of topics in the first chapters parallels those originally given in the earlier manuscripts. The majority of the material, however, supercedes the earlier versions and represents work through March, 1972. In recognition of Professor Nicolet's long interest in this text, and considering his role in the development of the field, it was felt that the title "Aeronomy," taken from the original manuscript, would be appropriate.

A work such as this cannot be prepared without extensive assistance. Our indebtedness to Professor Nicolet for his years of assistance and association is gratefully acknowledged. In addition, Dr. Eldon Ferguson provided valuable advice and data for Chapter 10 and subsequent calculations. Dr. J. R. Doupnik and Professor A. F. Nagy gave considerable time to reading and criticizing the manuscript in various stages of development. Dr. C. Y. Johnson and others provided many of the experimental data appropriately acknowledged in the various chapters.

The assistance provided by Miss Linda Simpson, Mrs. Carole Gonzales, and Mrs. Lynn Love at the University of California, San Diego was invaluable and helped to ease an otherwise impossible burden. In Belgium, Mrs. Marie-Thérèse Werpin and Miss Marguerite De Clercq typed the first versions of the older manuscripts. The numerous figures were carefully drawn by Arthur Simon and Jacques Schmitz while Maurice Jaumin prepared and checked the references cited in the text. Emile Falise, Jean-Claude Crop, and Marcel Rosseeuw wrote several computer programs necessary for the preparation of the tables and graphs. To the many others who helped in various ways we wish to express our sincere gratitude. In addition, we note that the responsibility for errors and omissions in the text must remain with us.

To our wives and children we must apologize for the long periods of time spent in preparation of the manuscript. Their understanding of our goals and needs provided us with the strength to continue a difficult task.

Contents of Part B

CHAPTER 1

General Introduction

1.1 Atmospheric Regions

Our present knowledge of the upper atmosphere leads to a nomenclature which is useful in the analysis of aeronomic phenomena (see *Figs. 1.1* and *1.2*). By analogy with the terms troposphere, tropopause, and stratosphere, S. Chapman developed an extended nomenclature based on the variation of temperature with altitude.

The *troposphere* is the lowest atmospheric region and begins at the earth's surface, which acts as a source of heat resulting from absorption of the visible sunlight. The atmospheric temperature decreases upwards to a level called the *tropopause*, the average altitude of which varies from about 18 km at the equator to 8 km at the geographical poles. The general distribution of temperature at the tropopause is such that a minimum value is found at the equator and a maximum value in the polar regions. Variations on the order of 5 km in the level of the tropopause in middle latitudes are related to changes in atmospheric structure associated with the development of high and low pressure systems.

The troposphere is characterized by its negative vertical temperature gradient (about 6° km^{-1}) which leads to temperatures between 190°K (equator) and 220°K (high latitudes) at the tropopause. The relative composition of the air in the troposphere is constant as far as the principal constituents are concerned.

The *stratosphere* lies above the tropopause and is a region of increasing temperature up to a maximum of about 270°K near 50 km; i.e., at the *stratopause*. It was widely thought after the discovery of the change in temperature gradient (about 1900, by deBort) that the stratosphere was a calm region and subject to molecular diffusion. It is now known that turbulence exists and atmospheric mixing to the extent of a constant relative composition for the principal constituents has been proved by various observations. The general circulation of the stratosphere is not completely known. The essential character of stratospheric winds corresponds to a change from westerly winds in winter to easterlies in summer. In contrast

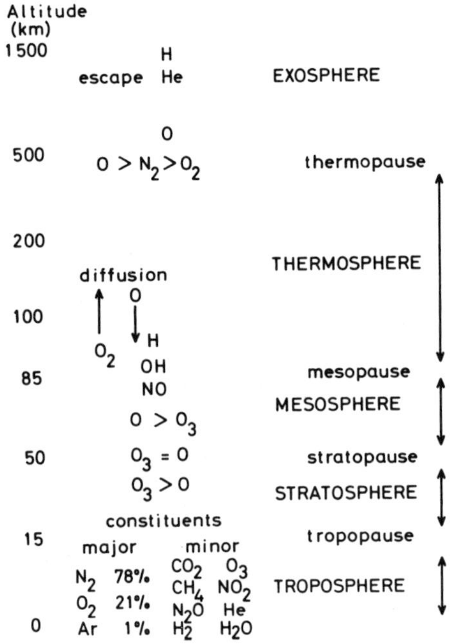

Fig. 1.1 Atmospheric nomenclature.

to the equatorial tropopause, the temperature of the equatorial stratopause is greater than that of the winter polar stratopause. For this reason stratospheric wind speeds increase above a height of 25 km. Since there is a seasonal variation in the temperature in the upper stratosphere at high latitudes, the altitude of the stratopause (50 ± 5 km) as well as its temperature (270 ± 20°K) must vary.

Although atmospheric mixing is present in the stratosphere, aeronomic reactions affecting minor constituents result in variations in concentrations of these gases which differ from the mixing distributions. Ozone is particularly important in this regard since it absorbs ultraviolet radiation harmful to life at the earth's surface.

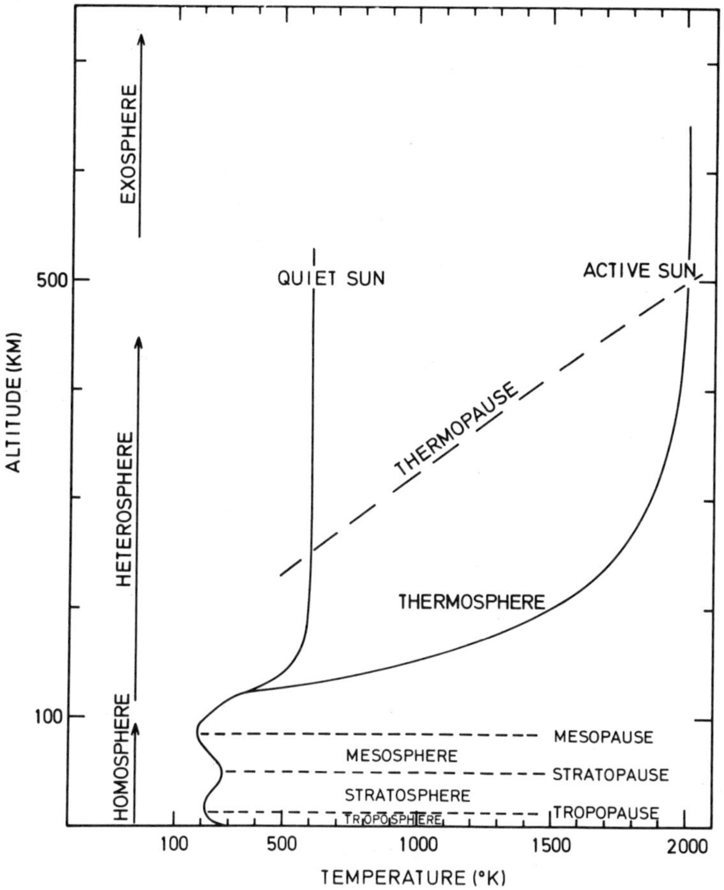

Fig. 1.2 Vertical distribution of temperature in the homosphere and heterosphere.

By analogy with the troposphere (negative temperature gradient) and stratosphere (positive temperature gradient) the *mesosphere* is a region of decreasing temperature between the stratopause and the *mesopause*, with a minimum of temperature found at 85 ± 5 km. Since the direction of the thermally driven wind in the mesosphere is opposite to that of the stratosphere, the speed of easterlies in summer and westerlies in winter must reach a maximum in the mesosphere. Furthermore, the atmospheric conditions at the mesopause are similar to those at the tropopause.

Because of oxidation processes and penetration of ultraviolet radiations which dissociate polyatomic molecules, the mesosphere is a more complex region than the stratosphere or troposphere. The chemical reactions which result are manifested in airglow emissions and the formation of a diverse collection of minor constituent gases which can be important in the thermal balance and ionization state of the mesosphere. Nevertheless, the three lowest regions of the terrestrial atmosphere have the same hydrodynamic character, and constitute the *homosphere* within which the mean molecular mass does not change significantly. Differences do exist, however, since the heat sources and losses are different in the three regions. Thus, while the earth's surface is the principal source of heat for the troposphere, the heat budget in the stratosphere is related to the absorption of solar ultraviolet radiation by ozone and to atmospheric infrared emission.

The region above the mesopause is called the *thermosphere*. Here, the behavior of the atmosphere changes dramatically (*Fig. 1.3*). Atomic oxygen becomes a permanent constituent of the atmosphere and there is a positive temperature gradient leading to temperatures greater than 1000°K. The large temperature gradient results from the absorption of solar ultraviolet energy and the lack of any effective means of thermal emission. Since the gradient of temperature is associated with the absorption of ultraviolet radiations by the atmosphere, the increase of temperature eventually becomes small. This limit is called the *thermopause*.

The fact that the gradient of temperature depends on the absorption of solar ultraviolet radiation has many consequences. The heat budget of the thermosphere is related to the variations of the solar energy and diurnal, day-to-day, seasonal, geographical, annual, and undecennial variations are necessarily involved. In addition, the fundamental processes of heat transport, convection, radiation, and conduction are of basic importance taken in conjunction with the change in molecular mass resulting from diffusion. Following the usual convention, the region of the upper atmosphere where mixing of the constituents by winds and dissipative turbulence does not greatly affect the vertical distribution of individual gases is called the *heterosphere* (see Fig. 1.3). The term *homosphere* is applied to the lower altitude regions where the rate of mixing of the atmospheric gases is sufficiently rapid to produce a uniform relative composition for the major constituents (N_2, O_2, Ar, CO_2).

Finally, it is necessary to distinguish between the possibilities by which the thermosphere may gain heat from the sun. These include ultraviolet radiation (the most important process), effects associated with solar disturbances, and heating resulting from external conduction and corpuscular radiation.

Since diffusion enhances the concentrations of light constituents (atoms) relative to the heavy constituents (molecules), one expects that the principal

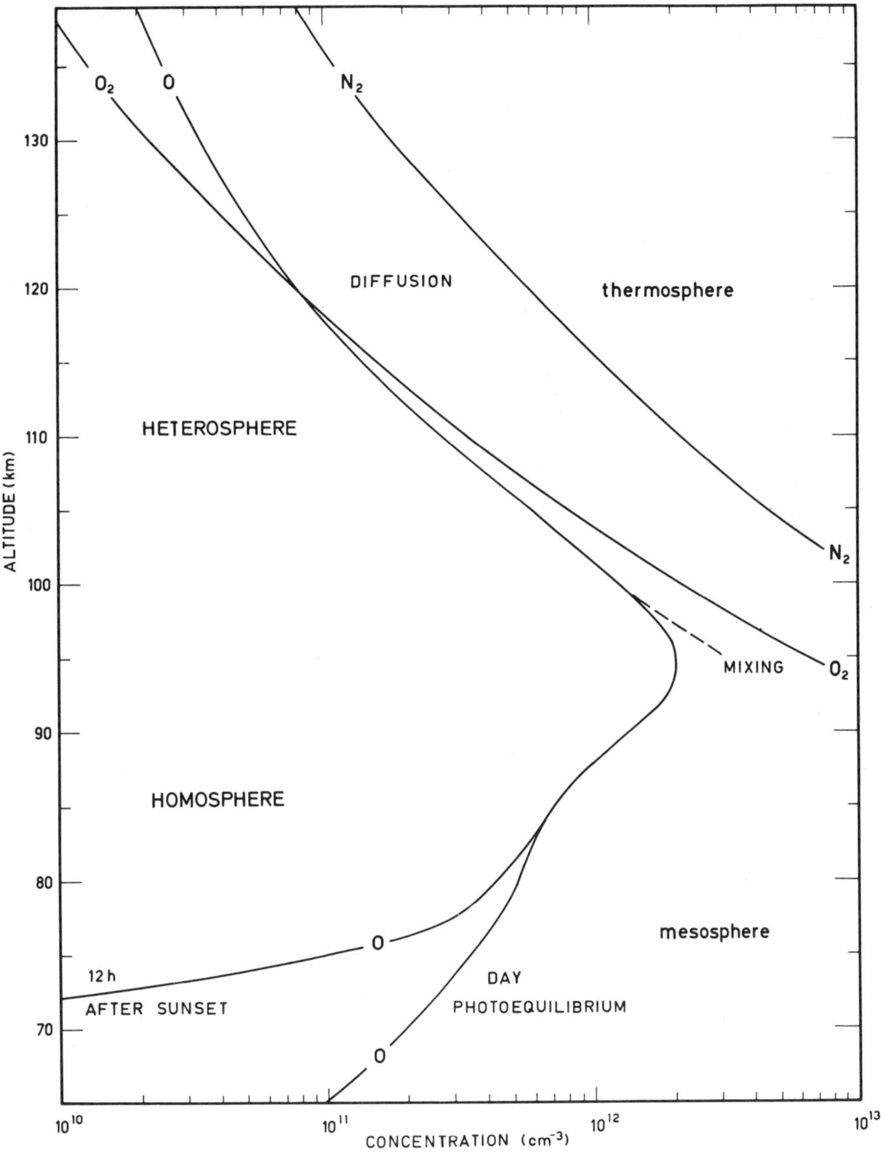

Fig. 1.3 Transition from the homosphere to the heterosphere. Dissociation of molecular oxygen and the transition from mixing to diffusion conditions are also shown.

atmospheric constituents above a certain altitude are atoms. If oxygen and nitrogen were the only constituents of the atmosphere, it would be possible to consider a simple model of the neutral atmosphere extending up to the limit given by the point at which gravity is balanced by centrifugal force. It is found, however, that atomic hydrogen and helium are present so that these gases, having a small atomic mass, predominate above a certain altitude with their concentrations decreasing much more slowly with height than those of oxygen and nitrogen. These atoms, through their thermal kinetic energy, are able to escape from the terrestrial atmosphere when they are in the *exosphere*, i.e., a region where collisions are not frequent. In this case, a hydrostatical model cannot be applied and dynamical processes must be introduced to take into account the effect of particle escape.

1.2 Atmospheric Density

Only a limited number of measurements of atmospheric density have been carried out above 100 km using rockets. Many studies, however, have been made of the orbital parameters of artificial satellites. Such data provide information on air drag between 200 and 1500 km and permit calculations of the atmospheric density up to 3000 km.

The atmospheric density follows a vertical distribution which depends on the hydrostatic distribution. Average values show that the variation of density with height is extremely different at low altitudes and at very high altitudes. The orders of magnitude for conditions of high solar activity are as follows (see also *Fig. 1.4*):

Altitude (km):	0	100	150	400	700	1500
Density (gm cm^{-3}):	10^{-3}	10^{-9}	10^{-12}	10^{-14}	10^{-16}	10^{-18}

These values show that the decrease of density over a 100 km region is small at several hundreds of kilometers when compared with the variation between ground level and 100 km. While the density is reduced by a factor of the order of 10^6 between 0 and 100 km, it decreases by a factor of only about 10^2 from 400 to 700 km and from 700 to 1500 km.

To a first approximation the atmospheric density ϱ is governed by the hydrostatic equation

$$\varrho g = - dp/dz \tag{1.1}$$

in which p denotes the atmospheric pressure and g is the acceleration of gravity at height z.

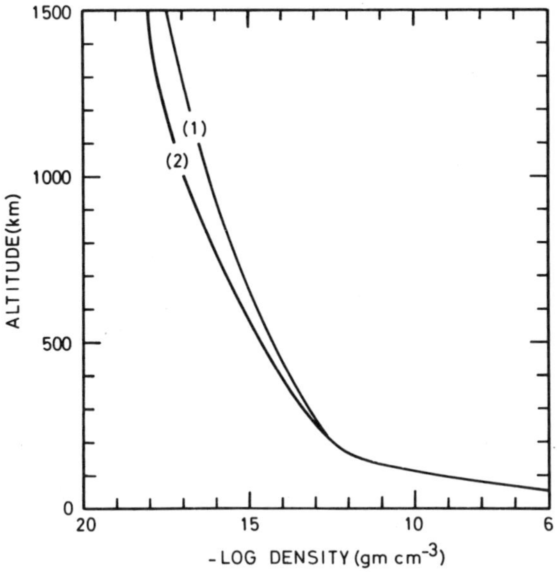

Fig. 1.4 Average daytime and nighttime atmospheric densities for conditions of high solar activity. Curve (1), day; curve (2), night.

For an ideal gas the pressure p is related to the neutral gas temperature T by

$$p = nkT \tag{1.2}$$

in which n is the molecular concentration (number density) and k is Boltzmann's constant ($k = 1.3803 \times 10^{-16}$ erg deg^{-1}).

With

$$\varrho = nm \qquad (m = \text{mean molecular mass}), \tag{1.3}$$

the hydrostatic equation can be written in the following form:

$$\frac{dp}{p} = \frac{dn}{n} + \frac{dT}{T} = \frac{d\varrho}{\varrho} + \frac{dT}{T} - \frac{dm}{m} = -\frac{dz}{kT/mg}. \tag{1.4}$$

This shows that a decrease in density of about 10^{-6} at a temperature of $T = 273°$K would require a change of altitude of about 100 to 110 km; i.e., this corresponds approximately to the first 100 km in the earth's atmosphere. However, the slower decrease in density found at higher altitudes requires higher temperatures, of the order of $T = 1500°$K and, in addition, at extreme altitudes, a low molecular mass m.

For example, a density of the order of 10^{-16} gm cm^{-3} near 700 km corresponds to a concentration of about 4×10^{6} oxygen atoms cm^{-3}. A density of the order of 10^{-18} gm cm^{-3} at 1500 km leads to the following concentrations (see Chapter 14):

hydrogen,	$n(\mathrm{H}) = 5\times10^{5}$ cm^{-3}
helium,	$n(\mathrm{He}) = 1.5\times10^{5}$ cm^{-3}
oxygen,	$n(\mathrm{O}) = 4\times10^{4}$ cm^{-3}.

Thus, when the total gas density is known it is possible to estimate the atmospheric temperatures associated with individual species concentrations required at the highest altitudes.

One of the most important aspects of the problem of atmospheric density is its variation from year to year. Density data deduced from orbital variations of satellites indicate a large decrease related to decreasing solar activity, for example, from 1958 to 1964 (see *Fig. 1.5*). Density determinations were made by various authors following the initial calculations of the acceleration of the first two satellites (Sputnik 1 and 2, 1957 α and β) whose perigees were below 225 km. The satellites Vanguard I (1958 β_2), Explorer I (1958 α), Explorer IV (1958 ε), and Sputnik III (1958 δ) also made possible

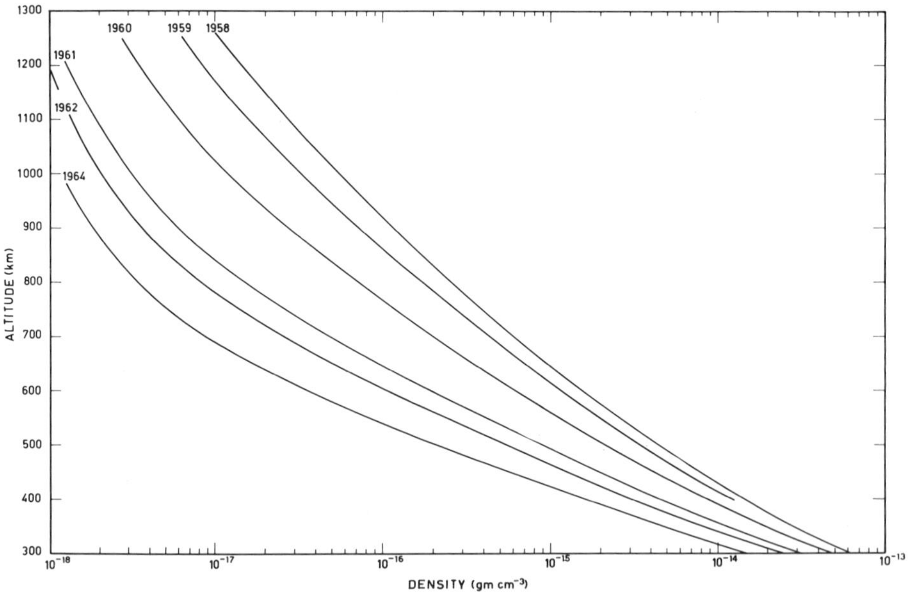

Fig. 1.5 Variation of atmospheric density for the years 1958–1964. Average daytime values.

an analysis of the densities at altitudes of approximately 650, 350, 260, and 220 km corresponding to the perigees of these satellites. Since the 1957–1958 maximum of solar activity, a large number of satellites have been launched and it is now possible to follow the variation of density over a complete solar cycle. The density between solar maximum and minimum solar activity (1958–1964) did not decrease by more than a factor of 2 at 200 km but declined by a factor of at least 100 at 750 km (see Fig. 1.5). Such a variation can be explained only by a temperature decrease of about 1000°K and a change in atmospheric composition. Between 1958 and 1964 the number density of oxygen atoms at 750 km decreased by a factor of more than 100 while the atomic hydrogen density *increased* by about the same factor. From Eq. (1.4) variations in composition and temperature such as these are required in order to explain the atmospheric density variation.

1.3 Temperature

A mean temperature–altitude graph up to several hundred kilometers is shown in Fig. 1.2. Since the direct measurement of atmospheric temperature is difficult above several hundred kilometers altitude, indirect measurements have frequently been used. For example, the temperature can be deduced from Eq. (1.4) using the *scale height*

$$H = kT/mg \tag{1.5}$$

which is the parameter (introduced by S. Chapman) adopted in all aeronomic problems relating to the logarithmic gradient of pressure. If the mean molecular mass m is known, the temperature T can be determined.

The acceleration of gravity is known from the relation

$$gr^2 = g_0 r_0^2 \tag{1.6}$$

in which g_0 and g are the values of the acceleration of gravity at the appropriate distances r_0 and r from the earth's center. Adoption of the conventional standard value of $g_0 = 980.655$ cm sec^{-2} leads to a value between the equator and pole of

$$g = \frac{980.6 \mp 2.6}{[1 + (1.573 \pm 0.005)\times 10^{-4} z]^2} \quad \text{cm sec}^{-2} \tag{1.7}$$

in which z is the altitude in kilometers. To a first approximation, Eq. (1.7) leads to

$$g/g_0 = 1 - 3.146\times 10^{-4} z. \tag{1.8}$$

For the same temperature and molecular mass, the decrease of g leads to an increase of the scale height by about 34% between ground level and 1000 km.

The variation of the mean molecular mass m is an important factor to consider when observational data are being analyzed. If there is complete dissociation of N_2 and O_2 in an oxygen–nitrogen atmosphere, the mean molecular mass may decrease by a factor of 2. But such a decrease is not sufficient to explain the slow decrease of density above 150 km; hence high temperatures and the presence of lighter constituents must be invoked.

In order to explain the absolute densities found between 400 and 500 km (see Fig. 1.5, years 1958–1964) atomic oxygen must be introduced because molecular nitrogen alone cannot explain the variation of the corresponding scale heights. The scale height of atomic oxygen, which varies by a factor of 2, i.e., from about 60 to 120 km, indicates a temperature variation of a factor of about 2 between 1958 and 1964. However, the atomic oxygen scale height adopted at 500 km cannot explain the absolute densities observed at 1000 km and above since the temperatures would be too high as compared with the temperature required at 500 km. The observational data can successively be interpreted in terms of a lower mean molecular mass at the highest altitudes involving helium and atomic hydrogen. An example of the variations of the scale height H is shown in *Fig. 1.6* for an atmosphere at a constant temperature $T = 750°K$. It is seen that the relative proportion of atomic hydrogen increases by a factor of 10 between 300

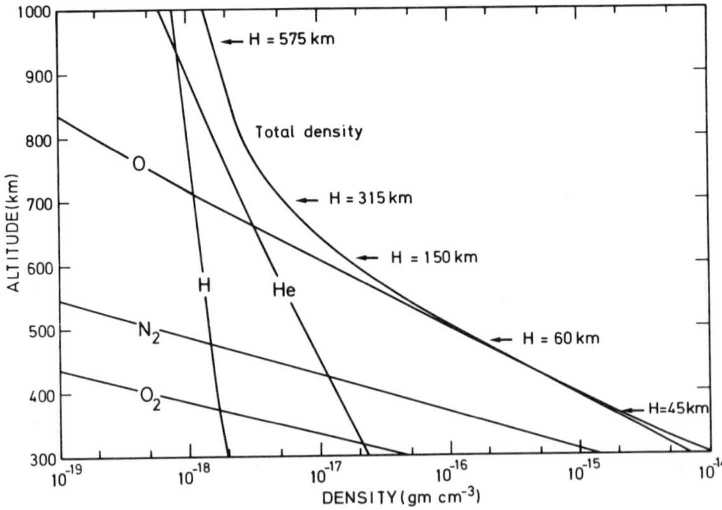

Fig. 1.6 Variation of the density in an atmosphere with constant temperature (750°K). Effects of N_2, O, He, and H scale heights are indicated.

and 400 km. The combined effect of the temperature and of the mean molecular mass properly explains the vertical distribution of density and its absolute value.

1.4 Molecular Mass

If the air composition does not change, the molecular mass m remains constant. It is possible to consider (see Fig. 1.2) an atmospheric region called the homosphere where the principal constituents (of molecular mass m_i), which effectively determine the mean molecular mass, maintain the same relative proportions per unit volume. In the homosphere, one has

$$m = \sum n_i m_i / \sum n_i = \text{constant} \tag{1.9}$$

and

$$H \propto T/g. \tag{1.10}$$

The first atmospheric constituent subject to changes in proportion per unit volume is molecular oxygen. Because O_2 undergoes photodissociation, atomic oxygen gradually becomes a principal atmospheric constituent and the mean molecular mass must therefore decrease with altitude. In the region where diffusion in the earth's gravitational field counterbalances mixing, the concentration of light constituents (atoms) increases relative to the heavy constituents (molecules). When (1.10) is replaced by

$$H \propto T/mg \tag{1.11}$$

with mg varying with height, the atmosphere is termed the *heterosphere* (see Fig. 1.2). Therefore, any observed increase of the atmospheric scale height H depends simultaneously on the increase of the temperature T and on the decrease of the molecular weight mg. As mentioned in Section 1.3, physical conditions between 750 and 1500 km require a certain molecular mass. If the temperature is less than 1500°K at 1500 km, the observed density can be explained with a molecular mass of the order of 4, i.e., helium, while at 750 km the molecular mass is of the order of 16, i.e., atomic oxygen. On the other hand a decrease of the atomic mass at 1500 km to 1, i.e., hydrogen, would require an excessively large concentration of such atoms. This illustrates that the analysis of the behavior of the upper atmosphere requires a knowledge of the vertical distribution of all constituents. A constituent which may be neglected in the homosphere can play a leading role in the heterosphere.

Figure 1.6 demonstrates how the effect of a decrease in the mean molecular mass affects the variation of the scale height. At a low temperature such as $T = 750°K$, which may be adopted for nighttime conditions corresponding to a very quiet sun, the scale height H increases with height. This increase is explained solely by successive transitions from an atmosphere with molecular nitrogen and atomic oxygen to atomic oxygen, helium, and atomic hydrogen belts. The scale heights of 45, 60, 150, 315, and 575 km indicated in Fig. 1.6 correspond to mean molecular masses 16 (atomic oxygen and helium), 5.4 (almost all helium), 2.5 (helium with hydrogen), and 1.8 (mainly atomic hydrogen), respectively.

1.5 The Ionosphere

Since atmospheric atoms and molecules are subject to ionization, the word *ionosphere* is used when the behavior of charged particles is studied. Conventional definitions of the ionosphere include that part of the atmosphere (*Fig. 1.7*) which comprises the D, E, and F regions, that is to say, roughly the atmosphere between 60 km, the beginning of the D region, and 400 km, above the peak of the F_2 region. The regions above the F_2 peak are usually referred to as the topside ionosphere.

The different regions which are identified in Figs. 1.7 and *1.8* correspond to an electron distribution increasing with height from about 10^3 cm^{-3} in the D region to about 10^5 in the E layer and to about 10^6 cm^{-3} at the F_2 peak. Above the F_2 peak the topside ionosphere is eventually bounded by the interplanetary medium.

The ionization caused by far ultraviolet radiations, x rays, and cosmic rays varies with the solar cycle and its behavior is related to changes in the solar radiation spectrum. The ionization of the D region is caused by the most penetrating radiation which is generally very sensitive to solar activity. The E region between 85 to 130 km is due to soft x rays and to long wavelength ultraviolet ionizing radiations. It is known that an ionization peak sometimes occurs in the E region corresponding to an electron concentration of 10^5 cm^{-3}, with NO^+ and O_2^+ as the dominant ionic constituents. In the D, E, and F_1 regions the electron concentration is highest near local noon. At night the electrons disappear from the D region and are reduced to a small concentration, of the order of 10^3 cm^{-3}, in the E and F_1 regions.

The behavior of the F_2 region is very complicated and depends on geomagnetic latitude. The vertical distribution of the electron density in and

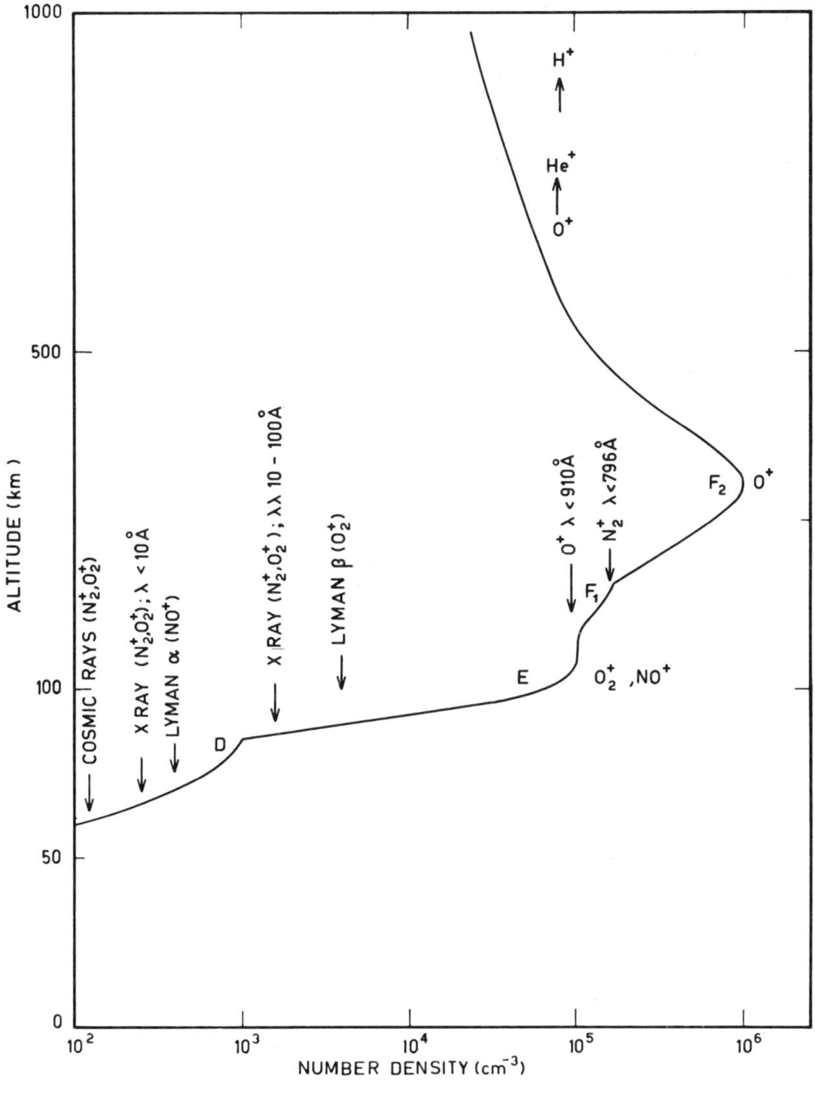

Fig. 1.7 Average daytime electron distribution with principal ions in the various layers of the ionosphere.

above the F_2 peak cannot be described in terms of the balance between ionization and recombination but rather as the effect of diffusion. Rocket and satellite observations show that the predominant ionic constituent near the F_2 peak is atomic oxygen, and it is found that there is finally a change with increasing altitude from O^+ to He^+ and H^+ in the topside region.

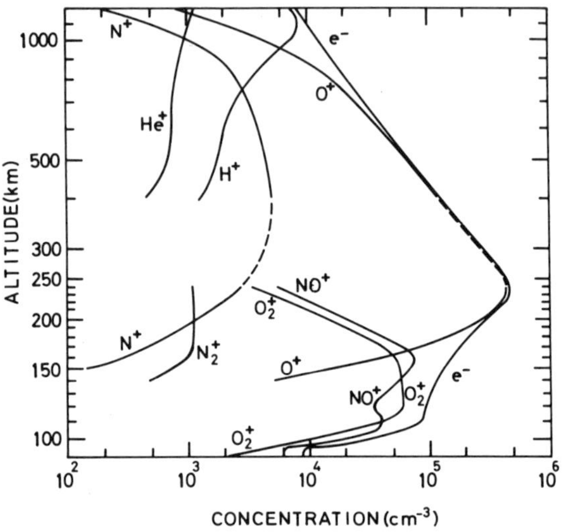

Fig. 1.8 Distribution of positive ions for a daytime ionosphere at solar minimum [C. Y. Johnson, *J. Geophys. Res.* **71**, 330 (1966)].

Even though the ionosphere is often characterized in terms of its electron density peaks it must be realized that it is the ions and their reactions which truly determine the behavior of the whole ionosphere and the different ionospheric regions. Figure 1.8 represents an example of a possible vertical distribution of atmospheric ions. The ionic composition of the ionosphere is related to ionic production which follows directly from the photoionization of the major constituents N_2, O_2, and O. The final composition can only be explained by various aeronomic reactions which can occur between the photoions and the neutral atmospheric constituents. At high altitudes diffusion processes between charged particles determine the ionospheric behavior. Under such conditions the potential energy of the plasma is the sum of that due to gravity, centrifugal force, and magnetic and electric fields for positive ions and electrons at their respective temperatures.

The increase in the ratio of charged particles to neutral particles with height suggests the importance of the earth's magnetic field in the upper ionosphere. The simplest magnetic field which corresponds most closely to the observed magnetic field of the earth is that of a central dipole whose axis is inclined at an angle of about 11° to the geographical axis.

Geomagnetic coordinates are related to the geomagnetic axis in the same way as geographical coordinates to the axis of rotation of the earth. Although an eccentric dipole is an improved approximation to the observed magnetic field, magnetic latitude as deduced from the observed magnetic inclination (dip) differs from the geomagnetic latitude obtained by the approximation of centered and eccentric dipoles.

The distribution of ionization in the topside ionosphere is strongly controlled by the geomagnetic field and the structure of the magnetosphere. As shown in *Fig. 1.9*, there are three regions where the behavior of thermal plasma (H^+, He^+) is determined by different effects.

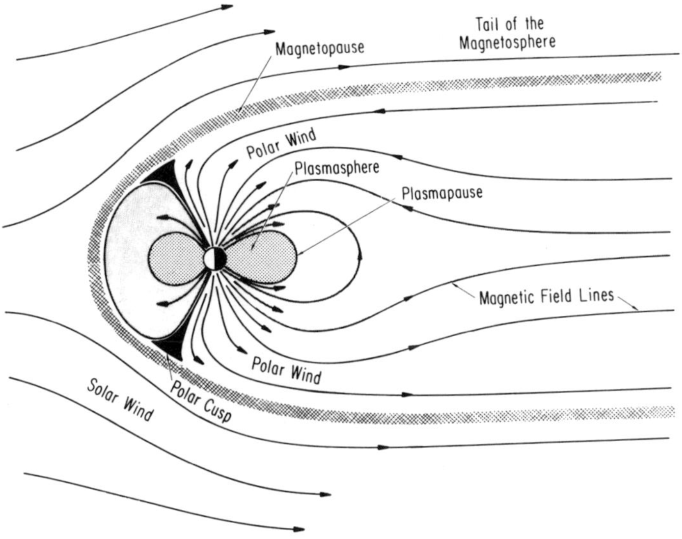

Fig. 1.9 Schematic view of the earth's plasma environment.

At low geomagnetic latitudes the *plasmasphere* represents a region of relatively high ion density (10^3 to 10^4 ions cm^{-3}) contained along the closed lines of geomagnetic force. The topside ionosphere in these regions is relatively quiescent and corotates with the earth, subject to small diurnal and sporadic electric fields which cause some lateral plasma motions at various times. Within the plasmasphere H^+ and He^+ are the principal ionic components above 1500 km and undergo diurnal flows along the magnetic field lines as a result of changes in the F_2 region (O^+) density and temperature.

The termination of the plasmasphere occurs at the *plasmapause*, a rather uneven surface defining the shape of the plasmasphere as an asymmetrical torus. The behavior of the topside ionosphere outside the plasmapause is very complex as a result of magnetospheric convection. Due to the interaction of the solar wind with the geomagnetic field, a large scale electric field is established within the magnetosphere, having a direction from dawn to dusk. This electric field drives thermal plasma at altitudes above 150 km toward the front of the magnetosphere in regions where the field lines are closed, and towards the tail of the magnetosphere in regions where field lines are open. Following the ideas of the theory of magnetospheric convection, it is possible to visualize the behavior of the thermal plasma in terms of motions of the magnetic field lines themselves. Thus, in the initial step of the convection cycle, the dipole-like field lines from the earth become connected to the interplanetary magnetic field, at the front of the magnetosphere. These field lines are then swept back over the polar cap into the tail of the magnetosphere where they eventually reconnect in the vicinity of the *neutral sheet*. During their traversal of the polar cap the field lines are open to the interplanetary medium and solar wind. Following reconnection in the tail of the magnetosphere, the field tubes migrate around the earth at relatively low geomagnetic latitudes. As the field tubes approach the front of the magnetosphere they begin to move towards higher magnetic latitudes and eventually disconnect to begin again the convection cycle.

The behavior of the topside ionosphere is strongly affected by the convection cycle. In the regions of open field lines, H^+ and He^+ are rapidly accelerated along the field tubes to supersonic speeds (10–20 km sec^{-1}) so that there is a net loss of plasma from the earth. As a result of this flow, called the *polar wind*, H^+ and He^+ densities are greatly reduced and F_2 region O^+ remains as the major ionic constituent to altitudes of 3000 to 5000 km.

In the regions where convection moves plasma on closed field lines, the upwards polar wind flow can continue only for periods of tens of hours. Since the time required for the convection cycle is usually less than this, polar wind flow is found over most of the regions where field tube convection is in progress. It appears that the plasmapause itself is a result of the thermal plasma distributions characteristic of the corotating plasma and that for convecting field tubes. Because of the polar wind flow the latter are greatly depleted of plasma, while the former have relatively high densities.

The location of the plasmapause is greatly dependent upon magnetic activity. The convection process itself is not steady, but subject to continual variation. As a result, the topside ionosphere in the vicinity of 45° to 70° magnetic is normally in a dynamic state with plasma inflow and outflow the usual state.

It is well known from satellite and spacecraft experiments that energetic particles are a very important component of the magnetosphere. Aurora, subvisual red arcs, and other airglow phenomena result from the interaction of these particles with the neutral atmosphere. This interaction is at times of great significance. For most conditions, however, the couplings implied are not of dominating importance so that the subjects of energetic particle behavior and the aeronomy of the neutral and ionized atmospheres can be discussed separately.

GENERAL REFERENCES

1. S. K. Mitra, *The Upper Atmosphere*. Asiatic Soc., Calcutta, 1952.
2. J. A. Ratcliffe (ed.), *Physics of the Upper Atmosphere*. Academic Press, New York, 1960.
3. J. W. Chamberlain, *Physics of the Aurora and Airglow*. Academic Press, New York, 1961.
4. R. G. Fleagle and J. A. Businger, *An Introduction to Atmospheric Physics*. Academic Press, New York, 1963.
5. M. Nicolet, La constitution et la composition de l'atmosphère supérieure, in *Geophysics: The Earth's Environment* (C. De Witt, J. Hieblot, and A. Lebeau, eds.), pp. 199–277. Gordon and Breach, New York, 1963.
6. R. A. Craig, *The Upper Atmosphere. Meteorology and Physics*. Academic Press, New York, 1965.
7. W. N. Hess (ed.), *Introduction to Space Science*. Gordon and Breach, New York, 1965.
8. W. L. Webb, *Structure of the Stratosphere and Mesosphere*. Academic Press, New York, 1966.
9. D. J. Williams and G. D. Mead (eds.), *Magnetospheric Physics*. Amer. Geophys. Union, Washington, D.C., 1969.
10. A. D. Danilov, *Chemistry of the Ionosphere*, Plenum Press, New York, 1970.
11. B. M. McCormac (ed.), *Particles and Fields in the Magnetosphere*. Reidel Publ., Dordrecht, Netherlands, 1970.
12. M. Nicolet, L'Aéronomie, in *Géophysique, Encyclopédie de la Pléiade*, pp. 447–553. Editions Gallimard, Paris, 1971.
13. R. C. Whitten and I. G. Poppoff, *Fundamentals of Aeronomy*. Wiley, New York, 1971.
14. S. I. Akasofu and S. Chapman, *Solar Terrestrial Physics*. Oxford Univ. Press (Claredon), London and New York, 1972.

CHAPTER 2

The Composition of the Homosphere

2.1 Introduction

Since absolute measurements of the concentrations of principal constituents of the homosphere, such as nitrogen and oxygen, have only been made over a rather short period of time, it is not possible to claim that the composition does not change at all. However, for periods of many years the mean molecular mass of the homosphere remains constant since the mixing, as far as the principal constituents are concerned, does not change in an important manner from the ground level up to the vicinity of the mesopause.

The homosphere is characterized basically as a gas composed of N_2, O_2, Ar, and CO_2 contributing to the mean molecular mass of the air in order of importance as: molecular nitrogen (78% by volume), molecular oxygen (21%), argon (1%), and carbon dioxide (0.03%).

2.2 Principal Permanent Constituents

The principal permanent constituents are given in *Table 2.1* taking into account the average isotopic composition. The total number density of molecules for standard pressure (760 Torr) and temperature (0°C) (STP) equals 2.69×10^{19} cm^{-3} (Loschmidt's number). For a vertical column such a number density n_0 leads to 2.15×10^{25} molecules cm^{-2}, i.e., to a thickness H_0 of about 8 km for a homogeneous atmosphere [1] in terms of the standard value of gravity of 980.66 cm sec^{-2}.

TABLE 2.1

PRINCIPAL CONSTITUENTS OF THE HOMOSPHERE

Constituent	Molecular mass (amu)	Molecular mass (gm $\times 10^{24}$)	Composition by volume (%)	Molecular concentration at STP (cm^{-3})	Atmospheric thickness (Atmos-cm)	Atmospheric column content (cm^{-2})
N_2	28.02	46.51	78.08	2.10×10^{19}	6.24×10^{5}	1.68×10^{25}
O_2	32.01	53.12	20.95	5.63×10^{18}	1.68×10^{5}	4.50×10^{24}
Ar	39.96	66.33	0.93	2.51×10^{17}	7.47×10^{3}	2.01×10^{23}
CO_2	44.02	73.06	0.03	8.87×10^{15}	2.64×10^{2}	7.09×10^{21}
Air	28.97	48.08	100.00	2.69×10^{19}	8.00×10^{5}	2.15×10^{25}

It should be pointed out that the mean molecular mass $M = 28.96$ is not much different from that of molecular nitrogen $M = 28$. The difference of about 3.5% between the molecular masses of the air and nitrogen is a measure of the difference in the behavior of N_2 in the homosphere and heterosphere. Therefore the vertical distribution of molecular nitrogen in the whole atmosphere is not greatly affected by diffusion; it depends mainly on the temperature distribution. With regard to oxygen and argon, their mass differs substantially from the mean molecular mass and their behavior in the heterosphere is quite different from that in the homosphere.

2.3 The Noble Gases

A study of the behavior of noble gases leads to interesting conclusions for aeronomic problems. Since the atoms of such gases are not subject to chemical reactions, they can be used as tracers in the homosphere and heterosphere. Essential data on the noble gases are given in *Table 2.2*. The cosmic abundances [2, 3], which decrease from helium to argon, are not related to terrestrial abundances. Argon is the most important of the terrestrial noble gases, but its mass $M = 40$ does not correspond to the normal argon mass $M = 36$ of the periodic table of elements since most of the atmospheric argon consists of the isotope produced by radioactivity in the earth's crust and mantle (see Table 2.6) [4–7]. Neon is more abundant than helium, which is also produced by radioactive sources. These various anomalies must be associated with atmospheric effects [8].

TABLE 2.2

Noble Gases

Atom	Ratio by volume in air	Atomic concentration at STP (cm^{-3})	Atmospheric column content (cm^{-2})	Atmospheric thickness (Atmos-cm)
He	5.24×10^{-6}	1.41×10^{14}	1.13×10^{20}	4.19
Ne	1.82×10^{-5}	4.89×10^{14}	3.89×10^{20}	14.56
Ar	9.34×10^{-3}	2.51×10^{17}	2.01×10^{23}	7.47×10^{3}
Kr	1.14×10^{-6}	3.06×10^{13}	2.45×10^{19}	0.91
Xe	8.7×10^{-8}	2.34×10^{12}	1.87×10^{18}	0.07

Helium is produced by the disintegration of uranium and thorium in the earth's crust and mantle [9]. It is possible to show that the generation of helium atoms (α particles) is about 2×10^6 cm^{-2} sec^{-1} at the earth's surface over geologic time [4]. Even if an important fraction of the atoms generated were unable to reach the atmosphere, it can be concluded that 10^6 atoms cm^{-2} sec^{-1} or 3.15×10^{13} atoms cm^{-2} year^{-1} represents the correct order of magnitude of helium atoms reaching the atmosphere. This is especially true when one considers the possibility of a ^{4}He contribution from an auroral source [10, 11]. Therefore, the total atmospheric content (see Table 2.2) of helium is obtained in a relatively short period $t(^4\text{He})$ given by

$$t(^4\text{He}) < (1.13 \times 10^{20})/(3.15 \times 10^{13}) = 3.5 \times 10^6 \quad \text{years},$$

i.e., assuming no change in the production rates, atmospheric helium has been completely renewed at least 1000 times since the earth's formation. One must therefore conclude that there is a rapid escape of helium at the top of the earth's atmosphere. Three principal modes of escape have been proposed to explain the loss of ^{4}He from the atmosphere: thermal evaporation from the exosphere [7, 12], ballistic escape of individual atoms using excess kinetic energy available in ^{4}He ion charge exchange reactions with N_2 [13], and loss via the polar wind [14, 15].

Computations of the average rate of thermal escape based on solar radiation fluxes during the 11-year period from 1951 to 1961 have shown that the escape flux was not more than 10^5 ^{4}He atoms cm^{-2} sec^{-1} [12]. This is substantially less than the best estimate of the present production rate, but recent measurements of atmospheric temperatures in high latitude regions indicate significant increases in temperature which are not related to variations in solar flux [16]. As a result, the actual thermal escape rate may be significantly higher than that computed using atmospheric models and solar data.

For the second process, laboratory measurements indicate exothermic ion-molecule reactions do not result in a release of adequate amounts of excess kinetic energy to permit ^{4}He escape [13]. Instead, internal energy states of the reactants appear to be excited with subsequent photon emission.

The polar wind loss of ^{4}He provides a continual loss of ^{4}He ions in regions outside the plasmapause. There the polar wind helium flux is approximately 2 to 4×10^6 ions cm^{-2} sec^{-1} [15]. The precise global loss rate due to ion escape is difficult to calculate, however, since the boundary between convection and corotation of magnetic field lines is continually changing. Nevertheless, it seems that the combination of thermal escape and polar wind loss is adequate to explain the terrestrial helium budget [14].

As pointed out by several authors [4, 10, 11] accretion of the noble gases could possibly occur from the solar wind via the aurora. Computations given in *Table 2.3* indicate the possible fluxes of various isotopes. It appears that the only species affected in any important way by this mechanism is ^3He [17].

There does not seem to be a radioactive production for neon, and it is difficult to conceive how the earth could have kept its primordial neon. If the neon present is due to the capture of cosmical neon, the total content of atmospheric neon corresponds to a flow of 3500 atoms cm^{-2} sec^{-1} during 3×10^9 years. Considering the cosmic ratio between the abundances of neon and hydrogen, the corresponding influx of hydrogen atoms would have to be not less than 10^7 hydrogen atoms cm^{-2} sec^{-1} at the top of the earth's atmosphere. Such fluxes are readily available in the solar wind, but the geomagnetic field deflects almost all the interplanetary plasma except in a few concentrated regions, such as the polar cusp and the auroral oval, where the solar wind could possibly gain access to the magnetosphere and earth's atmosphere. As a result, as discussed below, there is no simple explanation for the presence of Ne in the earth's atmosphere.

Terrestrial argon is produced from the 10% of radioactive potassium ^{40}K which decays by electron capture. The major part of ^{40}Ar in the atmosphere was supplied during a period when its production rate was much greater than the present rate [18]. Argon was certainly present before the beginning of the oxygen–nitrogen atmosphere. It is difficult, however, to make a correct estimate of the supply to the atmosphere during the early stages of development of the atmosphere since this requires a consideration of the differentiation between the crust and mantle. Nevertheless, the present atmospheric content of ^{40}Ar is probably not related to a primordial atmosphere but came from an early release of ^{40}Ar with additional amounts supplied during subsequent outgasing periods. An estimation of the relative contributions of ^{40}K to argon production during geological periods depends on the earth's composition and its evolution [19].

The analysis of the origin of atmospheric xenon leads to the conclusion that it does not correspond to a primordial xenon but is the result of accretion and decay of extinct radioactivities [20].

Extrapolations based on the cosmical abundances of the elements can be used to estimate the possible importance of solar wind accretion. Several authors have discussed the contribution such solar wind gases could make to the noble gas densities [2–4, 10]. Computations given in Table 2.3 have been based on measurements of the H^+ solar wind flux. It appears that the accumulation times for Ne, Ar, Kr, and Xe are far too long to permit

TABLE 2.3

Extraterrestrial Sources of the Elements

Element	Abundance in sun (Si = 10^6)	Particle kinetic energy in solar wind (keV)	Solar wind flux (cm^{-2} sec^{-1})	Atmospheric accretion rate (sec^{-1})	Atmospheric loss rate (sec^{-1})	Atmospheric column content (cm^{-2})	Accretion time τ (years)	Auroral flux (cm^{-2} sec^{-1})
^{1}H	2.6×10^{10}	0.85	2×10^{8}	3×10^{24}	2×10^{26}	—	—	3×10^{7}
^{2}H	3.9×10^{6}	1.7	1.7×10^{4}	3×10^{20}	$\sim 10^{22}$	—	—	3×10^{3}
^{3}He	6.3×10^{5}	2.6	2.7×10^{3}	3.5×10^{19}	4×10^{19}	1.41×10^{14}	7×10^{5}	350
^{4}He	2.1×10^{9}	3.4	9×10^{6}	1.4×10^{23}	5×10^{24}	1.13×10^{20}	1.3×10^{8}	1.4×10^{6}
^{12}C	1.35×10^{7}	11	6×10^{4}	9×10^{20}	—	—	—	9×10^{3}
^{14}N	2.44×10^{6}	13	1.1×10^{4}	1.7×10^{20}	—	—	—	1.7×10^{3}
^{16}O	2.36×10^{7}	14	1×10^{5}	1.5×10^{20}	—	—	—	1.5×10^{3}
^{20}Ne	2.36×10^{6}	17	1×10^{4}	1.5×10^{19}	—	3.84×10^{20}	4×10^{12}	150
^{36}Ar	2.28×10^{5}	34	1×10^{3}	1.5×10^{18}	—	1.93×10^{23}	2×10^{16}	15
^{84}Kr	64.4	71	$<3 \times 10^{-2}$	$<4.5 \times 10^{14}$	—	2×10^{19}	$>7 \times 10^{15}$	$<4.5 \times 10^{-3}$
^{129}Xe	7.1	110	$<3 \times 10^{-3}$	$<4.5 \times 10^{13}$	—	1.6×10^{18}	$>6 \times 10^{15}$	$<4.5 \times 10^{-4}$

such an explanation. Only the ^3He accretion rate is substantial enough to conceivably affect the global balance.

Another source of noble gases could arise in the form of cosmic dust grains surface saturated with solar wind gases. Upon entry into the atmosphere these gases would be released through frictional heating. Using current estimates of the rate of dust infall ($\sim 10^3$–2×10^3 tons per day) it appears that such a source may not be important [21, 22].

Since the concentrations of noble gases are remarkably constant in the homosphere, the mixing must be rapid [23]. When ratios such as He/N_2, Ne/N_2, and Ar/N_2 are determined above 100 km with sufficient precision, it is possible to reach important conclusions on the problem of diffusion and also on the effect of a production source with escape (He). Isotopic studies such as those of the ratios $^4He : ^3He$, $^{20}Ne : ^{21}Ne : ^{22}Ne$, $^{36}Ar : ^{38}Ar : ^{40}Ar$ and isotopes of krypton and xenon are important with relation to their production and possible variations.

2.4 Molecular Composition

Carbon dioxide has been treated in Section 2.2 as one of the principle constituents along with molecular nitrogen and oxygen since its volume ratio is about 3×10^{-4}. Other molecules are true minor constituents even though they may have important roles in the atmosphere. It is not yet possible to obtain a clear picture of the behavior of the molecular constituents in the homosphere since few spectroscopic observations have been made and systematic satellite observations are just beginning [24, 25]. *Table 2.4* gives a list of some of the molecules which should be present throughout the homosphere. Though industrial production affects the concentration of carbon dioxide, its general atmospheric distribution seems uniform [26, 27]. That is, CO_2 is effectively mixed during its lifetime in the atmosphere until its transfer to the oceans. A band of $^{13}CO_2$ detectable in the infrared spectrum [28] at 2 μm corresponds to a ratio $^{12}C/^{13}C$ of the order of 89. This ratio varies according to discrimination factors resulting from biological phenomena, and it would be very useful to determine the variations with height caused by the effect of chemical reactions.

Atmospheric methane is uniformly distributed [29–33] in the troposphere even though its production is due to biological phenomena. Its introduction into the atmosphere occurs at ground level and its lifetime is less than 100 years [34]. Its oxidation must be very rapid at the stratopause level. Application of an oxidation coefficient of the order of 10^{-6} sec^{-1} at 50 km,

TABLE 2.4

MOLECULAR CONTENT

Molecule	Ratio by volume	Remarks
CO_2	3×10^{-4}	Small variation
H_2O	10^{-2}–10^{-5}	Variable, particularly in troposphere
CH_4	1.5×10^{-6}	Fairly constant, mixed in troposphere
N_2O	2.5×10^{-7}	Approximately constant, mixed in troposphere
H_2	5×10^{-7}	Probably mixed in troposphere
O_3	10^{-8}–10^{-7}	Maximum in stratosphere
CO	5×10^{-8}–2×10^{-7}	Variable
NO_2	5×10^{-10}–2×10^{-8}	Variable, effect of NO

where the content of a vertical column should be 3×10^{16} CH_4 molecules cm^{-2}, yields an average lifetime of 40 years. However, such a loss process in the mesosphere for the removal of methane is unacceptable. It is impossible to sustain a constant mixing ratio throughout the stratosphere since it requires an upward flow of the order of 10^{10} CH_4 molecules cm^{-2} sec^{-1}. The maximum rate of diffusion flow of methane is of the order of only 7×10^6 molecules cm^{-2} sec^{-1} for mixing conditions in the lower stratosphere. Such a flow can maintain a constant mixing ratio in the lower stratosphere where the oxidation process is sufficiently slow. Consequently, there should be a continuous decrease of the mixing ratio of CH_4 in the stratosphere with height so that its concentration should be very small at the stratopause level and its lifetime only a few months. Spectroscopic observations made at or above the tropopause should detect such a decrease of the mixing ratio of methane.

N_2O is a molecule also produced at ground level [35]. However, such molecules are subject to a permanent photodissociation by solar ultraviolet radiation reaching the troposphere. The radiations penetrating into the stratosphere are able to dissociate [35, 36] about 3×10^9 molecules cm^{-2} sec^{-1} and the total atmospheric content of N_2O should be renewed after 70 years [37]. Again, such a high loss of N_2O molecules in the stratosphere cannot be compensated for by vertical transport or by any chemical action. Mixing occurs in the troposphere according to results obtained from ground level [38] and telluric [39] measurements. The tropospheric content (from ground to 16 km) corresponds to 90% of the total content of N_2O for a constant mixing ratio. But between 16 and 30 km it is almost certain that there is a continuous decrease of the mixing ratio with height [37]. The lifetime of N_2O molecules in the solar radiation field is of the order of 100 days at 50 km, and the actual content of a vertical column above 50 km is far below the content (10^{-2} of a vertical column at standard pressure and temperature) of a column having a constant mixing ratio.

The observed variations of NO_2 concentrations can be associated with the effect of industrial production. However, NO_2 can be produced by the dissociation of N_2 and O_2 by lightening, and its short term variations may be due to the presence or absence of solar radiation in the lower troposphere because the photochemical action of solar radiation on NO_2 is effective down to ground level.

Carbon monoxide is produced at the earth's surface both through anthropogenic processes (wood burning, automobile exhaust fumes, etc.) [40] and by surface waters of the oceans [41, 42]. A natural sink appears to exist above the troposphere such that there is a rapid decrease in CO density at the tropopause [43]. An oxidation for CO has been predicted [44, 45], but the measured decrease in density is far larger than that calculated [46]. The mean CO mixing ratio is variable in the range 4×10^{-8} ppm in the southern hemisphere to 1.8×10^{-7} ppm at 50° north latitude [46]. The atmospheric residence time is subject to uncertainty but appears to be less than one year [46]. The land source strength of CO is about 2.2×10^{9} molecules cm^{-2} sec^{-1}, while over water the value 0.6×10^{9} molecules cm^{-2} sec^{-1} has been found [46].

2.5 Isotopic Composition

Although the study of the isotopic constitution of the atmospheric constituents is of only recent development [47, 48], it already seems to have interesting possibilities for the solution of a whole series of aeronomic problems.

The hydrogen isotopes ^1H and $^2H \triangleq D$ found in air are present either as hydrogen in water vapor or as the component atoms of hydrogen molecules. The presence in air of the HDO molecule is known from spectroscopic analysis of the solar spectrum, in particular from identification of the fundamental ν_2 band at 3.67 μm. A numerical value of the order of 6700, which is adopted for the number concentration ratio H/D, is conventional, and is based on tap water determinations. Because water occurs in various forms and in all substances, the ratio H/D is variable. Indeed, the processes of condensation and evaporation cause isotopic separation proportional to the differences in vapor pressures. Specifically, the water vapor above the oceans will contain less deuterium than ocean surface water because H_2O evaporates more rapidly than HDO. In the same way, the continental water vapor suffers continuous diminution of its deuterium content, because rain removes more heavy than ordinary water from the moist air. Similarly, it is known that isotopic fractionation occurs when water is formed from ice, the latter gaining in deuterium content.

It must be pointed out that the problem of the ratio HD/H_2 in the hydrogen molecule is not the same as that for water vapor, and detailed analysis of the processes is very difficult. Considering that the abundance of deuterium in the stellar atmosphere is very low, it would appear interesting to review the whole question of H_2 and HD in the terrestrial atmosphere.

The investigation of ^3H (tritium) carried out after the discovery of deuterium showed that tritium was only present in the atmosphere in the very smallest amounts, but the conditions that affect the problem of tritium have been altered since the beginning of the era of thermonuclear explosions [48, 49].

In fact, the half-life of tritium is about 12 years, and its artificial production has disturbed the equilibrium of the tritium of pure cosmogenic origin. Specifically, the ratios $^1H\ ^3HO/^1H_2O$ and $^1H^3H/H_2$ are dependent on the rapidity of isotopic separation and variation in artificial production. Finally, the present distribution balance depends on direct exchange between the air and sea water [48, 50].

The disintegration of tritium through the emission of an electron leads to a stable isotope of helium, ^3He. Since thermal escape is important for ^4He, it is certain that ^3He has a shorter lifetime in the atmosphere than ^4He. Studies of the ^3He balance of the atmosphere have been made by many authors [2, 4, 17]. The essential problem is to find production processes adequate to account for the concentration of the ^3He isotope in the terrestrial atmosphere. The various loss processes appear to be well known. Considering only thermal escape over the period 1947–1968 the average ^3He thermal loss rate has been computed as 5.9 $cm^{-2}\ sec^{-1}$ [17]. In addition a polar wind loss rate of about 1.3 $cm^{-2}\ sec^{-1}$ seems appropriate during average geomagnetic activity. As a result, it is necessary to consider that the net production of ^3He in the earth's atmosphere must be about 7.2 atoms $cm^{-2}\ sec^{-1}$. Evaluation of various ^3He sources and losses is given in *Table 2.5*, taken from a recent study [17].

Although auroral precipitation of solar wind plasma is postulated [10] to account for the major portion of ^3He, an additional source may be found

TABLE 2.5

ESTIMATED[a] BUDGET FOR ATMOSPHERIC ^3He

Production (cm^{-2} sec^{-1})		Loss (cm^{-2} sec^{-1})	
Emission from earth's crust:	$\approx 0.10 \pm 0.05$	Nonthermal escape (polar wind):	≈ 1.3 cm^{-2} sec^{-1}
		Thermal escape:	≈ 5.90 cm^{-2} sec^{-1}
Accretion of galactic cosmic rays:	≤ 0.01	Total loss rate:[a]	≈ 7.2 cm^{-2} sec^{-1}
Accretion of solar cosmic rays:	≈ 0.01		
Galactic cosmic ray interactions, followed by tritium β decay:	$\approx 0.20 \pm 0.05$		
Cosmic ray interaction yielding ^3He directly:	$\approx 0.40 \pm 0.10$		
Solar cosmic-ray interactions:	≈ 0.01		
Meteorites:	<0.06		
Cosmic dust:	~ 4		
Release of primordial ^3He:	2		
Accretion of interstellar ^3He:	5		
Auroral precipitation of solar wind plasma:	≈ 4.0		
Total production of ^3He:	≈ 15.8		

[a] The net uncertainty in these quantities is probably of the order of $\pm 50\%$.

in deep sea regions where mantle convection occurs. Recent data [51] indicate substantial increases in seawater ^3He concentrations near crust upwellings which presumably expose ^3He accumulated during primeval periods. As a result, it is not clear whether the balance indicated by Table 2.5 is correct. It follows, however, that in some manner the atmospheric ^3He content is renewed every 400,000 years.

The solar spectrum indicates that the ^{13}C molecule is sufficiently abundant to be detected in the infrared 2 μm band of $^{13}CO_2$. In fact, its concentration corresponds to that of the air at ground level, since about 1% of the natural carbon is carbon-13. The actual $^{12}C/^{13}C$ ratio in the CO_2 of the air is of the order of 89. However, it is subject to notable variations [52], indicating that the relative variation of $^{13}C/^{14}C$ in limestone, and in bicarbonates in the sea is 6–7% greater than in the air. Other characteristic variations are due to the processes which discriminate between the isotopes. All variations occur as a result of the simultaneous interaction of the lithosphere, the atmosphere, and the biosphere. Nevertheless, we do not yet have any idea of their behavior in the regions above the troposphere.

Radioactive carbon ^{14}C, generated by cosmic radiation, is also produced by thermonuclear explosions. Since this isotope has a half-life of the order of 5730 years, it has been used for the age determination of biospheric products. The large proportion of ^{14}C in the oceans shows the transfer of CO_2 from the air into sea water [53]. Furthermore, its relative concentration in CH_4, corresponding to the normal distribution in living organic matter, shows that methane is a product of the biosphere.

The relative abundances of the three stable isotopes of oxygen in the air are quite well known [54]. The conventional value for the concentration ratio $^{16}O/^{17}O$ is 2670 ± 20 and for $^{16}O/^{18}O$ is 490 ± 10. The $^{16}O^{17}O$ and $^{16}O^{18}O$ bands were detected 25 years ago in the telluric spectrum; today the molecular constants are well known [55]. The ^{18}O isotope has also been discovered in the CO_2 spectrum at about 2 μm.

The relative concentration of oxygen-18 is subject to remarkable variations. Atmospheric oxygen is involved in the biospheric cycle where photosynthetic and respiratory processes play concurrent roles. For this reason the ^{18}O distribution must be analyzed in nature in the hydrosphere ($H_2{}^{16}O/H_2{}^{18}O$), in the air ($^{16}O^{16}O/^{16}O^{18}O$ and $C^{16}O^{16}O/C^{16}O^{18}O$), and in the lithosphere.

The relative ^{18}O concentration is a maximum in the air (0.2039%). Taking this percentage as a unit value the relative ^{18}O concentrations are shown in the accompanying tabulation [56]. It must be understood that these are

Air:	1.000
Fresh water:	0.972
Sea water:	0.978
Photosynthetic oxygen	
fresh water:	0.976 (15%)
sea water:	0.983 (85%)
average:	0.982

average values; for example the concentration of oxygen-18 in fresh water ($H_2{}^{16}O/H_2{}^{18}O$) varies by some 4‰. This depends on the temperature and the origin of the water vapor. Moreover there is a very clear correlation between the deuterium and oxygen-18 concentrations indicating that heavy isotopes are more sensitive, in comparison with light ones, to variations which are proportional to differences in vapor pressure. Thus, the oxygen-18 content in fresh water depends on the previous history of the water vapor.

Photosynthesis and respiration are, however, important processes which point to the existence of a very rapid cycle of reactions as shown by the relationships between the oxygen isotopes; the action of the biosphere must not be overlooked when considering the relation between the hydrosphere and the atmosphere.

TABLE 2.6

ISOTOPIC ABUNDANCES FOR THE ATMOSPHERIC CONSTITUENTS

Element	Symbol	Mass	Percentage	Abundance (cm^{-2})
Hydrogen	^1H	1.008	99.985	Variable
	^2H	2.015	0.015	H_2O, CH_4, H_2, ...
	^3H	3.017	5×10^{-8}	
Helium	^3He	3.017	1.25×10^{-4}	1.41×10^{14}
	^4He	4.004	100	1.13×10^{20}
Carbon	^{12}C	12.004	98.9	7.00×10^{21}
	^{13}C	13.0075	1.1	7.80×10^{19}
	^{14}C	14.008	1.2×10^{-10}	8.5×10^{9}
Nitrogen	^{14}N	14.0075	99.635	1.67×10^{25}
	^{15}N	15.005	0.365	6.12×10^{22}
Oxygen	^{16}O	16.000	99.759	4.49×10^{24}
	^{17}O	17.0045	0.0374	1.68×10^{21}
	^{18}O	18.00	0.2039	9.18×10^{21}
Neon	^{20}Ne	19.999	90.51	3.54×10^{20}
	^{21}Ne	20.999	0.27	1.00×10^{18}
	^{22}Ne	21.999	9.22	3.43×10^{19}
Argon	^{36}Ar	35.975	0.337	6.76×10^{21}
	^{38}Ar	37.975	0.063	1.26×10^{20}
	^{40}Ar	39.975	99.600	2.00×10^{23}

Since the ratio $^{18}O/^{16}O$ is not small, isotopic analysis should be carried out in the thermosphere to determine the variation due to diffusion in the earth's gravitational field. Starting from the ratio $^{18}O/^{16}O$ near 100 km it is important to know the vertical distribution of this ratio up to the highest altitude.

As far as the isotopic ratios of noble gases are concerned (see *Table 2.6*) several considerations have been put forward in the preceding paragraphs. Nevertheless, it should be pointed out that various studies can be made which could lead to the evaluation of new aeronomic problems. For a complete understanding of the behavior of the noble gases it would be necessary to study the atmospheric effects from the troposphere up to the thermosphere.

REFERENCES

1. E. Glückauf, in *Compendium of Meteorology* (T. F. Malone, ed.), pp. 3–10. Amer. Meteorolog. Soc., Boston, Massachusetts, 1951.
2. A. G. W. Cameron, in *The Origin and Evolution of the Atmosphere and Oceans* (P. J. Brancazio and A. G. W. Cameron, eds.), p. 235. Wiley, New York, 1964.
3. A. G. W. Cameron, in *Origin and Distribution of the Elements* (L. H. Ahrens, ed.), p. 125. Pergamon, Oxford, 1968.
4. G. J. F. MacDonald, *Rev. Geophys.* **1**, 305 (1963).
5. E. Glückauf and F. A. Paneth, *Proc. Roy. Soc.* **185A**, 89 (1946).
6. H. E. Suess and H. C. Urey, *Rev. Mod. Phys.* **28**, 53 (1956).
7. M. Nicolet, *Ann. Géophys.* **13**, 1 (1957).
8. P. Signer and H. E. Suess, in *Earth Sciences and Meteoritics* (J. Geiss and E. D. Goldberg, eds.), chapter 13. North-Holland Publ., Amsterdam, 1963.
9. P. Morrison and J. Pine, *Ann. N.Y. Acad. Sci.* **62**, 69 (1955).
10. W. I. Axford, in *Particles and Fields in the Magnetosphere* (B. M. McCormac, ed.), pp. 46–59. Reidel Publ., Dordrecht, The Netherlands, 1970.
11. W. I. Axford, in *Atmospheric Emissions* (B. M. McCormac and A. Omholt, eds.), pp. 317–325. Van Nostrand-Reinhold, Princeton, New Jersey, 1969.
12. G. Kockarts and M. Nicolet, *Ann. Géophys.* **18**, 269 (1962).
13. T. N. L. Patterson, *Rev. Geophys.* **6**, 553 (1968).
14. W. I. Axford, *J. Geophys. Res.* **73**, 6855 (1968).
15. P. M. Banks and T. E. Holzer, *J. Geophys. Res.* **74**, 6317 (1969).
16. J.-É. Blamont, Personal communication (1971).
17. H. E. Johnson and W. I. Axford, *J. Geophys. Res.* **74**, 2433 (1969).
18. M. Nicolet, *Bull. Cl. Sci., Acad. Roy. Belg.* **42**, 482 (1956).
19. G. J. Wasserburg, G. J. F. MacDonald, F. Hoyle, and W. A. Fowler, *Science* **143,** 465 (1964).
20. A. G. W. Cameron, *Icarus* **1**, 314 (1963).
21. D. Tilles, *Science* **148**, 1085 (1965).
22. S. F. Singer, *Science* **156**, 1080 (1967).

23. R. H. Bieri, M. Koide, E. A. Martell, and T. G. Scholz, *J. Geophys. Res.* **75**, 6731 (1970).
24. D. Q. Wark and D. T. Hilleary, *Science* **165**, 1256 (1969).
25. B. J. Conrath, R. A. Hanel, V. G. Kunde, and C. Prabhakara, *J. Geophys. Res.* **75**, 5831 (1970).
26. C. D. Keeling, *Tellus* **12**, 200 (1960).
27. B. Bolin and C. D. Keeling, *J. Geophys. Res.* **68**, 3899 (1963).
28. L. Goldberg, in *The Solar System*, Vol. II, *The Earth as a Planet* (G. P. Kuiper, ed.), pp. 434–490. Univ. of Chicago Press, Chicago, Illinois, 1954.
29. L. Goldberg, *Astrophys. J.* **113**, 567 (1951).
30. A. H. Nielsen and M. Migeotte, *Ann. Astrophys.* **15**, 134 (1952).
31. J. S. Seeley and J. T. Houghton, *Infrared Phys.* **1**, 116 (1961).
32. R. L. Bowman and J. H. Shaw, *Appl. Opt.* **3**, 176 (1963).
33. U. Fink, D. H. Rank, and T. A. Wiggins, *J. Opt. Soc. Amer.* **54**, 472 (1964).
34. G. E. Hutchinson, in *The Solar System*, Vol. II, *The Earth as a Planet* (G. P. Kuiper, ed.), pp. 371–433. Univ. of Chicago Press, Chicago, Illinois, 1954.
35. D. R. Bates and P. B. Hays, *Planet. Space Sci.* **15**, 189 (1967).
36. D. R. Bates and A. E. Witherspoon, *Mon. Notices Roy. Astronom. Soc.* **112**, 101 (1952).
37. K. Schütz, C. Junge, R. Beck, and B. Albrecht, *J. Geophys. Res.* **75**, 2230 (1970).
38. J. W. Birkcland and J. H. Shaw, *J. Opt. Soc. Amer.* **49**, 637 (1959).
39. D. H. Rank, A. F. Slomba, E. F. Gardner, and T. A. Wiggins, *J. Opt. Soc. Amer.* **52**, 858 (1962).
40. L. S. Jaffe, *J. Air Pollut. Contr. Ass.* **18**, 534 (1968).
41. J. W. Swinnerton, V. J. Linnenbom, and R. A. Lamontague, *Science* **167**, 984 (1970).
42. W. Seiler and C. Junge, *J. Geophys. Res.* **75**, 2217 (1970).
43. W. Seiler and C. Junge, *Tellus* **21**, 447 (1969).
44. J. Pressman and P. Warneck, *J. Atmos. Sci.* **27**, 155 (1970).
45. E. Hesstvedt, *Nature* (*London*) **225**, 50 (1960).
46. C. Junge, W. Seiler, and P. Warneck, *J. Geophys. Res.* **76**, 2866 (1971).
47. A. O. Nier, *Science* **121**, 737 (1955).
48. W. R. Schell, G. Sauzay, and B. R. Payne, *J. Geophys. Res.* **75**, 2251 (1970).
49. F. Bergman, in *Earth Science and Meteoritics* (J. Geiss and E. D. Goldberg, eds.), p. 169. North-Holland Publ., Amsterdam, 1963.
50. H. E. Suess, *Ann. Rev. Nucl. Sci.* **8**, 243 (1958).
51. W. F. Libby, *Res. Geochem.* **151**, 33 (1959).
52. W. B. Clarke, M. A. Beg, and H. Craig, *Earth Planet. Sci. Lett.* **6**, 213 (1969).
53. H. Craig, *J. Geol.* **62**, 115 (1954).
54. H. Craig, in *Earth Science and Meteoritics* (J. Geiss and E. D. Goldberg, eds.). Chapter 6, p. 103. North-Holland Publ., Amsterdam, 1963.
55. H. D. Babcock and L. Herzberg, *Astrophys. J.* **108**, 167 (1948).
56. G. A. Lane and M. Dole, *Science* **123**, 574 (1956).

CHAPTER 3

Physical Conditions of the Homosphere

3.1 Introduction

Thirty years ago it was generally accepted that the composition of the stratospheric air varied with altitude because it was believed that helium and oxygen samples taken above 20 km showed departures from a mixing distribution toward partial diffusive separation. Today one must conclude that the relative helium content is the same in the stratosphere as at ground level. Furthermore, it seems safe to conclude that gases which are sufficiently inert chemically do not exhibit changes in relative abundance below the thermosphere. Information as to the composition of the atmosphere up to 150 km indicates that the homosphere certainly extends to the mesopause. In other words, the mean molecular mass does not change up to the lower thermosphere, where it changes through the partial dissociation of molecular oxygen, and above 100 km, by virtue of diffusive separation.

By reference to the descriptive account of the atmosphere given in Section 1.1 and the atmospheric composition of the principal constituents given in Table 2.1 it is possible to determine the average conditions of the homosphere, and thus, to obtain orders of magnitude for various physical parameters.

Following the derivation of Eq. (1.4), the hydrostatic equation can be written as

$$\frac{dp}{p} = -\frac{dz}{H} = \frac{dn}{n} + \frac{dT}{T} \tag{3.1}$$

using previously defined quantities. The variation of pressure with altitude is given immediately as

$$p = p_0 \exp\left[-\int_{z_0}^{z} dz/H\right] \tag{3.2}$$

where the scale height is given by $H = kT/mg$ and p_0 is the gas pressure at the altitude z_0.

The gas concentration, n, is obtained from Eq. (3.2) using the perfect gas law, $p = nkT$:

$$n = n_0(T_0/T) \exp\left[-\int_{z_0}^{z} dz/H\right]. \tag{3.3}$$

Further reduction of (3.2) and (3.3) cannot be made without knowledge of the temperature profile of the atmosphere. To illustrate the general trends of pressure and density, however, three approximations to the variation of the scale height involving different assumptions about changes in gravity and temperature can be introduced: (1) constant temperature and gravity; (2) constant temperature with gravity dependent upon altitude, and (3) constant gradient of scale height.

(1) If both T and g are constants the scale height, H, is constant and (3.2) and (3.3) can be integrated to give

$$\frac{p}{p_0} = \frac{n}{n_0} = \frac{\varrho}{\varrho_0} = \exp\left[-\frac{z - z_0}{H_0}\right] \tag{3.4}$$

where $H_0 = kT_0/mg_0$, showing the exponential variation of pressure, concentration, and density with altitude above the reference point z_0.

The total number of particles in a square centimeter column lying above the altitude z is given by the column content

$$\int_{z_0}^{\infty} n(z)\, dz = n_0 H_0. \tag{3.5}$$

(2) For the second approximation, assume $T = T_0 \triangleq$ constant and g to be given by

$$g = g_0\left(\frac{r_0}{r}\right)^2 = \frac{g_0}{[1 + (z - z_0)/r_0]^2} \tag{3.6}$$

where g_0 is the acceleration of gravity at the geocentric distance r_0. Using Eqs. (3.2), (3.3), and (3.6), one obtains

$$\frac{p}{p_0} = \frac{n}{n_0} = \frac{\varrho}{\varrho_0} = \exp\left[-\frac{r_0}{H_0}\left(\frac{r - r_0}{r}\right)\right] \tag{3.7}$$

in terms of the geocentric distance r with $H_0 = kT_0/mg_0$. Noting that $r - r_0 = z - z_0$, it is possible to recast (3.7) in the form

$$\frac{p}{p_0} = \frac{n}{n_0} = \frac{\varrho}{\varrho_0} = \exp\left[-\frac{1}{H_0}\frac{(z-z_0)}{1+(z-z_0)/r_0}\right]. \tag{3.8}$$

If the *reduced altitude* or *geopotential height* is defined as

$$z' - z_0' = \frac{z - z_0}{1 + (z - z_0)/r_0}, \tag{3.9}$$

then Eq. (3.8) has the same form as Eq. (3.4):

$$\frac{p}{p_0} = \frac{n}{n_0} = \frac{\varrho}{\varrho_0} \exp\left[-\frac{z' - z_0'}{H_0}\right]. \tag{3.10}$$

In this manner the variation of scale height due to changes in gravity can be eliminated from the simple constant temperature model. Usually the geopotential height is defined with respect to ground level so that $z_0' = z_0 = 0$ and r_0 becomes the earth's radius. The column content of gas lying above the altitude r is not bounded for this case as a result of the ratio n/n_0 approaching the value $\exp[-r_0/H_0]$ as r becomes very large.

(3) If one introduces the gradient of the scale height $\beta \equiv dH/dz$, Eqs. (3.1) and (3.6) lead to the following equations:

$$\frac{dp}{p} = -\frac{1}{\beta}\frac{dH}{H} \tag{3.11}$$

and

$$\frac{d\varrho g}{\varrho g} = -\frac{1+\beta}{\beta}\frac{dH}{H}. \tag{3.12}$$

The procedure to be used in describing the vertical distribution of pressure and density depends on the form of variation assumed for β. It is necessary, however, to note that the scale height has a physical meaning, $H = kT/mg$, in order to avoid error when employing it as a parameter to follow a particular vertical distribution. Therefore, in approximating aeronomic conditions it is best to assume that the scale height gradient is practically constant within a specified altitude range. If β is constant, integration of (3.11) and (3.12) leads to

$$p/p_0 = (H/H_0)^{-1/\beta} \tag{3.13}$$

and

$$\varrho g/\varrho_0 g_0 = (H/H_0)^{-(1+\beta)/\beta} \tag{3.14}$$

if p_0, ϱ_0, g_0, and H_0 denote the values of p, ϱ, g, and H at height $z = z_0$.

The expansion of (3.13) and (3.14) in series enables a comparison to be made with the isothermal atmosphere model. These can be written as follows:

$$\frac{p}{p_0} = \exp\left\{- \frac{z - z_0}{\frac{1}{2}(H + H_0)} \left[1 + \frac{1}{3}\left(\frac{H - H_0}{H + H_0}\right)^2 + \cdots\right]\right\} \tag{3.15}$$

and

$$\frac{\varrho g}{\varrho_0 g_0} = \exp\left\{- \frac{(1 + \beta)(z - z_0)}{\frac{1}{2}(H + H_0)} \left[1 + \frac{1}{3}\left(\frac{H - H_0}{H + H_0}\right)^2 + \cdots\right]\right\}. \tag{3.16}$$

In the homosphere, the scale height gradient β is always less than 0.4. Expressions for p and ϱ, accurate to better than 1% for a height interval reaching one scale height, are obtained by keeping only the first term in brackets in (3.15) and (3.16), i.e.,

$$\frac{p}{p_0} = \exp\left\{- \frac{z - z_0}{\frac{1}{2}(H + H_0)}\right\} \tag{3.17}$$

and

$$\frac{\varrho g}{\varrho_0 g_0} = \exp\left\{- \frac{(1 + \beta)(z - z_0)}{\frac{1}{2}(H + H_0)}\right\}. \tag{3.18}$$

3.2 Numerical Values between 15 and 100 km

In recent years the density of the homosphere has been explored above 30 km by rockets which have given a general picture of the homosphere up to the mesopause. Variations are observed and it is difficult to use a standard model to represent the atmosphere with its seasonal and latitudinal variations [1–7]. Sources of heating due to the absorption of ultraviolet radiation in the spectral range 3000–2000 Å vary with height depending on latitudes and seasons and sources of cooling occur through infrared emission [8, 9] processes. The wind variation corresponding to the seasonal change from westerlies in winter to easterlies in summer is also an indication that there is not only a radiative balance but also important atmospheric dynamical effects. In any case, the establishment of rocket networks with consequent measurements of vertical profiles of temperature and density for

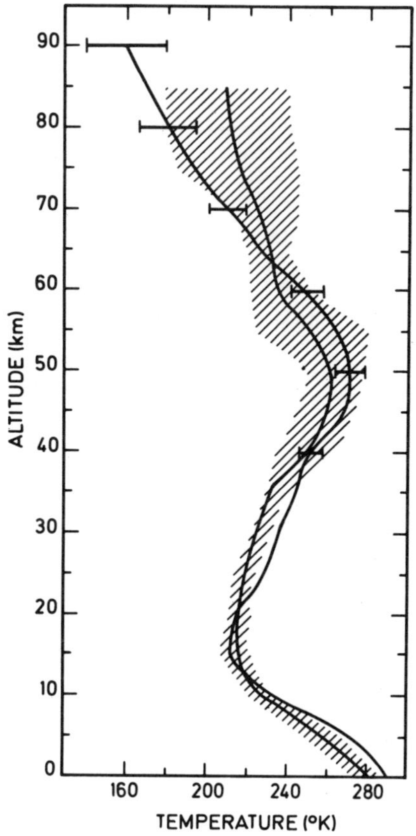

Fig. 3.1 Average profiles of temperature versus altitude for winter and summer over Wallops Island, Virginia (38°N) [10]. Maximum variation in temperature between sounding indicated by shading for winter and by horizontal bars for summer.

individual launching sites [10] (see *Fig. 3.1* for Wallops Island, Virginia) or for latitudinal cross sections [11] (see *Fig. 3.2*) leads to basic information [12] on the seasonal and latitudinal variations of temperature, density, and winds. In addition, direct profiles of temperature, humidity, and minor constituent concentration are available from satellite observations of infrared emissions [13]. However, in order to begin a study of aeronomic processes, it is necessary to use an atmospheric model for which the principal parameters are represented by average values. Computations of pressure p, density ϱ, and total concentration $n(M)$ have been made using constant gradients of the scale height, positive between 30 and 45 km, zero between 45 and 55 km, negative between 55 and 85 km, and positive from 85 to 100 km. The corresponding scale heights are given in the second column of *Table 3.1* showing, for steps of 1 km, pressure, density, and total concentrations.

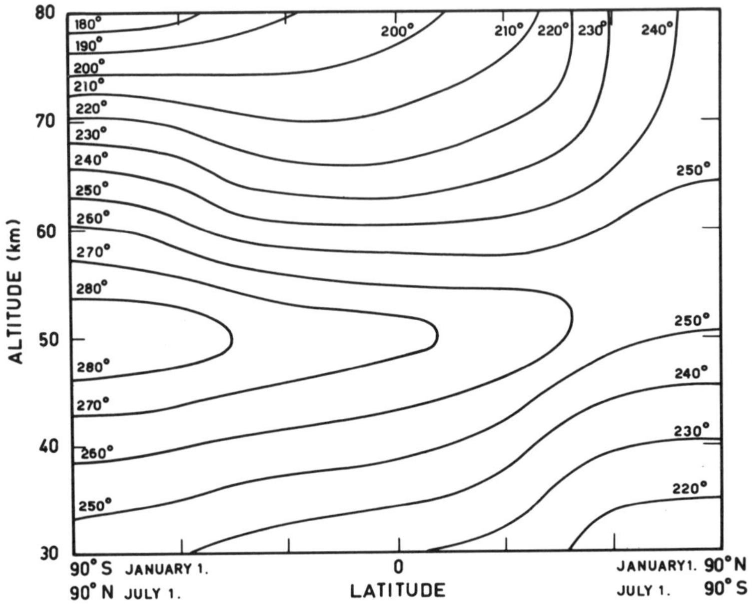

Fig. 3.2 Latitudinal temperature–height cross section; January or July [12].

A number of atmospheric models have been published which exhibit important differences. This arises from the fact that the computations have not been made in the same way and that time variations exist in the stratosphere and mesosphere as in the troposphere. For example, a density of 1.8×10^{-5} gm cm^{-3} adopted at 30 km can be accepted with a variation of $\pm 20\%$. A temperature of 273°K at 50 km may vary by about $\pm 10\%$ to reach extreme values. The stratopause level at 50 km may change by about ± 5 km considering all conditions. It can be seen that in Table 3.1 the density at 70 km is about 10^{-7} gm cm^{-3} while values of the order of $(7.5 \pm 2.5) \times 10^{-8}$ gm cm^{-3} are observed depending on mesospheric conditions. For this reason the pressure 3×10^{-4} Torr given in Table 3.1 for 100 km corresponds only to an order of magnitude estimate. In fact, the pressure at 100 km depends on mesospheric conditions and particularly on the temperature minimum at the mesopause level. If, for example, a variation of the order of 60°K occurs in the neighborhood of the mesopause, the pressure at 100 km can vary within a ratio of the order of 3 while the same conditions are maintained at 70 km.

Starting from data of Table 3.1 in which the density is about 10^{-7} gm cm^{-3} at 70 km, the general trend of the temperature at the mesopause and

TABLE 3.1

PRESSURE, DENSITY, CONCENTRATION, AND TEMPERATURE FROM 15 TO 100 KM

Altitude (km)	Scale height (km)	Temperature (°K)	Pressure (Torr)	Pressure (mb)	Density (gm cm^{-3})	Concentration (cm^{-3})
15	6.20	210.8	8.48×10^{1}	1.13×10^{2}	1.87×10^{-4}	3.89×10^{18}
16	6.25	212.4	7.22×10^{1}	9.63×10^{1}	1.58×10^{-4}	3.28×10^{18}
17	6.30	214.0	6.16×10^{1}	8.21×10^{1}	1.34×10^{-4}	2.78×10^{18}
18	6.35	215.7	5.26×10^{1}	7.01×10^{1}	1.13×10^{-4}	2.36×10^{18}
19	6.40	217.3	4.50×10^{1}	5.99×10^{1}	9.61×10^{-5}	2.00×10^{18}
20	6.45	218.9	3.85×10^{1}	5.13×10^{1}	8.16×10^{-5}	1.70×10^{18}
21	6.50	220.6	3.30×10^{1}	4.40×10^{1}	6.94×10^{-5}	1.44×10^{18}
22	6.55	222.2	2.83×10^{1}	3.77×10^{1}	5.91×10^{-5}	1.23×10^{18}
23	6.60	223.8	2.43×10^{1}	3.24×10^{1}	5.04×10^{-5}	1.05×10^{18}
24	6.65	225.4	2.09×10^{1}	2.78×10^{1}	4.30×10^{-5}	8.95×10^{17}
25	6.70	227.1	1.80×10^{1}	2.40×10^{1}	3.68×10^{-5}	7.65×10^{17}
26	6.75	228.7	1.55×10^{1}	2.07×10^{1}	3.15×10^{-5}	6.54×10^{17}
27	6.80	230.3	1.34×10^{1}	1.78×10^{1}	2.70×10^{-5}	5.61×10^{17}
28	6.85	231.9	1.16×10^{1}	1.54×10^{1}	2.31×10^{-5}	4.81×10^{17}
29	6.90	233.5	9.98×10^{0}	1.33×10^{1}	1.98×10^{-5}	4.13×10^{17}
30	6.95	235.2	8.64×10^{0}	1.15×10^{1}	1.71×10^{-5}	3.55×10^{17}
31	7.05	238.5	7.49×10^{0}	9.99×10^{0}	1.46×10^{-5}	3.03×10^{17}
32	7.15	241.8	6.51×10^{0}	8.68×10^{0}	1.25×10^{-5}	2.60×10^{17}
33	7.25	245.1	5.66×10^{0}	7.55×10^{0}	1.07×10^{-5}	2.23×10^{17}
34	7.35	248.4	4.94×10^{0}	6.58×10^{0}	9.23×10^{-6}	1.92×10^{17}
35	7.45	251.7	4.31×10^{0}	5.75×10^{0}	7.96×10^{-6}	1.66×10^{17}
36	7.55	255.0	3.78×10^{0}	5.03×10^{0}	6.88×10^{-6}	1.43×10^{17}
37	7.65	258.3	3.31×10^{0}	4.41×10^{0}	5.95×10^{-6}	1.24×10^{17}
38	7.75	261.6	2.91×10^{0}	3.88×10^{0}	5.16×10^{-6}	1.07×10^{17}
39	7.85	264.9	2.56×10^{0}	3.41×10^{0}	4.48×10^{-6}	9.32×10^{16}
40	7.95	268.2	2.25×10^{0}	3.00×10^{0}	3.90×10^{-6}	8.11×10^{16}
41	7.99	269.4	1.99×10^{0}	2.65×10^{0}	3.42×10^{-6}	7.12×10^{16}
42	8.03	270.7	1.75×10^{0}	2.34×10^{0}	3.01×10^{-6}	6.26×10^{16}
43	8.07	272.0	1.55×10^{0}	2.07×10^{0}	2.64×10^{-6}	5.50×10^{16}
44	8.11	273.2	1.37×10^{0}	1.82×10^{0}	2.33×10^{-6}	4.84×10^{16}
45	8.15	274.5	1.21×10^{0}	1.61×10^{0}	2.05×10^{-6}	4.26×10^{16}
46	8.15	274.4	1.07×10^{0}	1.43×10^{0}	1.81×10^{-6}	3.77×10^{16}
47	8.15	274.3	9.47×10^{-1}	1.26×10^{0}	1.60×10^{-6}	3.34×10^{16}
48	8.15	274.2	8.38×10^{-1}	1.12×10^{0}	1.42×10^{-6}	2.95×10^{16}
49	8.15	274.1	7.41×10^{-1}	9.88×10^{-1}	1.26×10^{-6}	2.61×10^{16}
50	8.15	274.0	6.56×10^{-1}	8.74×10^{-1}	1.11×10^{-6}	2.31×10^{16}
51	8.15	274.0	5.80×10^{-1}	7.73×10^{-1}	9.83×10^{-7}	2.04×10^{16}
52	8.15	273.9	5.13×10^{-1}	6.84×10^{-1}	8.70×10^{-7}	1.81×10^{16}

TABLE 3.1 (*Continued*)

Altitude (km)	Scale height (km)	Temperature (°K)	Pressure (Torr)	Pressure (mb)	Density (gm cm^{-3})	Concentration (cm^{-3})
53	8.15	273.8	4.54×10^{-1}	6.05×10^{-1}	7.69×10^{-7}	1.60×10^{16}
54	8.15	273.7	4.01×10^{-1}	5.35×10^{-1}	6.81×10^{-7}	1.42×10^{16}
55	8.15	273.6	3.55×10^{-1}	4.73×10^{-1}	6.02×10^{-7}	1.25×10^{16}
56	8.03	269.4	3.14×10^{-1}	4.18×10^{-1}	5.41×10^{-7}	1.12×10^{16}
57	7.91	265.3	2.77×10^{-1}	3.69×10^{-1}	4.84×10^{-7}	1.01×10^{16}
58	7.78	261.1	2.44×10^{-1}	3.25×10^{-1}	4.33×10^{-7}	9.01×10^{15}
59	7.66	256.9	2.14×10^{-1}	2.85×10^{-1}	3.87×10^{-7}	8.04×10^{15}
60	7.54	252.8	1.88×10^{-1}	2.50×10^{-1}	3.45×10^{-7}	7.17×10^{15}
61	7.42	248.6	1.64×10^{-1}	2.19×10^{-1}	3.07×10^{-7}	6.38×10^{15}
62	7.30	244.4	1.43×10^{-1}	1.91×10^{-1}	2.72×10^{-7}	5.66×10^{15}
63	7.17	240.2	1.25×10^{-1}	1.66×10^{-1}	2.41×10^{-7}	5.02×10^{15}
64	7.05	236.1	1.08×10^{-1}	1.44×10^{-1}	2.13×10^{-7}	4.43×10^{15}
65	6.93	231.9	9.40×10^{-2}	1.25×10^{-1}	1.88×10^{-7}	3.91×10^{15}
66	6.81	227.8	8.12×10^{-2}	1.08×10^{-1}	1.66×10^{-7}	3.44×10^{15}
67	6.69	223.6	7.00×10^{-2}	9.34×10^{-2}	1.45×10^{-7}	3.02×10^{15}
68	6.56	219.5	6.02×10^{-2}	8.03×10^{-2}	1.27×10^{-7}	2.65×10^{15}
69	6.44	215.3	5.16×10^{-2}	6.88×10^{-2}	1.11×10^{-7}	2.32×10^{15}
70	6.32	211.2	4.41×10^{-2}	5.88×10^{-2}	9.71×10^{-8}	2.02×10^{15}
71	6.28	209.8	3.77×10^{-2}	5.02×10^{-2}	8.34×10^{-8}	1.73×10^{15}
	6.22	207.8	3.76×10^{-2}	5.02×10^{-2}	8.41×10^{-8}	1.75×10^{15}
72	6.24	208.4	3.21×10^{-2}	4.28×10^{-2}	7.15×10^{-8}	1.49×10^{15}
	6.12	204.4	3.20×10^{-2}	4.27×10^{-2}	7.27×10^{-8}	1.51×10^{15}
73	6.20	207.0	2.73×10^{-2}	3.64×10^{-2}	6.13×10^{-8}	1.28×10^{15}
	6.02	201.0	2.71×10^{-2}	3.62×10^{-2}	6.27×10^{-8}	1.30×10^{15}
74	6.16	205.6	2.32×10^{-2}	3.10×10^{-2}	5.25×10^{-8}	1.09×10^{15}
	5.92	197.6	2.30×10^{-2}	3.06×10^{-2}	5.40×10^{-8}	1.12×10^{15}
75	6.12	204.2	1.98×10^{-2}	2.63×10^{-2}	4.49×10^{-8}	9.34×10^{14}
	5.82	194.2	1.94×10^{-2}	2.58×10^{-2}	4.63×10^{-8}	9.63×10^{14}
76	6.08	202.8	1.68×10^{-2}	2.24×10^{-2}	3.84×10^{-8}	7.99×10^{14}
	5.72	190.8	1.63×10^{-2}	2.17×10^{-2}	3.96×10^{-8}	8.24×10^{14}
77	6.04	201.4	1.42×10^{-2}	1.90×10^{-2}	3.28×10^{-8}	6.82×10^{14}
	5.62	187.4	1.36×10^{-2}	1.82×10^{-2}	3.38×10^{-8}	7.03×10^{14}
78	6.00	200.0	1.20×10^{-2}	1.61×10^{-2}	2.80×10^{-8}	5.82×10^{14}
	5.52	184.0	1.14×10^{-2}	1.52×10^{-2}	2.88×10^{-8}	5.98×10^{14}
79	5.96	198.6	1.02×10^{-2}	1.36×10^{-2}	2.38×10^{-8}	4.95×10^{14}
	5.42	180.6	9.50×10^{-3}	1.27×10^{-2}	2.44×10^{-8}	5.08×10^{14}
80	5.92	197.2	8.61×10^{-3}	1.15×10^{-2}	2.03×10^{-8}	4.22×10^{14}
	5.32	177.2	7.88×10^{-3}	1.05×10^{-2}	2.07×10^{-8}	4.30×10^{14}
81	5.88	195.8	7.27×10^{-3}	9.69×10^{-3}	1.72×10^{-8}	3.58×10^{14}

TABLE 3.1 (*Continued*)

Altitude (km)	Scale height (km)	Tempe-rature (°K)	Pressure (Torr)	Pressure (mb)	Density (gm cm^{-3})	Concentra-tion (cm^{-3})
	5.22	173.8	6.52×10^{-3}	8.70×10^{-3}	1.74×10^{-8}	3.62×10^{14}
82	5.84	194.4	6.13×10^{-3}	8.17×10^{-3}	1.46×10^{-8}	3.04×10^{14}
	5.12	170.4	5.38×10^{-3}	7.16×10^{-3}	1.46×10^{-8}	3.04×10^{14}
83	5.80	193.0	5.16×10^{-3}	6.88×10^{-3}	1.24×10^{-8}	2.58×10^{14}
	5.02	167.1	4.41×10^{-3}	5.88×10^{-3}	1.23×10^{-8}	2.55×10^{14}
84	5.76	191.6	4.34×10^{-3}	5.79×10^{-3}	1.05×10^{-8}	2.19×10^{14}
	4.92	163.7	3.61×10^{-3}	4.81×10^{-3}	1.02×10^{-8}	2.13×10^{14}
85	5.72	190.2	3.65×10^{-3}	4.86×10^{-3}	8.90×10^{-9}	1.85×10^{14}
	4.82	160.3	2.94×10^{-3}	3.92×10^{-3}	8.51×10^{-9}	1.77×10^{14}
86	5.76	191.5	3.06×10^{-3}	4.08×10^{-3}	7.43×10^{-9}	1.54×10^{14}
	4.92	163.6	2.39×10^{-3}	3.19×10^{-3}	6.79×10^{-9}	1.41×10^{14}
87	5.80	192.8	2.58×10^{-3}	3.44×10^{-3}	6.21×10^{-9}	1.29×10^{14}
	5.02	166.9	1.96×10^{-3}	2.61×10^{-3}	5.44×10^{-9}	1.13×10^{14}
88	5.84	194.1	2.17×10^{-3}	2.89×10^{-3}	5.19×10^{-9}	1.08×10^{14}
	5.12	170.1	1.61×10^{-3}	2.14×10^{-3}	4.38×10^{-9}	9.12×10^{13}
89	5.88	195.3	1.83×10^{-3}	2.44×10^{-3}	4.35×10^{-9}	9.04×10^{13}
	5.22	173.4	1.32×10^{-3}	1.76×10^{-3}	3.54×10^{-9}	7.37×10^{13}
90	5.92	196.6	1.54×10^{-3}	2.06×10^{-3}	3.65×10^{-9}	7.58×10^{13}
	5.32	176.7	1.10×10^{-3}	1.46×10^{-3}	2.88×10^{-9}	5.99×10^{13}
91	5.96	197.9	1.30×10^{-3}	1.74×10^{-3}	3.06×10^{-9}	6.37×10^{13}
	5.42	179.9	9.09×10^{-4}	1.21×10^{-3}	2.35×10^{-9}	4.88×10^{13}
92	6.00	199.1	1.10×10^{-3}	1.47×10^{-3}	2.57×10^{-9}	5.35×10^{13}
	5.52	183.2	7.57×10^{-4}	1.01×10^{-3}	1.92×10^{-9}	3.99×10^{13}
93	6.04	200.4	9.35×10^{-4}	1.25×10^{-3}	2.17×10^{-9}	4.51×10^{13}
	5.62	186.5	6.33×10^{-4}	8.44×10^{-4}	1.58×10^{-9}	3.28×10^{13}
94	6.08	201.7	7.93×10^{-4}	1.06×10^{-3}	1.82×10^{-9}	3.80×10^{13}
	5.72	189.7	5.30×10^{-4}	7.07×10^{-4}	1.30×10^{-9}	2.70×10^{13}
95	6.12	202.9	6.73×10^{-4}	8.97×10^{-4}	1.54×10^{-9}	3.20×10^{13}
	5.82	193.0	4.46×10^{-4}	5.95×10^{-4}	1.07×10^{-9}	2.23×10^{13}
96	6.16	204.2	5.72×10^{-4}	7.62×10^{-4}	1.30×10^{-9}	2.70×10^{13}
	5.92	196.2	3.76×10^{-4}	5.01×10^{-4}	8.90×10^{-10}	1.85×10^{13}
97	6.20	205.4	4.86×10^{-4}	6.48×10^{-4}	1.10×10^{-9}	2.29×10^{13}
	6.02	199.5	3.18×10^{-4}	4.24×10^{-4}	7.40×10^{-10}	1.54×10^{13}
98	6.24	206.7	4.14×10^{-4}	5.52×10^{-4}	9.30×10^{-10}	1.94×10^{13}
	6.12	202.7	2.70×10^{-4}	3.60×10^{-4}	6.18×10^{-10}	1.28×10^{13}
99	6.28	208.0	3.53×10^{-4}	4.71×10^{-4}	7.88×10^{-10}	1.64×10^{13}
	6.22	206.0	2.29×10^{-4}	3.06×10^{-4}	5.17×10^{-10}	1.08×10^{13}
100	6.32	209.2	3.01×10^{-4}	4.02×10^{-4}	6.68×10^{-10}	1.39×10^{13}
	6.32	209.2	1.96×10^{-4}	2.61×10^{-4}	4.34×10^{-10}	9.03×10^{12}

TABLE 3.2

CONCENTRATIONS AND TOTAL NUMBERS OF MOLECULES

	Concentrations		Total numbers		
Altitude (km)	$n(N_2)$ (cm^{-3})	$n(O_2)$ (cm^{-3})	$n(M)H$ (cm^{-2})	$n(N_2)H$ (cm^{-2})	$n(O_2)H$ (cm^{-2})
15	3.04×10^{18}	8.14×10^{17}	2.41×10^{24}	1.88×10^{24}	5.05×10^{23}
16	2.56×10^{18}	6.88×10^{17}	2.05×10^{24}	1.60×10^{24}	4.30×10^{23}
17	2.17×10^{18}	5.82×10^{17}	1.75×10^{24}	1.37×10^{24}	3.67×10^{23}
18	1.84×10^{18}	4.93×10^{17}	1.50×10^{24}	1.17×10^{24}	3.13×10^{23}
19	1.56×10^{18}	4.18×10^{17}	1.28×10^{24}	9.98×10^{23}	2.67×10^{23}
20	1.32×10^{18}	3.56×10^{17}	1.10×10^{24}	8.55×10^{23}	2.29×10^{23}
21	1.13×10^{18}	3.02×10^{17}	9.38×10^{23}	7.33×10^{23}	1.96×10^{23}
22	9.60×10^{17}	2.58×10^{17}	8.05×10^{23}	6.29×10^{23}	1.69×10^{23}
23	8.18×10^{17}	2.20×10^{17}	6.92×10^{23}	5.40×10^{23}	1.45×10^{23}
24	6.99×10^{17}	1.87×10^{17}	5.95×10^{23}	4.65×10^{23}	1.25×10^{23}
25	5.97×10^{17}	1.60×10^{17}	5.12×10^{23}	4.00×10^{23}	1.07×10^{23}
26	5.11×10^{17}	1.37×10^{17}	4.42×10^{23}	3.45×10^{23}	9.25×10^{22}
27	4.38×10^{17}	1.17×10^{17}	3.81×10^{23}	2.98×10^{23}	7.99×10^{22}
28	3.76×10^{17}	1.01×10^{17}	3.29×10^{23}	2.57×10^{23}	6.90×10^{22}
29	3.22×10^{17}	8.65×10^{16}	2.85×10^{23}	2.22×10^{23}	5.97×10^{22}
30	2.77×10^{17}	7.43×10^{16}	2.47×10^{23}	1.93×10^{23}	5.17×10^{22}
31	2.37×10^{17}	6.36×10^{16}	2.14×10^{23}	1.67×10^{23}	4.48×10^{22}
32	2.03×10^{17}	5.44×10^{16}	1.86×10^{23}	1.45×10^{23}	3.89×10^{22}
33	1.74×10^{17}	4.68×10^{16}	1.62×10^{23}	1.26×10^{23}	3.39×10^{22}
34	1.50×10^{17}	4.02×10^{16}	1.41×10^{23}	1.10×10^{23}	2.96×10^{22}
35	1.29×10^{17}	3.47×10^{16}	1.23×10^{23}	9.63×10^{22}	2.58×10^{22}
36	1.12×10^{17}	3.00×10^{16}	1.08×10^{23}	8.43×10^{22}	2.26×10^{22}
37	9.66×10^{16}	2.59×10^{16}	9.47×10^{22}	7.39×10^{22}	1.98×10^{22}
38	8.38×10^{16}	2.25×10^{16}	8.32×10^{22}	6.50×10^{22}	1.74×10^{22}
39	7.28×10^{16}	1.95×10^{16}	7.32×10^{22}	5.72×10^{22}	1.53×10^{22}
40	6.34×10^{16}	1.70×10^{16}	6.45×10^{22}	5.04×10^{22}	1.35×10^{22}
41	5.56×10^{16}	1.49×10^{16}	5.69×10^{22}	4.44×10^{22}	1.19×10^{22}
42	4.89×10^{16}	1.31×10^{16}	5.03×10^{22}	3.92×10^{22}	1.05×10^{22}
43	4.30×10^{16}	1.15×10^{16}	4.44×10^{22}	3.47×10^{22}	9.30×10^{21}
44	3.78×10^{16}	1.01×10^{16}	3.92×10^{22}	3.06×10^{22}	8.22×10^{21}
45	3.33×10^{16}	8.92×10^{15}	3.47×10^{22}	2.71×10^{22}	7.27×10^{21}
46	2.94×10^{16}	7.90×10^{15}	3.07×10^{22}	2.40×10^{22}	6.43×10^{21}
47	2.60×10^{16}	6.98×10^{15}	2.72×10^{22}	2.12×10^{22}	5.69×10^{21}
48	2.30×10^{16}	6.18×10^{15}	2.40×10^{22}	1.88×10^{22}	5.04×10^{21}
49	2.04×10^{16}	5.47×10^{15}	2.13×10^{22}	1.66×10^{22}	4.46×10^{21}
50	1.80×10^{16}	4.84×10^{15}	1.88×10^{22}	1.47×10^{22}	3.94×10^{21}
51	1.60×10^{16}	4.28×10^{15}	1.67×10^{22}	1.30×10^{22}	3.49×10^{21}
52	1.41×10^{16}	3.79×10^{15}	1.47×10^{22}	1.15×10^{22}	3.09×10^{21}

TABLE 3.2 (*Continued*)

	Concentrations		Total numbers		
Altitude (km)	$n(N_2)$ (cm^{-3})	$n(O_2)$ (cm^{-3})	$n(M)H$ (cm^{-2})	$n(N_2)H$ (cm^{-2})	$n(O_2)H$ (cm^{-2})
53	1.25×10^{16}	3.35×10^{15}	1.30×10^{22}	1.02×10^{22}	2.73×10^{21}
54	1.11×10^{16}	2.97×10^{15}	1.15×10^{22}	9.01×10^{21}	2.42×10^{21}
55	9.78×10^{15}	2.62×10^{15}	1.02×10^{22}	7.97×10^{21}	2.14×10^{21}
56	8.78×10^{15}	2.36×10^{15}	9.02×10^{21}	7.05×10^{21}	1.89×10^{21}
57	7.86×10^{15}	2.11×10^{15}	7.96×10^{21}	6.22×10^{21}	1.67×10^{21}
58	7.03×10^{15}	1.89×10^{15}	7.01×10^{21}	5.48×10^{21}	1.47×10^{21}
59	6.28×10^{15}	1.68×10^{15}	6.16×10^{21}	4.81×10^{21}	1.29×10^{21}
60	5.60×10^{15}	1.50×10^{15}	5.40×10^{21}	4.22×10^{21}	1.13×10^{21}
61	4.98×10^{15}	1.34×10^{15}	4.73×10^{21}	3.69×10^{21}	9.91×10^{20}
62	4.42×10^{15}	1.19×10^{15}	4.13×10^{21}	3.22×10^{21}	8.65×10^{20}
63	3.92×10^{15}	1.05×10^{15}	3.60×10^{21}	2.81×10^{21}	7.54×10^{20}
64	3.46×10^{15}	9.29×10^{14}	3.13×10^{21}	2.44×10^{21}	6.55×10^{20}
65	3.05×10^{15}	8.19×10^{14}	2.71×10^{21}	2.12×10^{21}	5.68×10^{20}
66	2.69×10^{15}	7.21×10^{14}	2.34×10^{21}	1.83×10^{21}	4.91×10^{20}
67	2.36×10^{15}	6.33×10^{14}	2.02×10^{21}	1.58×10^{21}	4.24×10^{20}
68	2.07×10^{15}	5.55×10^{14}	1.74×10^{21}	1.36×10^{21}	3.64×10^{20}
69	1.81×10^{15}	4.85×10^{14}	1.49×10^{21}	1.16×10^{21}	3.12×10^{20}
70	1.58×10^{15}	4.23×10^{14}	1.28×10^{21}	9.96×10^{20}	2.67×10^{20}
71	1.35×10^{15}	3.63×10^{14}	1.09×10^{21}	8.50×10^{20}	2.28×10^{20}
	1.37×10^{15}	3.66×10^{14}	1.09×10^{21}	8.50×10^{20}	2.28×10^{20}
72	1.16×10^{15}	3.12×10^{14}	9.28×10^{20}	7.25×10^{20}	1.94×10^{20}
	1.18×10^{15}	3.17×10^{14}	9.26×10^{20}	7.23×10^{20}	1.94×10^{20}
73	9.96×10^{14}	2.67×10^{14}	7.91×10^{20}	6.17×10^{20}	1.66×10^{20}
	1.02×10^{15}	2.73×10^{14}	7.85×10^{20}	6.13×10^{20}	1.64×10^{20}
74	8.53×10^{14}	2.29×10^{14}	6.73×10^{20}	5.25×10^{20}	1.41×10^{20}
	8.76×10^{14}	2.35×10^{14}	6.64×10^{20}	5.19×10^{20}	1.39×10^{20}
75	7.30×10^{14}	1.96×10^{14}	5.73×10^{20}	4.46×10^{20}	1.20×10^{20}
	7.52×10^{14}	2.02×10^{14}	5.60×10^{20}	4.38×10^{20}	1.17×10^{20}
76	6.24×10^{14}	1.67×10^{14}	4.86×10^{20}	3.79×10^{20}	1.02×10^{20}
	6.43×10^{14}	1.73×10^{14}	4.71×10^{20}	3.68×10^{20}	9.87×10^{19}
77	5.32×10^{14}	1.43×10^{14}	4.12×10^{20}	3.22×10^{20}	8.63×10^{19}
	5.49×10^{14}	1.47×10^{14}	3.95×10^{20}	3.09×10^{20}	8.28×10^{19}
78	4.54×10^{14}	1.22×10^{14}	3.49×10^{20}	2.72×10^{20}	7.31×10^{19}
	4.67×10^{14}	1.25×10^{14}	3.30×10^{20}	2.58×10^{20}	6.92×10^{19}
79	3.87×10^{14}	1.04×10^{14}	2.95×10^{10}	2.30×10^{20}	6.18×10^{19}
	3.97×10^{14}	1.06×10^{14}	2.75×10^{20}	2.15×10^{20}	5.77×10^{19}
80	3.29×10^{14}	8.83×10^{13}	2.50×10^{20}	1.95×10^{20}	5.23×10^{19}
	3.36×10^{14}	9.00×10^{13}	2.29×10^{20}	1.78×10^{20}	4.79×10^{19}
81	2.80×10^{14}	7.51×10^{13}	2.11×10^{20}	1.65×10^{20}	4.41×10^{19}

TABLE 3.2 (*Continued*)

Altitude (km)	Concentrations $n(N_2)$ (cm^{-3})	$n(O_2)$ (cm^{-3})	Total numbers $n(M)H$ (cm^{-2})	$n(N_2)H$ (cm^{-2})	$n(O_2)H$ (cm^{-2})
	2.83×10^{14}	7.59×10^{13}	1.89×10^{20}	1.48×10^{20}	3.96×10^{19}
82	2.38×10^{14}	6.38×10^{13}	1.78×10^{20}	1.39×10^{20}	3.72×10^{19}
	2.38×10^{14}	6.38×10^{13}	1.56×10^{20}	1.22×10^{20}	3.27×10^{19}
83	2.02×10^{14}	5.41×10^{13}	1.50×10^{20}	1.17×10^{20}	3.14×10^{19}
	1.99×10^{14}	5.34×10^{13}	1.28×10^{20}	1.00×10^{20}	2.68×10^{19}
84	1.71×10^{14}	4.58×10^{13}	1.26×10^{20}	9.84×10^{19}	2.64×10^{19}
	1.66×10^{14}	4.46×10^{13}	1.05×10^{20}	8.18×10^{19}	2.19×10^{19}
85	1.44×10^{14}	3.88×10^{13}	1.06×10^{20}	8.27×10^{19}	2.22×10^{19}
	1.38×10^{14}	3.71×10^{13}	8.53×10^{19}	6.66×10^{19}	1.79×10^{19}
86	1.21×10^{14}	3.24×10^{13}	8.90×10^{19}	6.95×10^{19}	1.86×10^{19}
	1.10×10^{14}	2.96×10^{13}	6.95×10^{19}	5.43×10^{19}	1.46×10^{19}
87	1.01×10^{14}	2.70×10^{13}	7.48×10^{19}	5.84×10^{19}	1.57×10^{19}
	8.84×10^{13}	2.37×10^{13}	5.68×10^{19}	4.44×10^{19}	1.19×10^{19}
88	8.43×10^{13}	2.26×10^{13}	6.31×10^{19}	4.92×10^{19}	1.32×10^{19}
	7.12×10^{13}	1.91×10^{13}	4.67×10^{19}	3.64×10^{19}	9.78×10^{18}
89	7.06×10^{13}	1.89×10^{13}	5.32×10^{19}	4.15×10^{19}	1.11×10^{19}
	5.76×10^{13}	1.54×10^{13}	3.85×10^{19}	3.00×10^{19}	8.06×10^{18}
90	5.92×10^{13}	1.59×10^{13}	4.49×10^{19}	3.51×10^{19}	9.40×10^{18}
	4.67×10^{13}	1.25×10^{13}	3.18×10^{19}	2.49×10^{19}	6.67×10^{18}
91	4.97×10^{13}	1.33×10^{13}	3.80×10^{19}	2.96×10^{19}	7.95×10^{18}
	3.81×10^{13}	1.02×10^{13}	2.64×10^{19}	2.06×10^{19}	5.54×10^{18}
92	4.18×10^{13}	1.12×10^{13}	3.21×10^{19}	2.51×10^{19}	6.73×10^{18}
	3.12×10^{13}	8.36×10^{12}	2.20×10^{19}	1.72×10^{19}	4.61×10^{18}
93	3.52×10^{13}	9.44×10^{12}	2.72×10^{19}	2.12×10^{19}	5.70×10^{18}
	2.56×10^{13}	6.86×10^{12}	1.84×10^{19}	1.44×10^{19}	3.86×10^{18}
94	2.96×10^{13}	7.95×10^{12}	2.31×10^{19}	1.80×10^{19}	4.84×10^{18}
	2.11×10^{13}	5.66×10^{12}	1.54×10^{19}	1.21×10^{19}	3.23×10^{18}
95	2.50×10^{13}	6.71×10^{12}	1.96×10^{19}	1.53×10^{19}	4.10×10^{18}
	1.74×10^{13}	4.67×10^{12}	1.30×10^{19}	1.01×10^{19}	2.72×10^{18}
96	2.11×10^{13}	5.66×10^{12}	1.67×10^{19}	1.30×10^{19}	3.49×10^{18}
	1.44×10^{13}	3.88×10^{12}	1.10×10^{19}	8.56×10^{18}	2.30×10^{18}
97	1.78×10^{13}	4.79×10^{12}	1.42×10^{19}	1.11×10^{19}	2.97×10^{18}
	1.20×10^{13}	3.22×10^{12}	9.27×10^{18}	7.24×10^{18}	1.94×10^{18}
98	1.51×10^{13}	4.05×10^{12}	1.21×10^{19}	9.43×10^{18}	2.53×10^{18}
	1.00×10^{13}	2.69×10^{12}	7.86×10^{18}	6.14×10^{18}	1.65×10^{18}
99	1.28×10^{13}	3.43×10^{12}	1.03×10^{19}	8.04×10^{18}	2.16×10^{18}
	8.49×10^{12}	2.25×10^{12}	6.69×10^{18}	5.22×10^{18}	1.40×10^{18}
100	1.08×10^{13}	2.91×10^{12}	8.78×10^{18}	6.86×10^{18}	1.84×10^{18}
	7.05×10^{12}	1.89×10^{12}	5.70×10^{18}	4.45×10^{18}	1.20×10^{18}

of the pressure at 100 km will be as follows (keeping T constant at 70 and 100 km):

Temperature at 85 km (°K):	130	140	150	160	170	180	190	200	210
Pressure at 100 km (10^{-4} Torr):	1.1	1.4	1.7	2.0	2.3	2.6	3.0	3.4	3.8

A pressure of the order of 10^{-4} Torr requires a very deep minimum of temperature which varies from 130° to 160°K for densities between 10^{-7} and 5×10^{-8} gm cm^{-3} at 70 km. There is, therefore, a possibility of an important variation of density and pressure at 100 km. With the adoption of a value

$$p = (2.5 \pm 1.5)\times 10^{-4} \quad \text{Torr at 100 km}$$

it is certain that the upper and lower limits correspond to real variations. Such variations are extremely important since they will affect the boundary conditions which are used for the analysis of the structure of the thermosphere. To illustrate the possible differences, Table 3.1 gives atmospheric parameters for two temperature profiles above 70 km. The first corresponds to a mesopause minimum of 190°K at 85 km, while in the second a deeper minimum of 160°K is reached at the same altitude.

It should also be pointed out that the variations of temperature in the mesosphere will play an important role in the chemosphere where chemical reactions depend on activation energies.

In order to have data available for the computation of various types of collisions and the determination of the absorption of the solar radiations, *Table 3.2* gives the concentrations of molecular oxygen and nitrogen and the total number of molecules of a vertical column. These values, which will also be useful for discussions of various aeronomic problems, must be considered only as first approximations to the true values.

3.3 Lower Thermosphere

When the number of oxygen atoms becomes sufficiently large compared with the number of oxygen molecules, the simple aspect of the homosphere disappears. The theoretical analysis of Chapter 4 shows that the lifetime or the recombination time of an oxygen atom increases rapidly between 85 and 100 km. Specifically, a concentration of the order of 10^{13} oxygen atoms cm^{-3}, equivalent to the total number of oxygen molecules dissociated by solar radiation in 1 sec, would have a mean lifetime of about 10 days at 95 km. However, a vertical distribution based on photochemical equilibrium would not correspond to a hydrostatic distribution and it is necessary to consider the possibility of downward transport of oxygen atoms to regions where recombination can occur. At the mesopause level, the lifetime of these oxygen atoms becomes small, typically of the order of 1 day. Under such conditions, a concentration peak is predicted between 85 and 100 km. The magnitude of this peak is related to the effectiveness of vertical transport and the varying aeronomic conditions in the lower thermosphere. Since the concentration of molecular oxygen near 100 km is given in Table 3.2 as being of the order of 10^{12} cm^{-3}, and since the total concentration is of the order of 10^{13} cm^{-3}, it is clear that an atomic oxygen concentration of the order of 10^{12} cm^{-3} shows that the heterosphere must begin somewhere above the mesopause and must be associated with the lower thermosphere. Consequently, the data of Tables 3.1 and 3.2 must be progressively changed from the altitude of the mesopause up to 100 km in order to introduce the effect of the increase of the atomic oxygen concentration compared with the other constituents.

It is known from observations of meteor trails and vapor releases that the lower thermosphere is in some kind of turbulent state which acts to prevent diffusive separation of the various atmospheric gases. As discussed in Chapter 15, partial pressure gradients of minor constituents tend to establish diffusive flows within the atmosphere. The properties of such flows are determined by the equations of continuity and momentum conservation. An important characteristic of minor constituent flow is the existence of a maximum value of the diffusive flux which is reached when the scale height of the minor constituent is the same as the scale height of the atmosphere. The magnitude of the limiting flux depends both upon atomic quantities and the parameters of the atmosphere. For a minor constituent mass less than the mean atmosphere mass, the flow is limited in an upward (+) sense, while heavier gases are limited in their downward (−) flow.

Using formulas developed in Chapter 15, the data of Table 3.2 can be used to obtain the limiting fluxes of argon and helium at 100 km:

$$F(\text{Argon}) = -3 \times 10^{10} \quad \text{cm}^{-2}\,\text{sec}^{-1}$$

$$F(\text{Helium}) \simeq 10^{8} \quad \text{cm}^{-2}\,\text{sec}^{-1}.$$

These values are equally valid down to the mesopause since the variation of temperature is small.

With regard to argon, it is possible to deduce the time required to modify an initial mixing distribution [H(argon) = H(atmosphere)] so that it is

50% of that represented by a diffusive distribution. Using the flux given above, one obtains:

85 km	90 km	95 km	100 km
2 months	1 month	10 days	2 days

Times between 2 days and 2 months show that a perfect mixing distribution of argon cannot be maintained above the mesopause, and it can be concluded that the lower thermosphere, even subject to turbulence, is a transition region between conditions of mixing and diffusive equilibrium. At the level where a perfect diffusion distribution can be applied for argon, its actual concentration is less than the normal mixing concentration.

Considering the conditions for an increase of helium by a factor of 2 times the normal mixing concentration, the following times are obtained:

85 km	90 km	95 km	100 km
3 months	1 month	2 weeks	1 week

Again it is clear that the lower thermosphere is a transition region of diffusive separation between conditions of perfect mixing and diffusion. At the level where diffusive equilibrium exists for helium, its actual concentration is greater than would be appropriate for mixing conditions. Present rocket information on diffusion in the thermosphere must be analyzed by taking into account the unknown and variable transition between regions of perfect mixing and diffusive equilibrium. In any atmospheric model the introduction of a diffusion level must always be below the actual level of diffusive equilibrium. Consequently, the heterosphere, which begins with the dissociation of molecular oxygen, is also subject in the same altitude range to departures from mixing and diffusive conditions. It is clear that if noble gases such as argon and helium are subject to vertical transport in the lower thermosphere, a principal constituent such as molecular oxygen must be affected by diffusive separation, and the mixing ratio $n(O_2)/n(N_2)$ is changed before reaching the diffusive equilibrium region.

REFERENCES

1. W. G. Stroud, W. Nordberg, W. R. Bandeen, F. L. Bartman, and P. Titus, *J. Geophys. Res.* **65**, 2307 (1960).
2. W. Nordberg and W. G. Stroud, *J. Geophys. Res.* **66**, 455 (1961).
3. L. M. Jones, J. W. Peterson, E. J. Schaefer, and H. F. Schulte, *J. Geophys. Res.* **64**, 2331 (1959).
4. A. E. Cole, *J. Geophys. Res.* **66**, 2773 (1961).

5. N. W. Spencer, R. L. Boggess, and D. R. Taeusch, *J. Geophys. Res.* **69**, 1367 (1964).
6. G. A. Faucher and J. F. Morrissey, *J. Geophys. Res.* **76**, 4145 (1971).
7. J. E. Salah, *J. Geophys. Res.* **72**, 5389 (1967).
8. R. J. Murgatroyd and R. M. Goody, *Quart. J. Roy. Meteorol. Soc.* **84**, 225 (1958).
9. R. J. Murgatroyd and F. Singleton, *Quart. J. Roy. Meteorol. Soc.* **87**, 125 (1961).
10. W. Nordberg, in Meteorological Observations above 30 Kilometers, pp. 37–57. NASA SP-49, Washington, D.C. (1964).
11. *CIRA 1965*. North-Holland Publ., Amsterdam, 1965.
12. S. Teweles, in Meteorological Observations above 30 Kilometers, pp. 15–35. NASA SP-49, Washington, D.C. (1964).
13. B. J. Conrath, R. A. Hanel, V. G. Kunde, and C. Prabhakara, *J. Geophys. Res.* **75**, 5831 (1970).

CHAPTER 4

The Transition from the Homosphere to the Heterosphere

4.1 Introduction

The transition from the homosphere to the heterosphere involves a competition between mixing and diffusive processes. The former tends to keep the mean molecular mass constant with altitude, while the latter acts to let each gas constituent assume a density and pressure distribution consistent with its own mass and temperature.

In the case of the earth's atmosphere there is an additional complication which arises from the photodissociation of molecular oxygen. As a result of this process, molecular oxygen diffuses upwards in the atmosphere to regions where dissociation occurs. The resulting oxygen atoms then diffuse downward to lower altitudes where recombination takes place. At sufficiently low altitudes the distribution of atomic oxygen may approximate a photochemical equilibrium, but above a certain height the photochemical distribution is not correct since the time for downward diffusion becomes shorter than the dissociation time.

The reduction in mean molecular mass which accompanies the dissociation of oxygen affects substantially models of the thermosphere. In this chapter the problem of oxygen dissociation and ozone formation are discussed as they affect atmosphere parameters near 100 km. More detailed studies of the behavior of atomic oxygen and ozone in the mesosphere and stratosphere are given in Chapter 11.

4.2 The Theoretical Problem

Dissociation of molecular oxygen is obtained by photodissociation due to solar radiation according to the process

$$(J_2):\quad O_2 + h\nu(\lambda \leqq \lambda_1) \rightarrow O + O \tag{4.1}$$

and the loss of atoms is due either to an association by a three-body collision

$$(k_1):\quad O + O + M \rightarrow O_2 + M \tag{4.2}$$

or it can be the result of a two-body collision corresponding to a radiative association

$$(k_4):\quad O + O \rightarrow O_2 + h\nu. \tag{4.3}$$

Reaction (4.3) can be neglected since it is not important compared with (4.2) in the lower thermosphere. However, the following reaction

$$(k_2):\quad O_2 + O + M \rightarrow O_3 + M \tag{4.4}$$

is important where a high concentration of molecular oxygen exists, i.e., in the stratosphere and mesosphere.

Due to the photolysis of ozone

$$(J_3):\quad O_3 + h\nu \rightarrow O_2 + O \tag{4.5}$$

and the reaction

$$(k_3):\quad O_3 + O \rightarrow 2O_2 \tag{4.6}$$

the concentrations of molecular oxygen, atomic oxygen, and ozone are not independent.

Considering processes (4.1), (4.2), and (4.4)–(4.6) to be of principal importance [1], it is possible to write the equations governing the rates of change of the concentrations $n(O_2)$, $n(O)$, and $n(O_3)$ when transport is ignored as

$$\begin{aligned} dn(O_2)/dt &+ n(O_2)J_2 + k_2 n(M)n(O)n(O_2) \\ &= k_1 n(M)n^2(O) + 2k_3 n(O)n(O_3) + n(O_3)J_3 \end{aligned} \tag{4.7}$$

$$\begin{aligned} dn(O)/dt &+ 2k_1 n(M)n^2(O) + k_2 n(M)n(O_2)n(O) + k_3 n(O_3)n(O) \\ &= 2n(O_2)J_2 + n(O_3)J_3 \end{aligned} \tag{4.8}$$

$$dn(O_3)/dt + n(O_3)J_3 + k_3 n(O)n(O_3) = k_2 n(M)n(O_2)n(O). \tag{4.9}$$

These three equations are not independent when the total number of oxygen atoms (free and attached) remains constant.

If we consider the situation for photochemical equilibrium [2], Eqs. (4.7) to (4.9) lead to

$$\frac{n(O_3)}{n(O)} = \frac{k_2 n(O_2) n(M)}{J_3 + k_3 n(O)} \tag{4.10}$$

and

$$n^2(O) = \frac{n(O_2) J_2}{k_1 n(M)[1 + \{k_2 k_3 n(O_2)/k_1[J_3 + k_3 n(O)]\}]}. \tag{4.11}$$

4.3 Ozone Formation

The conditions for the simultaneous variations of $n(O)$ and $n(O_3)$ can be conveniently written

$$\frac{dn(O)}{dt} + \frac{dn(O_3)}{dt} + 2k_1 n(M) n^2(O) + 2k_3 n(O_3) n(O) = 2n(O_2) J_2. \tag{4.12}$$

At sufficiently *low altitudes*, i.e., in the stratosphere, atomic oxygen is in equilibrium, and direct recombination is negligible. Equation (4.12) becomes

$$\frac{dn(O_3)}{dt} + n^2(O_3)\left[\frac{2J_3 k_3}{k_2 n(M) n(O_2) + k_3 n(O_3)}\right] = 2n(O_2) J_2 \left[\frac{k_2 n(M) n(O_2) - k_3 n(O_3)}{k_2 n(M) n(O_2) + k_3 n(O_3)}\right]. \tag{4.13}$$

Now, since

$$k_3 n(O_3) \ll k_2 n(M) n(O_2) \tag{4.14}$$

even during the night when J_2 and J_3 vanish, Eq. (4.13) can be simplified as

$$\frac{dn(O_3)}{dt} + n^2(O_3) \frac{2J_3 k_3}{k_2 n(M) n(O_2)} = 2n(O_2) J_2 \tag{4.15}$$

and the form used in studies of the main part of the ozone layer for photoequilibrium conditions is

$$n_{eq}(O_3) = [(k_2/k_3)(J_2/J_3) n(M) n^2(O_2)]^{1/2}. \tag{4.16}$$

Equations (4.15) and (4.16) show the degree to which the ozone content can vary in the stratosphere under photochemical conditions. The relevant solution to the differential equation (4.15) is simply

$$n_t(O_3) = n_{eq}(O_3) \cdot \tanh 2\left(\frac{k_3 J_2 J_3}{k_2 n(M)}\right)^{1/2} t, \tag{4.17}$$

t being the time measured from $t = 0$ for $n(O_3) = 0$.

Introducing the time, $\tau(O_3)$, necessary to attain 95% of the photochemical equilibrium value, the following relation is obtained:

$$\tau(O_3) = [k_2 n(M)/k_3 J_2 J_3]^{1/2}. \tag{4.18}$$

Lengthy computation leads to the numerical data shown in *Table 4.1*. In the neighborhood of the stratopause, there is an almost perfect photochemical equilibrium. But, it is also clear that below 30 km the time required to reach photochemical equilibrium becomes sufficiently long such that diurnal variations can lead to departures from equilibrium conditions. In

TABLE 4.1

Time Needed to Obtain Equilibrium Conditions of Ozone

Altitude (km)	Overhead sun time $\tau(O_3)$	Sun 15° above horizon time $\tau(O_3)$
50	6 hr	11 hr
45	12 hr	1 day
40	2 days	6 days
35	8 days	47 days
30	40 days	2 yr
25	170 days	38 yr
20	3 yr	3500 yr
15	28 yr	5×10^6 yr

the lower stratosphere a possibility of modifying a photochemical equilibrium exists. Such modifications may be related to the seasonal variation of stratospheric ozone and also with the exchange of ozone between the stratosphere and troposphere.

Using numerical values it can be shown that the ozone peak occurs in the stratosphere. But it is easy to show that the atomic oxygen concentration, which increases with altitude in the stratosphere, attains equality with the ozone concentration above the stratopause. The sunlit mesosphere is therefore characterized by atomic oxygen concentrations greater that that of ozone. In this case, Eq. (4.9) must be used and $n(\mathrm{O})$ can be considered constant during a sufficiently short interval of time. In a sunlit atmosphere for which $n(\mathrm{O}) > n(\mathrm{O}_3)$, an immediate equilibrium is reached and

$$n(\mathrm{O}_3) = k_2 n(\mathrm{O}) n(\mathrm{O}_2) n(\mathrm{M}) / [J_3 + k_3 n(\mathrm{O})]. \tag{4.19}$$

In a dark atmosphere, association of oxygen atoms with oxygen molecules [reaction (4.4)] is extremely rapid in the lower mesosphere. There is, therefore, a day–night variation of ozone in the mesosphere.

However, mesospheric conditions defined by Eq. (4.10) leading to

$$\frac{n(\mathrm{O}_3)}{n(\mathrm{O})} = \frac{k_2 n(\mathrm{M}) n(\mathrm{O}_2)}{J_3} \tag{4.20}$$

when $J_3 > k_3 n(\mathrm{O}_3)$ in a sunlit atmosphere, and to

$$n(\mathrm{O}_3) = \frac{k_2 n(\mathrm{M}) n(\mathrm{O}_2)}{k_3} \tag{4.21}$$

for a nighttime atmosphere where $J_3 = 0$, do not represent actual conditions. They would be correct for a pure oxygen–nitrogen atmosphere but cannot be kept in an atmosphere where atomic hydrogen plays a role. In other words, the vertical distribution of ozone in the mesosphere depends in aeronomic processes involving other minor constituents (see Chapter 13).

4.4 Lifetime of Atomic Oxygen

Since $n(\mathrm{O}_3) < k_2 n(\mathrm{M}) n(\mathrm{O}_2)/k_3$ the differential equation for atomic oxygen becomes

$$dn(\mathrm{O})/dt + n^2(\mathrm{O})[2k_1 n(\mathrm{M}) + 2k_2 k_3 n(\mathrm{M}) n(\mathrm{O}_2)/J_3] = 2n(\mathrm{O}_2) J_2. \tag{4.22}$$

The relevant solution of this equation is

$$n_t(\mathrm{O})/n^*(\mathrm{O}) = \tanh\{4n(\mathrm{O}_2) J_2 k_1 n(\mathrm{M}) [1 + k_2 k_3 n(\mathrm{O}_2)/k_1 J_3]\}^{1/2} t, \tag{4.23}$$

t being the time measured from $t = 0$ when $n(\mathrm{O}) = 0$ and $n^*(\mathrm{O})$ the photochemical equilibrium value. Introducing the time $\tau_{\mathrm{eq}}(\mathrm{O})$ necessary to attain 50% of the photochemical equilibrium value, the following relation is obtained:

$$\tau_{eq}(O) = 0.275/\{J_2 k_1 n(M) n(O_2)[1 + k_2 k_3 n(O_2)/k_1 J_3]\}^{1/2}. \quad (4.24)$$

At sufficiently high altitudes, Eq. (4.24) can be reduced to

$$\tau_{eq}(O) = 0.275/\{J_2 k_1 n(M) n(O_2)\}^{1/2}. \quad (4.25)$$

The increase of $n_t(O)$ is not rapid. It reaches 80% of the photochemical value after $2\tau_{eq}(O)$, 93% after $3\tau_{eq}(O)$, and 97.5% after $4\tau_{eq}(O)$.

At altitudes of the mesopause and higher, Eq. (4.25) may be used instead of Eq. (4.24). In fact, from (4.8) the rate of change of atomic oxygen is, when the terms related to ozone concentrations are neglected,

$$dn(O)/dt + 2k_1 n(M) n^2(O) = 2n(O_2)J_2. \quad (4.26)$$

Using the following numerical data at the mesopause level: $n(O_2)J_2 = 7\times 10^{-5}$ cm^{-3} sec^{-1} and $k_1 n(M) = 5\times 10^{-19}$ cm^3 sec^{-1}, the time required to reach an atomic oxygen concentration of about 50% of its equilibrium value is about 5 days according to Eq. (4.25).

In the mesosphere at 70 km for example, where the atomic oxygen loss rate coefficient is of the order of 10^{-16} cm^3 sec^{-1} and its production rate is about 4×10^6 cm^{-3} sec^{-1}, only 5 hr are required according to Eq. (4.24) to attain 50% of the photochemical value. Thus, it can be assumed that conditions of photochemical equilibrium of atomic oxygen can be reached in the mesosphere.

After sunset $J_2 = J_3 = 0$, and Eq. (4.8) becomes

$$dn(O)/dt + 2k_1 n(M) n^2(O) + k_2 n(M) n(O_2) n(O) + k_3 n(O_3) n(O) = 0. \quad (4.27)$$

In the thermosphere, the following approximation may be used

$$dn(O)/dt + 2k_1 n(M) n^2(O) = 0 \quad (4.28)$$

and the time of recombination of an oxygen atom can be deduced from the following expression

$$n_t(O) = ([1/n_0(O)] + 2k_1 n(M) t)^{-1} \quad (4.29)$$

where $n_0(O)$ is the concentration when $t = 0$.

Since $2k_1 n(M) = 5\times 10^{-20}$ and 1×10^{-18} cm^3 sec^{-1} at 100 and 70 km, respectively, it is clear that $n_t(O)$ does not differ from $n_0(O)$ even after 5×10^4 sec; i.e., after a night. Atomic oxygen is not affected by recombination in the thermosphere but is subject to transport processes.

In the mesosphere the following approximation of Eq. (4.27) may be used:

$$dn(\mathrm{O})/dt + k_2 n(\mathrm{M}) n(\mathrm{O_2}) n(\mathrm{O}) = 0 \tag{4.30}$$

which leads to

$$n_t(\mathrm{O}) = n_0(\mathrm{O}) \exp[-k_2 n(\mathrm{M}) n(\mathrm{O_2}) t]. \tag{4.31}$$

Assuming a numerical value $k_2 = 3.3 \times 10^{-35} e^{950/T}$ and the introduction of data for $n(\mathrm{M})$ and $n(\mathrm{O_2})$ taken from Table 3.2, lifetimes of an oxygen atom after sunset in the mesosphere can be obtained; examples are given in *Table 4.2*. From this table we conclude that atomic oxygen cannot be maintained during the night in the mesosphere. At the mesopause level the atomic oxygen concentration is reduced to almost 10% of its daytime value after about 12 hr.

TABLE 4.2

LOSS OF OXYGEN ATOMS IN THE MESOSPHERE[a]

z (km)	$t = 1$ hr	$t = 3$ hr	$t = 6$ hr	$t = 12$ hr	$t = 24$ hr
85	0.9	0.7	0.5	0.2	5×10^{-2}
80	0.6	0.2	4×10^{-2}	1×10^{-3}	2×10^{-6}
75	1×10^{-1}	1×10^{-3}	1×10^{-6}	1×10^{-12}	1×10^{-24}
70	1×10^{-4}	1×10^{-12}	1×10^{-24}	0	0
65	1×10^{-10}	1×10^{-30}	0	0	0
60	1×10^{-24}	0	0	0	0

[a] Ratio $n_t(\mathrm{O})/n_{t=0}(\mathrm{O})$ with $t = 0$ corresponds to sunset conditions.

In any case, it can be concluded that atomic oxygen must be a permanent constituent in the lower thermosphere. Its concentration will be variable and will depend on the aeronomic conditions in the lower thermosphere. Its concentration peak will be subject to the penetration of solar radiation dissociating O_2 in this region (latitudinal and seasonal effects), and to dynamic effects related to vertical or horizontal transport. One must conclude that the altitude of the concentration peak of atomic oxygen will vary in the lower thermosphere between 85 and 100 km and that the maximum concentration may reach $10^{13}\ \mathrm{cm^{-3}}$ without difficulty and will not be less that $10^{11}\ \mathrm{cm^{-3}}$.

4.5 Atomic Oxygen at 100 km

Taking into account such observational results as the measurement of pressure and density of molecular oxygen, some idea can be obtained about physical conditions in the neighborhood of 100 km.

When one considers that photoequilibrium conditions do not exist at 100 km and that mixing is the most important process in the analysis of dissociation, the following relation can be written

$$n(\mathrm{O}) = \tfrac{1}{2}n(\mathrm{N_2}) - 2n(\mathrm{O_2}). \tag{4.32}$$

Any departure from mixing ratios would lead to an increase of atomic oxygen–nitrogen ratio; if there is in addition a negligible O_2 concentration, the result would be an atomic oxygen concentration greater than 50% of that of molecular nitrogen. If we consider observational data [3] such as direct measurement of O_2 absorption and atmospheric density from x-ray transmission data, it is clear that molecular oxygen at 100 km remains an important fraction of the total density. Recent observations [4–7] have confirmed such a conclusion. Results obtained from rocket-borne mass spectrometers [8] show that $n(\mathrm{N_2}) > n(\mathrm{O_2}) > n(\mathrm{O})$ at 100 km.

Taking the value of the pressure given in Table 3.1, i.e., $p = 3 \times 10^{-4}$ Torr at 100 km, it is possible to use formula (4.32) in order to introduce the atomic oxygen concentration. However, one should also take into account the downward transport of atomic oxygen which opposes photoequilibrium conditions. Actually, the combination of turbulent transport with time dependent photodissociation of O_2 can be responsible for a certain variation of O around 100 km. Nevertheless, if the degree of dissociation were greater than 35% the concentration of atomic oxygen would be greater than that of molecular oxygen and would not differ sufficiently from the values for photochemical equilibrium. It is probably more realistic to consider that near 100 km the concentration of molecular oxygen is still greater than that of atomic oxygen and that a percentage of the order of 25% may represent the best working figure for the degree of dissociation. Since a choice is unavoidable when numerical data are required to study the behavior of the thermosphere, we have adopted three sets of parameters between 100 and 120 km which are shown in detail in *Table 4.3*. In any case, it seems that the lifetime of an oxygen atom must be of the order of several months at 100 km and increases with height. Numerical values for molecular association given in Table 4.2 also lead to such a conclusion. It can be assumed therefore that the oxygen atoms have a sufficiently long lifetime to follow

TABLE 4.3a

FIRST EXAMPLE OF PHYSICAL PARAMETERS BETWEEN 100 AND 120 KM

Altitude (km)	H (km)	p (Torr)	T (°K)	ϱ (gm cm^{-3})	$n(N_2)$ (cm^{-3})	$n(O_2)$ (cm^{-3})	$n(O)$ (cm^{-3})	$n(M)$ (cm^{-3})
100	6.37	3.0×10^{-4}	200	6.6×10^{-10}	1.1×10^{13}	2.2×10^{12}	1.4×10^{12}	1.4×10^{13}
105	6.87	1.4×10^{-4}	216	2.9×10^{-10}	4.7×10^{12}	9.5×10^{11}	6.1×10^{11}	6.3×10^{12}
110	7.37	6.9×10^{-5}	230	1.3×10^{-10}	2.2×10^{12}	4.4×10^{11}	2.8×10^{11}	2.9×10^{12}
115	7.87	3.7×10^{-5}	246	6.5×10^{-11}	1.0×10^{12}	2.1×10^{11}	1.4×10^{11}	1.4×10^{12}
120	8.37	2.0×10^{-5}	262	3.3×10^{-11}	5.3×10^{11}	1.1×10^{11}	7.0×10^{10}	7.1×10^{11}

TABLE 4.3b

Second Example of Physical Parameters between 100 and 120 km

Altitude (km)	H (km)	p (Torr)	T (°K)	ϱ (gm cm^{-3})	$n(N_2)$ (cm^{-3})	$n(O_2)$ (cm^{-3})	$n(O)$ (cm^{-3})	$n(M)$ (cm^{-3})
100	6.37	3.0×10^{-4}	200	6.6×10^{-10}	1.1×10^{13}	2.2×10^{12}	1.4×10^{12}	1.4×10^{13}
105	7.37	1.4×10^{-4}	231	2.7×10^{-10}	4.5×10^{12}	9.2×10^{11}	5.9×10^{11}	6.0×10^{12}
110	8.37	7.6×10^{-5}	262	1.3×10^{-10}	2.1×10^{12}	4.3×10^{11}	2.7×10^{11}	2.8×10^{12}
115	9.37	4.3×10^{-5}	293	6.5×10^{-11}	1.1×10^{12}	2.2×10^{11}	1.4×10^{11}	1.4×10^{12}
120	10.37	2.6×10^{-5}	324	3.5×10^{-11}	5.8×10^{11}	1.2×10^{11}	7.6×10^{10}	7.8×10^{11}

TABLE 4.3c

THIRD EXAMPLE OF PHYSICAL PARAMETERS BETWEEN 100 AND 120 KM

Altitude (km)	H (km)	p (Torr)	T (°K)	ϱ (gm cm^{-3})	$n(N_2)$ (cm^{-3})	$n(O_2)$ (cm^{-3})	$n(O)$ (cm^{-3})	$n(M)$ (cm^{-3})
100	6.37	3.0×10^{-4}	200	6.6×10^{-10}	1.1×10^{13}	2.2×10^{12}	1.4×10^{12}	1.4×10^{13}
105	7.87	1.5×10^{-4}	247	2.6×10^{-10}	4.3×10^{12}	8.8×10^{11}	5.6×10^{11}	5.8×10^{12}
110	9.37	8.2×10^{-5}	293	1.2×10^{-10}	2.0×10^{12}	4.1×10^{11}	2.6×10^{11}	2.7×10^{12}
115	10.87	5.0×10^{-5}	340	6.5×10^{-11}	1.1×10^{12}	2.2×10^{11}	1.4×10^{11}	1.4×10^{12}
120	12.37	3.3×10^{-5}	386	3.7×10^{-11}	6.1×10^{11}	1.2×10^{11}	8.0×10^{10}	8.1×10^{11}

TABLE 4.3d

ATMOSPHERIC PARAMETERS INVOLVING ATOMIC OXYGEN CONCENTRATION BETWEEN 90 AND 120 KM

z (km)	T (°K)	H (km)	M ($^{12}C = 12$)	$n(N_2)$ (cm^{-3})	$n(O_2)$ (cm^{-3})	$n(O)$ (cm^{-3})	$n(Ar)$ (cm^{-3})	$\sum n$ (cm^{-3})	ϱ (gm cm^{-3})	p (Torr)
90	176.7	5.34	28.87	4.67×10^{13}	1.25×10^{13}	3.00×10^{11}	5.59×10^{11}	6.01×10^{13}	2.88×10^{-9}	1.10×10^{-3}
91	179.9	5.44	28.83	3.81	1.02	3.10	4.56	4.91	2.35	9.14×10^{-4}
92	183.2	5.56	28.81	3.12	8.36×10^{12}	3.15	3.73	4.02	1.92	7.62
93	186.5	5.66	28.77	2.56	6.86	3.20	3.06	3.31	1.58	6.39
94	189.7	5.73	28.80	2.11	5.66	3.25	2.52	2.73	1.31	5.36
95	193.0	5.85	28.88	1.74	4.67	3.30	2.08	2.26	1.08	4.51
96	196.2	5.97	28.73	1.45	3.88	3.28	1.73	1.89	9.00×10^{-10}	3.84
97	199.5	6.07	28.70	1.20	3.23	3.26	1.44	1.57	7.48	3.24
98	202.7	6.17	28.69	1.00	2.69	3.24	1.20	1.31	6.24	2.75
99	206.0	6.31	28.55	8.40×10^{12}	2.25	3.22	1.01	1.11	5.26	2.37
100	209.2	6.41	28.54	7.05	1.90	3.20	8.43×10^{10}	9.35×10^{12}	4.43	2.02
101	213.5	6.57	28.42	5.91	1.50	3.15	7.21	7.80	3.68	1.72
102	217.9	6.73	28.35	4.96	1.20	3.10	6.19	6.53	3.07	1.47
103	222.2	6.90	28.18	4.19	1.00	3.05	5.34	5.55	2.60	1.28

104	226.6	7.09	27.97	3.55	8.00×10^{11}	3.00	4.62	4.70	2.18	1.10
105	230.9	7.24	27.94	3.03	6.50	2.90	4.01	4.01	1.86	9.58×10^{-5}
106	237.1	7.46	27.86	2.59	5.60	2.80	3.21	3.46	1.60	8.49
107	243.3	7.73	27.60	2.22	4.70	2.70	2.59	2.99	1.37	7.53
108	249.5	7.97	27.44	1.92	3.95	2.55	2.10	2.59	1.18	6.69
109	255.7	8.21	27.38	1.66	3.35	2.45	1.71	2.26	1.02	5.98
110	261.9	8.42	27.26	1.44	2.85	2.30	1.40	1.97	8.91×10^{-11}	5.34
111	268.1	8.66	27.16	1.26	2.35	2.10	1.15	1.72	7.74	4.77
112	274.3	8.90	27.05	1.10	1.95	1.92	9.51×10^{9}	1.50	6.72	4.26
113	280.6	9.15	26.92	9.63×10^{11}	1.60	1.76	7.89	1.31	5.84	3.80
114	286.8	9.40	26.81	8.45	1.30	1.62	6.58	1.14	5.09	3.38
115	293.0	9.63	26.72	7.41	1.10	1.49	5.50	1.01	4.46	3.06
116	299.2	9.88	26.63	6.52	8.75×10^{10}	1.36	4.62	8.80×10^{11}	3.89	2.73
117	305.4	10.14	26.48	5.75	7.25	1.26	3.89	7.78	3.42	2.46
118	311.6	10.40	26.36	5.08	6.00	1.17	3.29	6.88	3.01	2.22
119	317.8	10.64	26.28	4.50	4.90	1.08	2.79	6.10	2.66	2.01
120	324.0	10.87	26.21	4.00	4.00	1.00	2.38	5.42	2.36	1.82

a vertical distribution not too different from that of other constituents in the region of 100 to 120 km. Under these conditions, there is some justification in presenting the data of Table 4.3 as an example of real physical conditions which provide boundary conditions for an analysis of the structure of the middle and upper thermosphere. It must be pointed out that such a ratio of $n(O_2)/n(O)$ leads to a mean molecular mass for the mixing region $M = 27$, which does not differ too much from $M(N_2) = 28$.

It should be remembered that the viewpoint in the studies of the problem of dissociation has changed a great deal during the last twenty years. After the earlier strict application of photochemical equilibrium based on solar emission corresponding to a blackbody at 6000°K, a departure from photo-equilibrium condition was considered [9] as a better representation of oxygen dissociation. Such a departure leads to an increase of the molecular oxygen concentration as compared with that resulting from photochemical equilibrium conditions, and therefore implies an atomic oxygen atmosphere at only the highest altitudes.

A detailed analysis of the data of Tables 4.3 shows that various conclusions can be reached when one introduces variations of pressure at 100 km. First, it must be pointed out that an almost constant density is obtained at 120 km, namely $(3.5 \pm 0.2) \times 10^{-11}$ gm cm^{-3}, for various gradients of the scale height, $0.1 \leqq \beta \leqq 0.3$, or for different temperature gradients, $3°\text{K km}^{-1} \leqq dT/dz \leqq 10°\text{K km}^{-1}$. This density value corresponds to $\varrho = 6.6 \times 10^{-10}$ gm cm^{-3} and $p = 3 \times 10^{-4}$ Torr at 100 km. Furthermore, the ratio $n(O_2)/n(O) = 1.5$ at 120 km is obtained when the total density is $\varrho = 3.5 \times 10^{-11}$ gm cm^{-3} with temperature variation of the order of 125°K.

If the pressure at 100 km were reduced to $\frac{2}{3}$ and $\frac{1}{3}$ of the value given in Tables 4.3 the resulting density at 120 km would become $\varrho = 2.3 \times 10^{-11}$ gm cm^{-3} and $\varrho = 1.2 \times 10^{-11}$ gm cm^{-3}, respectively.

The average vertical distribution of density is shown in the accompanying tabulation.

Altitude (km):	100	105	110	115	120	
ϱ (Table 4.3b):	6.6	2.7	1.3	0.65	0.35×10^{-10}	gm cm^{-3}
$\frac{2}{3}\varrho$:	4.4	1.8	0.87	0.53	0.23×10^{-10}	gm cm^{-3}
$\frac{1}{3}\varrho$:	2.2	9.0	0.43	0.22	0.12×10^{-10}	gm cm^{-3}

These various figures show the need of a precise knowledge of all parameters by direct observation in the transition region. Nevertheless, such differences correspond to real variation in the lower thermosphere. A reduction of density from 3.5×10^{-11} gm cm^{-3} at 120 km to 2.3×10^{-11} and 1.2×10^{-11} gm cm^{-3} may be considered as the total variation acceptable for all conditions occurring during a solar cycle. It can be said that such density decreases are equivalent to a change of altitude from 120 to 115 km and about 110 km, respectively. In other words, considering a temperature of the order of 290°K which corresponds to 115 km in Table 4.3b and to 110 km in Table 4.3c, it can be said that a density of the order of 4×10^{-11} gm cm^{-3} is a numerical value representing thermospheric conditions between 110 and 120 km. In the same way, the ratio $n(O_2)/n(O) = 1$ occurs in the altitude range 120–130 km depending on the variation of the total density. Consequently, when using Tables 4.3 it must be taken into account that extreme conditions are equivalent to values such as

$$\varrho = 3.5 \times 10^{-11} \quad \text{gm cm}^{-3} \quad \text{at} \quad 115 \pm 5 \quad \text{km}$$

and

$$n(O)/n(O_2) = 1 \quad \text{at} \quad 125 \pm 5 \quad \text{km}.$$

Measured values of molecular oxygen near 100 km do not give any indication of the atomic oxygen concentration. Values which are shown in Tables 4.3 are examples indicating, in fact, that the scale height gradient has a minor effect and that the temperature is not known. The absolute value of the density at 100 km is the most important parameter. Since there are several mass spectrometer measurements [10–15] of the ratio $n(O)/n(O_2)$ near 120 km, it is possible to suggest vertical distributions of N_2, O_2, and O in the lower thermosphere if turbulent mixing is introduced as a transport process in addition to molecular diffusion.

Another way to determine the vertical distribution of atomic oxygen and its absolute concentration should be the analysis of the chemiluminescent reactions involving oxygen atoms.

For example, three oxygen atoms must be involved in the production of the green line (5577 Å) radiation in the lower thermosphere airglow. If the rate coefficient of this luminescent radiation of the order of 1.5×10^{-34} cm^6 sec^{-1}, as deduced from laboratory measurements [16], is accepted, 10^{12} oxygen atoms cm^{-3} are required to produce the observed [17] emissions of the green line (150 photons cm^{-3} sec^{-1}) at 100 km. A concentration of about 1.5×10^{12} cm^{-3} at 95 km would correspond to an emission peak of 350 photons cm^{-3} sec^{-1}. At the mesopause level and also near 110 km, at least 3×10^{11} oxygen atoms would be required to give rise to an emission of not less than 15 photons cm^{-3} sec^{-1}.

If a comparison is made between the various possible values of atomic oxygen concentration deduced from transport effects [18] and the value deduced from airglow observations [17], and with use of laboratory measurement [16], the values are inconsistent unless a possible error of a factor of 5 is accepted.

A more recent rocket observation of the 5577 Å emission has been analyzed [19] using a rate coefficient of 1.5×10^{-33} cm^6 sec^{-1} deduced from new quenching rates [20] of O(^{1}S). The results are in good agreement with time-dependent transport computations [21].

Thus, it is not easy to determine an exact distribution of atomic oxygen in the lower thermosphere. A turbulent effect does not remain constant and errors in experimental rate coefficients are possible. Nevertheless, it is certain that the best method available for an exact determination of the vertical distribution of atomic oxygen and of its absolute value is the analysis of the airglow. A clear idea of various excitation processes of atomic and molecular emissions involving atomic oxygen will lead to the solution of such a problem. A special effort should be made in order to obtain more airglow observations in the lower thermosphere and to develop other experimental methods [22–25] for the study of the oxygen luminescence.

For the discussion in the following chapter of the temperature gradient in the thermosphere, the values at 120 km in Table 4.3 can be used to define the lower boundary conditions. But, in order to have a detailed model of the lower thermosphere connecting the values given in Tables 3.1 and 3.2 with the atmospheric models which will be used in the study of the upper part of the thermosphere, the values shown in Table 4.3d are adopted. They are almost identical with the adopted values for an atmosphere subject to photodissociation with turbulent and molecular diffusion. An error of a factor of 2 is possible, since all aeronomic parameters are poorly determined in the neighborhood of 100 km.

REFERENCES

1. S. Chapman, *Mem. Roy. Met. Soc.* **3**, 103 (1930).
2. D. R. Bates and M. Nicolet, *J. Geophys. Res.* **55**, 301 (1950).
3. H. Friedman, in *Physics of the Upper Atmosphere* (J. S. Ratcliffe, ed.), pp. 175, 210. Academic Press, New York, 1960.
4. A. A. Pokhunkov, *Planet. Space Sci.* **9**, 269 (1962).
5. A. S. Jursa, M. Nakamura, and Y. J. Tanaka, *J. Geophys. Res.* **68**, 6145 (1963).
6. L. H. Weeks and L. G. Smith, *J. Geophys. Res.* **73**, 4835 (1968).
7. P. J. Brannon and J. H. Hoffman, *J. Geophys. Res.* **76**, 4630 (1971).

8. E. J. Schaefer and M. H. Nichols, *Space Res.* **4**, 205 (1964).
9. M. Nicolet and P. Mange, *J. Geophys. Res.* **59**, 16 (1954).
10. A. O. Nier, *Mass spectromet.* (*Japan*) **85**, 67 (1967).
11. A. E. Hedin and A. O. Nier, *J. Geophys. Res.* **71**, 4121 (1966).
12. E. J. Schaefer, *Space Res.* **8**, 959 (1968).
13. U. von Zahn, *J. Geophys. Res.* **72**, 59 (1967).
14. K. Mauersberger, D. Muller, D. Offermann, and U. von Zahn, *J. Geophys. Res.* **73**, 1071 (1968).
15. W. T. Kasprzak, D. Krankowsky, and A. O. Nier, *J. Geophys. Res.* **73**, 6765 (1968).
16. R. A. Young and G. Black, *J. Chem. Phys.* **44**, 3741 (1966).
17. I. S. Gulledge, D. M. Packer, S. G. Tilford, and J. T. Vanderslice, *J. Geophys. Res.* **73**, 5535 (1968).
18. D. F. Colegrove, W. B. Hanson, and F. S. Johnson, *J. Geophys. Res.* **71**, 2227 (1966).
19. B. S. Dandekar and J. P. Turtle, *Planet. Space Sci.* **19**, 949 (1971).
20. G. Black, T. G. Slanger, G. A. St.John, and R. A. Young, *J. Chem. Phys.* **51**, 116 (1969).
21. T. J. Keneshea and S. P. Zimmerman, *J. Atmos. Sci.* **27**, 831 (1970).
22. W. R. Henderson and H. I. Schiff, *Planet. Space Sci.* **18**, 1527 (1970).
23. W. R. Henderson, *J. Geophys. Res.* **76**, 3166 (1971).
24. D. Golomb, N. W. Rosenberg, C. Aharonian, J. A. F. Hill, and H. L. Alden, *J. Geophys. Res.* **70**, 1155 (1965).
25. D. Golomb and R. E. Good, *Space Res.* **12**, 675 (1972).

CHAPTER 5

Temperature Gradient in the Thermosphere

5.1 Introduction

It has been pointed out on several occasions that an essential characteristic of the thermosphere is its very slow decrease of density with altitude. According to Eq. (3.1) the only possible explanation is an important increase of the scale height with altitude. A transition from a molecular atmosphere ($M = 29$) to an atomic atmosphere ($M = 16$ or 14) would double the value of the scale height $H = kT/mg$, but this is not sufficient to explain the high atmospheric densities in the upper thermosphere. A great increase in the temperature is therefore necessary [1, 2] to explain the observed densities [3–6]. Two possibilities could be considered: an increase of the temperature with height with a temperature gradient present up to the highest altitudes, or a sharp increase of the temperature below 200 km, followed by a decrease of the temperature gradient above 200 km, changing to an isothermal region at very high altitudes.

When a temperature gradient is assumed, physical conditions are such that an external heat flow is needed to maintain such a gradient. Even if such a temperature gradient occurs, it can be shown (see Chapter 14) that it is not possible to consider a gradient which increases with height; any atmospheric model requiring such a permanent gradient must be rejected. In the thermosphere a temperature gradient must always decrease with height, for there is no other transport of heat available to counterbalance heat conduction [7].

It can be shown that it is not possible to reach the densities observed near 700 km if the temperature gradient near 150 km is too small, even though the densities observed near 200 km are similar to those obtained by calculation. The reason for this is that the absolute value of density near 200 km is not very sensitive to the gradient of temperature near 150 km, but depends on the boundary conditions near 100 to 120 km.

Considering again Eqs. (3.13) and (3.14) in which the local scale height H is a physical parameter:

$$H = \frac{kT}{mg} = \frac{kT}{mg_a}\left(\frac{a+z}{a}\right)^2 \tag{5.1}$$

we must remember that

$$\frac{d\varrho g}{\varrho g} = \frac{dn}{n} + \frac{dm}{m} + \frac{dg}{g} = -\frac{1+\beta}{\beta}\frac{dH}{H} \tag{5.2}$$

is a general equation, since

$$\frac{dH}{H} = \frac{dT}{T} - \left(\frac{dm}{m} + \frac{dg}{g}\right). \tag{5.3}$$

A parameter such as H_n chosen to represent the variation of the number density

$$H_n = -\left(\frac{d(\log n)}{dz}\right)^{-1} \tag{5.4}$$

can be used when T is a constant. In other cases, it may lead to an incorrect conclusion since

$$\frac{dH_n}{H_n} = \frac{dT}{T} - \left[\frac{dm}{m} + \frac{dg}{g} + \frac{kT}{mg}\frac{d(dT/dz)}{T}\right]\left[1 + \frac{kT}{mg}\frac{dT/dz}{T}\right]^{-1}. \tag{5.5}$$

When this is compared with (5.3), it shows how complicated the relationship of H_n is to such parameters as T, m, and g; thus, H_n should not be used where there is a strong gradient of temperature.

When there is mixing, (5.2) is subject to the following condition:

$$\frac{d\varrho_i g}{\varrho_i g} = -\frac{1+\beta}{\beta}\frac{dH}{H} \tag{5.6}$$

for each molecule of the ith type. When there is diffusion, (5.2) is subject to another condition:

$$\frac{d\varrho_i g}{\varrho_i g} = -\frac{1+\beta_i}{\beta_i}\frac{dH_i}{H_i} \tag{5.7}$$

in which

$$\beta_i m_i \simeq \beta m \tag{5.8}$$

with

$$\beta_i = \frac{k}{m_i} \frac{d(T/g)}{dz} \tag{5.9}$$

and

$$\beta = k \frac{d(T/mg)}{dz}. \tag{5.10}$$

When T is constant, Eqs. (5.9) and (5.10) become

$$\beta_i = \frac{kT}{m_i} \frac{d(1/g)}{dz} = \frac{2(a+z)}{a^2} H_{ia} \tag{5.11}$$

and

$$\beta = kT \frac{d(1/gm)}{dz} = \frac{2(a+z)}{a^2} H_a \left[1 - \frac{a+z}{2m} \frac{dm}{dz}\right]. \tag{5.12}$$

5.2 Conditions at 200 km

In order to obtain numerical data, the physical parameters given in Table 4.3 are used to represent constant boundary conditions at 120 km. Since the object of the calculation is to show that the density at 150 km is not affected by the gradient of the scale height between 120 and 150 km, a constant gradient can be used in Eq. (5.6) or (5.7). Various values between $\beta = 0.3$ and $\beta = 0.7$ have been adopted and any of these values leads to practically the same value of the density at 150 km. *Table 5.1* shows the values of physical parameters such as temperature and concentrations. When the conditions are known at 120 km, the density at 150 km, $\varrho(150 \text{ km})$, is practically constant, namely

$$\varrho(150 \text{ km}) = (2.42 \pm 0.02) \times 10^{-12} \quad \text{gm cm}^{-3} \tag{5.13}$$

when

$$\varrho(120 \text{ km}) = 3.54 \times 10^{-11} \quad \text{gm cm}^{-3} \tag{5.14}$$

with

$$600^\circ\text{K} \leq T(150 \text{ km}) \leq 970^\circ\text{K}. \tag{5.15}$$

In other words, any temperature gradient between 120 and 150 km leads to the same density at 150 km, and density measurements in a small range of heights are not sufficient to lead to a knowledge of physical parameters.

TABLE 5.1

EXAMPLES OF PHYSICAL PARAMETERS BETWEEN 120 AND 150 KM

Altitude (km)					
			Scale height and temperature		
120	10.37	10.37	10.37	10.37	10.37 km
	324	324	324	324	324°K
130	13.37	14.37	15.37	16.37	17.37 km
	416	447	478	509	540°K
140	16.37	18.37	20.37	22.37	24.37 km
	508	570	632	694	756°K
150	19.37	22.37	25.37	28.37	31.37 km
	599	692	784	877	970°K
			Density (gm cm^{-3})		
120	3.54×10^{-11}	3.54×10^{-11}	3.54×10^{-11}	3.54×10^{-11}	3.54×10^{-11}
130	1.18×10^{-11}	1.13×10^{-11}	1.09×10^{-11}	1.05×10^{-11}	1.02×10^{-11}
140	4.93×10^{-12}	4.81×10^{-12}	4.70×10^{-12}	4.58×10^{-12}	4.47×10^{-12}
150	2.38×10^{-12}	2.42×10^{-12}	2.44×10^{-12}	2.44×10^{-12}	2.43×10^{-12}
			Density with diffusion beginning at 120 km (gm cm^{-3})		
150	2.37×10^{-12}	2.41×10^{-12}	2.42×10^{-12}	2.42×10^{-12}	2.40×10^{-12}
			Scale height with diffusion beginning at 120 km (km)		
150	20.65	23.70	26.74	29.78	32.82
			Atomic oxygen concentration without and with diffusion (cm^{-3})		
150	5.1×10^{9}	5.2×10^{9}	5.2×10^{9}	5.2×10^{9}	5.2×10^{9}
	1.2×10^{10}	1.2×10^{10}	1.1×10^{10}	1.1×10^{10}	1.0×10^{10}

Furthermore, it is clear that diffusion or mixing does not affect the density at 150 km. Table 5.1 shows, however, that atomic oxygen concentrations are different when mixing or diffusion is involved, but such differences cannot be detected since measurements cannot be made with sufficient accuracy to give a value of the scale height with the precision needed.

TABLE 5.2

EXAMPLE OF PHYSICAL PARAMETERS AT 200 AND 220 KM

Altitude (km)						Average values
			Temperature (°K)			
150	599	692	784	877	970	800 ± 200
			ϱ = density (gm cm^{-3}) × 10^{12} [±0.01]			
150	2.37	2.41	2.43	2.42	2.41	2.40 ± 0.03
			$\varrho \times 10^{13}$ [±≦10%]			
200	2.10	2.70	3.35	3.90	4.45	3.3 ± 1.2
			$\varrho \times 10^{13}$ [±≦5%]			
220	1.10	1.45	1.90	2.30	2.70	1.9 ± 0.8
			$\varrho H^{1/2} \times 10^{10}$ [±≦10%] diffusion at 150 km			
220	1.9	3.0	4.1	5.3	6.5	(4.2 ± 2.3)
			$\varrho H^{1/2} \times 10^{10}$ diffusion at 120 km			
220	2.4	3.5	4.7	5.9	7.0	(4.7 ± 2.3)
			Mass M—without diffusion			
220	27.43	—	—	—	27.43	27.43
			Diffusion at 150 km (±1%)			
220	25.2	25.6	25.9	26.1	26.2	25.7 ± 0.5
			Diffusion at 120 km (±1%)			
220	22.5	23.2	23.7	24.1	24.4	23.5 ± 1.0
			$\varrho(O) \times 10^{14}$ [≦±10%] without diffusion			
220	0.55	0.80	1.00	1.25	1.50	1.0 ± 0.5
			Diffusion at 150 km			
220	1.60	2.00	2.30	2.65	3.00	2.3 ± 0.7
			$\varrho(O) \times 10^{14}$ [≦±10%] diffusion at 120 km			
220	3.85	4.50	5.00	5.35	5.80	4.8 ± 1.0
			Temperature $\beta_{<150\text{ km}} \geq \beta_{>150\text{ km}}$			
220	1000	1100	1200	1300	1400	1200 ± 100°K
220	1200	1500	1800	2100	2200	1700 ± 500°K
			Temperature $\beta_{<150\text{ km}} \leq \beta_{>150\text{ km}}$			
220	1800	1900	2000	2100	2200	2000 ± 200°K

An arbitrary total density such as $\varrho_{150} = 2.4 \times 10^{-12}$ gm cm^{-3} (see Table 5.1) can be diminished by a factor of two in order to follow various observational results. But such a decrease may correspond to a change of boundary conditions at 120 km (see discussion of Tables 4.3). In other words, if 2.4×10^{-12} gm cm^{-3} is adopted at 145 ± 5 km, it is possible to explain all observational data with an *isopycnic level* defined for a large range of scale heights. For this reason, it is very convenient to refer to such a level in order to discuss the physical conditions determined by the temperature and its gradient. However, it is necessary to know the vertical distribution of the principal constituents N_2, O_2, and O in order to obtain a clear view of the situation at 150 km.

In order to show that it is practically impossible to give an exact physical description of the thermosphere below 200 km based on the measurement of only one parameter, one can determine the density at 200 and 220 km and show its functional dependence on various temperature gradients associated with temperature values between 600 and 1000°K at 150 km. With the same values for the temperature gradient between 150 and 220 km, it can be shown (*Table 5.2*) that densities at 200 km (or 220 km) are not sensitive to the gradient adopted. Differences are less than $\pm 10\%$ at 200 km and $\pm 5\%$ at 220 km. Variations between 2×10^{-13} and 4.5×10^{-13} gm cm^{-3} at 200 km are essentially due to the effect of the temperature gradient below 150 km. *Figure 5.1* shows the vertical distribution of the density from 120 to 220 km according to conditions given in Table 5.2, namely, a density defined at 120 km and temperatures varying between 600° and 970°K at 150 km for any kind of gradient above 150 km, with mixing or with diffusion. But, the decrease of the mean molecular mass M from 27.4 to 23.5 is not detectable [8]; i.e., the variation of the concentration of atomic oxygen by a factor of 5 cannot be discussed from measurement of the density between 100 and 200 km. In other words, one cannot determine the temperature at 200 km without an additional physical guide. In fact, the data of Tables 5.1 and 5.2 show why such different conclusions have been obtained on the behavior of the thermosphere between 100 and 200 km. An arbitrary hypothesis may explain the observational data but anomalies introduced in the gradient of the scale height have little effect on the vertical distribution of the density. On the other hand, any anomaly introduced in the density curve is easily explained by an ad hoc hypothesis.

There is only one clear conclusion which can be reached with certainty from the data of Table 5.2: The density at 200 to 220 km is strongly dependent on the gradient near or below 150 km. *Figure 5.2* shows that there is an almost linear relation between ϱ(200 km) and T(150 km) with no

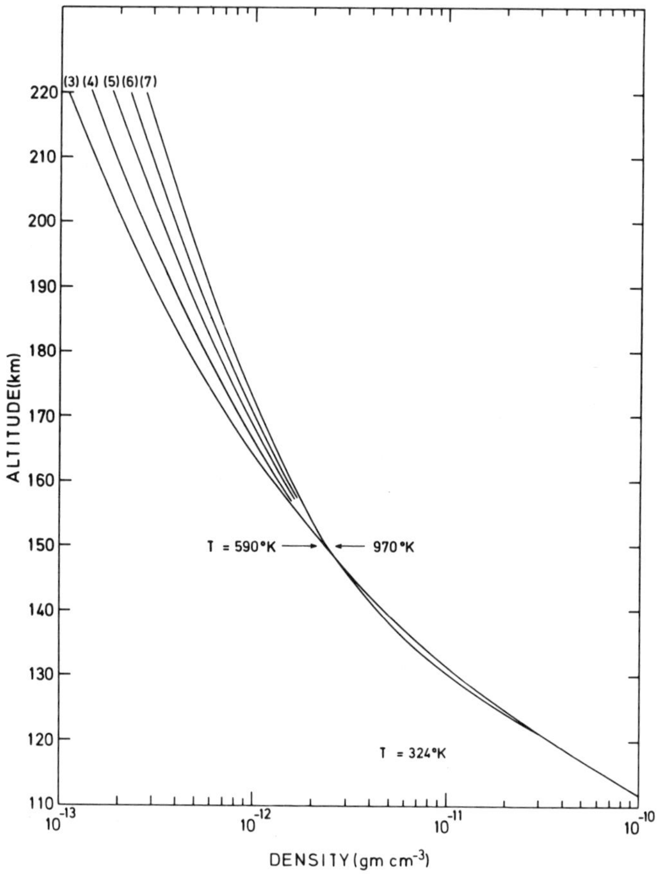

Fig. 5.1 Atmospheric densities as a function of the temperature gradient between 120 and 200 km with constant boundary conditions at 120 km. The boundary conditions are $\varrho(120\text{ km}) = 3.5 \times 10^{-11}$ gm cm^{-3} and $T(120\text{ km}) = 324°\text{K}$. The values for T at 150 km are: (3) 590°K, (4) 692°K, (5) 784°K, (6) 877°K, and (7) 970°K.

variations of $\varrho(150\text{ km})$ when $\varrho(120\text{ km})$ is constant. It can be concluded that densities at 200 and 220 km vary as follows:

$$\varrho(200\text{ km}) = (4 \pm 2) \times 10^{-13} \quad \text{gm cm}^{-3} \tag{5.16}$$

$$\varrho(220\text{ km}) = (2.4 \pm 1.2) \times 10^{-13} \quad \text{gm cm}^{-3} \tag{5.17}$$

for

$$T(150\text{ km}) = 900 \pm 300°\text{K} \tag{5.18}$$

when

$$\varrho(120\ \text{km}) \sim 3.5 \times 10^{-11} \quad \text{gm cm}^{-3} \tag{5.19}$$

and

$$T(120\ \text{km}) \sim 325°\text{K}. \tag{5.20}$$

For high values of density at 200 km, it is necessary that a strong temperature gradient exist at 150 km. An exact value cannot be given since boundary conditions at 120 km are not necessarily constant.

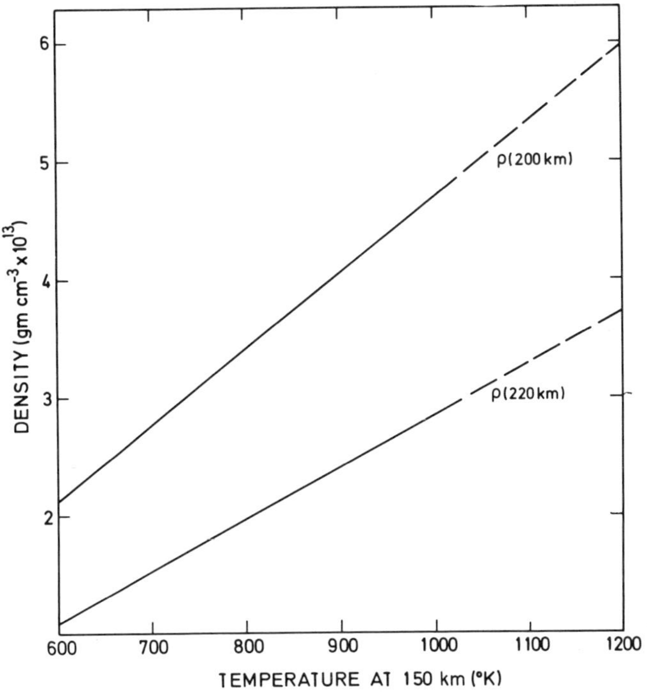

Fig. 5.2 The atmospheric density at 200 and 220 km as a function of the temperature at 150 km with constant boundary conditions at 120 km. $T(120\ \text{km}) = 324°\text{K}$; $\varrho(120\ \text{km}) = 3.5 \times 10^{-11}$ gm cm^{-3}. $T(150\ \text{km}) = 900° \pm 300°\text{K}$; $\varrho(200\ \text{km}) = (4 \pm 2) \times 10^{-13}$ gm cm^{-3}; $\varrho(220\ \text{km}) = (2.4 \pm 1.3) \times 10^{-13}$ gm cm^{-3}.

If the average value of the density $\varrho(200\ \text{km}) = 4 \times 10^{-13}$ gm cm^{-3} is accepted, it cannot be assumed that the adopted model is always correct and can be used as a standard model able to represent all physical conditions of the thermosphere. It may be chosen to represent physical conditions of a thermosphere in which the temperature gradient varies in a certain range.

A small variation of the lower boundary conditions modifies the conclusions.

As an example, we adopt the density ϱ(120 km) given by (5.19), with a very small variation ($\pm 10\%$) of the pressure, i.e., p(120 km) = $(2.6 \pm 0.26) \times 10^{-5}$ Torr. *Figure 5.3* shows the result which is obtained when the temperature reaches 1225°K at 275 km. The average densities at 200, 250, and 275 km are 3.7×10^{-13}, 1.0×10^{-13}, and 6×10^{-14} gm cm^{-3}, respectively.

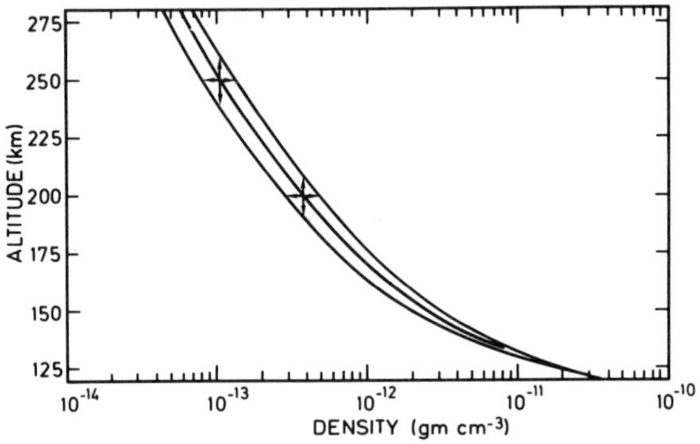

Fig. 5.3 Atmospheric densities for a constant temperature of 1225°K at 275 km and a pressure variation of $\pm 10\%$ at 120 km.

However, the variation represented by horizontal bars in Fig. 5.3 is not less than $\pm 20\%$. There is, therefore, an appreciable variation of the density when there is a small change of the boundary conditions, i.e., the temperature at 120 km varying by $\pm 10\%$ with constant densities. Furthermore, such conditions lead to a corresponding height change of about ± 10 km represented by vertical bars in Fig. 5.3. Thus, when observational and theoretical data are compared, it is difficult to select thermospheric parameters which are correct.

A very great precision is required at 150 km. In Fig. 5.3 the density at 150 ± 5 km which is ϱ(150 km) = $(2.5 \pm 0.5) \times 10^{-12}$ gm cm^{-3} with $T = 764 \pm 27$°K shows the difficulty in obtaining a final solution from observational or theoretical results. The adoption of ϱ(150 km) = $(1.5 \pm 0.3) \times 10^{-12}$ gm cm^{-3} for the same temperature conditions leads to ϱ(200 km) = $(2.8 \pm 0.6) \times 10^{-13}$ gm cm^{-3} and ϱ(250 km) = $(8.3 \pm 2.0) \times 10^{-14}$ gm cm^{-3} instead (see Fig. 5.3) of 3.7×10^{-13} and 1.0×10^{-14} gm cm^{-3}, respectively.

It can be concluded, when $T = 1225$°K at 275 km, that the following densities vary by $\pm 20\%$ if a pressure variation of only $\pm 10\%$ is introduced in the lower boundary conditions at 120 km:

$$\varrho(200\text{ km}) = 3.7\times10^{-13} \quad\text{or}\quad 2.8\times10^{-13}\text{ gm cm}^{-3};\quad \pm20\% \tag{5.21}$$

$$\varrho(250\text{ km}) = 1.0\times10^{-13} \quad\text{or}\quad 8.3\times10^{-14}\text{ gm cm}^{-3};\quad \pm20\% \tag{5.22}$$

$$\varrho(275\text{ km}) = 6.0\times10^{-14} \quad\text{or}\quad 5.0\times10^{-14}\text{ gm cm}^{-3};\quad \pm20\% \tag{5.23}$$

when

$$\varrho(150\text{ km}) = (2.5 \pm 0.5)\times10^{-12} \quad\text{or}\quad (1.5 \pm 0.3)\times10^{-12}\text{ gm cm}^{-3} \tag{5.24}$$

For lower temperature ($T < 1000°$K) the variations are more important. An analysis [9] of the density at 150 km obtained from the growth of chemiluminescent clouds leads to an average value of 2.4×10^{-12} gm cm^{-3}. Mass spectrometric data [9] and extreme ultraviolet extinction measurements give, however, an average value [10, 11] of 1.96×10^{-12} gm cm^{-3}. Actually, the discussion [9] of experimental data prior to 1970 clearly shows the fundamental importance of a good knowledge of the boundary conditions between 100 and 120 km. No attempt is made in this chapter to explain in detail the observations, but emphasis is placed on the effects produced by small variations of the boundary conditions.

5.3 Conditions above 250 km

The slow decrease of the density above 200 km must be explained by invoking a sufficiently high temperature. But, although there is a large increase in the scale height H between 200 and 700 km, the explanation is not necessarily to be found by an increase of the temperature: It can result from a decrease of the molecular weight mg. For example, a change in the scale height from $H = 50$ km near 200 km to $H = 100$ km near 750 km can take place in an isothermal atmosphere if atomic oxygen is in diffusive equilibrium in an oxygen–nitrogen thermosphere.

Isothermal models of the high atmosphere must be associated with a lower thermosphere in which the temperature gradient results from the absorption of solar ultraviolet radiation. A diurnal variation of temperature will occur which will be the result of differences in the heat transport between sunlit and dark atmospheres. One must know the relative proportions of atomic oxygen and molecular nitrogen at each altitude. The scale height gradient at sufficiently great heights is due to the decrease of mg. This cannot be ascertained with sufficient precision since an equilibrium does not exist. The actual state is variable and dependent on a variety of physical conditions. To show the effect of diffusion, the data of Tables 5.1 and 5.2 have been used and three levels at which diffusion equilibrium begins have been chosen, namely 220, 150, and 120 km. *Figure 5.4* shows that the vertical distributions of density are almost identical between 250 and 700 km in isothermal atmospheres for temperatures between about 1200 and 1375°K. These curves fit observational data. However, they exhibit large differences in the ratios of the atomic oxygen and molecular nitrogen densities. In the neighborhood of 500 km, a temperature variation less than ±100°K corresponds to a variation of a factor of 10 in the density ratios.

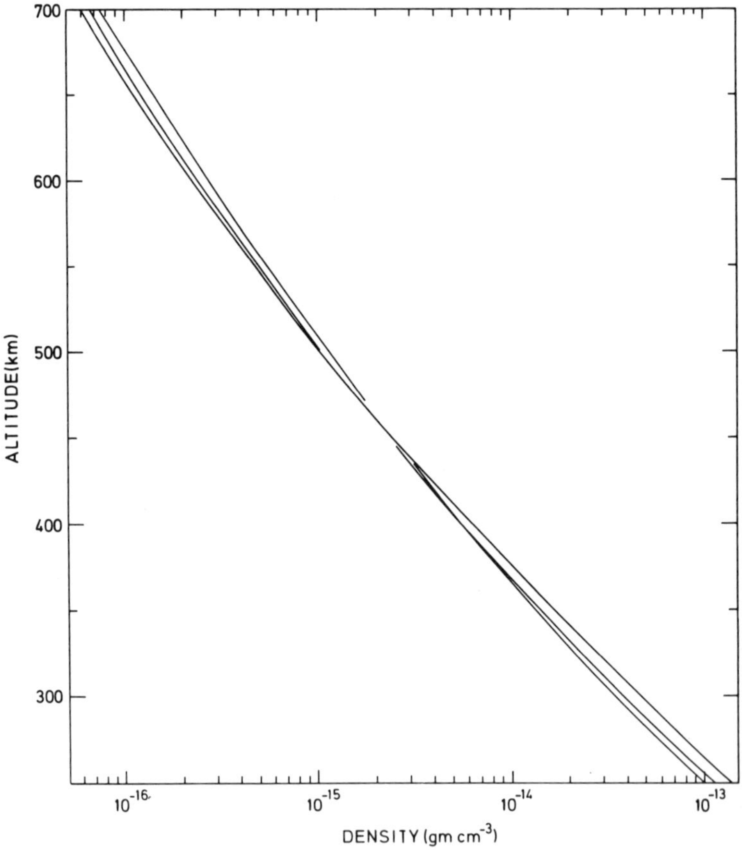

Fig. 5.4 The atmospheric density between 250 and 700 km in isothermal atmospheres for various values of the ratio $n(O)/n(N_2)$. The temperature is 1282 ± 91°K. The values of the ratio $n(O)/n(N_2)$ at 500 km are 17.5 for $T = 1373$°K, 52.5 for $T = 1282$°K, and 157.5 for $T = 1191$°K.

Such numerical data show that hasty conclusions must be avoided when parameters such as temperature, mean molecular mass, or atomic and molecular concentrations are employed. Furthermore, since variations are expected, the uncritical use of observational data obtained in different places or at different times may lead to inconsistencies.

5.4 Energies in the Thermosphere

In order to determine conditions for which a temperature gradient can exist and be maintained we must discover whether or not a permanent heat flow can exist. Since it is only necessary to estimate within orders of magnitude, the kinetic energy of an atmospheric vertical column is a useful parameter.

At 100 km, the total kinetic energy (Table 3.2) cannot be less than 3×10^5 erg cm^{-2} while it is only of the order of 10^4 erg cm^{-2} at 200 km. If an energy of the order of 1 erg cm^{-2} sec^{-1} is introduced in the atmosphere at 100 km, the thermosphere will not be subject to great changes below 100 km. But, since such amounts of energy are absorbed near 200 km, it is clear that the thermosphere will receive a total energy in one day corresponding to the kinetic energy of the vertical column. Thus, one can see that a temperature gradient may result. In any case, the existence of a strong temperature gradient requires an assessment of the heat conduction in the presence of heat sources and sinks. It will be shown in Chapter 14 that the thermal gradient associated with an energy flux (E erg cm^{-2} sec^{-1}) of heat transport is given by the following expression, $300 \leqq T \leqq 1000$°K,

$$dT/dz = (18 \pm 7)E \quad {}^\circ\text{K km}^{-1} \tag{5.25}$$

for an atomic oxygen atmosphere and by

$$dT/dz = (25 \pm 10)E \quad {}^\circ\text{K km}^{-1} \tag{5.26}$$

for an undissociated atmosphere.

Therefore, an energy of the order of 1 erg cm^{-2} sec^{-1} leads to a temperature gradient of more than 10°K km^{-1} in the thermosphere and can explain the high temperatures attained within a layer of some tens of kilometers, or a variation of the scale height from 10 to 50 km over an altitude interval of less than 100 km.

Since the scale height gradient at sufficiently high altitudes is due to the decrease of the mean molecular mass, scale heights of the order of 100 km are easily obtained in an isothermal layer.

REFERENCES

1. M. Nicolet, *The Earth as Planet* (G. P. Kuiper, ed.), p. 644. Univ. of Chicago Press, Chicago, Illinois, 1954.
2. D. R. Bates, *Proc. Roy. Soc.* **253**, 451 (1959).
3. G. V. Groves, *Nature* (*London*) **181**, 1055 (1958).
4. D. G. King-Hele and D. C. M. Leslie, *Nature* (*London*) **181**, 1761 (1958).
5. I. Harris and R. Jastrow, *Planet. Space Sci.* **1**, 20 (1959).
6. L. G. Jacchia, *Nature* (*London*) **183**, 526 (1959).
7. M. Nicolet, *Planet. Space Sci.* **5**, 1 (1961).
8. P. Mange, *Space Res.* **2**, 1002 (1961).
9. U. von Zahn, *J. Geophys. Res.* **75**, 5517 (1970).
10. L. A. Hall, W. Schweiger, and H. E. Hinteregger, *J. Geophys. Res.* **70**, 105 (1965).
11. L. A. Hall, C. W. Chagnon, and H. E. Hinteregger, *J. Geophys. Res.* **72**, 3425 (1967).

CHAPTER 6

Photoabsorption

6.1 Introduction

Less than 25 years ago, the ultraviolet spectrum of the sun was not known and the importance of solar x rays was unsuspected. At that time it was assumed that the solar emission in the unobservable ultraviolet wavelength regions corresponded to a blackbody with a temperature of 6000°K; i.e., a temperature approximating the radiation flux found at visible wavelengths. From aeronomical studies [1] it was deduced that below 3000 Å the ultraviolet radiation was less than that corresponding to a 6000°K blackbody. For chromospheric radiation below 1000 Å, the blackbody temperature increases again [2]. After the identification of coronal rays [3, 4] between 1942 and 1945 and the later (1950) theoretical study of ultraviolet emissions [5], the importance of coronal radiation was established and it became clear that a detailed analysis of the solar spectrum with wavelengths shorter than 1000 Å was necessary to determine the extent of ionospheric and atmospheric effects [6].

The first calculation [7] of the emission of the corona in the far ultraviolet and x-ray regions (1952) verified their importance to atmospheric ionization. Before this, the problem of the origin of the ionospheric regions was unresolved. Forty-five years ago [8] it was thought that the ionization arose from both solar ultraviolet radiation and energetic particles. Ionospheric observations made during eclipses, however, indicated [9–11] that the E region at an altitude of 100 km was not a result of corpuscular bombardment but probably arose from ionization produced by solar x rays having a large coefficient of absorption. This idea was developed later [12–

14] and definitely accepted following the first rocket observations of solar x rays [15–17].

Our knowledge of the ultraviolet spectrum has been obtained only in the past two decades [18, 19]. This information has been adequate to determine the general features of the most important spectral regions affecting the ionization of atoms and molecules in the earth's atmosphere. However, it is important to recognize that the absolute values of the photon fluxes and energies measured outside the atmosphere are subject to systematic errors of various types as well as variations due to solar activity.

In this chapter principal interest lies not in the solar spectrum, but in the cross sections [20, 21] for absorption and ionization by the atmospheric gases. In general, these data are rather uncertain, even with there having been a number of detailed laboratory and space experiments.

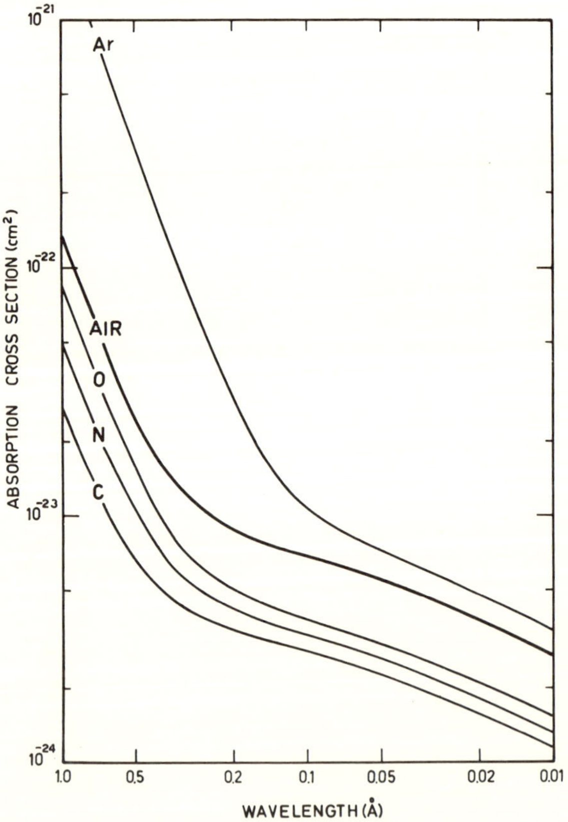

Fig. 6.1 Absorption between 0.01 and 1 Å for carbon, nitrogen, oxygen, and argon. The total absorption cross section by air is shown.

6.2 Absorption of x Rays between 1 and 10 Å

The absorption of x rays is more dependent on the concentration than on the nature of the elements if their atomic masses are not too different. For this reason, the absorption by the principal constituents must be considered in the homosphere. *Figures 6.1* and *6.2* show the absorption cross section between 0.01 and 1 Å and 1 and 15 Å, respectively. Between 0.01 and 15 Å the cross sections vary from about 3×10^{-24} to 10^{-19} cm^2. *Table 6.1* gives numerical values of the absorption cross section from 0.01 to 10 Å related to the altitude at which the unit optical depth occurs [22]. It can be seen that the absorption peak for wavelengths between 0.01 and 0.7 Å occurs in the stratosphere from 27.5 to 50 km. The mesospheric absorption corresponds to a spectral range between 0.7 and 7 Å. *Figure 6.3* shows the

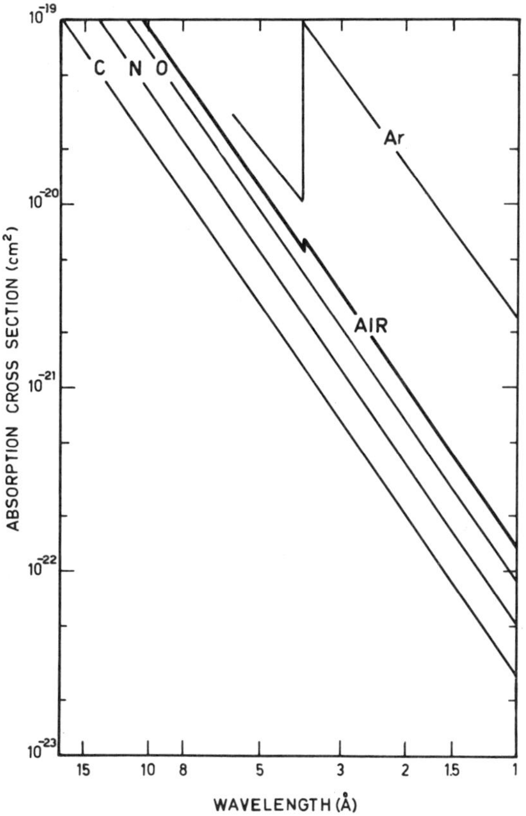

Fig. 6.2 Absorption cross section of the principal atmospheric constituents between 1 and 10 Å.

TABLE 6.1

ABSORPTION CROSS SECTIONS AND ALTITUDE OF THE UNIT OPTICAL DEPTH OF X RAYS BETWEEN 0.01 AND 10 Å

λ (Å)	$K(N_2)$ (cm^2)	$K(O_2)$ (cm^2)	K(Ar) (cm^2)	K(air) (cm^2)	Altitude $\tau_{air} = 1$ (km)
0.01	2.66×10^{-24}	3.03×10^{-24}	3.42×10^{-24}	2.74×10^{-24}	27
0.02	3.69	4.22	4.77	3.81	30
0.05	5.28	6.05	7.11	5.46	32
0.10	6.64	7.65	1.09×10^{-23}	6.89	34
0.20	8.40	1.01×10^{-23}	2.96	8.95	36
0.40	1.45×10^{-23}	2.03	1.66×10^{-22}	1.71×10^{-23}	41
0.60	2.91	4.59	5.23	3.72	47
0.80	5.71	9.51	1.20×10^{-21}	7.56	53
1.00	1.03×10^{-22}	1.76×10^{-22}	2.27	1.38×10^{-22}	58
1.20	1.71	2.95	3.83	2.31	62
1.50	3.23	5.63	7.18	4.37	66
2.00	7.47	1.31×10^{-21}	1.59×10^{-20}	1.01×10^{-21}	72
2.50	1.44×10^{-21}	2.57	2.90	1.92	76
3.00	2.46	4.29	4.67	3.26	79
4.00	5.73	9.93	1.10	6.55	83
5.00	1.10×10^{-20}	1.90×10^{-20}	1.93	1.26×10^{-20}	87
6.00	1.87	3.20	3.09	2.13	90
8.00	4.27	7.22	6.23	4.85	95
10.00	8.05×10^{-20}	1.34×10^{-19}	1.05×10^{-19}	9.09×10^{-20}	99

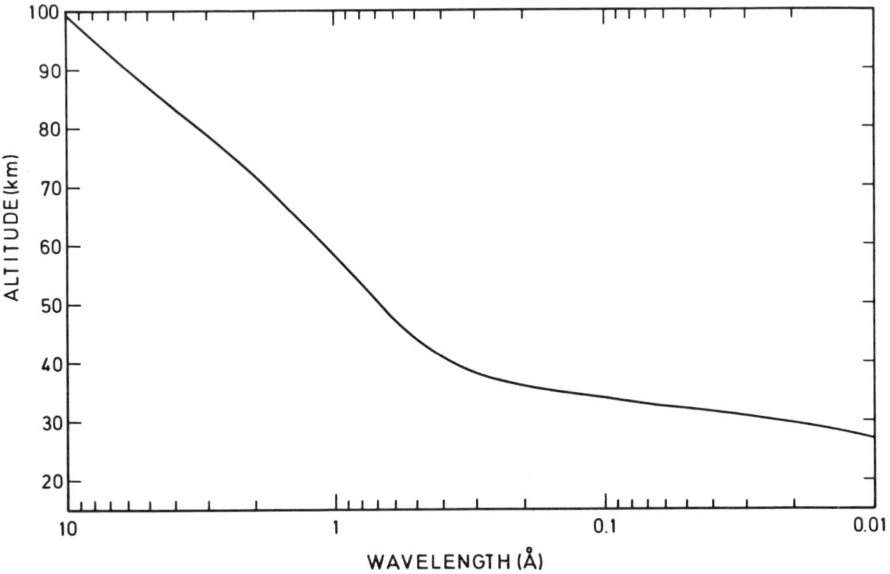

Fig. 6.3 Altitudes of the unit optical depth for $0.01 \leq \lambda \leq 10$ Å.

absorption of x rays versus height by means of a plot of the altitude of the unit optical depth versus wavelength. *Figure 6.4* shows the contributions of the ionization and scattering cross sections for atomic oxygen.

Absorption of x rays causes several effects which must be analyzed for the particular atmospheric constituents invoked. The short wavelength radiation which is absorbed initially causes photoionization. The K-limit ionization thresholds for oxygen and nitrogen are 23.3 and 31.0 Å, respectively, and the resulting photoelectric electrons produce other ionization effects if their energies are sufficiently high. For radiation with wavelengths shorter than 0.5 Å, Compton scattering must be added to the photoelectric effect and the decrease of the total cross section with wavelength becomes less important than at longer wavelengths. Nevertheless, the final result of any absorption process is the ionization of oxygen and nitrogen. In fact, the number of electrons may be estimated by assuming an energy utilization of 35 eV per ion pair [23]. It is therefore possible to determine the number of electrons produced per solar photon. The approximate values for $\lambda > 1$ Å are

Wavelength (Å):	6	4	2
Production (electrons/photon):	45	75	165

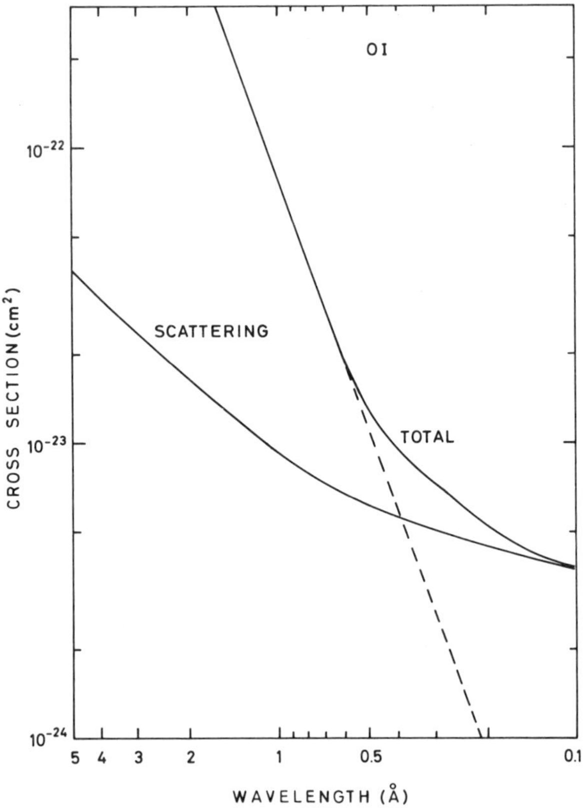

Fig. 6.4 Absorption cross section for atomic oxygen involving the scattering effect.

As far as the shortest wavelengths are concerned, the efficiency is very high:

Wavelength (Å):	0.62	0.31	0.124
Quantum energy (keV):	20	40	100
Production (electrons/photon):	570	1140	2860

One may conclude that the ionization by x rays of wavelength shorter than 10 Å will occur in the D region of the ionosphere. The precise ionization effect cannot be determined without a knowledge of the solar radiation involved in this spectral range and also without assessing other possible ionization processes to be discussed later. Finally, it may be pointed out that argon shows an important absorption cross section in this spectral range, but its effect on the general ionization is not important. The discontinuity near 3.65 Å in Fig. 6.2 is the effect of the K threshold of argon.

6.3 Absorption of x Rays between 10 and 100 Å

The K limits of carbon, nitrogen, and oxygen lie in the spectral range between 10 and 100 Å. *Figures 6.5* and *6.6* show that the spectral distribution of the absorption by these three elements and the atmospheric absorption corresponds to cross sections between 10^{-19} and 10^{-18} cm^2. The unit

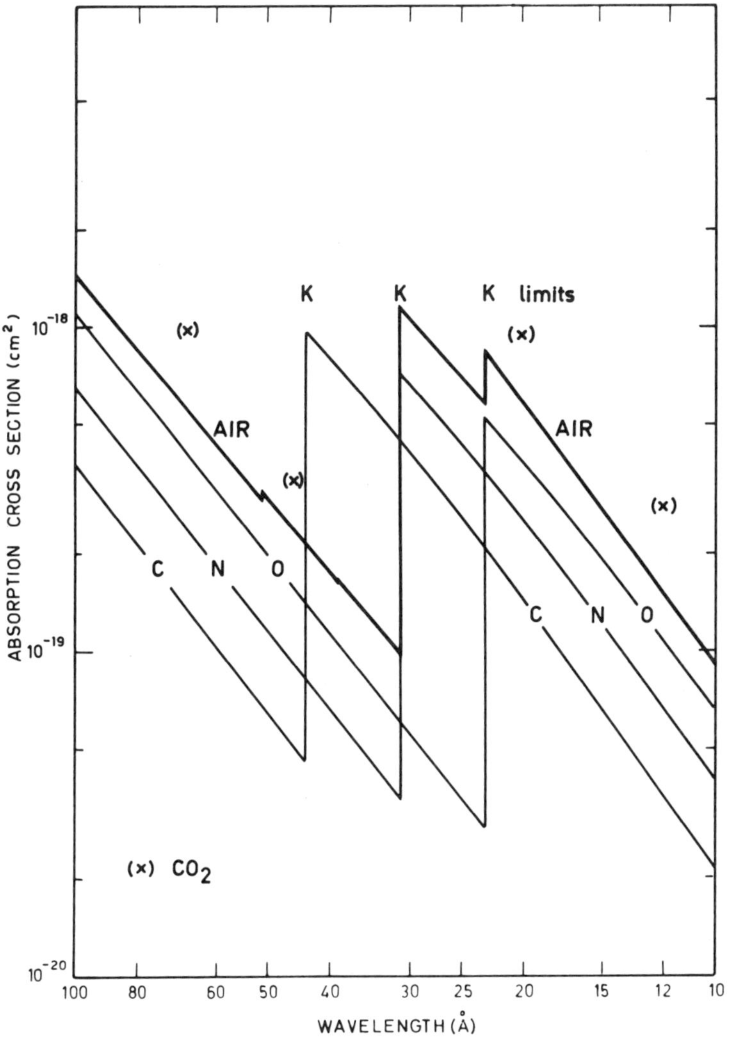

Fig. 6.5 Absorption of x rays between 10 and 100 Å. The effect of the K thresholds leads to two spectral ranges: 100–31 Å and 31–10 Å, with almost identical cross sections.

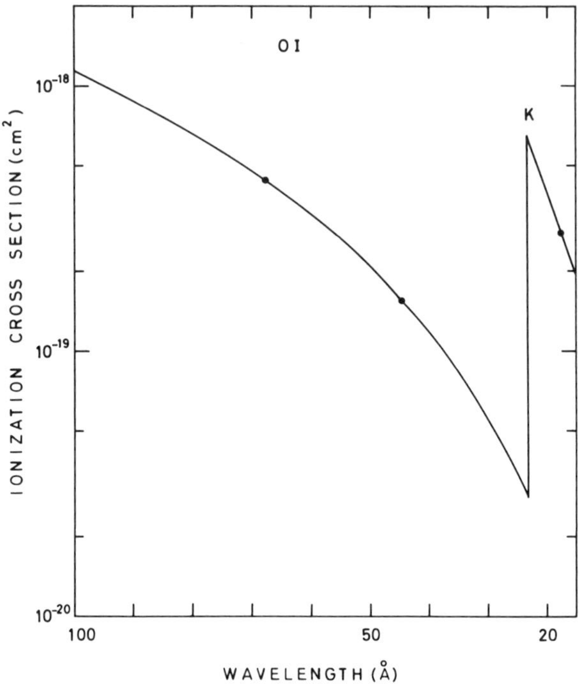

Fig. 6.6 Ionization cross section of atomic oxygen. The curve shows empirical values; the points, experimental ones.

optical depth for the whole spectral range is concentrated in the layer about 20 km thick centered near 110 km. One can therefore deduce a close association with the E layer of the ionosphere. It must be pointed out that two spectral ranges must be considered; these relate to the absorption for wavelengths greater or less than 31 Å. In fact, ionization for $\lambda > 31$ Å occurs in the L continuum and for $\lambda < 23$ Å corresponds to the K continuum (see *Table 6.2*). The first spectral range will be the more efficient for the atmospheric ionization because solar energy in the range $\lambda > 31$ Å is greater than that at $\lambda < 31$ Å. Nevertheless, other processes such as ionization by ultraviolet radiation and the various recombination effects must be considered before attempting to assess the nature of the ionization equilibrium of the E layer.

In order to simplify the calculation of x-ray ionization, the spectral ranges from 100 to 31 Å and from 31 to 10 Å may be divided into three parts for each of which a mean value for the cross section is appropriate. *Table 6.3* shows how a comparison may be made of different cross sections for wavelengths in the whole range from 100 to 10 Å.

TABLE 6.2

ABSORPTION CROSS SECTIONS OF X RAYS BETWEEN 10 AND 100 Å

λ (Å)	$K(N_2)$ (cm^2)	$K(O_2)$ (cm^2)	K(Ar) (cm^2)	K(air) (cm^2)
10	8.05×10^{-20}	1.34×10^{-19}	1.05×10^{-19}	9.09×10^{-20}
12	1.34×10^{-19}	2.20	1.60	1.51×10^{-19}
15	2.47	3.94	3.00	2.75
20	5.26	7.95	6.16	5.77
23.3	7.4	1.3×10^{-18}		8.5
23.3	7.4	5.6×10^{-20}		5.80
25	8.96	7.00×10^{-20}	1.00×10^{-18}	7.10
30	1.38×10^{-18}	1.12×10^{-19}	1.70×10^{-18}	1.10×10^{-18}
31	1.40	1.21	1.79	1.15
31	0.71×10^{-19}	1.21	1.79	0.98×10^{-19}
35	0.97	1.64	2.50	1.33
40	1.35	2.29	2.50	1.76
50	2.35	4.00	3.80×10^{-18}	3.03
60	3.71	6.31	4.20×10^{-19}	4.22
70	5.45	9.28	5.60	6.20
80	7.61	1.30×10^{-18}	7.10	8.66
90	1.12×10^{-18}	1.74	8.50×10^{-19}	1.16×10^{-18}
100	1.33	2.26	1.00×10^{-18}	1.57

TABLE 6.3

ABSORPTION CROSS SECTIONS OF X RAYS BETWEEN 10 AND 100 Å

Designation (Å):	70	50	35
Spectral range (Å):	80–60	60–41	41–31
Cross sections (cm^2)			
total:	6.2×10^{-19}	3.0×10^{-19}	1.3×10^{-19}
$K(N_2)$:	5.5	2.4	0.97
$K(O_2)$:	9.3	4.0	1.6
Spectral range (Å):	25–19	19–12.8	12.8–10.3
Designation (Å):	20.5	15.5	11.5

6.4 Absorption of Ultraviolet Radiation Shorter than 1000 Å

The study of *ultraviolet absorption* is more complicated than that of x rays since it must be analyzed for each constituent. Atoms and molecules have very different cross sections even in the same spectral range. Nevertheless, it is necessary to attempt an analysis of the absorption behavior in order to reach some conclusions of a general nature.

When one compares ionization potentials of atoms and molecules which are greater than that of molecular oxygen (see *Table 6.4*), one can see that O_2 absorption limits the penetration of solar radiation capable of ionizing polyatomic molecules whose ionization potential is between 1010 and 940 Å.

TABLE 6.4

IONIZATION THRESHOLDS OF ATOMS AND MOLECULES FOR $\lambda \leq 1028$ Å

Constituent	λ (Å)	Constituent	λ (Å)	Constituent	λ (Å)	Constituent	λ (Å)
O_2	1028	N_2O	961	O	910	N_2	796
NO_2	1008	CH_4	954	CO_2	899	Ar	787
H_2O	985	OH	940	CO	885	Ne	575
O_3	969	H	911	N	852	He	504

Molecules such as CH_4, O_3, H_2O, NO_2, and N_2O can be neglected in the atmospheric ionization since photodissociation processes will play the leading role, and no radiation will be available to ionize these molecules in lower altitudes where they are sufficiently abundant.

In the same way, CO and CO_2 and inert gases such as Ar and Ne will not participate sufficiently in the atmospheric ionization since atomic oxygen is photoionized at $\lambda \leq 910$ Å and molecular nitrogen at $\lambda \leq 796$ Å. These two constituents together with molecular oxygen will limit the photoionization of all other constituents in the thermosphere.

Difficulties also arise from the presence of multiple ionization states. Molecular oxygen, for example, has several ionization potentials corresponding to electronic states of the ionized molecule:

State:	$^2\Pi_g$	$^4\Pi_u$	$^2\Pi_u$	$^4\Sigma_g^-$	$^4\Sigma_u^-$
Wavelength (Å):	1027.8	769.2	738.5	682.5	504.7
Energy (eV):	12.06	16.12	16.79	18.17	24.56

The ionization cross section is always less than 4×10^{-18} cm^2 (*Fig. 6.7*) in the continuum corresponding to the absorption spectral range of the first ionization potential (1028 Å $\leq \lambda \leq$ 910 Å). One can see, however, that the ionization cross section does not follow the continuum distribution but depends on the structure [24, 25] of absorption bands which are, in fact, preionization bands in several cases.

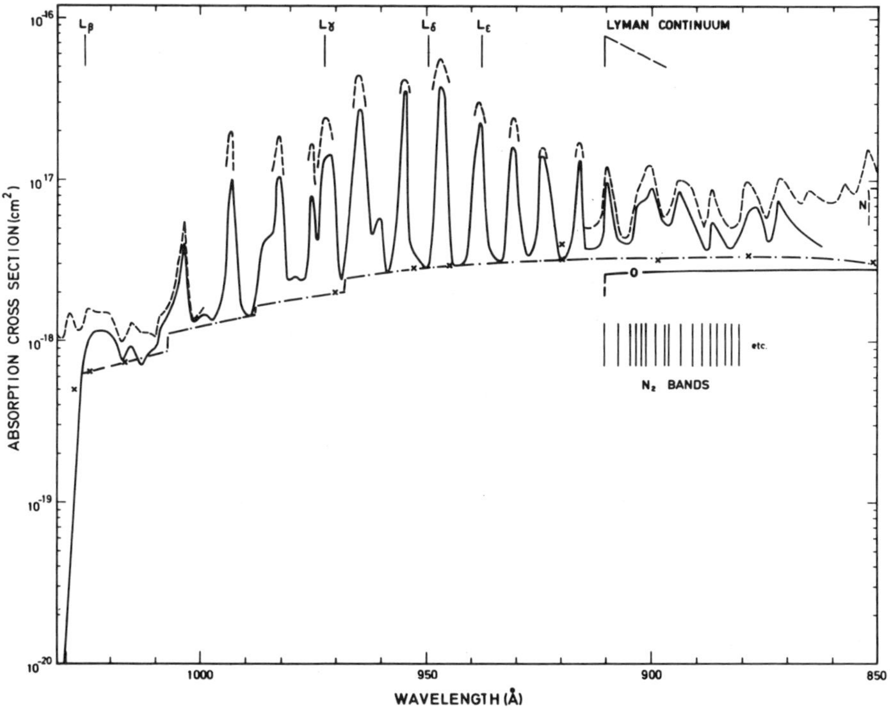

Fig. 6.7 Absorption and ionization cross section of molecular oxygen between 1030 and 850 Å. Atomic oxygen has its threshold near 910 Å. Dashed line, absorption cross section (Watanabe–Marmo [25]), curve, ionization cross section (Watanabe–Marmo [101]), crossed line, minimum ionization cross section (Lee [24]), circled line, atomic oxygen (Bates–Seaton [29]).

As a result, the ionization cross section has a very complicated structure. The example of the Lyman series of hydrogen shows how greatly the ionization efficiency is related to the detailed structure of the absorption spectrum. It is very difficult to obtain exact values of ionization coefficients when the spectrum of oxygen with superposed absorption bands is associated with the solar spectrum which also has a complicated structure over that spectral range. Furthermore, when absorption occurs in a particular band, it is necessary to know the exact fraction associated with the ionization cross section. Lyman β at 1025.7 Å is a typical example. Its ionization cross section is equal to 9.6×10^{-19} cm^2 and its total absorption cross section is 1.55×10^{-18} cm^2. Such values of the absorption of Lyman β show that an absorption peak must occur at a level within the E layer, where x rays of $\lambda > 10$ Å are also absorbed.

Experimental data, based on the absorption of monochromatic radiation within molecular bands, may vary since rotational structure is a function of temperature. It must therefore be stressed that when experimental data are used for aeronomic purposes special care must be taken. When measurements are related to a continuum, results obtained with sufficient resolution can be applied to the atmosphere; but, when bands are involved, results should be considered only as indications leading to semiquantitative values. This remark is of special importance in the study of the thermosphere where temperatures greater than 1000°K exist. Rotational structure and even vibrational distribution may modify experimental data obtained at lower temperatures.

Figure 6.7 indicates the beginning of the ionization continuum of atomic oxygen in order to show the simultaneous absorption by another constituent in the atmosphere and, consequently, the complexity of the total absorption. The effect of the absorption bands of molecular nitrogen should also be introduced in this spectral region. In fact, the whole spectrum from 1030 to 850 Å includes nitrogen bands with complicated structure. Although their position in the spectrum is known, it is not yet possible to give reliable limits for the absorption cross section. Experimental determinations made at too low dispersion are not useful for this purpose unless the rotational distribution is known. In fact, variations of several orders of magnitude must be expected over very short spectral intervals. In other words, two solar radiations in the same spectral range may correspond to a region in the atmosphere where the absorption is strong and to an atmospheric window with deep atmospheric penetration, respectively. As an example, Lyman γ at 972.5 Å and Lyman δ at 949.7 Å have absorption cross sections which differ by a factor of 10 in molecular oxygen (see Fig. 6.7). The exact values, however, depend on the temperature of the absorbing gas.

6.4.1 Atomic Oxygen

Recent laboratory measurements [26] have been sufficiently precise to determine the spectral structure leading to the ionization of atomic oxygen.

Nine Rydberg series have been observed up to their series limits corresponding to the three lower levels of O^+; namely $^4S^0_{3/2}$, $^2D^0_{5/2,3/2}$, and $^2P^0_{3/2,1/2}$.

The ground state of O(II), the level $^4S^0_{3/2}$, is reached from the ground state O(I)($^3P_{2,1,0}$) by a photon of wavelength $\lambda = 910.440$ Å corresponding to an ionization potential of 13.618 eV. The two higher O^+ states ^2D and ^2P correspond to wavelengths $\lambda = 731.82$ and $\lambda = 665.32$ Å, respectively.

Several more energetic ion states corresponding to the L shell limits have configurations $2p^4\ ^4P$ and $2p^4\ ^2P$. Finally, the extreme limit of photoionization corresponds to the ejection of an electron from the K shell by radiation of wavelength shorter than 23.32 Å. Data for the different states of O(II) are given in *Table 6.5*. Outer shell ejection leads to the ion ground

TABLE 6.5

IONIZATION THRESHOLDS OF ATOMIC OXYGEN

		cm^{-1}	Å	eV
First ionization: O(II) $2s^2 2p^2 2p^3$	$^4S^0_{3/2}$	109837.03	910.440	13.618
Second ionization: O(II)	$^2D^0_{5/2}$	136645.4	731.821	16.941
	$^2D^0_{3/2}$	136666.4	731.709	16.944
Third ionization: O(II)	$^2P^0_{3/2}$	150303.9	665.319	18.635
	$^2P^0_{1/2}$	150305.4	665.312	18.635
Fourth ionization: O(II) $2s2p^4$	$^4P_{5/2}$	229674.7	435.40	28.48
	$^4P_{3/2}$	229838.1	435.09	28.50
	$^4P_{1/2}$	229920.5	434.93	28.51
Fifth ionization: O(II) $2s2p^4$	$^2P_{5/2}$	322430.2	310.14	39.98
	$^2P_{3/2}$	322599.4	309.98	40.00
Ionization K: K shell		428800.0	23.32	531.7

state O(II)($^4S^0$) and to the metastable states $^2D^0$ and $^2P^0$ for wavelengths of 910, 732, and 665 Å. When an electron is removed from internal shells, the $2p^4\ ^4P$ and ^2P transitions occur at 435 and 310 Å, respectively. Radiation with wavelengths shorter than 23.3 Å leads to K-shell ionization most important for x rays.

A number of theoretical calculations have been made of the cross sections for absorption and ionization of O(I). The first measurements have given an accuracy estimated by the experimenters [27] as $\pm 30\%$ but there can be no greater precision claimed for the calculated values. The first calcula-

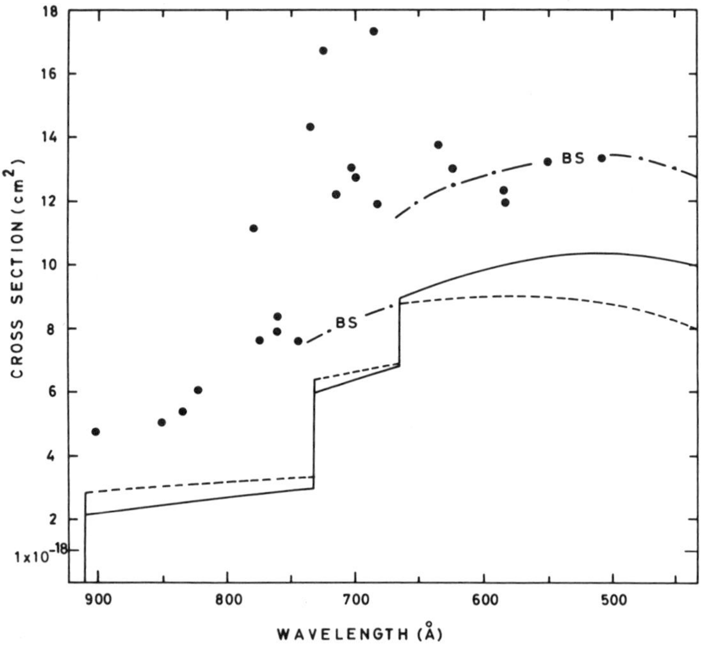

Fig. 6.8 Comparison of theoretical and experimental absorption cross sections for atomic oxygen. Points, experimental values (Cairns and Samson [27]); curve, dipole length, short dashes, dipole velocity (Henry [32]); dotted line (Bates–Seaton [29]).

tions indicated [28, 29] that the total cross sections for absorption were about 3×10^{-18} cm^2 for $910 > \lambda > 732$ Å, 8×10^{-18} cm^2 for $732 > \lambda > 655$ Å, and reached $(1.2 \pm 0.1) \times 10^{-17}$ cm^2 for $665 > \lambda > 400$ Å. More refined methods [30–32] led to the same values (see *Fig. 6.8*). It is noted that some very large experimental values for the cross section at 685.5 and 735.9 Å (see Fig. 6.8) result from oxygen absorption lines which lead to greater absorption than is appropriate to continuum ionization [33].

The most recent experimental data [34], as well as new theoretical results [35], shown in *Figs. 6.8–6.10*, give cross section values which are substantially smaller than the first experimental results.

Considering the various values for the oxygen atom ionization cross section corresponding to the first ionization potential, the theoretical cross section is $(3 \pm 0.3) \times 10^{-18}$ cm^2 with a possible error of $\pm 30\%$. The experimental values in contrast, are between 5×10^{-18} cm^2 and 3×10^{-18} cm^2. At wavelengths in the range 732–665 Å, the theoretical value of 6 to 7×10^{-18} cm^2 is intermediate between the experimental values of 5×10^{-18} cm^2 and 10^{-17} cm^2. Finally, in the continuum below 665 Å the cross section

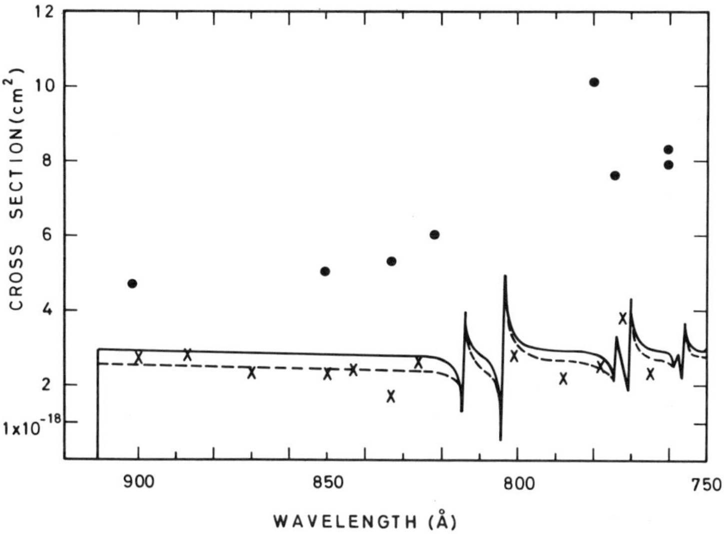

Fig. 6.9 Recent theoretical and experimental absorption data for atomic oxygen in the range 910–750 Å. Experimental values, dots (Cairns and Samson [27]); crosses (Comes *et al.* [34]); theory (Henry [39]), line—dipole velocity, dashes—dipole length.

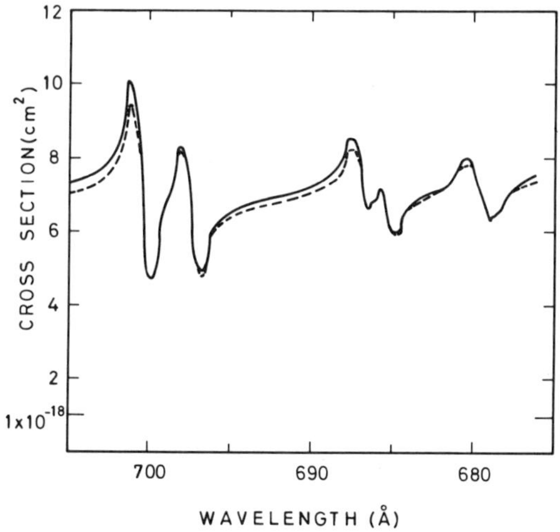

Fig. 6.10 Theoretical absorption cross section for atomic oxygen between 705 and 605 Å showing the influence of autoionizing states. The theory (Henry [39]) gives dipole velocity (solid curve) and dipole length (dashed curve).

TABLE 6.6

IONIZATION CROSS SECTIONS FOR ATOMIC OXYGEN ATOMS 910 TO 375 Å

Wavelengths (Å)	Ionization cross section (cm^2)
910–732	4.0×10^{-18}
732–665	7.5×10^{-18}
665–375	1.1×10^{-17}

reaches 10^{17} cm². Taking account of the different results, it is possible to accept the value $(1 \pm 0.1) \times 10^{-17}$ cm².

Summarizing the various experimental and theoretical data, the cross sections of *Table 6.6* for ionization of O(I) can be adopted as maximum values. Minimum values can be found through a 30% reduction of the listed cross sections.

Examination of the detailed theoretical cross sections [35] leading to the ejection of inner shell electrons O(II)($2p^4\ {}^4P$, ${}^2P$) shows that these values are a factor of 10 smaller than those for the outer shell states. Detailed theoretical results are given in *Table 6.7* for wavelengths less than 435 Å.

TABLE 6.7

ATOMIC OXYGEN IONIZATION CROSS SECTIONS FOR THE ION STATES $O^+({}^4P)$ AND $O^+({}^2P)$

Wavelength (Å)	$O^+({}^4P)$		$O^+({}^2P)$	
	Dipole length	Dipole velocity	Dipole length	Dipole velocity
435	7.1×10^{-19}	5.4×10^{-19}	—	—
400	8.3	6.1	—	—
350	9.2	6.7	—	—
310	—	—	3.8×10^{-19}	2.2×10^{-19}
300	9.0	6.3	4.3	2.5
250	8.0	5.9	5.6	2.9
200	6.3	5.0	5.0	3.0
150	4.7	3.8	3.8	2.9
100	2.8	2.2	2.0	1.1

TABLE 6.8

RATIOS OF $O^+(^2D)/O^+(^4S)$ AND $O^+(^2P)/O^+(^4S)$

Ion ratios	Wavelength bands (Å)				
	732–665	665–375	375–275	275–150	150–80
$O^+(^2D)/O^+(^4S)$	1.25 ± 0.1	1.55 ± 0.1	1.45 ± 0.5	1.30 ± 0.05	1.25 ± 0.5
$O^+(^2P)/O^+(^4S)$	0	0.90 ± 0.1	0.90 ± 0.05	0.80 ± 0.05	0.75 ± 0.5

It can be seen that the effect on the ion production will be such that the number of excited states O(II)($2p^4\,^4P$, ^2P) will be about one-tenth the number of metastable ions ^2D and ^4P. With regard to the outer shell states, the ground state O(I)(^4S) and the two metastable states ^2D and ^2P will be produced in approximately equal numbers. The most recent calculations [32] lead to cross-section ratios which depend upon wavelength but which can be approximated as shown in *Table 6.8* (see also *Figs. 6.11* and *6.12*).

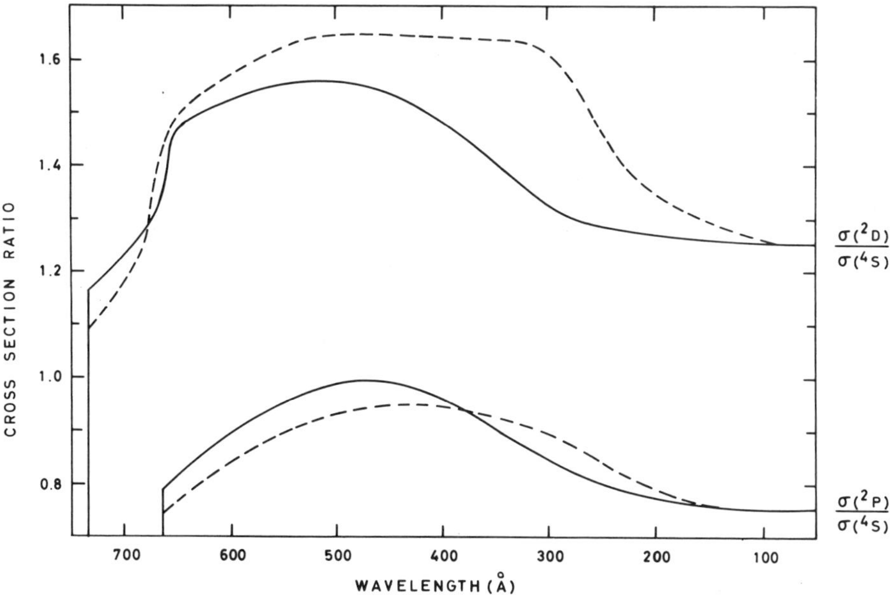

Fig. 6.11 Theoretical ratios for photoionization to the states $O^+(^2D)$ and $O^+(^2P)$ as compared with $O^+(^4S)$. The values from Henry [32] give dipole length (solid curve) and dipole velocity (dashed curve).

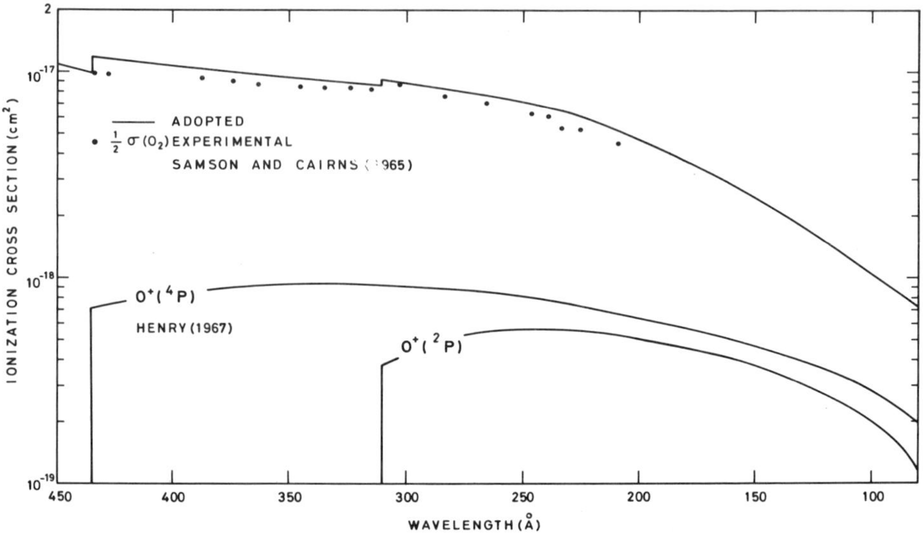

Fig. 6.12 Adopted photoionization cross sections for atomic oxygen in the range 450–100 Å.

For the spectral regions below 450 Å it is possible to make extrapolations based on the absorption of molecular oxygen [36, 37]. These empirical values, shown in Fig. 6.12, are summarized in *Table 6.9*.

At very small wavelengths it is necessary to take account of the fact that the absorption of photons leads to energetic electrons which can cause further ionization. For example, in the spectral ranges from 150 to 80 Å,

TABLE 6.9

Mean Ionization Cross Sections for Atomic Oxygen Atoms from 910 to 80 Å

Wavelengths (Å)	Ionization cross section (cm^2)
910–732	4.0×10^{-18}
732–665	7.5×10^{-18}
665–375	1.1×10^{-17}
375–275	8.0×10^{-18}
275–150	3.8×10^{-18}
150–80	1.8×10^{-18}

TABLE 6.10

NUMBER OF IONIZATIONS PER ABSORBED PHOTON FOR ATOMIC OXYGEN ATOMS

Wavelength band (Å)	Mean wavelength (Å)	Yield
80–60	62	3
60–40	50	5
40–31	31	7

one can adopt a mean value of two ionizations per photon. For smaller wavelengths the values shown in *Table 6.10* are appropriate.

For $\lambda < 31$ Å, one can adopt cross sections used previously as shown in Figs. 6.4–6.6. From 100 to 23.3 Å the absorption cross section is based on two experimental points [38] but below the K limit more data are available. It should be noted (see Fig. 6.4) that at very short wavelengths the photoelectric effect represents only a part of the absorption cross section due to the effect of scattering.

TABLE 6.11

ATOMIC OXYGEN ABSORPTION CROSS SECTIONS FROM 80 TO 10.3 Å

Wavelength band (Å)	Cross section (cm^2)
80–60 or 23.3–19	4.6×10^{-19}
60–41 or 19–12.8	2.0×10^{-19}
41–23.3 or 12.8–10.3	8.0×10^{-20}

In the spectral regions of wavelength greater than 23.3 Å, it is possible to adopt three domains with mean cross sections given in *Table 6.11*. The ionization rates are given by the product of the number of photons times the absorption cross section and the factors 3, 5, and 7 for the wavelengths 70, 50, and 35 Å, respectively.

The final adopted cross sections are listed in *Table 6.12*.

TABLE 6.12

ATOMIC OXYGEN PHOTOIONIZATION CROSS SECTIONS FROM 100 TO 2 Å

Wavelength (Å)	Cross section (cm^2)	Wavelength (Å)	Cross section (cm^2)
100	1.1×10^{-18}	23.3	6.5×10^{-19}
90	8.7×10^{-19}	20	4.0×10^{-19}
80	6.5×10^{-19}	15	2.0×10^{-19}
70	4.6×10^{-19}	12	1.1×10^{-19}
60	3.2×10^{-19}	10	6.7×10^{-20}
50	2.0×10^{-19}	8	3.6×10^{-20}
40	1.2×10^{-19}	6	1.6×10^{-20}
35	8.2×10^{-20}	5	9.5×10^{-21}
30	5.6×10^{-20}	4	5.0×10^{-21}
25	3.5×10^{-20}	3	2.2×10^{-21}
23.3	2.8×10^{-20}	2	6.5×10^{-22}

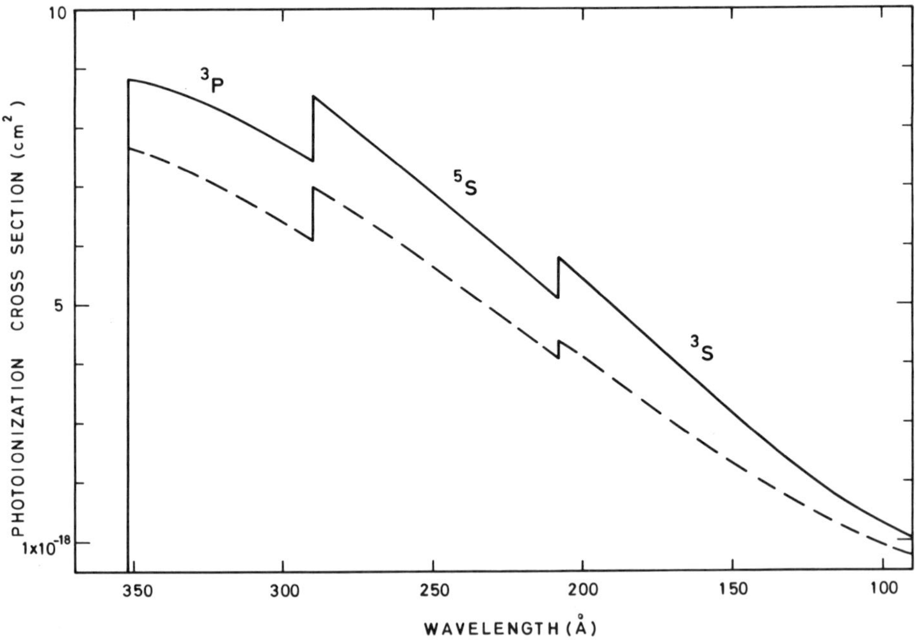

Fig. 6.13 Theoretical photoionization cross sections for atomic oxygen ions leading to O (III). The values from Henry [39] give dipole length (solid curve) and dipole velocity (dashed curve).

The ion O(II) can be photoionized by ultraviolet radiation with wavelengths smaller than 352 Å. The ground state of O(III) is the ^3P state but removal of other electrons leads [39] to the ^5S and ^3S states at greater energies (see *Fig. 6.13*). Accurate cross sections leading to these states are difficult to compute but a total photoionization cross section of O(II) can be adopted as shown in *Table 6.13*.

TABLE 6.13

PHOTOIONIZATION CROSS SECTIONS OF $O^+(^4S)$ LEADING TO O(III)

Wavelength band (Å)	Cross section (cm^2)
350–275	8×10^{-18}
275–150	5×10^{-18}
150–80	2.5×10^{-18}

The excited states of O(II) can spontaneously emit characteristic radiations. The wavelengths of the ^2P and ^4P transitions [40] are listed in *Table 6.14*.

TABLE 6.14

EMISSIONS FROM THE STATES $O^+(^2P)$ AND $O^+(^4P)$

Transition	Wavelength (Å)
$2p^4\ ^2P_{3/2}$–$2p^3\ ^2D^0_{5/2}$	538.256
$^2P_{1/2}$–$^2D^0_{3/2}$	537.830
$^2P_{3/2}$–$^2D^0_{3/2}$	538.318
$2p^4\ ^2P_{3/2}$–$2p^3\ ^2P^0$	580.967
$^2P_{1/2}$–$^2P^0$	580.400
$2p^4\ ^4P_{5/2}$–$2p^3\ ^4S^0_{3/2}$	834.467
$^4P_{3/2}$–$^4S^0_{3/2}$	833.332
$^4P_{1/2}$–$^4S^0_{3/2}$	832.762

The ratio of the multiplets 538 Å/581 Å is about 3.3 according to calculations [41]. The radiation emitted at 832 Å in the transition $^4P \rightarrow {}^4S$ is reabsorbed in the terrestrial atmosphere by molecular oxygen. The supplementary ionization due to this secondary process cannot be as important as direct ionization by radiation of wavelength $\lambda < 435$ Å, however.

As to other emissions leading to ground state transitions of O(II), there are contributions from the two metastable levels ^2D and ^2P. The energy level diagram of *Fig. 6.14* indicates that the lifetime of $^2P^0$ is relatively short (2.98 sec) corresponding to the emission probabilities $^2P^0 \rightarrow {}^2D^0$

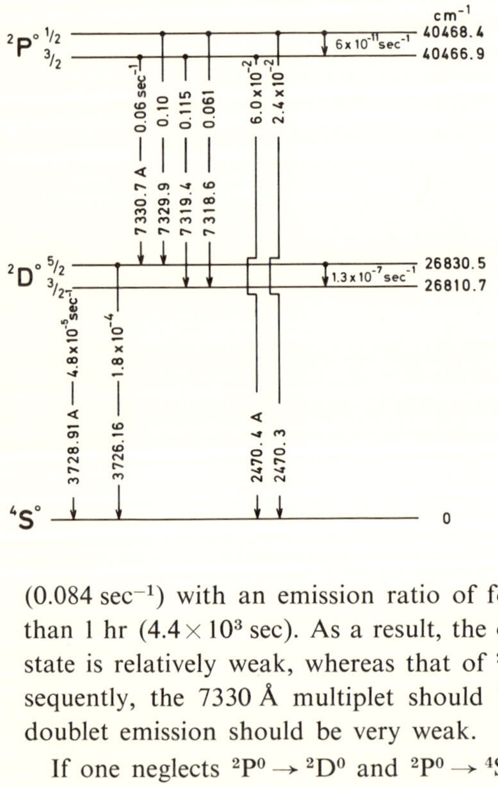

Fig. 6.14 Metastable states of O(II) with wavelengths of forbidden emissions.

(0.084 sec^{-1}) with an emission ratio of four. The lifetime of $^2D^0$ is more than 1 hr (4.4×10^3 sec). As a result, the collisional deactivation of the $^2P^0$ state is relatively weak, whereas that of $^2D^0$ can be very important. Consequently, the 7330 Å multiplet should be observable, while the 3728 Å doublet emission should be very weak.

If one neglects $^2P^0 \rightarrow {}^2D^0$ and $^2P^0 \rightarrow {}^4S^0$ transitions (see Fig. 6.14) and uses the production ratios in the states $^2D^0$, $^2P^0$, and $^4S^0$ (see Fig. 6.11), one finds that $^2D^0$ ions account for about 40% of the total O^+ ions resulting from photoionization. In introducing the emissions starting from $^2P^0$ with no deactivation, the production ratio of $^2D^0/^4S^0$ is about two. Thus, the ions O(II)($^2D^0$) constitute about two-thirds of the total initial ions produced by photoionization. Further discussion of this effect is given in Chapter 20.

6.4.2 Atomic Nitrogen

Recent laboratory measurements [42] have detected, after the first ionization limit at $\lambda = 852$ Å, the Rydberg series leading to a convergence corresponding to the $^5S^0$ state at a wavelength $\lambda = 610$ Å. The lines of the Rydberg series are broadened, indicating autoionization.

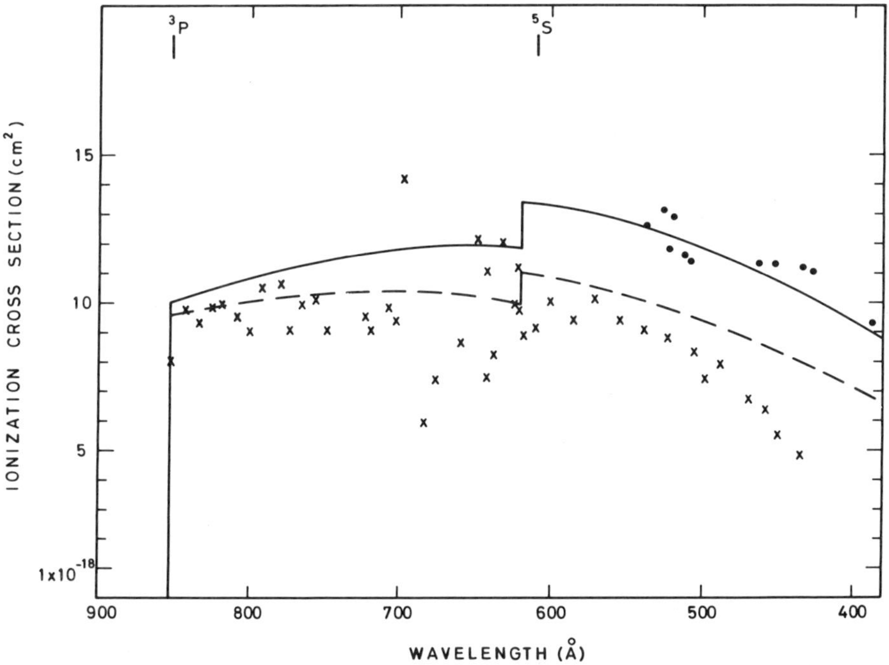

Fig. 6.15 Photoionization cross section of atomic nitrogen from ionization threshold to 250 Å. The values from Henry [39] give dipole length (solid curve) and dipole velocity (dashed curve). Other values are shown as crosses (Comes and Elzer [45]) and dots (Samson and Cairns, [44]), for $\frac{1}{2}\sigma(N_2)$.

The first experimental results [43], as well as initial calculations [29], indicated that the ionization cross section was of the order $(1.3 \pm 0.1) \times 10^{-17}$ cm² between 700 and 500 Å. The most recent theoretical calculations [39] (see *Fig. 6.15*) lead to values which are difficult to apply to the calculations of ionization. In order to make the computations for the ultraviolet limit it is necessary to make an extension of values taken from x-ray data using the values [44] of $\frac{1}{2}\sigma(N_2)$. On the long wavelength side the choice is still more difficult if one compares the most recent data [45].

TABLE 6.15

IONIZATION THRESHOLDS OF ATOMIC NITROGEN

		(cm^{-1})	(Å)	(eV)
First ionization: N(II)	$2s^2 2p^2\ {}^3P$	117,345	852.19	14.549
Second ionization: N(II)	$2s2p^3\ {}^5S$	163,999	609.76	20.333
Third ionization: N(II)	$2s2p^3\ {}^3S$	272,479	367.00	33.783
K ionization: K shell		10	30.99	400.0

A list of the different ionization states is given in *Table 6.15*. The adopted cross sections are listed in *Table 6.16*. One sees that the principal cross section is that which leads to the ionization state N(II)(^3P) with only a small contribution from N(II)(^5S) below 610 Å. In apportioning the ionization continuum of atomic nitrogen following the wavelength subdivisions adopted for atomic oxygen, one obtains the mean cross sections given in *Table 6.17*.

In the x-ray region molecular binding energies are unimportant and it is possible to adopt one-half the molecular nitrogen cross section (see *Fig. 6.16*). Values of these ionization cross sections are given in *Table 6.18*.

TABLE 6.16

PHOTOIONIZATION CROSS SECTIONS FOR ATOMIC NITROGEN ATOMS FROM 850 TO 150 Å

Wavelength (Å)	Cross section (cm^2) $(\times 10^{-17})$	Wavelength (Å)	Cross section (cm^2) $(\times 10^{-17})$
850	1.0	400	0.8
800	1.0	367	0.8
700	1.1	350	0.7
610	1.2	300	0.6
600	1.2	250	0.4
500	1.1	200	0.3
		150	0.2

TABLE 6.17

MEAN PHOTOIONIZATION CROSS SECTIONS FOR ATOMIC NITROGEN ATOMS FROM 852 TO 80 Å

Wavelength band (Å)	Cross section (cm^2)
852–732	1.0×10^{-17}
732–665	1.1×10^{-17}
665–375	1.0×10^{-17}
375–275	6.5×10^{-18}
275–150	2.5×10^{-18}
150–80	1.1×10^{-18}

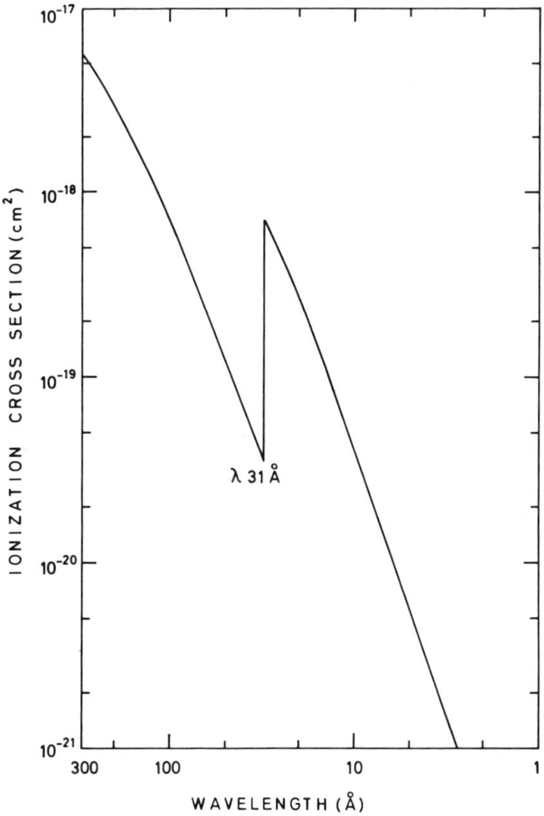

Fig. 6.16 Photoionization cross section of atomic nitrogen from 300 Å to the K shell discontinuity at 31 Å.

TABLE 6.18

PHOTOIONIZATION CROSS SECTIONS FOR ATOMIC NITROGEN FROM 100 TO 2 Å

Wavelength (Å)	Cross section (cm^2)	Wavelength (Å)	Cross section (cm^2)
100	6.7×10^{-19}	31	7.0×10^{-19}
90	5.1×10^{-19}	25	4.5×10^{-19}
80	3.8×10^{-19}	20	2.6×10^{-19}
70	2.7×10^{-19}	15	1.2×10^{-19}
60	1.9×10^{-19}	10	4.1×10^{-20}
50	1.2×10^{-19}	8	2.1×10^{-20}
40	6.7×10^{-20}	6	9.4×10^{-21}
35	4.8×10^{-20}	4	2.9×10^{-21}
31	3.5×10^{-20}	2	3.7×10^{-22}

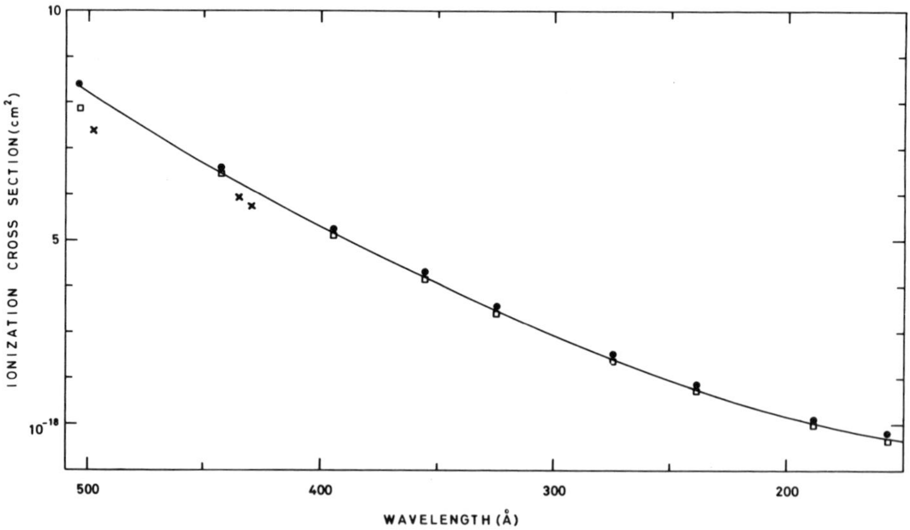

Fig. 6.17 Photoionization cross section of helium from ionization threshold to 150 Å. Experimental values, dots (Lowry *et al.* [49]); theoretical values, squares (Stewart and Webb [47]); experimental values, crosses (Cairns and Samson [27]).

6.4.3 Helium

The first theoretical photoionization cross sections for helium were computed in 1933, whereas the first successful experiments were not made [46] until 1955. Recent theoretical [47, 48] and experimental results [49, 50] give ionization cross sections from the limit at 504 Å down to 100 Å. *Figures 6.17* and *6.18* present the various values, along with the adopted variation,

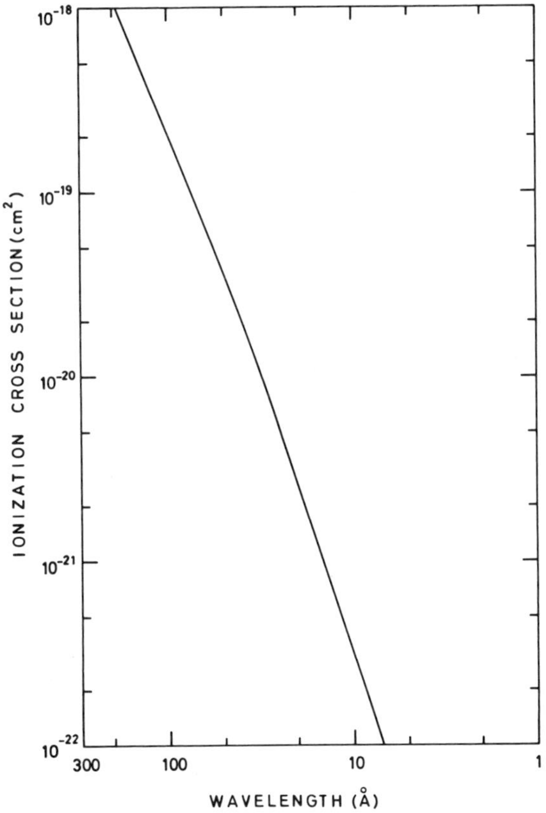

Fig. 6.18 Photoionization cross section of helium from 300 to 7 Å.

over the range $504 > \lambda > 1$ Å. Values of the cross section are given for selected wavelengths in *Table 6.19* while the mean values over the selected wavelength ranges are given in *Table 6.20*.

For wavelengths shorter than 189.577 Å photoionization of helium can be accompanied by simultaneous excitation [51]. In effect, it is energetically possible at energies above 65.399 eV to simultaneously ionize and excite

TABLE 6.19

PHOTOIONIZATION CROSS SECTIONS FOR HELIUM

Wavelength (Å)	Cross section (cm^2)	Wavelength (Å)	Cross section (cm^2)
500	8.2×10^{-18}	150	6.0×10^{-19}
450	6.2×10^{-18}	100	2.0×10^{-19}
400	5.3×10^{-18}	50	3.8×10^{-20}
350	4.3×10^{-18}	25	5.0×10^{-21}
300	2.9×10^{-18}	20	2.6×10^{-21}
250	2.0×10^{-18}	15	1.1×10^{-21}
200	1.2×10^{-18}	10	3.3×10^{-22}

the atom to the state $n = 2$. From this there results the resonance line emission of He(II) at 303.781 and 301.786 Å. The cross section for photoionization leading to an emission of He(II) (304 Å) is of the order of 10% of the total photoionization cross section.

TABLE 6.20

MEAN PHOTOIONIZATION CROSS SECTIONS FOR HELIUM

Wavelength band (Å)	Cross section (cm^2)
500–375	6×10^{-18}
375–275	3×10^{-18}
275–150	1×10^{-18}
150–80	3×10^{-19}

6.4.4 *Atomic Hydrogen*

The results for atomic hydrogen have been well determined through both theoretical calculations and experiment. Including the Gaunt factor, g, the ionization cross section can be written [52] as (see also *Fig. 6.19*)

$$\sigma(\mathrm{H}) = 6.31 \times 10^{-18} g(\lambda/\lambda_0)^3 \quad \mathrm{cm}^2 \tag{6.1}$$

where $\lambda_0 = 911.75$ Å. Results are listed in *Table 6.21*. The mean cross sections for the selected wavelength intervals are given in *Table 6.22*.

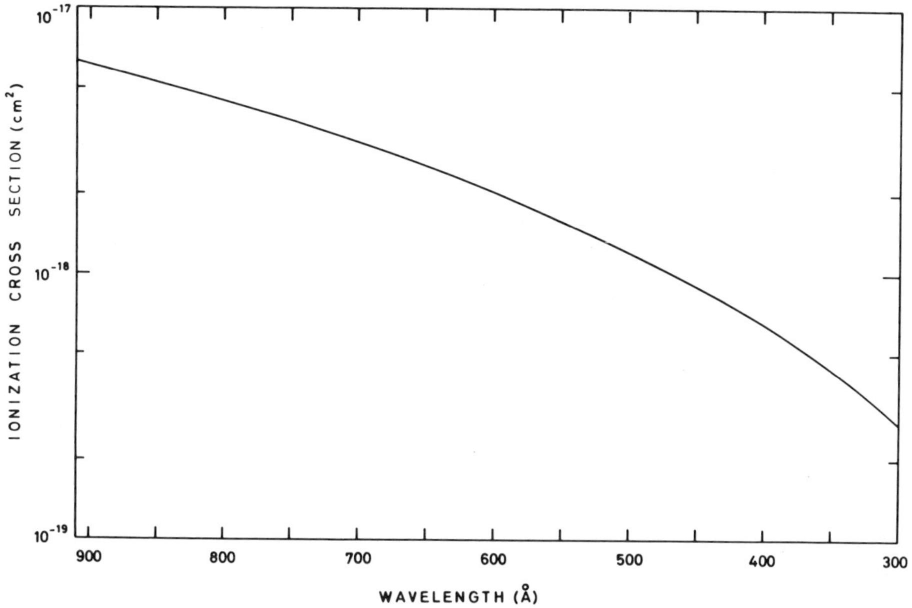

Fig. 6.19 Photoionization cross section of atomic hydrogen from ionization threshold to 300 Å.

TABLE 6.21

PHOTOIONIZATION CROSS SECTIONS FOR ATOMIC HYDROGEN

Wavelength (Å)	Cross section (cm^2)	Wavelength (Å)	Cross section (cm^2)
900	6.12×10^{-18}	450	9.01×10^{-19}
850	5.26×10^{-18}	400	6.43×10^{-19}
800	4.46×10^{-18}	350	4.36×10^{-19}
750	3.74×10^{-18}	300	2.78×10^{-19}
700	3.10×10^{-18}	250	1.62×10^{-19}
650	2.53×10^{-18}	200	0.83×10^{-19}
600	2.02×10^{-18}	150	0.35×10^{-19}
550	1.59×10^{-18}	100	0.10×10^{-19}
500	1.22×10^{-18}	50	$(0.01) \times 10^{-19}$

TABLE 6.22

MEAN PHOTOIONIZATION CROSS SECTIONS FOR ATOMIC HYDROGEN

Wavelength band (Å)	Cross section (cm^2)
910–796	5.5×10^{-18}
796–732	4.0×10^{-18}
732–665	3.2×10^{-18}
665–375	1.5×10^{-18}
375–275	3.0×10^{-19}
275–150	1.0×10^{-19}
150–80	3.0×10^{-20}

6.4.5 Argon

Although the ionization of argon is unimportant in the ionospheric reaction balance, it is worthwhile to know the ionization cross section in connection with studies of the primeval terrestrial atmosphere.

The different ionization potentials of argon are listed in *Table 6.23.*

Recent experimental [53, 54] studies provide a detailed knowledge of the cross section down to 300 Å (see *Fig. 6.20*). The structure appearing at $\lambda = 465.5$ Å before the ionization at 424 Å corresponds to the ejection

TABLE 6.23

IONIZATION THRESHOLDS FOR ARGON

Limits	Wavelength (Å)	Energy (eV)
M3	786.72	15.759
M2	777.96	15.937
M1	424.03	29.239
L(III)	49.5	248.5 ± 0.3
L(II)	49.9	250.6 ± 0.3
L(I)	38.1	326.5 ± 0.5
K	3.865	3202.9

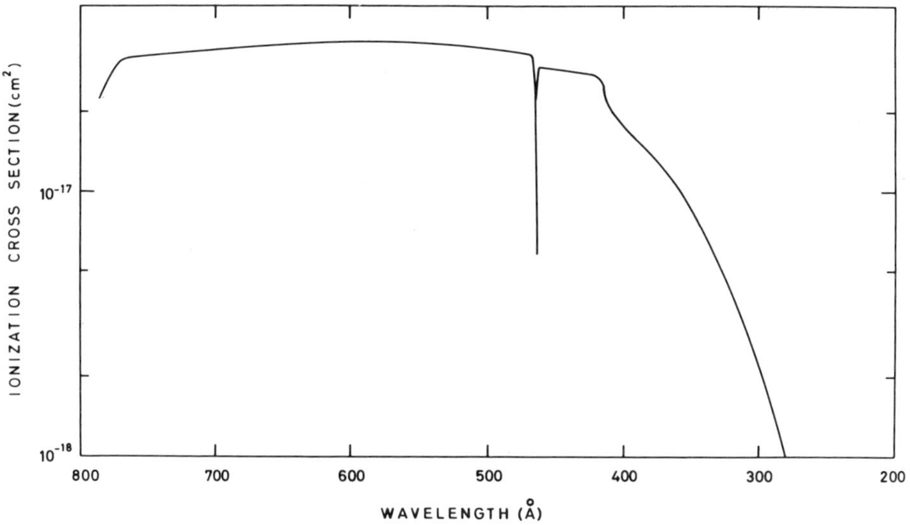

Fig. 6.20 Photoionization cross section of argon from ionization threshold to 300 Å. The curve is from Samson [54].

of an s electron. The values adopted here are given in *Table 6.24* where it is seen that in the range $778 > \lambda > 375$ Å the cross section is greater than 10^{-17} cm^2. The mean cross sections are given in *Table 6.25*.

Below 300 Å the spectrum of argon passes through several maxima [55]. The L and K ionization heads are well marked by thin discontinuities, as shown in *Fig. 6.21*. The mean cross section in the range $275 > \lambda > 80$ Å can be adopted as 1×10^{-18} cm^2.

TABLE 6.24

PHOTOIONIZATION CROSS SECTIONS FOR ARGON FROM 783 TO 300 Å

Wavelength (Å)	Cross section (cm^2)	Wavelength (Å)	Cross section (cm^2)
783	2.5×10^{-17}	450	3.0×10^{-17}
750	3.2×10^{-17}	425	2.75×10^{-17}
700	3.4×10^{-17}	400	2.0×10^{-17}
650	3.6×10^{-17}	375	1.35×10^{-17}
600	3.7×10^{-17}	350	8.50×10^{-18}
550	3.6×10^{-17}	325	4.75×10^{-18}
500	3.4×10^{-17}	300	2.15×10^{-18}

TABLE 6.25

MEAN PHOTOIONIZATION CROSS SECTIONS FOR ARGON

Wavelength band (Å)	Cross section (cm^2)
786–732	3.2×10^{-17}
732–665	3.4×10^{-17}
665–375	3.4×10^{-17}
375–275	3.0×10^{-18}

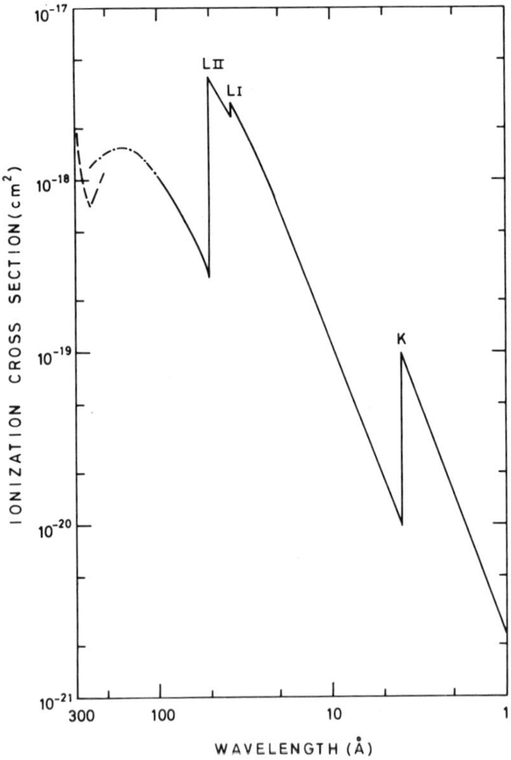

Fig. 6.21 Photoionization cross section of argon from 300 to 1 Å. The curve is from Nicolet [37], the dotted line from Lukirskii and Zimkina [55], and the dashed line from Samson [54].

TABLE 6.26

PHOTOIONIZATION CROSS SECTIONS FOR ARGON FROM 100 TO 10 Å

Wavelength (Å)	Cross section (cm^2)	Wavelength (Å)	Cross section (cm^2)
100	1.0×10^{-18}	31	1.8×10^{-18}
90	8.5×10^{-19}	30	1.7×10^{-18}
80	7.1×10^{-19}	25	1.0×10^{-18}
60	4.2×10^{-19}	20	6.2×10^{-19}
51	3.0×10^{-19}	15	3.0×10^{-19}
51	4.0×10^{-18}	10	1.0×10^{-19}
50	3.8×10^{-18}	8	6.2×10^{-20}
40	2.5×10^{-18}		

In the region where the L(II) and K limits are found (see *Fig. 6.21*) the cross section given in *Table 6.26* can be used while the mean cross sections of *Table 6.27* correspond to the previous spectral regions.

Below 3.87 Å the absorption leading to direct photoionization can be approximated as [56, 57]

$$\sigma(\mathrm{Ar}) = 1.0 \times 10^{-19} (3.87/\lambda)^{2.75} \quad \mathrm{cm}^2. \tag{6.2}$$

TABLE 6.27

MEAN PHOTOIONIZATION CROSS SECTIONS FOR ARGON FROM 80 TO 10.3 Å

Wavelength bands (Å)	Cross sections (cm^2)
80–60	5.6×10^{-19}
60–40	4.0×10^{-19}
40–31	2.3×10^{-19}
31–19	1×10^{-19}
19–12.8	3×10^{-19}
12.8–10.3	1.3×10^{-19}

6.4.6 *Molecular Oxygen*

Photoionization of molecular oxygen [58] in the ground state starts at (1027.8 ± 0.1) Å. Observations of Rydberg series and emission or absorption structure show the different processes involved in the O_2 photoionization. *Table 6.28* gives some of the energy levels represented in *Fig. 6.22*. There exist, however, other levels at shorter wavelengths which have not yet been analyzed. In this respect, one should actually consider the levels corresponding to the 435, 310, and 23.3 Å ionization limits of atomic oxygen. Dissociative ionization can occur for wavelengths shorter than 658, 598, and 530 Å giving $O^+(^4S)$ and oxygen atoms in the states ^3P, ^1D, and ^1S, respectively.

TABLE 6.28

ENERGY LEVELS OF O_2^+ STATES

Ion state	Wavelength (Å)	Energy (eV)	Wave number (cm^{-1})	Dissociation
$X\,^2\Pi_g$	1027.8 ± 0.1	12.063 ± 0.001	97,297	$O(^3P) + O^+(^4S^0)$
$a\,^4\Pi_u$	769.2	16.11	130,000	$O(^3P) + O^+(^4S^0)$
$A\,^2\Pi_u$	738.5	16.788	135,405	$O(^3P) + O^+(^4S^0)$
$b\,^4\Sigma_g^-$	682.5	18.170	146,556	$O(^1D) + O^+(^4S^0)$
$c\,^2\Sigma_g^-$	610.8	20.296	163,700	$O(^3P) + O^+(^2D^0)$
$c\,^4\Sigma_u^-$	504.7	24.564	198,125	

Several experimental determinations [59–62] of the absorption and ionization cross sections have been made and *Fig. 6.23* shows the O_2 spectrum between 2000 and 600 Å. Several absorption bands are superimposed on the ionization continuum starting at 1027.8 Å. These bands produce a perturbation in the ionization continuum down to 700 Å. The absorption cross section is always larger than the ionization cross section. The largest contribution to the ionization results from absorption in the continuum and also from preionization bands.

For aeronomical applications, it is useful to have mean cross sections over chosen wavelength intervals. Between the first O_2 ionization potential (1028 Å) and the first atomic ionization potential (910 Å), two solar spectral lines are of major importance: Lyman β at 1025.7 Å and C(III) at 977.0 Å. At 1025.7 Å, the O_2 absorption cross section is $\sigma_A = 1.55 \times 10^{-18}$ cm^2 and

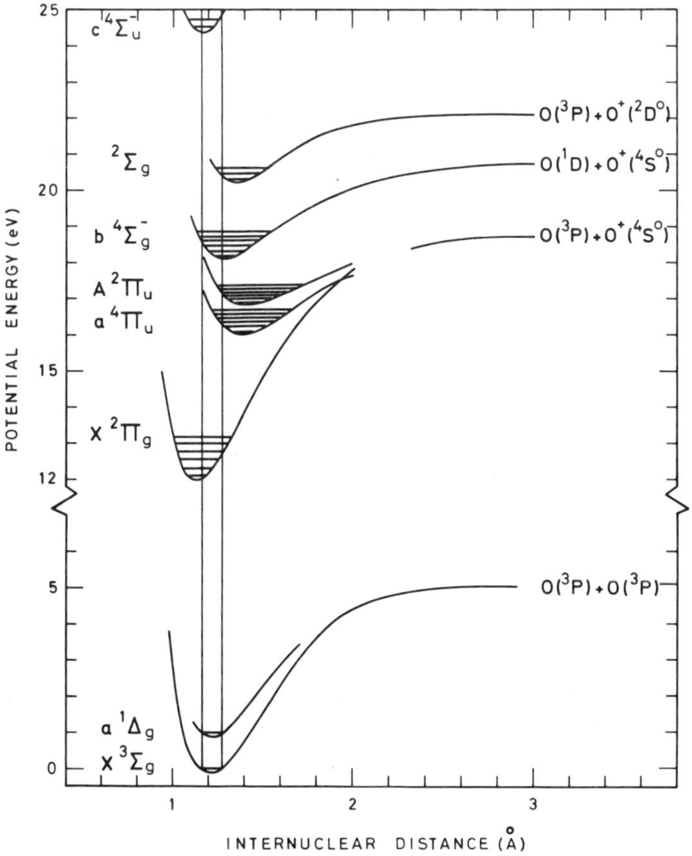

Fig. 6.22 Potential energy diagram for O_2. The Franck–Condon principle indicates which vibration states are most probable.

the ionization cross section is $\sigma_I = 1.0 \times 10^{-18}$ cm^2. For the C(III) line the values are respectively $\sigma_A = 4.0 \times 10^{-18}$ cm^2 and $\sigma_I = 2.5 \times 10^{-18}$ cm^2. The other mean values which can be adopted are shown in *Table 6.29*. In the spectral region 910–796 Å, characterized by bands and by a continuum, the adopted value can only be an approximation. Between 275 and 150 Å the cross sections are interpolated between ultraviolet measurements and x-ray determinations (see Section 6.3). Measurements [44] below 400 Å are shown in *Fig. 6.24*.

Dissociative photoionization has been observed for molecular oxygen. Below 719 Å, the following process [63]

$$O_2 + h\nu\ (>17.25\ \text{eV}) \rightarrow O^+ + O + e \tag{6.3}$$

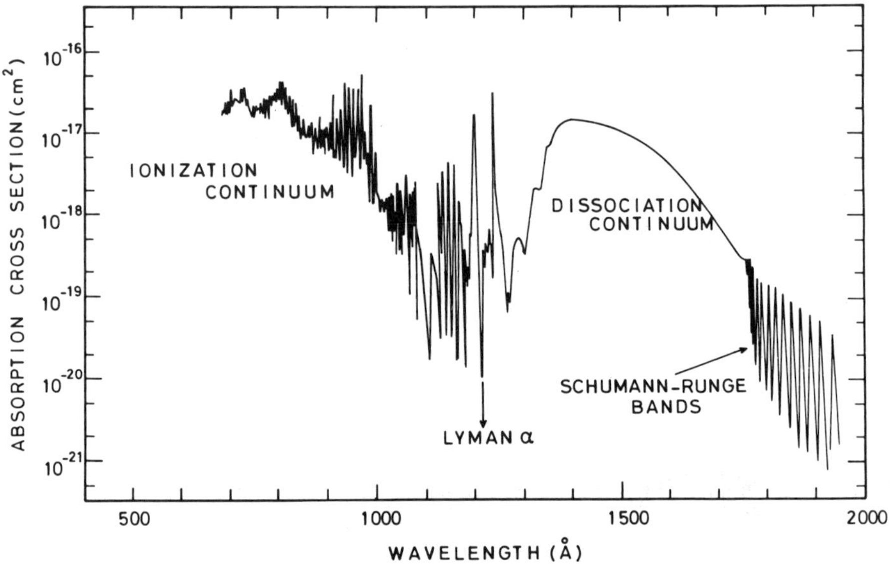

Fig. 6.23 Absorption cross section for O_2 showing the Schumann–Runge bands and the ionization continuum.

TABLE 6.29

MEAN PHOTOIONIZATION AND ABSORPTION CROSS SECTIONS FOR MOLECULAR OXYGEN

Wavelength band (Å)	Absorption cross section (cm²)	Ionization cross section (cm²)
1026	1.55×10^{-18}	1.0×10^{-18}
977	4.0×10^{-18}	2.5×10^{-18}
910–796	7.5×10^{-18}	5.0×10^{-18}
796–732	2.0×10^{-17}	1.0×10^{-17}
732–665	2.0×10^{-17}	2.0×10^{-17}
665–375	2.5×10^{-17}	2.5×10^{-17}
375–275	1.7×10^{-17}	1.7×10^{-17}
275–150	7.5×10^{-18}	7.5×10^{-18}
150–80	3.5×10^{-18}	3.5×10^{-18}

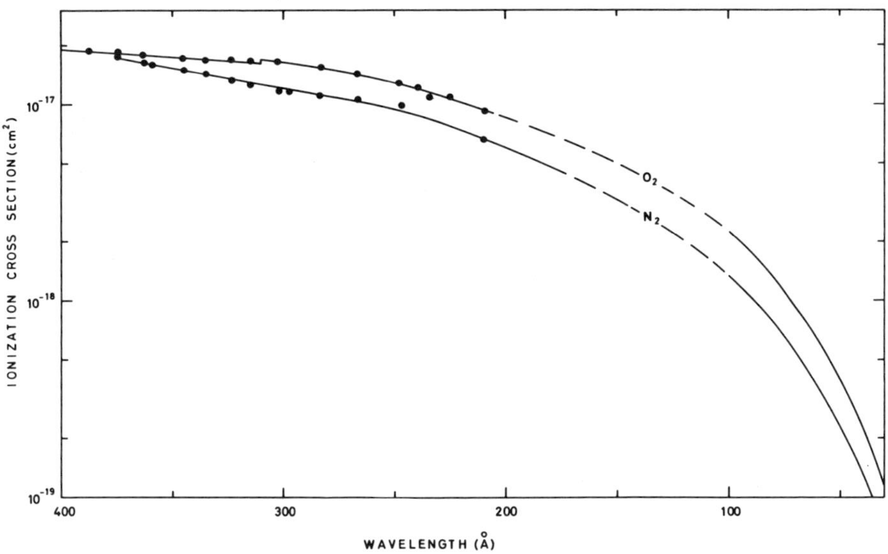

Fig. 6.24 Photoionization cross sections for molecular oxygen and nitrogen from 400 to 31 Å. The curve shows the adopted values; the points are from Samson and Cairns [82].

can lead to a ratio $n(O^+)/n(O_2^+) = 0.03$. Recent measurements [34] below 660 Å indicate that the process

$$O_2(^3\Sigma_g^-) + h\nu\ (>18.73\ \text{eV}) \rightarrow O^+(^4S^0) + O(^3P) + e \qquad (6.4)$$

has a cross section changing from 1×10^{-19} cm² at 650 Å to 1×10^{-18} cm² at 610 Å. Therefore, a cross section of $(3 \pm 1) \times 10^{-18}$ cm² can be used to estimate the dissociative photoionization. Below 600 Å, the following process is also possible:

$$O_2(^3\Sigma_g^-) + h\nu\ (>20.7\ \text{eV}) \rightarrow O^+(^4S^0) + O(^1D) + e. \qquad (6.5)$$

Measurements [64] at 584 Å, which indicate that 15% of O_2 is subject to dissociative ionization, show that energetic ions are produced. The aeronomical importance of those ions cannot be established before the energy spectrum is known.

It should also be mentioned that a fluorescence has been observed [65] which begins around 730 Å (O_2^+ in the A $^2\Pi_u$ state) and increases around 680 Å (O_2^+ in the b $^4\Sigma_g^-$ state). In any case, it is necessary to determine which fractions of O_2 molecules are excited to the different ionization states as a function of wavelength. The fraction of ionized molecules able to emit a fluorescence spectrum as well as the energy of the photoelectrons could then be determined.

Statistical weights of the positive ions have been used [66] to determine ionization cross sections for individual states. Such a procedure is, however, not compatible with experimental results [67, 68]. At the present time it is very difficult to determine the fluorescence conditions, the ionization in different states, and the energy of the photoelectrons. A very crude approximation of the population in the different states of O_2^+ is given as a function of wavelength in *Table 6.30*. In such a way, the excited O_2^+ ions can lead to ultraviolet emissions such as the Hopfield bands c $^4\Sigma_u^-$ → b $^4\Sigma_g^-$ or to visible emission such as the first negative system b $^4\Sigma_g^-$ → a $\Sigma^4\Pi_u$ and the second negative system A $^2\Pi_u$ → X $^2\Pi_g$.

TABLE 6.30

FRACTIONS OF IONS IN DIFFERENT STATES PRODUCED BY PHOTOIONIZATION OF MOLECULAR OXYGEN

Wavelength band (Å)	$O_2^+(X\,^2\Pi)$	$O_2^+(^4\Pi, ^2\Pi)$	$O_2^+(^4\Sigma)$
1027.8–796	1.00	—	—
796–665	0.45	0.55	—
665–80	0.45	0.35	0.20

6.4.7 Molecular Nitrogen

The spectrum of N_2 is characterized by numerous band systems superimposed on the ionization continuum [69] which starts at 796 Å. The principal states of N_2^+ are shown in *Fig. 6.25*, and the ionization limits are given in *Table 6.31*. Several experimental determinations [44, 53, 60, 70, 71] of the absorption spectrum have been made, but the spectrum is not well known below 500 Å. In *Fig. 6.26* it is seen that for $\lambda < 910$ Å (corresponding to the Lyman continuum of solar emission) the spectrum is dominated by numerous absorption bands. As a consequence, the ionization of atomic oxygen can be significantly influenced by N_2 absorption. The absorption of the solar Lyman γ line by N_2 is a typical example showing how a poor knowledge of the absorption cross section [72] can lead to erroneous con-

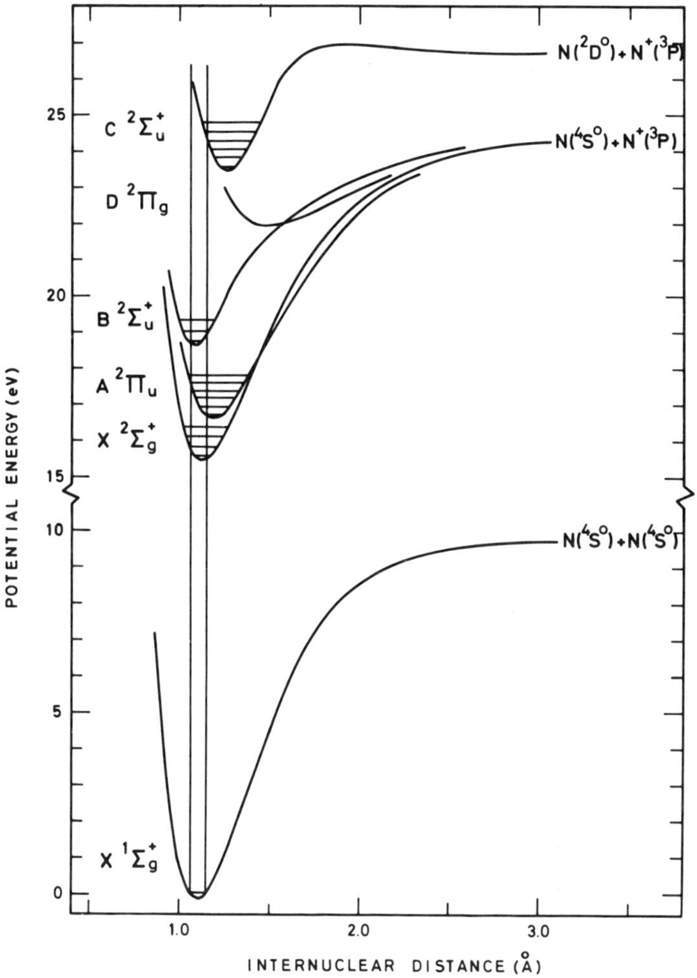

Fig. 6.25 Potential energy diagram for molecular nitrogen showing various N_2^+ states along the paths of vibrational excitation.

clusions. Measurements at specific wavelengths are the best guide for checking absorption and ionization cross sections.

For the Lyman β radiation at 1025.7 Å, the N_2 absorption cross section is about 1×10^{-21} cm^2 whereas for the C(III) line at 977 Å a value of 8.2×10^{-20} cm^2 can be adopted.

In the Lyman continuum, Fig. 6.26 shows that rather arbitrary values have to be adopted. The absorption effect of N_2 between 910 and 796 Å can be estimated in the way indicated in *Table 6.32*. For the other wave-

TABLE 6.31

IONIZATION THRESHOLDS OF N_2^+

Ion state	Wavelength (Å)	Energy (eV)	Wave number (cm^{-1})	Dissociation
$X\,^2\Sigma_g^+$	795.76	15.580	125,666	$N(^4S^0) + N^+(^3P)$
$A\,^2\Pi_u$	742.22	16.704	134,731	$N(^4S^0) + N^+(^3P)$
$B\,^2\Sigma_u^+$	661.23	18.750	151,233	$N(^4S^0) + N^+(^3P)$
$D\,^2\Pi_g$	563.94	21.993	177,324	$N(^4S^0) + N^+(^3P)$
$C\,^2\Sigma_u^+$	525.74	23.582	190,208	$N(^2D^0) + N^+(^3P)$

length intervals the adopted values are given in *Table 6.33*. These values cannot take into account the detailed structure shown in *Fig. 6.27*. As in the case of O_2, an interpolation is necessary between 200 and 68 Å where no measurements exist at the present time.

A fluorescence phenomena has been observed [73] when N_2 is ionized to the $B\,^2\Sigma_u^+$ state by radiation with $\lambda < 661.2$ Å. By observing the fluorescence [74] at 3914 Å, the cross section between 661 and 460 Å seems to be practically constant. A value of 2×10^{-18} cm^2 represents about 10% of

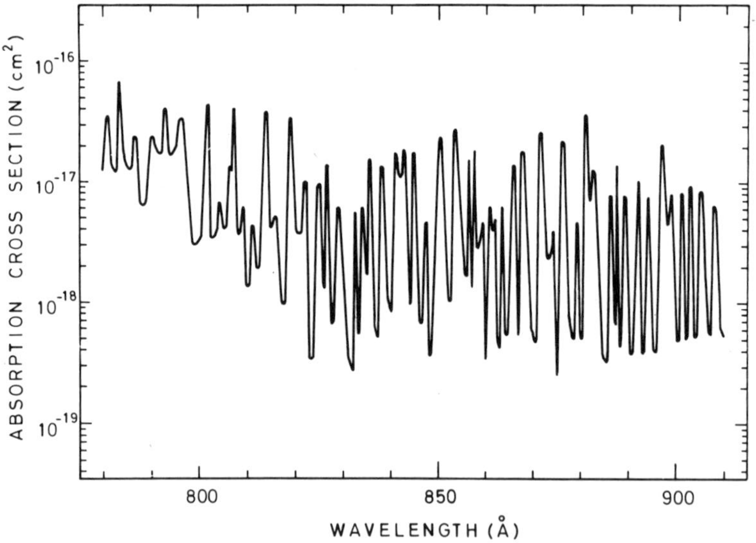

Fig. 6.26 Absorption structure of molecular nitrogen in the Lyman continuum ($\lambda < 910$ Å) and first photoionization continuum beginning at 795.6 Å.

TABLE 6.32

MOLECULAR NITROGEN ABSORPTION CROSS SECTIONS FOR 910 TO 796 Å

Fraction of 910–796 Å photon flux (%)	Absorption cross section (cm^2)
5	1.0×10^{-19}
5	2.5×10^{-19}
10	5.0×10^{-19}
20	1.0×10^{-18}
20	5.0×10^{-18}
20	1.0×10^{-17}
20	2.0×10^{-17}

the total photoionization cross section. By spontaneous emission from the $B\ ^2\Sigma_u^+$ state of N_2 the following radiations of the first negative system $B\ ^2\Sigma_u^+ \rightarrow X\ ^2\Sigma_g^+$ can be observed with the transition probabilities A_{ij}:

$$\lambda = 3914\ \text{Å}: \quad A_{0-0} = 1.07 \times 10^7 \quad \text{sec}^{-1}$$
$$\lambda = 4278\ \text{Å}: \quad A_{0-1} = 3.50 \times 10^6 \quad \text{sec}^{-1}$$
$$\lambda = 4709\ \text{Å}: \quad A_{0-2} = 7.99 \times 10^5 \quad \text{sec}^{-1}$$
$$\lambda = 5228\ \text{Å}: \quad A_{0-3} = 1.67 \times 10^5 \quad \text{sec}^{-1}.$$

TABLE 6.33

MEAN MOLECULAR NITROGEN ABSORPTION CROSS SECTIONS FROM 796 TO 80 Å

Wavelength band (Å)	Absorption cross section (cm^2)	Ionization cross section (cm^2)
796–732	2.0×10^{-17}	1.0×10^{-17}
732–665	2.3×10^{-17}	1.0×10^{-17}
665–375	2.3×10^{-17}	2.3×10^{-17}
375–275	1.4×10^{-17}	1.4×10^{-17}
275–150	5.0×10^{-18}	5.0×10^{-18}
150–80	2.2×10^{-18}	2.2×10^{-18}

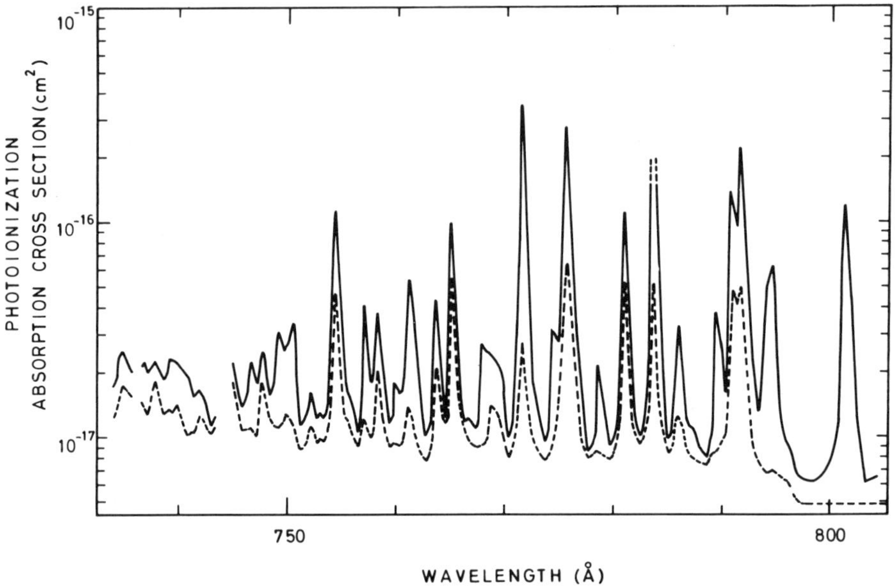

Fig. 6.27 Absorption and photoionization cross sections of molecular nitrogen in the region of the ionization continuum. Solid curve, absorption; dashed curve, ionization.

The observation of these bands [75, 76] in the dayglow or at twilight is an indication of the direct photoionization to the B $^2\Sigma_u^+$ state. However, absorption of solar radiation at 3914 Å also contributes to the fluorescence emission. For twilight conditions 80% of the emission from the first negative system results from fluorescence whereas 20% is due to direct photoionization in the B $^2\Sigma_u^+$ state. The total emission at 3914 Å corresponds to 4×10^{-2} photon sec^{-1} per N_2^+ ion. Other mechanisms such as electron impact ionization and charge transfer can, however, also affect the ionization states of molecular nitrogen.

Dissociation photoionization has been observed [77] near 510 Å with a cross section of 2×10^{-19} cm^2, i.e., 1/100 of the total photoionization cross section. Recent measurements [78] indicate a value of $(5\pm1)\times10^{-19}$ cm^2 between 500 and 450 Å, i.e., 1/40 of the total cross section. An analysis of atomic nitrogen ionization states [79] seems to indicate a cross section corresponding to approximately 20% of the total photoionization cross section.

The approximate distribution of the various states of N_2^+ can be evaluated [67, 68] with the same restrictions as for O_2. Possible values are shown in *Table 6.34*.

TABLE 6.34

FRACTIONS OF IONS IN DIFFERENT STATES PRODUCED BY PHOTOIONIZATION OF MOLECULAR NITROGEN

Wavelength band (Å)	$X\ ^2\Sigma_g^+$	$A\ ^2\Pi_u$	$B\ ^2\Sigma_u^+$
796–732	1.0	—	—
732–665	0.4	0.6	—
665–375	0.45	0.45	0.10

6.5 Absorption of Ultraviolet Radiation Longer than 1000 Å

Some indication can be given of the possible ionization processes in the spectral range of $\lambda > 1027.8$ Å, the limit associated with the first ionization potential of molecular oxygen. A list of atoms and molecules which are involved is given in *Table 6.35*. Nothing is known about the cross sections of CH_3 and CH. Very precise measurements have been made for nitric oxide [80]. At Lyman α, the absorption cross section is 2.42×10^{-18} cm^2 and the ionization cross section equals 2.02×10^{-18} cm^2.

Sodium and calcium cannot be neglected since their ionization thresholds occur at wavelengths for which solar radiation is important. Their ionization coefficient should be of the order of 10^{-5} and 10^{-6} sec^{-1}, respectively [81]. Other meteor atoms such as magnesium and silicon have ionization thresholds in the Schumann–Runge continuum.

TABLE 6.35

IONIZATION THRESHOLDS OF ATOMS AND MOLECULES FOR $\lambda > 1028$ Å

Atom	λ (Å)	Molecule	λ (Å)
Na	2412	NO	1340
Al	2071	CH_3	1260
Ca	2028	NH_3	1221
Mg	1622	CH	1117
Si	1521		
C	1100		

Another possibility is ionization from excited states of atoms and molecules by solar radiation of wavelengths greater than $\lambda = 1028$ Å (corresponding to the photoionization of O_2). Possibly the most important metastable state is the electron state $^1\Delta_g$ of O_2 which leads [82, 83] to an ionization threshold at 1118 Å. Since $O_2(^1\Delta_g)$ molecules are present in the lower thermosphere and mesosphere [84] this process can contribute to lower E-region ionization [85]. However, absorption by CO_2 must also be considered.

Above 1000 Å ozone and molecular oxygen are the principal atmospheric constituents responsible for the absorption of solar radiation, even though the mixing ratio of ozone is not higher than 10^{-6}.

6.5.1 Ozone

Ozone begins to absorb in the visible part of the spectrum with a maximum absorption cross section around 6000 Å of the order of 5×10^{-21} cm² in the Chappuis bands. Between 3100 and 3500 Å the Huggins bands are important and, finally, there is a large absorption in the Hartley bands with a maximum

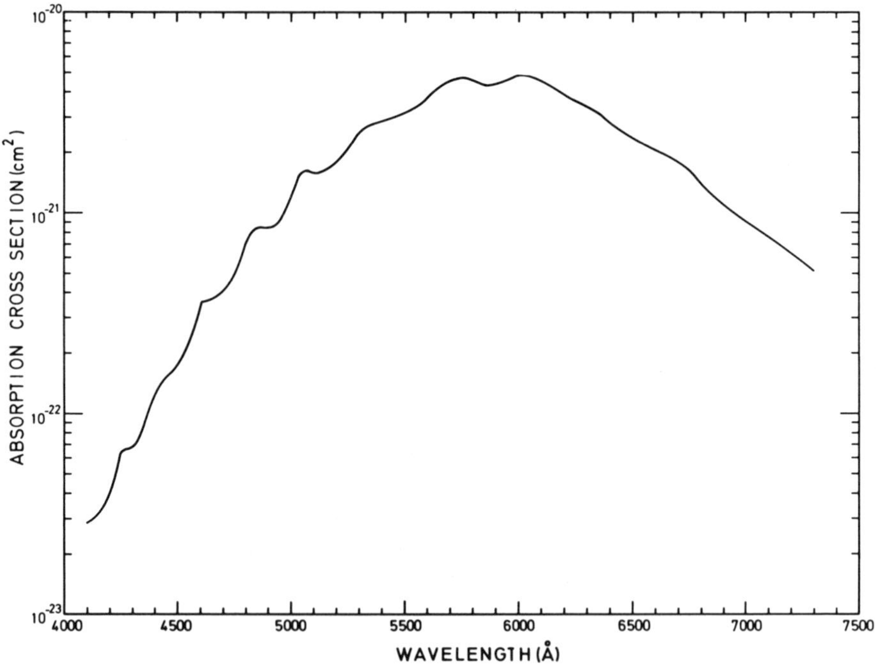

Fig. 6.28 Absorption cross section [86] of ozone in the Chappuis bands.

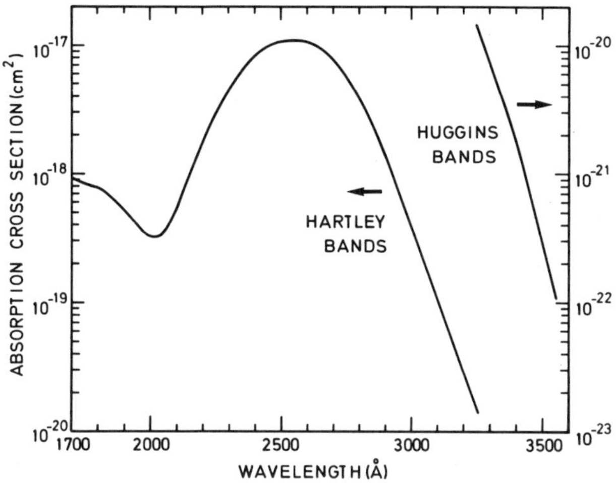

Fig. 6.29 Absorption cross section [86] of ozone in the Hartley and Huggins bands.

cross section of the order of 1×10^{-17} cm² around 2500 Å. The absorption cross sections shown in *Figs. 6.28–6.30* have recently been compiled and analyzed in detail [86].

The dissociation processes of ozone are summarized in *Table 6.36*. It is important to note that O_3 photodissociation leads to the production of excited atoms and molecules. These species play a fundamental part in the

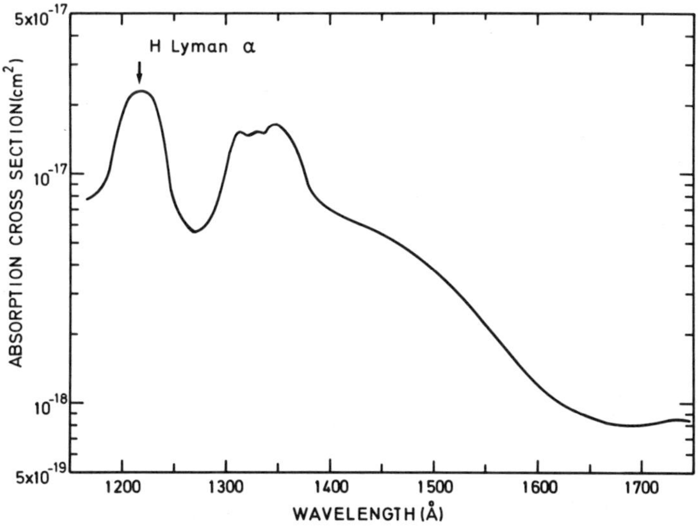

Fig. 6.30 Absorption cross section [86] of ozone from 1750 to 1100 Å.

TABLE 6.36

DISSOCIATION PROCESSES OF OZONE

Products	Threshold wavelength (Å)
$O_3 \rightarrow O_2(X\,^3\Sigma_g^-) + O(^3P)$	11843
$O_3 \rightarrow O_2(a\,^1\Delta_g) + O(^3P)$	6112
$O_3 \rightarrow O_2(b\,^1\Sigma_g^+) + O(^3P)$	4621
$O_3 \rightarrow O_2(X\,^3\Sigma_g^-) + O(^1D)$	4113
$O_3 \rightarrow O_2(a\,^1\Delta_g) + O(^1D)$	3103
$O_3 \rightarrow O_2(A\,^1\Sigma_g^+) + O(^1D)$	2666
$O_3 \rightarrow O_2(X\,^3\Sigma_g^-) + O(^1S)$	2368
$O_3 \rightarrow O_2(a\,^1\Delta_g) + O(^1S)$	1994
$O_3 \rightarrow O_2(A\,^1\Sigma_g^+) + O(^1S)$	1804
$O_3 \rightarrow O_2(A\,^3\Sigma_u^+) + O(^3P)$	2245
$O_3 \rightarrow O_2(B\,^3\Sigma_u^-) + O(^3P)$	1717

chemospheric processes discussed in Chapter 11. For instance, the optical window around 2000 Å is of special importance, since excited $O_2(^1\Delta_g)$ molecules are observed [87].

6.5.2 *Molecular Oxygen*

Absorption by molecular oxygen has to be considered below 2424 Å, its first dissociation limit. In the Herzberg continuum the absorption cross section is of the order of 10^{-24} cm^2 near the threshold and becomes 10^{-23} cm^2 near 2050 Å. In the Schumann–Runge band system, absorption cross sections have been measured for more than fifty years [88–90]. *Figure 6.31* indicates large variations as a function of wavelength [21, 91–96]. An analysis of the fine structures can only be made with high resolution measurements. Analysis of the line broadening in the Schumann–Runge system shows that predissociation occurs for vibrational levels with $v' > 3$. Using total line widths ranging between 0.5 and 3.7 cm^{-1} and band integrated absorption cross sections, it has been possible to compute the theoretical cross sections [97] shown in Fig. 6.32 with a wave number resolution of 0.5 cm^{-1} over the entire Schumann–Runge system. These computations include the effect of the Herzberg continuum and the rotational lines belonging to the $v'' = 1$ sequence. The overlapping of the rotational lines

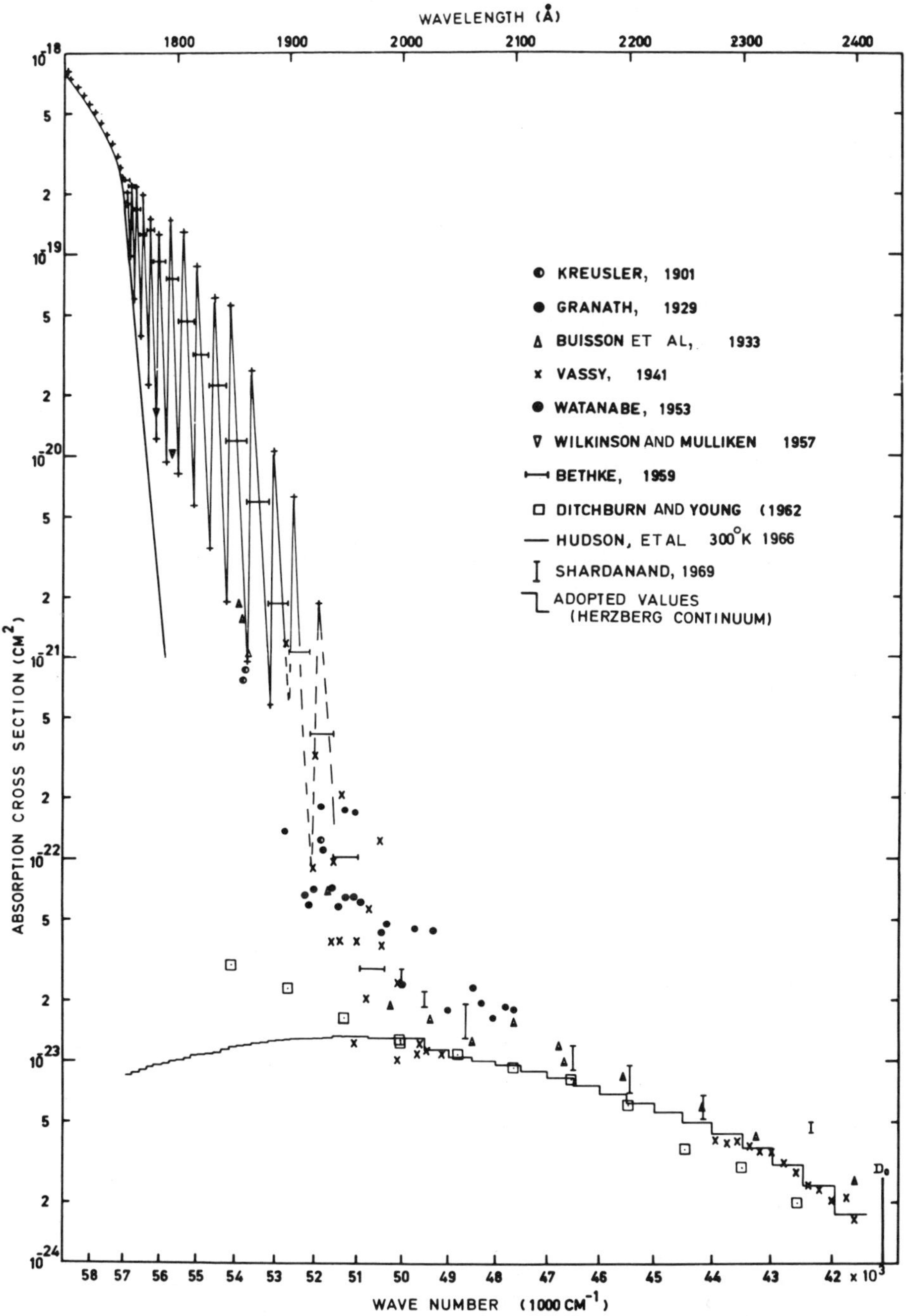

Fig. 6.31 Absorption cross section of molecular oxygen in the Schumann–Runge bands.

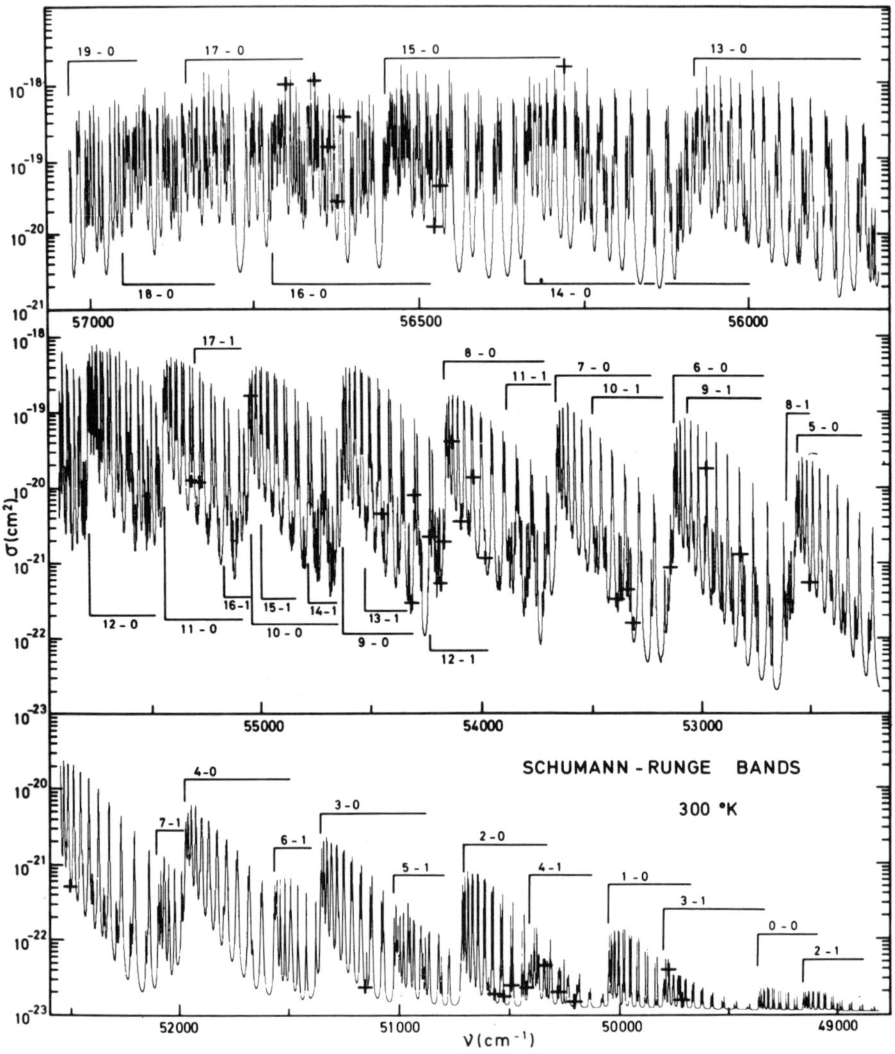

Fig. 6.32 Comparison of theoretical and experimental results for the absorption cross section of molecular oxygen in the Schumann–Runge bands.

within a band has to be taken into account in order to explain experimental values obtained at precisely known wavelengths of silicon emission lines. For example, in the 4–0 band the contribution of the rotational lines is taken into account up to 20 line widths from the center of the line. Recent measurements [98] are also explained by the results of *Fig. 6.32*. When the cross sections are averaged over approximately 500 cm^{-1} intervals, there is practically no temperature effect. However, at great resolution the temperature effect can be quite large. Variations of a factor of two are possible when the temperature changes from 300° to 200°K. Such an effect is important for the penetration of the solar radiation in the mesosphere and stratosphere. Therefore, temperature dependent cross sections should be adopted for detailed calculations in the thermosphere [99]. The predissociation in the Schumann–Runge bands is responsible for a large fraction of the atomic oxygen production between 60 and 90 km altitude.

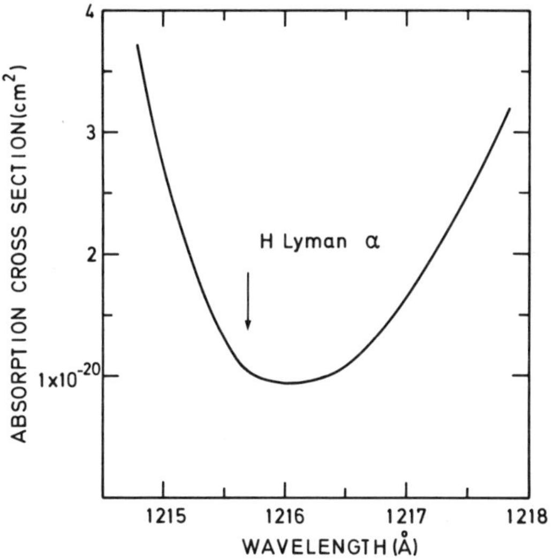

Fig. 6.33 Absorption cross section of molecular oxygen in the vicinity of Lyman α. From Ogawa [105].

The continuum of the Schumann–Runge system has an absorption cross section [100–104] varying from about 2×10^{-19} cm^2 at 1750 Å with a peak of 1.5×10^{-17} cm^2 at 1425 $\pm$ 25 Å to 3×10^{19} cm^2 near 1220 Å. The penetration of solar radiation in regions where the total O_2 content is greater than 10^{19} molecules cm^{-2} occurs only at certain windows at 1108, 1143, 1157, 1167, 1187, and 1216 Å where the absorption cross section is of the order of 10^{-20} cm^2. The window at Lyman α (1215.7 Å) is of special importance since the solar line has an energy of several ergs cm^{-2} sec^{-1} and it can ionize nitric oxide in the D region. The absorption cross section at Lyman α can be taken equal to 1×10^{-20} cm^2. The variation as a function of wavelength [105] is shown on *Fig. 6.33.*

REFERENCES

1. M. Nicolet and P. Mange, *J. Geophys. Res.* **59**, 16 (1954).
2. M. Nicolet, *Le Rayonnement Solaire*, Chapter II, p. 38. Mém. Inst. Roy. Météorol. Belg., 1945.
3. B. Edlen, *Z. Astrophys.* **22**, 30 (1942).
4. B. Edlen, *Mon. Notices Roy. Astronom. Soc.* **105**, 323 (1945).
5. R.v.d.R. Woolley and C. W. Allen, *Mon. Notices Roy. Astronom. Soc.* **110**, 358 (1950).
6. M. Nicolet, in *Physics and Medicine of the Upper Atmosphere* (C. S. White and O. O. Benson, eds.). Chapter XXII, dp. 182–225. Univ. of New Mexico Press, Albuquerque, New Mexico, 1952.
7. G. Elwert, *Z. Naturforsch.* **7a**, 202 (1952).
8. P. O. Pedersen, *The Propagation of Radio Waves*. Gad, Copenhagen, 1927.
9. E. A. W. Müller, *Nature* (*London*) **135**, 187 (1935).
10. E. A. W. Müller, *Z. Astrophys.* **10**, 52 (1935).
11. E. A. W. Müller, *Z. Geophys.* **11**, 126 (1935).
12. E. O. Hulburt, *Phys. Rev.* **53**, 344 (1938).
13. L. Vegard, *Geophys. Publik.* **12**, 5 (1938).
14. F. Hoyle and D. R. Bates, *J. Geophys. Res.* **53**, 51 (1948).
15. T. R. Burnight, *Phys. Rev.* **76**, 165 (1949).
16. H. Friedman, S. W. Lichtman, and E. T. Byram, *Phys. Rev.* **83**, 1025 (1951).
17. R. Tousey, K. Watanabe, and J. D. Purcell, *Phys. Rev.* **83**, 792 (1951).
18. R. Tousey, *Space Sci. Rev.* **2**, 3 (1963).
19. H. E. Hinteregger, *Space Sci. Rev.* **4**, 461 (1965).
20. G. L. Weissler, in *Handbuch der Physik*, Vol. XXI, pp. 304–382. Springer, Berlin, 1956.
21. K. Watanabe, *Advan. Geophys.* **5**, 153 (1958).
22. M. Nicolet and A. Aikin, *J. Geophys. Res.* **65**, 1469 (1960).
23. J. M. Valentine and S. C. Curran, *Rep. Progr. Phys.* **21**, 1 (1958).
24. P. Lee, *J. Opt. Soc. Amer.* **45**, 703 (1955).
25. K. Watanabe and F. F. Marmo, *J. Chem. Phys.* **25**, 965 (1956).
26. R. E. Huffman, J. C. Larrabee, and Y. Tanaka, *J. Chem. Phys.* **46**, 2213 (1967).
27. R. B. Cairns and J. A. R. Samson, *Phys. Rev.* **139A**, 1403 (1965).
28. D. R. Bates, *Mon. Notices Roy. Astronom. Soc.* **100**, 25 (1939).
29. D. R. Bates and M. J. Seaton, *Mon. Notices Roy. Astronom. Soc.* **109**, 698 (1949).
30. A. Dalgarno and D. Parkinson, *J. Atmos. Terr. Phys.* **18**, 335 (1960).
31. A. Dalgarno, R. J. W. Henry, and A. L. Stewart, *Planet. Space Sci.* **12**, 235 (1964).
32. R. J. W. Henry, *Planet. Space Sci.* **15**, 1747 (1967).
33. R. E. Huffman, J. C. Larrabee, and Y. Tanaka, *Phys. Rev. Lett.* **16**, 1033 (1966).
34. F. J. Comes, F. Speier, and A. Elzer, *Z. Naturforsch.* **23A**, 125 (1968).
35. R. J. W. Henry, *Planet. Space Sci.* **16**, 1503 (1968).
36. J. A. R. Samson and R. B. Cairns, *J. Opt. Soc. Amer.* **55**, 1035 (1965).
37. M. Nicolet, *Mém. Soc. Roy. Sci., Liège, 5e sér.* **4**, 319 (1961).
38. R. H. Messner, *Z. Phys.* **85**, 727 (1933).
39. R. J. W. Henry, *J. Chem. Phys.* **48**, 3635 (1968).
40. R. L. Kelly, "Atomic Emission Lines below 2000 Å, Hydrogen through Argon, NRL." US Govt. Printing Office, Washington, D.C., 1968.

41. M. Cohen and A. Dalgarno, *Proc. Roy. Soc.* **A280**, 258 (1964).
42. P. K. Carrol, R. E. Huffman, J. C. Larrabee, and Y. Tanaka, *Astrophys. J.* **146**, 553 (1960).
43. A. W. Ehler and G. L. Weissler, *J. Opt. Soc. Amer.* **45**, 1035 (1955).
44. J. A. R. Samson and R. B. Cairns, *J. Opt. Soc. Amer.* **55**, 1035 (1965).
45. F. J. Comes and A. Elzer, *Z. Naturforsch.* **23a**, 133 (1968).
46. P. Lee and G. L. Weissler, *Phys. Rev.* **99**, 540 (1953).
47. A. L. Stewart and T. G. Webb, *Proc. Phys. Soc.* **82**, 532 (1963).
48. K. E. Bell and A. E. Kingston, *Proc. Phys. Soc.* **90**, 31 (1967).
49. J. F. Lowry, D. H. Tomboulian, and D. L. Ederer, *Phys. Rev.* **137A**, 1054 (1965).
50. R. B. Cairns and J. A. R. Samson, *J. Geophys. Res.* **70**, 99 (1965).
51. J. A. R. Samson, *Phys. Rev. Lett.* **22**, 693 (1969).
52. J. A. R. Samson, *Advan. At. Mol. Phys.* **2**, 178 (1966).
53. R. E. Huffman, Y. Tanaka, and J. C. Larrabee, *J. Chem. Phys.* **39**, 902 (1963).
54. J. A. R. Samson, *J. Opt. Soc. Amer.* **54**, 420 (1964).
55. A. P. Lukirskii and T. M. Zimkina, *Bull. Acad. Sci. USSR Phys. Ser.* **27**, 808 (1963).
56. F. Wuilleumier, *J. Phys.* **26**, 776 (1965).
57. F. Wuilleumier, *C.R. Acad. Sci. Paris* **270**, 272 (1970).
58. J. A. R. Samson, *J. Opt. Soc. Amer.* **56**, 769 (1966).
59. J. A. R. Samson and R. B. Cairns, *J. Geophys. Res.* **69**, 4583 (1964).
60. G. R. Cook and P. H. Metzger, *J. Chem. Phys.* **41**, 321 (1964).
61. R. E. Huffman, J. C. Larrabee, and Y. Tanaka, *J. Chem. Phys.* **40**, 356 (1964).
62. L. De Reilhac and N. Damany-Astoin, *C.R. Acad. Sci. Paris* **258**, 519 (1964).
63. V. H. Dibeler and J. A. Walker, *J. Opt. Soc. Amer.* **57**, 1007 (1967).
64. P. H. Doolittle, R. I. Schoen, and K. E. Schubert, *J. Chem. Phys.* **49**, 5108 (1968).
65. G. R. Cook, P. H. Metzger, M. Ogawa, R. A. Becker, and B. K. Ching, Aerospace Corp. Rep. TDR-469 (9260-01)-4 (1965).
66. A. Dalgarno and M. B. McElroy, *Planet. Space Sci.* **13**, 947 (1965).
67. R. I. Schoen, *J. Chem. Phys.* **40**, 1830 (1964).
68. A. J. Blake and J. H. Carver, *J. Chem. Phys.* **47**, 1038 (1967).
69. R. E. Worley, *Phys. Rev.* **64**, 207 (1943).
70. G. L. Weissler, P. Lee, and E. I. Mohr, *J. Opt. Soc. Amer.* **42**, 84 (1952).
71. M. Astoin and J. Granier, *C.R. Acad. Sci. Paris* **224**, 1350 (1957).
72. M. Ogawa and R. B. Cairns, *Planet. Space Sci.* **12**, 656 (1964).
73. R. E. Huffman, Y. Tanaka, and J. G. Larrabee, *J. Chem. Phys.* **38**, 1920 (1963).
74. D. L. Judge and G. L. Weissler, *J. Chem. Phys.* **48**, 4590 (1968).
75. A. L. Broadfoot, *Planet. Space Sci.* **15**, 1801 (1967).
76. A. L. Broadfoot, *Planet Space Sci.* **16**, 693 (1968).
77. G. L. Weissler, J. A. R. Samson, M. Ogawa, and G. R. Cook, *J. Opt. Soc. Amer.* **49**, 338 (1959).
78. F. J. Comes and W. Lessman, *Z. Naturforsch.* **23a**, 125 (1968).
79. M. B. McElroy, *Planet. Space Sci.* **15**, 457 (1967).
80. K. Watanabe, *J. Chem. Phys.* **22**, 1564, 1954.
81. M. Nicolet, in *Meteors* (T. R. Kaiser, ed.), p. 99. Pergamon, Oxford, 1956.
82. J. A. R. Samson and R. B. Cairns, *J. Opt. Soc. Amer.* **56**, 169 (1966).
83. G. R. Cook and R. J. McNeal, *J. Chem. Phys.* **47**, 5385 (1967).
84. W. F. J. Evans, D. M. Hunten, E. J. Llewellyn, and A. Vallance-Jones, *J. Geophys. Res.* **73**, 2885 (1968).

85. D. M. Hunten and M. B. McElroy, *J. Geophys. Res.* **73**, 2421 (1968).
86. M. Ackerman, in *Mesospheric Models and Related Experiments* (G. Fiocco, ed.), p. 149. Reidel Publ., Dordrecht, The Netherlands, 1971.
87. J. F. Noxon and A. Vallance-Jones, *Nature* (*London*) **196**, 157 (1962).
88. H. Kreusler, *Ann. Phys.* **6**, 412 (1901).
89. L. P. Granath, *Phys. Rev.* **34**, 1045 (1929).
90. H. Buisson, G. Jausseran, and P. Rouard, *Rev. Opt.* **12**, 70 (1933).
91. A. Vassy, *Ann. Phys.* **16**, 145 (1941).
92. G. W. Bethke, *J. Chem. Phys.* **31**, 669 (1959).
93. P. G. Wilkinson and R. S. Mülliken, *Astrophys. J.* **125**, 594 (1957).
94. R. W. Ditchburn and P. Young, *J. Atmos. Terr. Phys.* **24**, 127 (1962).
95. R. D. Hudson, V. L. Carter, and J. A. Stein, *J. Geophys. Res.* **71**, 2295 (1966).
96. Shardanand, *Phys. Rev.* **186**, 5 (1969).
97. M. Ackerman, F. Biaumé, and G. Kockarts, *Planet. Space Sci.* **18**, 1639 (1970).
98. M. Ogawa, *J. Chem. Phys.* **54**, 2550 (1971).
99. G. Kockarts, in *Mesospheric and Related Experiments* (G. Fiocco, ed.), p. 160. Reidel Publ., Dordrecht, The Netherlands, 1971.
100. R. W. Ditchburn and D. W. O. Hedlle, *Proc. Roy. Soc.* **A226**, 509 (1954).
101. K. Watanabe and F. F. Marmo, *J. Chem. Phys.* **25**, 967 (1956).
102. P. H. Metzger and G. R. Cook, *J. Quant. Spectrosc. Radiat. Transfer* **4**, 107 (1964).
103. R. Goldstein and F. N. Mastrup, *J. Opt. Soc. Amer.* **56**, 765 (1966).
104. A. J. Blake, J. H. Carver, and G. N. Haddad, *J. Quant. Spectrosc. Radiat. Transfer* **6**, 451 (1966).
105. M. Ogawa, *J. Geophys. Res.* **73**, 6759 (1968).

CHAPTER 7

Solar Radiation and Photoionization

7.1 Introduction

Our knowledge of solar emission in the ultraviolet and x-ray regions is due to extensive rocket and satellite data. The first spectrum recorded above the ozone layer was obtained [1] in 1946 (*Fig. 7.1*) and was extended to shorter wavelengths [2] at altitudes above 100 km in 1949. At wavelengths greater than 2900 Å the solar radiation temperature is found to be between 5500 and 6000°K. Descending in wavelength from 2900 Å, the effective radiation temperature decreases until it reaches about 5000°K. Below 2400 Å, where molecular oxygen begins to absorb in its first con-

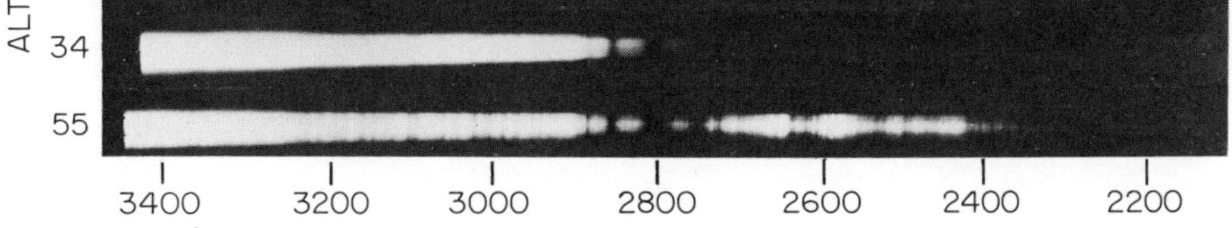

Fig. 7.1 First solar spectrum recorded above the ozone layer by the Naval Research Laboratory [1] on October 10, 1946 using a V-2 rocket. At 55 km, the spectrum is visible throughout the Hartley bands.

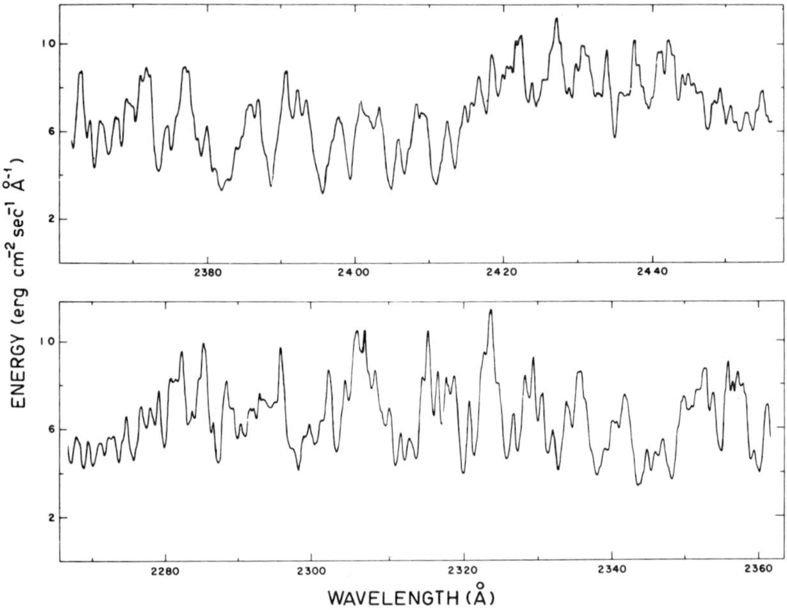

Fig. 7.2 The solar energy distribution at the top of the earth's atmosphere between 2450 and 2260 Å obtained by the Naval Research Laboratory [3].

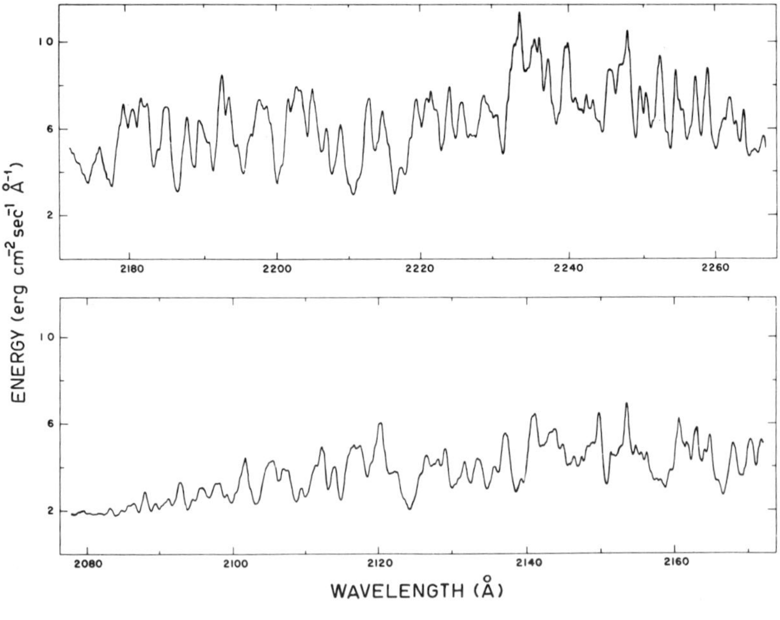

Fig. 7.3 The solar energy distribution at the top of the earth's atmosphere between 2260 and 2080 Å obtained by the Naval Research Laboratory [3].

tinuum, the solar spectrum still provides an energy greater than 50 erg cm^{-2} sec^{-1} for $\Delta\lambda = 10$ Å. This radiation energy decreases slowly with wavelength [3] and is of the order of 2 erg cm^{-2} sec^{-1} for $\Delta\lambda = 1$ Å near 2100 Å (see *Figs. 7.2* and *7.3*). The average spectrum follows the spectral distribution of a blackbody having a temperature lower than 5000°K down to 1300 Å. However, the structure of the absorption spectrum modifies the smooth distribution of the continuum.

Since the Fraunhofer spectrum progressively becomes less important below 1300 Å monochromatic radiation becomes more important in comparison with the solar continuum. Below $\lambda < 1220$ Å the solar spectrum is essentially represented by monochromatic radiation [4] which is not uniform over the solar surface. *Figure 7.4*, which shows the solar spectrum between 1800 and 500 Å, illustrates the importance of line emissions. In the Lyman continuum near 900 Å the radiation temperature is of the order of 6000°K and may reach about 7000°K near 800 Å.

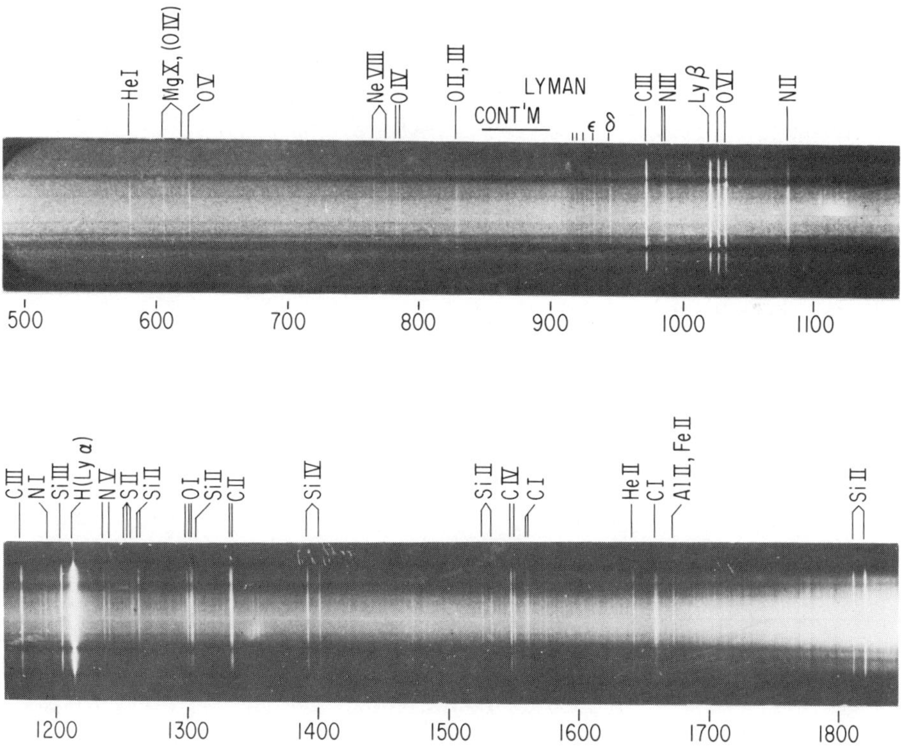

Fig. 7.4 Solar spectrum with its prominent lines between 1800 and 500 Å. Photographed by the Naval Research Laboratory [6], March 13, 1959, at 198 km.

7.2 Blackbody Emissions and Solar Data

In following the distribution of solar radiation with wavelength we must know how the emission varies with the assumed radiation temperature. It is most convenient to compute the emissions in terms of number of the photons cm^{-2} sec^{-1} available at the top of the terrestrial atmosphere. This is found from the radiation density, $\varrho(\nu)$, for a blackbody using the formula

$$\varrho(\nu) = (8\pi h\nu^3/c^3)(e^{h\nu/kT} - 1)^{-1} \tag{7.1}$$

where $h = 6.62\times 10^{-27}$ erg sec, $k = 1.38\times 10^{-16}$ erg deg^{-1}, $c = 2.998\times 10^{10}$ cm sec^{-1}, and ν denotes the frequency.

At the top of the earth's atmosphere, the radiation density $\varrho(\nu)$, is reduced by the dilution factor β_s

$$\beta_s = R^2/4r^2 = 5.4\times 10^{-6} \tag{7.2}$$

where R represents the solar radius and r represents the mean earth-to-sun distance. The variation with the earth-to-sun distance corresponds to $\pm 3.3\%$, i.e., $\beta_s(\max) = 5.59\times 10^{-6}$ and $\beta_s(\min) = 5.23\times 10^{-6}$.

The number of photons cm^{-2} sec^{-1} of frequency ν to $\nu + d\nu$ available at the top of the earth's atmosphere is $q(\nu)\, d\nu$, with

$$q(\nu) = (8\pi\beta_s\nu^2/c^2)(e^{h\nu/kT} - 1)^{-1}. \tag{7.3}$$

Thus, the total number of photons, Q, available at frequencies greater than ν, is expressed as

$$Q = \int_{\nu_1}^{\infty} q(\nu)\, d\nu = \frac{8\pi\beta_s}{c^2}\left(\frac{kT}{h}\right)^3 \int_x^{\infty} x^2(e^x - 1)^{-1}\, dx \tag{7.4}$$

where $x = h\nu/kT$. When $x \gg 1$, which is normal for the spectral regions in the visible and ultraviolet at typical solar disk temperatures in the range 4000° to 6000°K, (7.6) can be written as

$$Q = \frac{8\pi\beta_s}{c^2}\frac{k^3}{h^3}T^3 e^{-h\nu/kT}\left[\left(\frac{h\nu}{kT} + 1\right)^2 + 1\right]. \tag{7.5}$$

Figure 7.5 shows the values of Q computed for various temperatures between 6000° and 4000°K from 2500 to 1000 Å. Numerical data are given in *Table 7.1*. The divergence between Q for a constant temperature and the actual solar emission is shown by photon flux intensities obtained from rocket data [3] represented by a dotted line for the continuum and by circles for

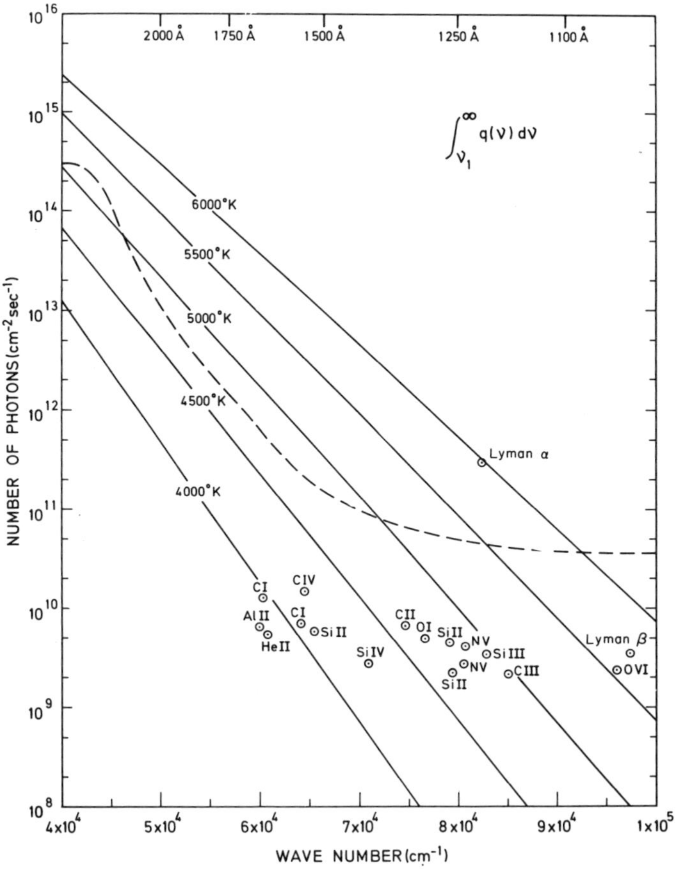

Fig. 7.5 Solar emission from 2500 to 1000 Å. Number of photons available at the top of the earth's atmosphere between the wave number indicated on abscissa and infinity. Blackbody curves between 4000 and 6000°K. Solar values obtained from Naval Research Laboratory data [6]. Solid curve, blackbody; dashed curve, solar continuum; circles, lines.

monochromatic radiation. The solar radiation level follows that expected for a 5000°K blackbody for the spectral region of absorption by molecular oxygen and ozone. Near 1400 Å, the radiation level increases up to that similar to a 5250°K blackbody and reaches a radiation temperature of the order of 5500°K near 1250 Å. For shorter wavelengths the effect of monochromatic radiation such as Lyman α and Lyman β must be recognized as very important even though the continuum is characterized by higher radiation temperatures. In fact Lyman α, corresponding to 5 erg cm^{-2} sec^{-1}, i.e., about 3×10^{11} photons cm^{-2} sec^{-1} at the top of the earth's atmosphere,

TABLE 7.1

NUMBER OF PHOTONS[a] $Q(\nu)$ AT THE TOP OF THE EARTH'S ATMOSPHERE

ν (cm^{-1})	$T = 4000°K$	$T = 4500°K$	$T = 4900°K$	$T = 5000°K$	$T = 5250°K$	$T = 5500°K$	$T = 6000°K$
5.0×10^4	4.9×10^{11}	4.1×10^{12}	1.7×10^{13}	2.3×10^{13}	4.8×10^{13}	9.5×10^{13}	3.1×10^{14}
5.5	9.8×10^{10}	1.0×10^{12}	4.6×10^{12}	6.5×10^{12}	1.5×10^{13}	3.1×10^{13}	1.1×10^{14}
6.0	1.9×10^{10}	2.4×10^{11}	1.3×10^{12}	1.8×10^{12}	4.4×10^{12}	9.8×10^{12}	4.0×10^{13}
6.5	3.7×10^{9}	5.6×10^{10}	3.4×10^{11}	5.0×10^{11}	1.3×10^{12}	3.1×10^{12}	1.4×10^{13}
7.0	7.0×10^{8}	1.3×10^{10}	8.9×10^{10}	1.4×10^{11}	3.8×10^{12}	9.6×10^{11}	4.8×10^{12}
7.5	1.3×10^{8}	3.0×10^{9}	2.4×10^{10}	3.7×10^{10}	1.1×10^{11}	3.0×10^{11}	1.7×10^{12}
8.0	2.5×10^{7}	6.9×10^{8}	6.1×10^{9}	1.0×10^{10}	3.2×10^{10}	9.0×10^{10}	5.7×10^{11}
8.5	4.7×10^{6}	1.6×10^{8}	1.6×10^{9}	2.7×10^{9}	9.0×10^{9}	2.7×10^{10}	1.9×10^{11}
9.0	8.6×10^{5}	3.6×10^{7}	4.1×10^{8}	7.1×10^{8}	2.6×10^{9}	8.3×10^{9}	6.4×10^{10}
9.5	1.6×10^{5}	8.0×10^{6}	1.0×10^{8}	1.9×10^{8}	7.2×10^{8}	2.5×10^{9}	2.2×10^{10}
10.0	2.9×10^{4}	1.8×10^{6}	2.7×10^{7}	4.9×10^{7}	2.0×10^{8}	7.4×10^{8}	7.2×10^{9}

[a] $Q(\nu)$ is measured in $cm^{-2}\ sec^{-1}$.

provides an energy equivalent to that of a blackbody at 6000°K for $\lambda < 1216$ Å.

Chromospheric radiations can play an important role in the spectral range of wavelength shorter than that of Lyman α. For example, O(VI) near 1035 Å, and especially Lyman β, ionizing molecular oxygen, contribute more than 2×10^9 photons cm^{-2} sec^{-1} and are important chromospheric radiations superposed on the solar continuum.

Not only are precise observations of the spectral distribution of the energy in the region 2500 to 1300 Å needed to study the dissociation of oxygen, but also absorption and dissociation cross section measurements between 2400 to 1750 Å are necessary to compute correct dissociation coefficients. Table 7.1 shows how the number of photons is dependent on the assumed temperature. In fact, prior to the use of the rocket as a research tool, it was customary to assume the radiation was that characteristic of a 6000°K blackbody except in a few cases where 5000°K was adopted [5]. When a comparison is made between blackbody curves and observational data (*Fig. 7.6*) there is no difficulty in selecting an average value of the

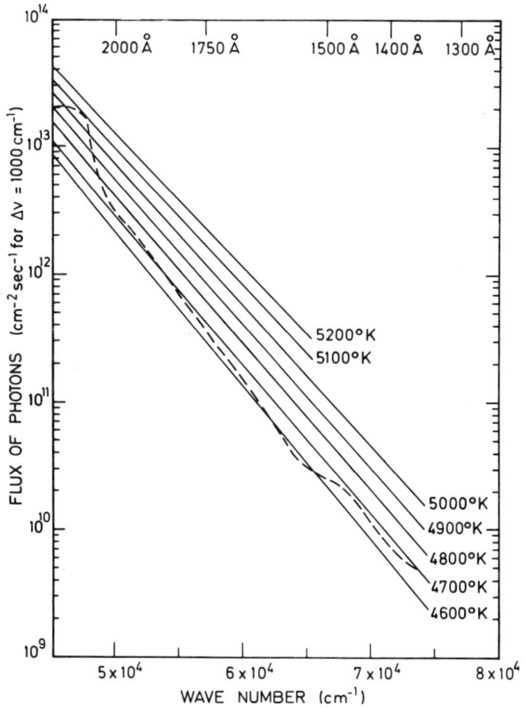

Fig. 7.6 Number of photons available at the top of the earth's atmosphere between 2200 and 1300 Å.

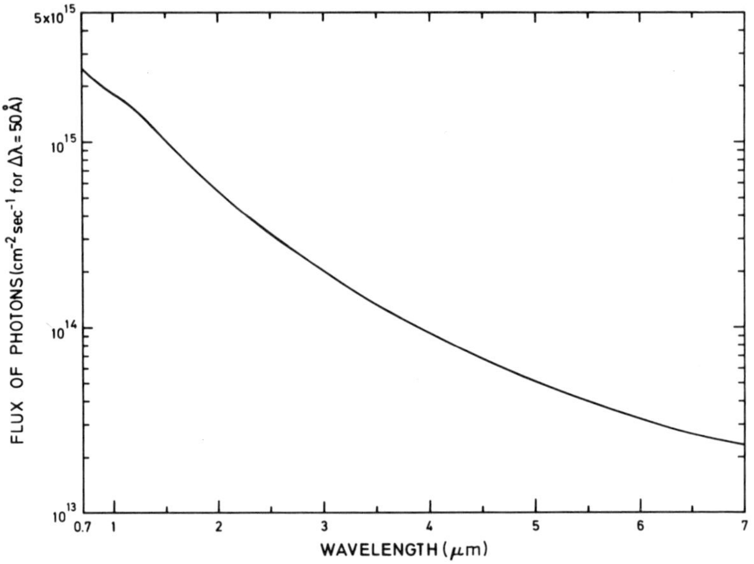

Fig. 7.7 Infrared solar radiation: Number of photons cm^{-2} sec^{-1} for $\Delta\lambda = 50$ Å available [6] at the top of the earth's atmosphere between 0.7 and 7 μm.

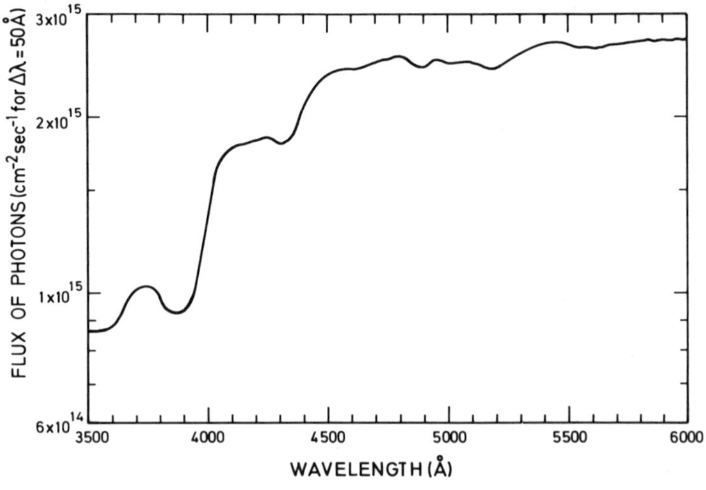

Fig. 7.8 Visible solar radiation: Number of photons cm^{-2} sec^{-1} for $\Delta\lambda = 50$ Å available [6] at the top of the earth's atmosphere between 3500 and 6500 Å.

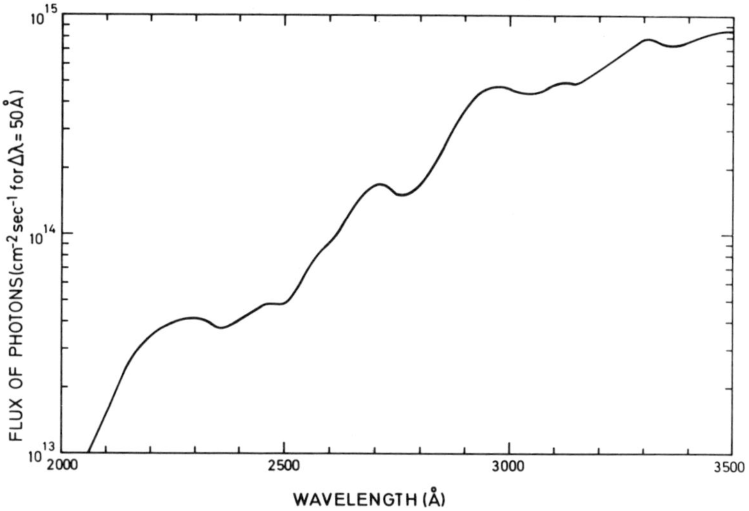

Fig. 7.9 Ultraviolet solar radiation: Number of photons cm^{-2} sec^{-1} for $\Delta\lambda = 50$ Å available [6] at the top of the earth's atmosphere between 2050 and 3500 Å.

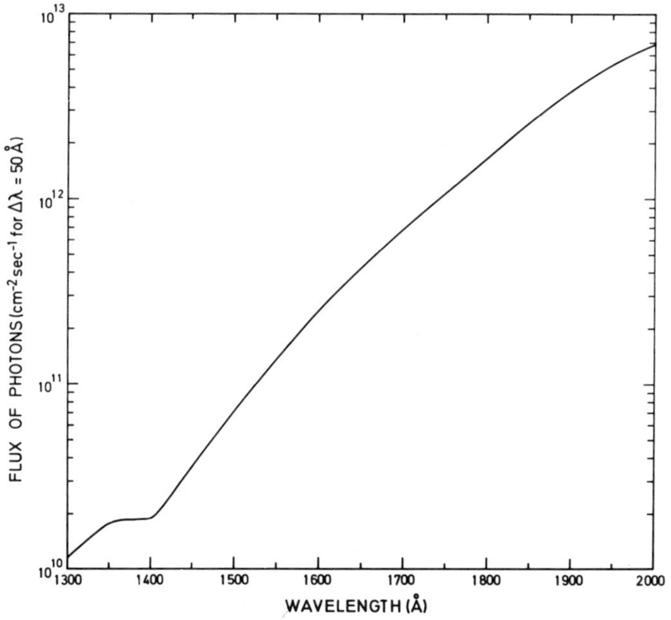

Fig. 7.10 Ultraviolet solar radiation: Number of photons cm^{-2} sec^{-1} for $\Delta\lambda = 50$ Å available [6] at the top of the earth's atmosphere between 1300 and 2050 Å.

TABLE 7.2

NUMBER OF PHOTONS AT THE TOP OF THE EARTH'S ATMOSPHERE[a]

λ (Å)	Photons	λ (Å)	Photons	λ (μm)	Photons
3050	4.45×10^{14}	4800	2.59×10^{15}	0.7	2.52×10^{15}
3100	4.90	4850	2.46	0.8	2.26
3150	4.97	4900	2.44	0.9	2.03
3200	5.70	4950	2.53	1.0	1.81
3250	6.65	5000	2.48	1.1	1.67
3300	7.96	5050	2.49	1.2	1.50
3350	7.36	5100	2.50	1.3	1.32
3400	7.74	5150	2.43	1.4	1.15
3450	8.40	5200	2.43	1.5	1.00
3500	8.65	5250	2.52	1.6	8.80×10^{14}
3550	8.65	5300	2.58	1.7	7.74
3600	8.68	5350	2.64	1.8	6.84
3650	9.79	5400	2.67	1.9	6.05
3700	1.02×10^{15}	5450	2.70	2.0	5.40
3750	1.03	5500	2.68	2.1	4.81
3800	9.83×10^{14}	5550	2.66	2.2	4.32
3850	9.32	5600	2.66	2.3	3.89
3900	9.27	5650	2.67	2.4	3.51
3950	1.02×10^{15}	5700	2.67	2.5	3.18
4000	1.34	5750	2.69	2.6	2.88
4050	1.67	5800	2.71	2.7	2.63
4100	1.77	5850	2.71	2.8	2.40
4150	1.79	5900	2.71	2.9	2.21
4200	1.83	5950	2.72	3.0	2.01
4250	1.85	6000	2.72	3.1	1.88
4300	1.80	6050	2.71	3.2	1.71
4350	1.84	6100	2.70	3.3	1.58
4400	2.12	6150	2.70	3.4	1.45
4450	2.27	6200	2.70	3.5	1.34
4500	2.38	6250	2.69	3.6	1.25
4550	2.42	6300	2.68	3.7	1.16
4600	2.41	6350	2.66	3.8	1.08
4650	2.45	6400	2.66	3.9	1.00
4700	2.50	6450	2.65	4.0	9.50×10^{13}
4750	2.56	6500	2.63		

[a] Values are deduced from Naval Research Laboratory data. The number of photons is measured in $cm^{-2}\ sec^{-1}$ for $\Delta\lambda = 50$ Å.

solar radiation temperature of the order of 4700°K for the spectral range 2000 to 1400 Å which contributes primarily to the total dissociation rate.

The solar intensity distribution is now fairly well known even though there may be systematic errors in a few spectral regions. The spectral curves obtained after reduction of observational data [6] are shown in *Figs. 7.7–7.10*. However, absolute values in the ultraviolet region of the solar spectrum depend on the calibration [7, 8] used below 4000 Å. The value of the solar constant is 1.97 cal cm^{-2} min^{-1} according to a recent analysis [9] while the first direct measurement made [10] at an altitude of 82 km gave a value of 1.95 cal cm^{-2} min^{-1}. Thus, the numbers of photons cm^{-2} sec^{-1} at the top of the earth's atmosphere have been adjusted in *Table 7.2* to correspond to a solar constant of 1.96 cal cm^{-2} min^{-1}.

Data obtained by the Naval Research Laboratory at wavelengths less than 3000 Å and given in *Table 7.3* should be correct to within ±20%. A very great precision is required in the spectral region 2400–1800 Å where ozone and molecular oxygen are subject to photodissociation. For example, a decrease of about 25% in the solar flux at 2000 Å reduces the radiation temperature by about 100°K.

TABLE 7.3

NUMBER OF SOLAR PHOTONS[a] BETWEEN 3000 AND 1450 Å

λ	$\Delta\lambda = 50$ Å	Total	λ	$\Delta\lambda = 50$ Å	Total
3000	4.58×10^{14}	2.69×10^{15}	2200	3.41×10^{13}	1.08×10^{14}
2950	4.65	2.23	2150	2.58	7.34×10^{13}
2900	3.77	1.76	2100	1.52	4.76
2850	2.42	1.39	2050	9.23×10^{12}	3.24
2800	1.68	1.14	2000	7.00	2.32
2750	1.51	9.77×10^{14}	1950	5.36	1.63
2700	1.69	8.26	1900	3.90	1.09
2650	1.33	6.57	1850	2.59	7.01×10^{12}
2600	9.10×10^{13}	5.24	1800	1.71	4.42
2550	7.14	4.33	1750	1.05	2.71
2500	4.75	3.62	1700	6.97×10^{11}	1.66
2450	4.78	3.15	1650	4.13	9.65×10^{11}
2400	4.08	2.67	1600	2.56	5.52
2350	3.76	2.26	1550	1.32	2.96
2300	4.14	1.88	1500	7.13×10^{10}	1.65
2250	3.94	1.47	1450	3.63	9.32×10^{10}

[a] Values are measured in cm^{-2} sec^{-1}.

Recent data [11–15] actually show lower values of the flux around 2000 Å as compared to the first NRL data [6]. The reduction in the blackbody temperature can be seen [16] in *Fig. 7.11.*

The spectral distribution represented in Fig. 7.10 for a spectral range of 50 Å shows a gradual decrease of the number of photons available at the top of the earth's atmosphere in the important region leading to the dissociation of molecular oxygen. A small systematic error may change the radiation temperature by more than 100°K.

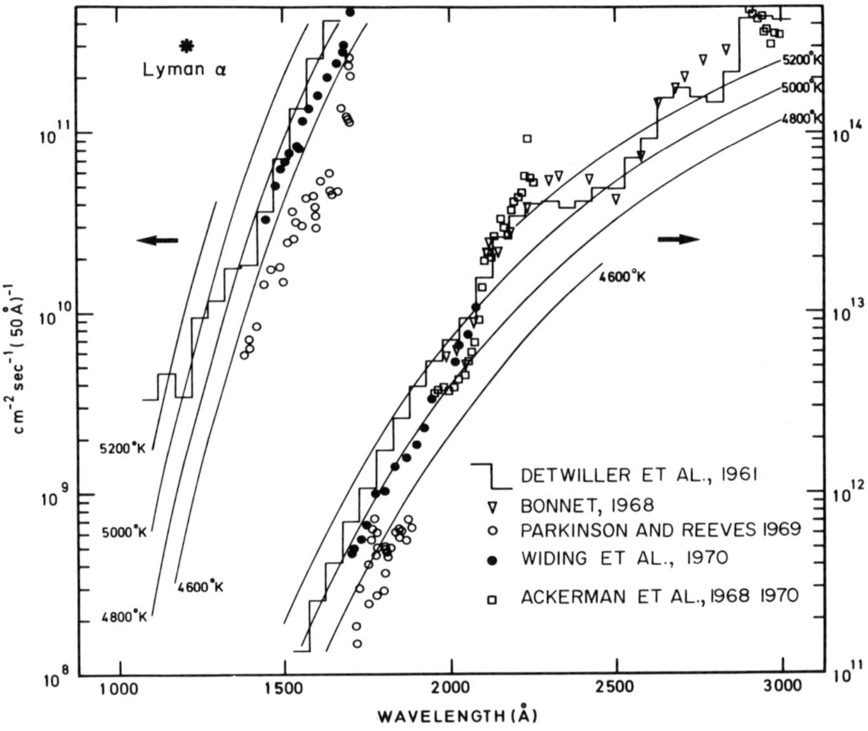

Fig. 7.11 Solar ultraviolet flux as a function of wavelength.

The precision required for a detailed analysis of the atmospheric absorption must be very high. Using more recent observational data it is possible to improve the accuracy of the photon flux values of Table 7.3 below 2000 Å. We have adopted the values which are given in *Table 7.4* in wave number intervals of 500 and 1000 cm^{-1}. Such values are used in the next chapter to compute the photodissociation rate of O_2.

TABLE 7.4

Flux of Solar Photons below 2000 Å for $\Delta\nu = 500$ and $1000\ cm^{-1}$

Wave number (cm^{-1})	Photon flux ($cm^{-2}\ sec^{-1}$)	Wave number (cm^{-1})	Photon flux ($cm^{-2}\ sec^{-1}$)	Wave number (cm^{-1})	Photon flux ($cm^{-2}\ sec^{-1}$)	Wave number (cm^{-1})	Photon flux ($cm^{-2}\ sec^{-1}$)
$\Delta\nu = 500\ cm^{-1}$		$\Delta\nu = 500\ cm^{-1}$		$\Delta\nu = 1000\ cm^{-1}$		$\Delta\nu = 1000\ cm^{-1}$	
5.00×10^4	1.5×10^{12}	5.40×10^4	4.4×10^{11}	5.8×10^4	2.3×10^{11}	6.7×10^4	2.3×10^{10}
5.05×10^4	1.4×10^{12}	5.45×10^4	3.8×10^{11}	5.9×10^4	1.8×10^{11}	6.8×10^4	1.8×10^{10}
5.10×10^4	1.3×10^{12}	5.50×10^4	3.1×10^{11}	6.0×10^4	1.2×10^{11}	6.9×10^4	1.3×10^{10}
5.15×10^4	9.9×10^{12}	5.55×10^4	2.9×10^{11}	6.1×10^4	5.5×10^{10}	7.0×10^4	1.0×10^{10}
5.20×10^4	7.3×10^{11}	5.60×10^4	2.6×10^{11}	6.2×10^4	6.4×10^{10}	7.1×10^4	7.6×10^9
5.25×10^4	6.6×10^{11}	5.65×10^4	2.3×10^{11}	6.3×10^4	4.8×10^{10}	7.2×10^4	5.8×10^9
5.30×10^4	5.9×10^{11}	5.70×10^4	1.8×10^{11}	6.4×10^4	3.6×10^{10}	7.3×10^4	5.6×10^9
5.35×10^4	5.0×10^{11}	5.75×10^4	1.4×10^{11}	6.5×10^4	2.9×10^{10}	7.4×10^4	5.1×10^9
				6.6×10^4	2.7×10^{10}		

Recent measurements [17] indicate a variation with solar activity of ultraviolet flux between 1200 and 3000 Å. The amplitude of the variation, however, decreases with increasing wavelength. At 1750 Å, a possible variation of 50% has been observed. This would lead to a significant variation in the production rate of atomic oxygen in the lower thermosphere. Therefore, the O_2 photodissociation rate may be changed by 50% as a result of solar activity.

7.3 x Rays

Direct observations [18, 19] of x rays at wavelengths shorter than 10 Å show how greatly this radiation varies with solar activity. In order to relate the emission to the absorption conditions defined in Chapter 6 we adopt the spectral distribution of *Table 7.5* to follow the principal observational results. These values correspond to the following spectral ranges:

Indication (Å):	6	4	2
Range (Å):	8–5	5–3.33	3.33–1.5
Wave number (10^7 cm^{-1}):	1.25–2	2–3	3–6.67

A completely quiet sun corresponds to the minimum of the sunspot cycle while the slightly disturbed sun and disturbed sun are to apply to the maximum of the sunspot cycle. Flares are represented by two cases in which the energy is 10^4–10^5 times greater than that for the very quiet sun.

TABLE 7.5

ENERGY[a] OF SOLAR RADIATIONS AT 2, 4, AND 6 Å

Solar conditions	2 Å	4 Å	6 Å
Completely quiet	10^{-8}	10^{-7}	10^{-6}
Quiet	10^{-7}	10^{-6}	10^{-5}
Lightly disturbed	10^{-6}	10^{-5}	10^{-4}
Disturbed	10^{-5}	10^{-4}	10^{-3}
Special events (flares)	10^{-4}	10^{-3}	10^{-2}
Strong flares	10^{-3}	10^{-2}	10^{-1}

[a] The units are erg cm^{-2} sec^{-1}

TABLE 7.6

IONIZATION RATE COEFFICIENTS[a] AT 2, 4, AND 6 Å

Solar conditions	2 Å	4 Å	6 Å
Completely quiet	1.6×10^{-19}	9.8×10^{-18}	2.8×10^{-16}
Quiet	1.6×10^{-18}	9.8×10^{-17}	2.8×10^{-15}
Lightly disturbed	1.6×10^{-17}	9.8×10^{-16}	2.8×10^{-14}
Disturbed	1.6×10^{-16}	9.8×10^{-15}	2.8×10^{-13}
Special events	1.6×10^{-15}	9.8×10^{-14}	2.8×10^{-12}
Strong flares	1.6×10^{-14}	9.8×10^{-13}	2.8×10^{-11}

[a] The units are sec^{-1}.

Introducing the ionization cross sections of Chapter 6 for x rays of 2, 4, and 6 Å, one obtains the ionization rate coefficients [20] given in *Table 7.6.*

If one considers x rays of energy greater than 10 keV ($\lambda < 1.2$ Å), the ionization produced by galactic cosmic rays must be taken into account. In fact, unusual solar conditions must develop before the ionization produced in the D region by cosmic rays is modified to any great extent.

Figure 7.12 illustrates the changes in the x-ray spectrum for various solar conditions [21]. The energy distribution is given for sunspot minimum,

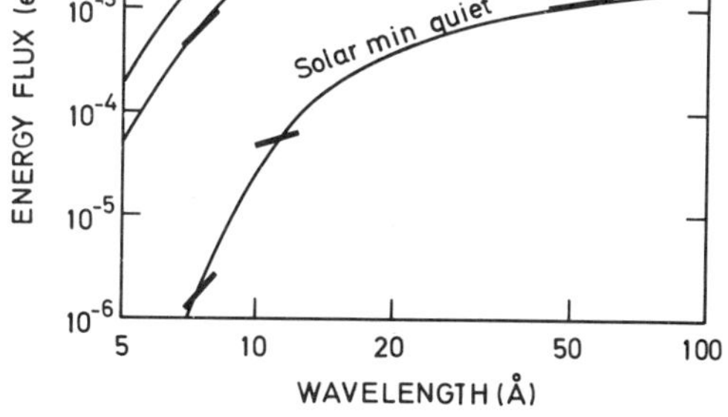

Fig. 7.12 Solar x-ray emission for various solar conditions [21].

sunspot maximum, and solar flare conditions. The measurements were made in three wavelength bands as indicated by the heavy bar segments. The slopes of the curves cannot be compared with the spectral slope used in Table 7.5 which illustrates the energy ranges by factors of 10. Observations [22] providing the value of 5×10^{-5} erg cm^{-2} sec^{-1} at $\lambda < 8$ Å for a "quiet sun" indicate that the ionization rate coefficient is only about 10^{-14} sec^{-1}. In such conditions the x rays do not significantly contribute to the electron concentration in the D region. However, the development of the x-ray spectra associated with flares even at sunspot minimum suggests an important enhancement of radiation of $\lambda < 15$ Å which affects the ionization rate (see Table 7.6). *Figure 7.13* illustrates the spectrum obtained [22] during

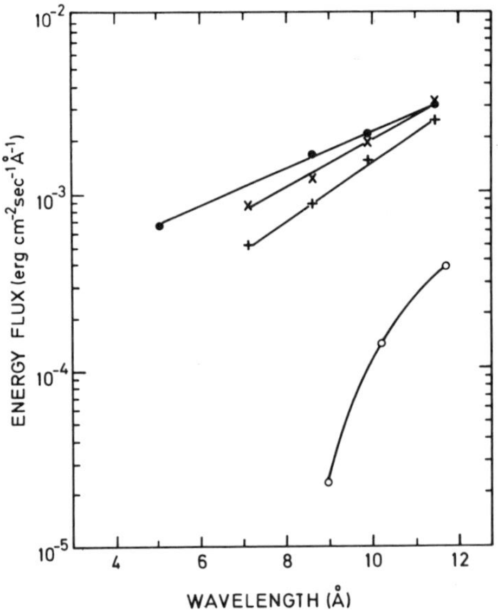

Fig. 7.13 Solar x ray during the decay phase of a solar flare [22], May 1, 1962. ○, 1727–1733 U.T.; ●, 1923 U.T.; ×, 1925 U.T.; and +, 1927 U.T.

the decay phase of a flare of importance 1. The flare emission dominates the entire spectral range and has a spectral profile completely different from the preflare distribution, represented by the lower curve. Thus, it is clear that the variation of the solar x-ray emission is accompanied by the hardening or softening of the spectrum and is extremely sensitive to solar conditions which cannot be determined by visual observations.

Spectral observations made in 1966 [23] and 1967 [24] show that the solar spectrum below 8 Å can be observed when the sun is not completely quiet. Rapid variations appear in the emission spectra of individual active regions. A group of lines [24] near 1.9 Å seem to correspond to lines of Fe(XXV), with the Lyman α line of Fe(XXVI) on the ultraviolet side and FeKα on the other side. Emission lines from the highest stages of ionization of iron are prominent during solar flares. Other lines in the band between 2 and 3 Å correspond to Ca(XX). Thus, it is clear that the hardening of the spectrum near 2 Å is due to the initial ionization of atoms of very high ionization potential. Emissions in the region between 3.3 and 5 Å may be due [23] to Ca(XX) and Ar(XVIII). In the spectral region between 5 and 8 Å, identified radiations belong to Si(XIV), Al(XI), Al(XII), and Na(XI).

Exact predictions cannot be made for the variable distribution of the intensity in this part of the solar spectrum. For example, the quiet and slightly disturbed states of the sun in Table 7.5 correspond to values for a "quiet sun" with low and high average activity, i.e., for solar conditions with a microwave emission at $\lambda = 10$ cm between 100 and 150 W m^{-2} Hz^{-1}. In other words, typical values for the spectral ranges are: 2 Å, between 10^{-7} and 10^{-6} erg cm^{-2} sec^{-1}, 4 Å between 10^{-6} and 10^{-5} erg cm^{-2} sec^{-1}, and 6 Å, between 10^{-5} and 10^{-4} erg cm^{-2} sec^{-1}. A solar flare may lead to increases of these by a factor of 10^4 at 2 Å, of 10^3 at 4 Å, and of 10^2 at 6 Å but during a flare the relative enhancements do not maintain the same ratios. Thus variable ionization conditions in the D region imply temporal variations in solar activity.

Another important spectral range of x-ray emission corresponds to wavelengths greater than 31 Å, according to the values of absorption cross section in air. Energies involved in the spectral range centered near 60 Å vary between 0.1 and 0.7 erg cm^{-2} sec^{-1} during a sunspot cycle and values greater than 1 erg cm^{-2} sec^{-1} may be reached during solar flares [18]. Due to the variation of the absorption cross section it is necessary to divide the spectral region into at least three groups.

Table 7.7 shows such a division for which 70, 50, and 35 Å represent the following associated spectral ranges:

Mean value (Å):	70	50	35
Associated ranges (Å):	80–60	60–41	41–31
	31–19	19–13	13–10

The ionization rate coefficients were computed by assuming the following fluxes for completely quiet conditions: 0.05 erg cm^{-2} sec^{-1} at 70 Å, 0.04 erg cm^{-2} sec^{-1} at 50 Å, and 0.01 erg cm^{-2} sec^{-1} at 35 Å. The justification for such values can be obtained from an analysis of the spectrum.

The principal lines in the spectral range 31–41 Å are given in *Table 7.8* for conditions of very low solar activity. The total energy in this spectral

TABLE 7.7

IONIZATION CONDITIONS BETWEEN 10 AND 80 Å FOR A COMPLETELY QUIET SUN

	70 Å	50 Å	35 Å
Energy flux (erg cm^{-2} sec^{-1})	0.05	0.04	0.01
Photon flux (cm^{-2} sec^{-1})	1.75×10^{8}	1.0×10^{8}	1.75×10^{7}
Ionization efficiency	3	5	7
O_2 absorption (cm^2)	9.28×10^{-19}	4.40×10^{-19}	1.64×10^{-19}
O_2 ionization (sec^{-1})	4.87×10^{-10}	2.2×10^{-10}	2.0×10^{-11}
N_2 absorption (cm^2)	5.45×10^{-19}	2.35×10^{-19}	9.67×10^{-20}
N_2 ionization (sec^{-1})	2.86×10^{-10}	1.20×10^{-10}	1.18×10^{-11}
O absorption (cm^2)	4.64×10^{-19}	2.0×10^{-19}	0.82×10^{-19}
O ionization (sec^{-1})	2.44×10^{-10}	1.0×10^{-10}	1.00×10^{-11}
Air absorption (cm^2)	6.12×10^{-19}	2.72×10^{-19}	1.08×10^{-19}
Air ionization (sec^{-1})	3.25×10^{-10}	1.5×10^{-10}	1.65×10^{-11}

range, involving lines and continuum regions, is of the order of 0.01 erg cm^{-2} sec^{-1} for a completely quiet sun. With the variation of solar activity there is a variation of the ratio of C(V)/C(VI) lines; with the enhancement of solar activity $\lambda = 33.73$ Å increases as compared with $\lambda = 40.27$ Å [25, 26].

TABLE 7.8

PRINCIPAL SOLAR LINES IN THE SPECTRAL REGION 31–41 Å

Solar wavelength (Å)	Energy flux (10^{-3} erg cm^{-2} sec^{-1})	Atom	Transition	λ (Å)
33.73	2.5	C(VI)	$1s^2\ ^2S$–$2p\ ^2P^0$	33.73
35.1	2.5	C(VI)	$1s^2\ ^1S$–$1s3p\ ^1P^0$	34.97
36.2	7.0			
40.27	1.3	C(V)	$1s^2\ ^1S_0$–$1s2p\ ^1P^0$	40.27
40.72 }	1.6		$1s^2\ ^1S_0$–$1s2p\ ^3P_1{}^0$	40.73
40.92 }		Si(XII)	$2s^2S$–$3p\ ^2P^0$	40.91

The emission lines in the spectral range 41–60 Å are listed in *Table 7.9*. The spectrum includes a number of important lines of silicon [Si(XI) and Si(XII)] and iron [Fe(XIV), Fe(XV), and Fe(XVI)]. The total intensity of this spectral range, including the continuum, for a completely quiet sun is about 0.04 erg cm^{-2} sec^{-1}.

TABLE 7.9

PRINCIPAL SOLAR LINES IN THE SPECTRAL REGION 41–60 Å

Solar wavelength (Å)	Energy flux (10^{-3} erg cm^{-2} sec^{-1})	Atom	Transition	λ (Å)
42.54		S(X)	$2s^22p^3\ ^4S^0 - 2s^22p^23d\ ^4P$	42.54
43.76	0.9	Si(XI)	$2s^2\ ^1S^0 - 2s3p\ ^1P^0$	43.76
44.18	1.2	Si(XII)	$2p\ ^2P^0 - 3d\ ^2D$	44.16
45.49 }	0.6	Si(XII)	$2p\ ^2P^0_{1/2} - 3s\ ^2S_{1/2}$	45.48
45.73 }			$2p\ ^2P^0_{3/2} - 3s\ ^2S_{3/2}$	45.66
46.37		Si(XI)	$2s2p\ ^3P^0 - 2s3d\ ^3P$	46.3
49.22	1.4	Si(XI)	$2s2p\ ^1P^0 - 2s3d\ ^1D_2$	49.22
50.51 }	3.5	Si(X)	$2p\ ^2P^0_{1/2} - 3d\ ^2D_{3/2}$	50.52
50.58 }			$2p\ ^2P^0_{3/2} - 3d\ ^2D_{5/2}$	50.69
52.30	0.7	Si(XI)	$2s2p\ ^1P_1{}^0 - 2s3s\ ^1S_0$	52.30
52.97	0.7	Fe(XV)	$3s^2\ ^1S_0 - 3s4p\ ^1P_1{}^0$	52.91
54.14	0.8	Fe(XVI)	$3p\ ^2P^0_{1/2} - 4d\ ^2D_{3/2}$	54.14
54.67	0.8	Fe(XVI)	$3p\ ^2P^0_{3/2} - 4d\ ^2D_{5/2}$	54.73
55.10 }	2.7	Si(IX)	$2p^2\ ^3P - 2p3d\ ^3P^0$	55.10
55.31 }			$2p^2\ ^3P - 2p3d\ ^3D^0$	55.30
57.87	1.7	Mg(X)	$2s\ ^2S - 3p\ ^2P^0$	57.90
58.97		Fe(XIV)	$3s^23p\ ^2P^0_{1/2} - 3s^24d\ ^2D_{3/2}$	58.90
59.61		Fe(XIV)	$3s^23p\ ^2P^0_{3/2} - 3s^24d\ ^2D_{5/2}$	59.58

Table 7.10 lists the emission lines of the quiet sun for the spectral range 60–80 Å. The lines reported belong to magnesium [Mg(VIII), Mg(IX), and Mg(X)] and to iron [Fe(XII), Fe(XIII), Fe(XIV), Fe(XV), and Fe(XVI)]. An absolute energy flux of 0.05 erg cm^{-2} sec^{-1} has been adopted for the spectral range 60–80 Å when the sun is completely quiet. Thus, a total flux between 31 and 81 Å of the order of 0.1 erg cm^{-2} sec^{-1} is adopted for a completely quiet sun.

TABLE 7.10

PRINCIPAL SOLAR LINES BETWEEN 60 AND 80 Å

Solar wavelength (Å)	Energy flux (10^{-3} erg cm^{-2} sec^{-1})	Atom	Transition	λ (Å)
61.05	2.5	Si(VIII)	$2p^3\ {}^4S^0 - 2p^2 3d\ {}^4P$	61.03
62.74	1.4	Mg(IX)	$2s^2\ {}^1S_0 - 2s3p\ {}^1P^0$	62.75
62.98	0.7	Fe(XVI)	$3p\ {}^2P^0_{1/2} - 4s\ {}^2S_{1/2}$	62.88
63.72			$3p\ {}^2P^0_{3/2} - 4s\ {}^2S_{1/2}$	63.72
65.67	1.4	Mg(X)	$2p\ {}^2P^0_{1/2} - 3s\ {}^2S_{1/2}$	65.67
65.89			$2p\ {}^2P^0_{3/2} - 3s\ {}^2S_{1/2}$	65.85
66.35	1.6	Fe(XVI)	$3d\ {}^2D_{3/2} - 4f\ {}^2F^0_{5/2}$	66.26
			$3d\ {}^2D_{5/2} - 4f\ {}^2F^0_{7/2}$	66.37
67.20	1.3	Mg(IX)	$2s2p\ {}^3P^0 - 2s3d\ {}^3D$	67.2
69.96	2.8	Fe(XV)	$3s3d\ {}^3D - 3s4f\ {}^3P^0$	70.0
71.00	0.5	Fe(XIV)	$3s^2 3p\ {}^2P^0_{1/2} - 3s^2 4s\ {}^2S_{1/2}$	71.00
71.95	1.0		$3s^2 3p\ {}^2P^0_{3/2} - 3s^2 4s\ {}^2S_{1/2}$	71.96
72.31	1.3	Mg(IX)	$2s2p\ {}^1P^0 - 2s3d\ {}^1D_2$	72.31
74.85	1.4	Mg(VIII)	$2s^2 2p\ {}^2P_{1/2} - 2s^2 3d\ {}^2D_{3/2}$	74.86
75.03			$2s^2 2p\ {}^2P_{3/2} - 2s^2 3d\ {}^2D_{5/2}$	75.03
76.01	1.8	Fe(XIII)	$3s^2 2p^2\ {}^3P_1 - 3s^2 3p\ {}^3P_1{}^0$	76.01
76.38			$3s^2 2p^2\ {}^3P_0 - 3s^2 3p\ {}^3P_1{}^0$	76.48
76.48			$3s^2 2p^2\ {}^3P_2 - 3s^2 3p\ {}^3P_1$	76.55
77.74	1.4	Mg(IX)	$2s2p\ {}^1P_1{}^0 - 2s3s\ {}^1S_0$	77.74
79.48	2.0	Fe(XII)	$3p^3\ {}^4S^0_{3/2} - 3p^2\ 4s\ {}^4P_{5/2}$	79.48
80.03			$3p^3\ {}^4S^0_{3/2} - 3p^2\ 4s\ {}^4P_{3/2}$	80.00
80.50			$3p^3\ {}^4S_{3/2} - 3p^2\ 4s\ {}^4P_{1/2}$	80.49

In order to follow the varying conditions during a solar cycle, the absolute energy fluxes which are adopted for the spectral range 31–80 Å are given in the accompanying tabulation.

Completely quiet sun:	0.1 erg cm^{-2} sec^{-1}
Quiet sun	
low average:	0.2
high average:	0.4
Active sun (maximum):	0.6
Disturbed sun (flares):	1.0
Exceptional:	>1.0

Such conditions correspond closely to solar conditions in Table 7.5 for $\lambda < 10$ Å.

The spectrum between 25 and 10 Å includes a number of important high-temperature emission lines of iron [Fe(XI)–Fe(XXIV)] which appear in plages and in local x-ray emission regions of the solar corona. The absolute energy flux of this spectral region is negligible during quiet solar conditions as compared with the total flux of the 31–80 Å region. It is known that the enhancement during very important solar flares can increase up to 0.5 erg cm^{-2} sec^{-1}. Thus, the spectral region 10–31 Å is not so important from an aeronomic viewpoint as the 31–80 Å region.

7.4 Ultraviolet Radiation at Wavelengths Shorter than 1050 Å

Rocket results [27, 28] such as those shown in *Fig. 7.14* give a clear idea of the solar spectrum below 1050 Å. It is essentially a spectrum of chromospheric and coronal radiations [29].

Lines of the Lyman series of hydrogen and its continuum are observed.

Below 1300 Å the observational data are uncertain by as much as a factor of 2 and the values which will be discussed are only approximate. Furthermore, it is not possible to give an exact value of the energies since they are subject to solar activity effects [29]. Finally, the ionization rate coefficients are only approximate because of the uncertainty in the absorption and ionization cross sections.

Since the precision necessary for a detailed analysis of ionization cross sections has not yet been reached, one may try to represent the energy of the ultraviolet spectrum by taking into account the observational results [28–33] and absorption cross section of oxygen and nitrogen in the various spectral ranges.

The total number of photons in the spectral range 1026–910 Å is of the order of 1×10^{10} cm^{-2} sec^{-1}; 5×10^{9} photons cm^{-2} sec^{-1} should be accepted for the 1026–990 Å range with about 50% for Lyman β.

For wavelengths shorter than 911.7 Å, the solar emission is characterized by the Lyman continuum. Atomic oxygen with an ionization continuum beginning at 910.4 Å can be ionized by this radiation. Figure 7.14 shows [31] that the Lyman continuum covers the 900–800 Å range. For wavelengths shorter than 800 Å numerous lines of ionized atoms are observed. *Figure 7.15* illustrates the extension [31] of the spectrum to 160 Å. It is clear that many lines are present below 250 Å which appear with intensities greater than those of lines of $\lambda > 300$ Å. Figure 7.15, presenting two scans of the

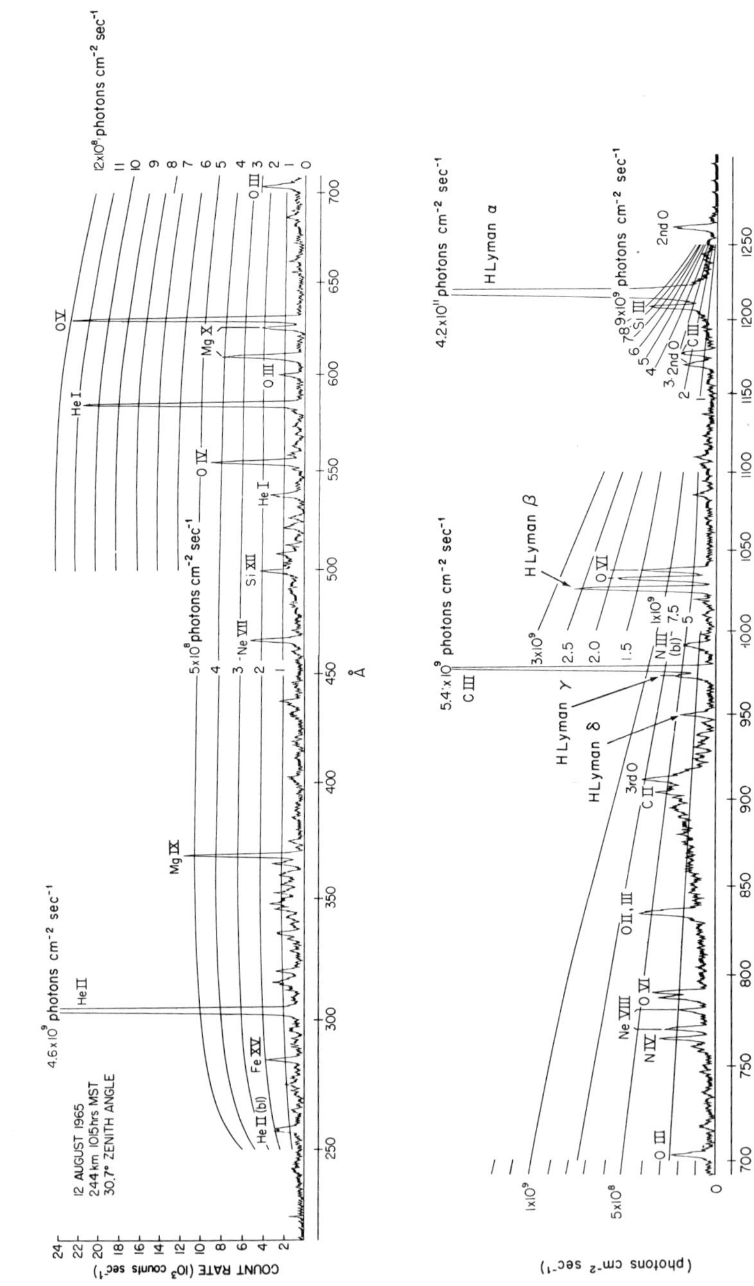

Fig. 7.14 Principal solar radiation [40] between 1250 and 250 Å.

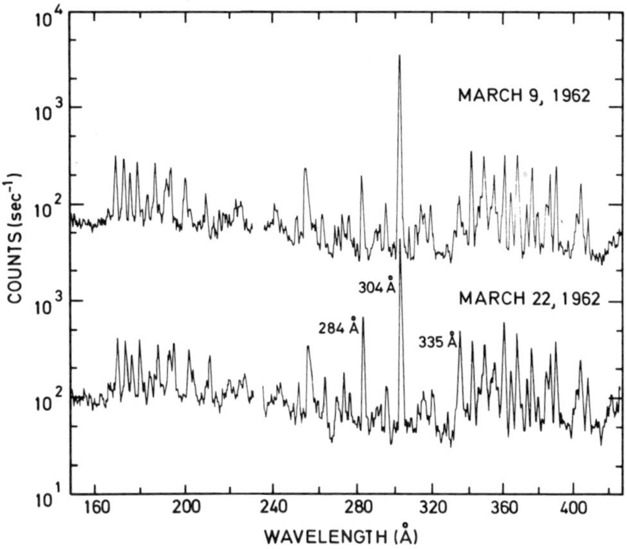

Fig. 7.15 Solar spectrum between 400 and 160 Å for quiet and active conditions [31].

spectrum, shows the variation of the flux with solar activity. An increase of at least 50% was observed from 9 March to 23 March 1962. Approximate values of the number of photons are given in *Table 7.11* for various spectral ranges as an aid to the calculation of ionization rates at zero optical depth. These values encompass the variation during a sunspot cycle and a possible

TABLE 7.11

NUMBER OF SOLAR PHOTONS[a] AT THE TOP OF THE EARTH'S ATMOSPHERE

Spectral range (Å)	Photons (cm^{-2} sec^{-1})	Spectral range (Å)	Photons (cm^{-2} sec^{-1})
Lyman β at 1025 Å	$(5 \pm 1) \times 10^9$	665–375	$(8 \pm 4) \times 10^9$
C(III) at 977 Å	$(4 \pm 2) \times 10^9$	375–275	$(1.4 \pm 0.6) \times 10^{10}$
910–796	$(1.1 \pm 0.4) \times 10^{10}$	275–150	$(1.4 \pm 0.6) \times 10^{10}$
796–732	$(1.5 \pm 0.5) \times 10^9$	150–80	$(1.25 \pm 0.75) \times 10^9$
732–665	$(7.5 \pm 2.5) \times 10^8$		

[a] Values are measured in cm^{-2} sec^{-1}.

error of a factor of the order of 2. It is noted that the total number of photons of wavelength shorter than 500 Å may vary by an important factor during a sunspot cycle; lower and upper limits in Table 7.11 should be accepted as possible values, since both real variations and systematic errors are involved.

Considering again that photoionization of molecular nitrogen occurs for $\lambda \leq 796$ Å, $\lambda \leq 743$ Å, and $\lambda \leq 661$ Å, that of atomic oxygen for $\lambda < 910$ Å, $\lambda \leq 732$ Å, and $\lambda \leq 665$ Å we may divide the ultraviolet spectrum below 1050 Å into spectral ranges from 910 to 796 Å, from 796 to 732 Å, from 732 to 665 Å, from 665 to 375 Å, from 375 to 275 Å, and from 275 to 150 Å. The number of photons for these various spectral ranges is of the order of 1×10^{10} photons cm^{-2} sec^{-1} with the exception of the 796–665 Å ranges which correspond to an approximate flux of 2×10^{9} photons cm^{-2} sec^{-1}.

7.4.1 1026–910 Å

Using the data given in Table 7.11 and the cross section of Chapter 6, the ionization rate coefficient at zero optical depth by the solar Lyman β line at 1026 Å is

$$I_{1026}(O_2) = (5 \pm 1) \times 10^{-9} \quad \text{sec}^{-1}.$$

The absorption by N_2 is certainly much smaller than that by O_2, since the cross section is less than 10^{-19} cm^{-2}. The principal N_2 absorption is under the influence of rotational structure of a band with a head at 972.2 Å and a rotational line at 976.98 Å. Since the optical depth at 977 Å is not less than twice that for Lyman β, its effect in the lower part of the E layer is less important than that of 1025 Å. Thus, Lyman β is the principal ultraviolet radiation which must be considered with solar x rays in the formation of the E layer.

At 977 Å, there is also an important chromospheric line (see Fig. 7.14) due to C(III). The ionization coefficient for O_2 is

$$I_{977}(O_2) = (1.1 \pm 0.4) \times 10^{-8} \quad \text{sec}^{-1}.$$

7.4.2 910–796 Å

The ionization rate coefficient $I_{910-796}(O)$ of atomic oxygen for zero optical depth in its continuum between the threshold at 910 Å and the first ionization potential of N_2 at 796 Å is

$$I_{910-796}(O) = (4.5 \pm 1.5) \times 10^{-8} \quad \text{sec}^{-1}.$$

In the same spectral range, molecular oxygen is ionized at a slightly higher rate and one obtains

$$I_{910-796}(O_2) = (5.7 \pm 2) \times 10^{-8} \quad \text{sec}^{-1}.$$

The real difficulty in the analysis of the atmospheric ionization produced by the Lyman continuum is the presence of N_2 bands which show very large variations of the absorption coefficient with wavelength. The nitrogen absorption spectrum consists of at least 30 bands, i.e., not less than an average of one band every 2 Å. Since almost all bands have a rotational structure, atmospheric absorption will depend on the variation of rotational lines with altitude from the E layer to the F region. Generally, the absorption cross section will be very great near the head of the bands. Since the rotational lines cover the spectral interval between two heads, it is practically impossible to follow the absorption structure between 910 and 850 Å. Laboratory measurements are far from being sufficient to clarify the uncertainty.

Between 850 and 796 Å the solar spectrum is represented by multiplets of O(II) and O(III) between 833 and 835 Å. Their total intensity is only a fraction of that between 910 and 850 Å and it is permissible to neglect in a first approximation the effect of this spectral range compared with the principal part of the Lyman continuum.

The photoionization rate coefficient of atomic hydrogen for zero optical depth is about

$$I_{910-796}(H) = (6.2 \pm 2) \times 10^{-8} \quad \text{sec}^{-1}.$$

7.4.3 796–732 Å

For the spectral region between 796 Å (first ionization potential of N_2) and 732 Å (second ionization potential of O), the following photoionization coefficients at zero optical depth may be considered:

$$I_{796-732}(N_2) = (1.5 \pm 0.5) \times 10^{-8} \quad \text{sec}^{-1}$$

$$I_{796-732}(O) = (1.5 \pm 0.5) \times 10^{-8} \quad \text{sec}^{-1}$$

$$I_{796-732}(O_2) = (1.5 \pm 0.5) \times 10^{-8} \quad \text{sec}^{-1}$$

$$I_{796-732}(H) = (6 \pm 2) \times 10^{-9} \quad \text{sec}^{-1}.$$

7.4.4 732–665 Å

Between the second and third ionization potential of O, similar values can be adopted:

$$I_{732-665}(N_2) = (1.6 \pm 0.5)\times 10^{-8} \quad sec^{-1}$$

$$I_{732-665}(O) = (1.5 \pm 0.5)\times 10^{-8} \quad sec^{-1}$$

$$I_{732-665}(O_2) = (1.5 \pm 0.5)\times 10^{-8} \quad sec^{-1}$$

$$I_{732-665}(H) = (2.4 \pm 0.8)\times 10^{-9} \quad sec^{-1}.$$

7.4.5 665–375 Å

The spectral range of wavelengths shorter than 660 Å is difficult to analyze for a whole solar cycle. Nevertheless, the ionization rate coefficients for zero optical depth are

$$I_{665-375}(N_2) = (1.9 \pm 0.9)\times 10^{-7} \quad sec^{-1}$$

$$I_{665-375}(O_2) = (2 \pm 1)\times 10^{-7} \quad sec^{-1}$$

$$I_{665-375}(O) = (8.7 \pm 4.3)\times 10^{-8} \quad sec^{-1}$$

$$I_{665-375}(H) = (1.2 \pm 0.6)\times 10^{-9} \quad sec^{-1}.$$

Helium is also ionized at $\lambda \leq 504.3$ Å, and

$$I_{665-375}(He) = (2.4 \pm 1.2)\times 10^{-8} \quad sec^{-1}.$$

7.4.6 375–275 Å

From 375 to 275 Å the solar spectrum has several groups of radiations, and the most important is certainly He(II) at $\lambda = 303.8$ Å (see Fig. 7.13) which represents about 25% of the total energy of that spectral range. The approximate ionization rate coefficients at zero optical depth are

$$I_{375-275}(N_2) = (1.9 \pm 0.9)\times 10^{-7} \quad sec^{-1}$$

$$I_{375-275}(O_2) = (2.3 \pm 1)\times 10^{-7} \quad sec^{-1}$$

$$I_{375-275}(O) = (1.1 \pm 0.5)\times 10^{-7} \quad sec^{-1}$$

$$I_{375-275}(He) = (4.2 \pm 1.8)\times 10^{-8} \quad sec^{-1}$$

$$I_{375-275}(H) = (4.2 \pm 1.9)\times 10^{-9} \quad sec^{-1}.$$

7.4.7 275–150 Å

Between 275 and 150 Å, the solar emission is still important [34] (see Fig. 7.14). Unfortunately, the uncertainty is almost complete as far as the absorption and ionization cross sections are concerned. Ionization resulting from solar emission in this spectral range will characterize the thermospheric region between the E layer, where x rays and Lyman β play an important role, and the F layer, where radiation between 800 and 300 Å undergoes absorption. If one compares the ionization cross sections of this spectral region with those of the Lyman continuum, one can see that they are of the same order of magnitude. This means that the vertical distribution of the atmospheric absorption between 275 and 150 Å will be mixed with that of the Lyman continuum. Approximate values are

$$\begin{aligned} I_{275-150}(N_2) &= (7.0 \pm 3.0)\times 10^{-8} \quad \text{sec}^{-1} \\ I_{275-150}(O_2) &= (1.0 \pm 0.5)\times 10^{-7} \quad \text{sec}^{-1} \\ I_{275-150}(O) &= (5.2 \pm 2.4)\times 10^{-8} \quad \text{sec}^{-1} \\ I_{275-150}(He) &= (1.4 \pm 0.6)\times 10^{-8} \quad \text{sec}^{-1} \\ I_{275-150}(H) &= (1.4 \pm 0.6)\times 10^{-10} \quad \text{sec}^{-1}. \end{aligned}$$

In conclusion, substantial uncertainty is due to a lack of sufficient experimental knowledge of absorption and ionization cross sections in which there are differences of a factor of 2 in certain spectral regions. Furthermore, values of the solar energy depend on solar activity and are also subject to some uncertainty in observational data [32, 33]. Finally, it should be pointed out that the absorption cross section is extremely variable with wavelength and that the ionization rate coefficients will be subject to strong variations in the thermosphere where the optical depth varies. The values of *Table 7.12* are order of magnitude estimates at zero optical depth. They give the total photoionization rate coefficients which may be taken as follows:

$$\begin{aligned} I_{\infty}(O_2) &= (7.0 \pm 3.0)\times 10^{-7} \quad \text{sec}^{-1} \\ I_{\infty}(N_2) &= (4.9 \pm 2.3)\times 10^{-7} \quad \text{sec}^{-1} \\ I_{\infty}(O) &= (3.2 \pm 1.3)\times 10^{-7} \quad \text{sec}^{-1} \\ I_{\infty}(He) &= (8.0 \pm 4.0)\times 10^{-8} \quad \text{sec}^{-1} \\ I_{\infty}(H) &= (8.8 \pm 3.3)\times 10^{-8} \quad \text{sec}^{-1}. \end{aligned}$$

When individual ion states are taken into account, a more detailed listing is needed using the ion ratios discussed in Chapter 6. *Tables 7.13–7.15* indicate the magnitude of the effect.

TABLE 7.12

IONIZATION COEFFICIENTS[a] BY SOLAR ULTRAVIOLET RADIATION

Spectral range (Å)	O_2	O	H	N_2	He
Lyman β at 1026 Å	$(5 \pm 1) \times 10^{-9}$				
C(III) at 977 Å	$(1.1 \pm 0.4) \times 10^{-8}$				
910–796	$(5.7 \pm 2) \times 10^{-8}$	$(4.5 \pm 1.5) \times 10^{-8}$	$(6.2 \pm 2) \times 10^{-8}$		
796–732	$(1.5 \pm 0.5) \times 10^{-8}$	$(1.5 \pm 0.5) \times 10^{-8}$	$(6 \pm 2) \times 10^{-9}$	$(1.5 \pm 0.5) \times 10^{-8}$	
732–665	$(1.5 \pm 0.5) \times 10^{-8}$	$(1.5 \pm 0.5) \times 10^{-8}$	$(2.4 \pm 0.8) \times 10^{-9}$	$(1.6 \pm 0.5) \times 10^{-8}$	
665–375	$(2 \pm 1) \times 10^{-7}$	$(8.7 \pm 4.3) \times 10^{-8}$	$(1.2 \pm 0.6) \times 10^{-8}$	$(1.9 \pm 0.9) \times 10^{-7}$	$(2.4 \pm 1.2) \times 10^{-8}$
375–275	$(2.3 \pm 1) \times 10^{-7}$	$(1.1 \pm 0.5) \times 10^{-7}$	$(4.2 \pm 1.9) \times 10^{-9}$	$(1.9 \pm 0.9) \times 10^{-7}$	$(4.2 \pm 1.8) \times 10^{-8}$
275–150	$(1.0 \pm 0.5) \times 10^{-7}$	$(5.2 \pm 2.4) \times 10^{-8}$	$(1.4 \pm 0.6) \times 10^{-9}$	$(7.0 \pm 3.0) \times 10^{-8}$	$(1.4 \pm 0.6) \times 10^{-8}$
150–80	$(9.0 \pm 5) \times 10^{-8}$	$(4.4 \pm 2.6) \times 10^{-9}$	$(7.5 \pm 4.5) \times 10^{-11}$	$(5.5 \pm 3.3) \times 10^{-9}$	$(7.5 \pm 4.5) \times 10^{-10}$
Total	$(7.0 \pm 3.0) \times 10^{-7}$	$(3.2 \pm 1.3) \times 10^{-7}$	$(8.8 \pm 3.3) \times 10^{-8}$	$(4.9 \pm 2.3) \times 10^{-7}$	$(8.0 \pm 4.0) \times 10^{-8}$

[a] Measurements are in sec^{-1}. The values are for all ion states.

TABLE 7.13

ATOMIC OXYGEN PHOTOIONIZATION COEFFICIENTS[a]

Spectral range	Ion state		
λ (Å)	$O^+(^4S)$	$O^+(^2D)$	$O^+(^2P)$
1026	—	—	—
977	—	—	—
910–796	4.5×10^{-8}	—	—
796–732	1.5×10^{-8}	—	—
732–665	6.7×10^{-9}	8.4×10^{-9}	—
665–375	2.5×10^{-8}	3.9×10^{-8}	2.3×10^{-8}
375–275	3.2×10^{-8}	4.6×10^{-8}	2.8×10^{-8}
275–150	1.7×10^{-8}	2.2×10^{-8}	1.3×10^{-8}
150–80	1.5×10^{-9}	1.8×10^{-9}	1.1×10^{-9}
Total	1.4×10^{-7}	1.2×10^{-7}	6.5×10^{-8}

[a] Values are in sec^{-1}.

TABLE 7.14

MOLECULAR OXYGEN PHOTOIONIZATION COEFFICIENTS[a]

Spectral range	Ion state		
λ (Å)	$O_2^+(X\,^2\Pi)$	$O_2^+(^4\Pi, ^2\Pi)$	$O_2^+(^4\Sigma)$
1026	5×10^{-9}	—	—
977	1.1×10^{-8}	—	—
910–796	5.7×10^{-8}	—	—
796–732	6.7×10^{-9}	8.2×10^{-9}	—
732–665	6.7×10^{-9}	8.2×10^{-9}	—
665–375	9.0×10^{-8}	7.0×10^{-8}	3.0×10^{-8}
375–275	1.0×10^{-7}	8.0×10^{-8}	4.0×10^{-8}
275–150	5.0×10^{-8}	3.0×10^{-8}	1.0×10^{-8}
150–80	4.0×10^{-8}	3.2×10^{-8}	1.8×10^{-8}
Total	3.7×10^{-7}	2.3×10^{-7}	1.0×10^{-7}

[a] Values are in sec^{-1}.

TABLE 7.15

MOLECULAR NITROGEN PHOTOIONIZATION COEFFICIENTS[a]

Spectral range	Ion state		
λ (Å)	$N_2^+(X\ ^2\Sigma^+)$	$N_2^+(A\ ^2\Pi_u)$	$N_2^+(B\ ^2\Sigma_u)$
1026	—	—	—
977	—	—	—
910–796	—	—	—
796–732	1.5×10^{-8}	—	—
732–665	6.4×10^{-9}	9.6×10^{-9}	—
665–375	8.5×10^{-8}	8.5×10^{-8}	2.0×10^{-8}
375–275	8.5×10^{-8}	8.5×10^{-8}	2.0×10^{-8}
275–150	3.1×10^{-8}	3.1×10^{-8}	8.0×10^{-9}
150–80	2.5×10^{-9}	2.5×10^{-9}	5.0×10^{-10}
Total	2.3×10^{-7}	2.1×10^{-7}	4.9×10^{-8}

[a] Values are in sec^{-1}.

7.5 Lyman α and Lyman β

The first line of the Lyman series of hydrogen [35–37] at 1215.7 Å can lead to about 3×10^{11} photons cm^{-2} sec^{-1} at the top of the earth's atmosphere. Its absorption in the mesosphere is primarily dependent on molecular oxygen absorption even though the cross section for absorption by CH_4 and H_2O is very high. Absorption by CO_2 is less important.

The penetration of Lyman α in the mesosphere is governed by molecular oxygen (absorption), but leads to the ionization of nitric oxide inside the D region. However, since Lyman α emission is not emitted uniformly by the solar disk [38] it may vary with the solar cycle [39]. A solar flux of $(3 \pm 1) \times 10^{11}$ photons cm^{-2} sec^{-1} corresponds to a variation between 3.5 and 6.5 ergs cm^{-2} sec^{-1} at the top of the earth's atmosphere. The ionization rate coefficient of nitric oxide I_{NO} (Lyman α) reaches such values as

$$I_{NO}(\text{Lyman } \alpha) = (6 \pm 2) \times 10^{-7} \quad sec^{-1}$$

which corresponds to a very efficient photoionization process at zero optical depth. Unit optical depth occurs at an altitude of 75 km for an overhead sun (*Table 7.16*). Below 70 km the photoionization rate coefficient of nitric

TABLE 7.16

IONIZATION RATE COEFFICIENT OF NO BY LYMAN α

	Rate coefficient (sec^{-1})		
Altitude (km)	Overhead sun $\chi = 0°$	Horizon $\chi = 90°$	Below horizon $\chi = 95°$
60	7.0×10^{-12}		
65	2.0×10^{-9}		
70	4.1×10^{-8}		
75	1.9×10^{-7}	3.0×10^{-28}	
80	3.7×10^{-7}	5.4×10^{-16}	
85	5.0×10^{-7}	2.1×10^{-10}	
90	5.6×10^{-7}	3.6×10^{-8}	
95	5.9×10^{-7}	2.0×10^{-7}	
100	5.9×10^{-7}	4.0×10^{-7}	
105	6.0×10^{-7}	5.1×10^{-7}	2.7×10^{-24}
110	6.0×10^{-7}	5.6×10^{-7}	1.5×10^{-13}
115	6.0×10^{-7}	5.8×10^{-7}	2.7×10^{-9}
120	6.0×10^{-7}	5.9×10^{-7}	7.5×10^{-8}
125	6.0×10^{-7}	6.0×10^{-7}	2.2×10^{-7}
130	6.0×10^{-7}	6.0×10^{-7}	4.4×10^{-7}
140	6.0×10^{-7}	6.0×10^{-7}	5.7×10^{-7}
150	6.0×10^{-7}	6.0×10^{-7}	6.0×10^{-7}

oxide decreases rapidly and any exceptional ionization at 60 km cannot be due to Lyman α. There is no doubt that ionization is produced in the mesosphere even though the nitric oxide concentration is a small fraction of the total concentration.

As far as Lyman β is concerned, its emission at 1025.7 Å is at least of the order of 3×10^9 photons cm^{-2} sec^{-1} during the maximum of the sunspot cycle. It leads to the following ionization rate of molecular oxygen, I_{O_2} (Lyman β) at zero optical depth:

$$I_{O_2}(\text{Lyman } \beta) \geq (5 \pm 1) \times 10^{-9} \quad \text{sec}^{-1}.$$

It can be easily seen from the absorption cross section for Lyman β and the data of Chapter 6 that unit optical depth for an overhead sun is in the neighborhood of 105 km. With an absorption cross section of about 1.6×10^{-18} cm^{-2}, we may write

$$I_{O_2}(\text{Lyman } \beta) = (5 \pm 1) \times 10^{-9} \exp[-1.6 \times 10^{-18} n(O_2) H(O_2)] \quad \text{sec}^{-1}.$$

Lyman β is an important radiation occurring near the ionization threshold of molecular oxygen. It penetrates into the E region and its effect must be introduced in determining the ionization equilibrium of the E layer.

7.6 Variation of Solar Emission

Since all solar ultraviolet radiations do not vary with the same amplitude during a complete solar cycle, we have adopted average solar conditions as follows:

1. completely quiet sun: minimum,
2. quiet sun: low average,
3. quiet sun: high average,
4. active sun: maximum
5. extreme conditions: flares and x ray.

In the E region, ultraviolet radiations between 1028 and 911 Å involving Lyman β at 1025 Å and C(III) at 977 Å ionize O_2. The variation from a completely quiet sun to an active sun should be of the order of 2. x Rays between 80 and 35 Å also ionize in the E region, but their variation from a completely quiet to an active sun corresponds to at least a factor of 6, and for extreme conditions can reach a factor of 10.

The transition from the E layer to the F_1 layer depends on the absorption by O_2 and O of the solar radiation in the Lyman continuum (910–796 Å) and also on the absorption by N_2, O_2, and O of ultraviolet radiations of $\lambda < 275$ Å. Factors of 2 and 3 may be chosen for the variation from a completely quiet to an active sun for 910 to 796 Å and 275 to 150 Å, respectively.

Solar radiations between 796 and 275 Å, which are absorbed by N_2, O, and O_2 in the F layer, also vary with solar activity. Detailed results are given in the analysis of ionization processes of the F region. A factor of 3 may be adopted for the variations from a completely quiet to an active sun.

Table 7.17 shows possible variations of the ultraviolet fluxes over a whole solar cycle. Although a general correlation [29] has been established between the solar ultraviolet fluxes and the 10.7 cm solar radioelectric flux, it cannot be concluded that a strict correlation exists for typical chromospheric emission lines such as 1025, 977, 584, and 304 Å.

The 910–796 Å region can be subdivided into seven intervals representing respectively 5, 5, 10, 20, 20, 20, and 20% of the values in Table 7.14. These fluxes have to be used in the computation of the absorption by the N_2 bands with the cross sections given in Chapter 6.

TABLE 7.17

Solar Flux for Various Conditions of Solar Activity

Range (Å)	Minimum (photons cm^{-2} sec^{-1} and erg cm^{-2} sec^{-1})	Low average (photons cm^{-2} sec^{-1} and erg cm^{-2} sec^{-1})	High average (photons cm^{-2} sec^{-1} and erg cm^{-2} sec^{-1})	Maximum (photons cm^{-2} sec^{-1} and erg cm^{-2} sec^{-1})
Lyman β 1027	4×10^{9}	5×10^{9}	5×10^{9}	6×10^{9}
	7.7×10^{-2}	9.7×10^{-2}	9.7×10^{-2}	1.2×10^{-1}
C(III) 997	3×10^{9}	4×10^{9}	4×10^{9}	6×10^{9}
	6.1×10^{-2}	8.1×10^{-2}	8.1×10^{-2}	1.2×10^{-1}
910–796	7.5×10^{9}	1.0×10^{10}	1.25×10^{10}	1.5×10^{10}
	1.8×10^{-1}	2.3×10^{-1}	2.9×10^{-1}	3.5×10^{-1}
796–732	1.0×10^{9}	1.25×10^{9}	1.5×10^{9}	2.0×10^{9}
	2.6×10^{-2}	3.2×10^{-2}	3.9×10^{-2}	5.2×10^{-2}
732–665	5.0×10^{8}	6.0×10^{8}	8.0×10^{8}	1.0×10^{9}
	1.4×10^{-2}	1.7×10^{-2}	2.3×10^{-2}	2.8×10^{-2}
665–375	4.0×10^{9}	6.0×10^{9}	8.0×10^{9}	1.2×10^{10}
	1.7×10^{-1}	2.5×10^{-1}	3.3×10^{-1}	5.0×10^{-1}
375–275	7.5×10^{9}	1.0×10^{10}	1.5×10^{10}	2.0×10^{10}
	4.7×10^{-1}	6.3×10^{-1}	9.4×10^{-1}	1.2
275–150	7.5×10^{9}	1.0×10^{10}	1.5×10^{10}	2.0×10^{10}
	7.7×10^{-1}	1.0	1.5	2.0
150–80	5.0×10^{8}	7.5×10^{8}	1.0×10^{9}	2.0×10^{9}
	9.5×10^{-2}	1.4×10^{-1}	1.9×10^{-1}	3.8×10^{-1}
80–60	1.75×10^{8}	3.5×10^{8}	7×10^{8}	1.05×10^{9}
	5.1×10^{-2}	1.0×10^{-1}	2.0×10^{-1}	3.0×10^{-1}
60–41	1.0×10^{8}	2.0×10^{8}	4.0×10^{8}	6.0×10^{8}
	4.1×10^{-2}	8.2×10^{-2}	1.6×10^{-1}	2.4×10^{-1}
41–31	1.75×10^{7}	3.5×10^{7}	7.0×10^{7}	1.05×10^{8}
	9.8×10^{-3}	2.0×10^{-2}	3.9×10^{-2}	5.9×10^{-2}
6	2.9×10^{2}	2.9×10^{3}	2.9×10^{4}	2.9×10^{5}
	9.6×10^{-7}	9.6×10^{-6}	9.6×10^{-5}	9.6×10^{-4}
4	2.0×10^{1}	2.0×10^{2}	2.0×10^{3}	2.0×10^{4}
	9.9×10^{-8}	9.9×10^{-7}	9.9×10^{-6}	9.9×10^{-5}
2	1.0	1.0×10^{1}	1.0×10^{2}	1.0×10^{3}
	9.9×10^{-9}	9.9×10^{-8}	9.9×10^{-7}	9.9×10^{-6}

TABLE 7.18

POSSIBLE FLUXES IN THE 6, 4, AND 2 Å RANGES DURING SOLAR FLARES

Range (Å)	Solar conditions (photons cm^{-2} sec^{-1} and erg cm^{-2} sec^{-1})			
	Completely quiet sun	Low average	High average	Active sun
6	6.0×10^4	3.0×10^5	6.0×10^5	3.0×10^6
	2.0×10^{-4}	1.0×10^{-3}	2.0×10^{-3}	1.0×10^{-2}
4	4.0×10^2	4.0×10^4	1.2×10^5	8.1×10^5
	2.0×10^{-6}	2.0×10^{-4}	6.0×10^{-4}	4.0×10^{-3}
2	2.0×10^1	3.0×10^3	2.0×10^4	2.0×10^5
	2.0×10^{-7}	3.0×10^{-5}	2.0×10^{-4}	2.0×10^{-3}

Table 7.18 gives extreme variations in the 2, 4, 6 Å ranges during solar flare conditions. The wavelength region between 31 and 10 Å has not been included since it has practically no effect on the D region ionization.

The final result for the entire ultraviolet spectrum is shown in *Table 7.19* in which the ionization rate coefficients for the principal thermospheric

TABLE 7.19

IONIZATION COEFFICIENTS (10^{-7} sec^{-1}) FOR VARIOUS CONDITIONS OF SOLAR ACTIVITY

Component	Quiet	Low	High	Active
O_2	3.6	4.9	6.9	9.3
N_2	2.5	3.6	4.9	7.1
O	1.7	2.4	3.3	4.5
He	0.42	0.58	0.84	1.2
H	0.55	0.75	0.96	1.2

constituents are given for the conditions of (1) completely quiet sun, (2) quiet sun: low average, (3) quiet sun: high average, and (4) active sun. It may be pointed out that the range of the variations does not differ much from values given in Table 7.12.

REFERENCES

1. W. A. Baum, F. S. Johnson, J. J. Oberly, C. C. Rockwood, C. V. Strain, and R. Tousey, *Phys. Rev.* **70**, 781 (1946).
2. F. S. Johnson, J. D. Purcell, R. Tousey, and K. Watanabe, *J. Geophys. Res.* **57**, 157 (1952).
3. H. H. Malitson, J. D. Purcell, R. Tousey, and C. E. Moore, *Astrophys. J.* **132**, 746 (1960).
4. C. R. Detwiler, J. D. Purcell, and R. Tousey, *Mém. Soc. Roy. Sci. Liège 5e sér.* **4**, 253 (1961).
5. D. R. Bates and M. Nicolet, *J. Geophys. Res.* **55**, 301 (1950).
6. C. R. Detwiler, D. L. Garrett, J. D. Purcell, and R. Tousey, *Ann. Géophys.* **17**, 263 (1961).
7. M. Nicolet, *Arch. Meteorol. Geophys. Bioklimatol.* **3B**, 209 (1951).
8. F. S. Johnson, *J. Meteorol.* **11**, 431 (1954).
9. D. Labs and H. Neckel, *Z. Astrophys.* **69**, 1 (1968).
10. E. G. Lane and A. J. Drummond, *Science* **161**, 888 (1968).
11. R.-M. Bonnet, J.-É. Blamont, and P. Gildwarg, *Astrophys. J.* **148**, L115 (1967).
12. M. Ackerman, D. Frimout, and R. Pastiels, *Ciel et Terre* **84**, 408 (1968).
13. R.-M. Bonnet, *Space Res.* **8**, 458 (1968).
14. W. H. Parkinson and E. M. Reeves, *Solar Phys.* **10**, 342 (1969).
15. K. G. Widing, J. D. Purcell, and G. D. Sandlin, *Solar Phys.* **12**, 52 (1970).
16. M. Ackerman, in *Mesospheric Models and Related Experiments* (G. Fiocco, ed.), p. 149. Reidel Publ., Dordrecht, The Netherlands, 1971.
17. D. F. Heath, NASA Report X-631-71-116, Goddard Space Flight Center, Greenbelt, Maryland (1971).
18. R. W. Kreplin, *Ann. Géophys.* **17**, 151 (1961).
19. R. W. Kreplin, T. A. Chubb, and H. Friedman, *J. Geophys. Res.* **67**, 2231 (1962).
20. M. Nicolet and A. C. Aikin, *J. Geophys. Res.* **65**, 1469 (1960).
21. T. A. Chubb and E. T. Byram, *Space Res.* **3**, 1046 (1963).
22. K. A. Pounds, *Ann. Astrophys.* **28**, 132 (1965).
23. G. Fritz, R. W. Kreplin, J. F. Meckins, A. E. Unzicker, and H. Friedman, *Astrophys. J.* **148**, L133 (1967).
24. W. M. Neupert, W. Gates, M. Swartz, and R. Young, *Astrophys. J.* **149**, L79 (1967).
25. J. E. Manson, *Astrophys. J.* **147**, 703 (1967).
26. K. C. Widing and G. D. Sandlin, *Astrophys. J.* **152**, 545 (1968).
27. H. E. Hinteregger, *J. Geophys. Res.* **66**, 2367 (1961).
28. R. Tousey, *Space Sci.*n*Rev.* **2**, 3 (1963).
29. L. A. Hall, J. E. Higgis, C. W. Chagnon, and H. E. Hinteregger, *J. Geophys. Res.* **74**, 4181 (1969).
30. J. C. Lindsay, *Planet. Space Sci.* **12**, 379 (1964).
31. L. A. Hall, K. R. Damon, and H. E. Hinteregger, *Space Res.* **3**, 745 (1963).
32. H. E. Hinteregger, *Ann. Géophys.* **26**, 547 (1970).
33. L. A. Hall and H. E. Hinteregger, *J. Geophys. Res.* **75**, 6959 (1970).
34. W. M. Neupert, W. E. Behring, and J. C. Lindsay, *Space Res.* **4**, 719 (1964).
35. E. T. Byram, T. A. Chubb, H. Friedman, J. E. Kupperian Jr., and R. W. Kreplin, *Astrophys. J.* **128**, 738 (1958).

36. A. I. Yefremov, A. L. Podmoshensky, O. N. Yefimov, and A. A. Lebedev, *Space Res.* **3**, 843 (1963).
37. H. E. Hinteregger, L. A. Hall, and G. Schmidtke, *Space Res.* **5**, 1175 (1965).
38. R. Tousey, *Astrophys. J.* **149**, 239 (1967).
39. L. H. Weeks, *Astrophys. J.* **147**, 1203 (1967).
40. H. E. Hinteregger, in *Annals of the IQSY* (A. C. Strickland, ed.), Vol. 5, p. 305. MIT Press, Cambridge, Massachusetts, 1969.

CHAPTER 8

Photodissociation

8.1 Molecular Oxygen

The first dissociation energy of molecular oxygen is 5.11 eV and corresponds to a continuum beginning at 2424 Å according to the process

$$O_2 + h\nu(\lambda \leq 2424\ \text{Å}) \rightarrow O(^3P) + O(^3P). \tag{8.1}$$

The various electronic transitions corresponding to the dissociation into two normal oxygen atoms are as follows [1, 2]:

(i) The infrared atmospheric system a $^1\Delta_g$–X $^3\Sigma_g^-$ is a magnetic dipole intercombination transition with an important band (0–0). The emission probability [3] of this system is 2.6×10^{-4} sec^{-1}.

(ii) The red atmospheric oxygen bands b $^1\Sigma_g^+$–$^3\Sigma_g^-$ have a transition probability of the order of 10^{-1} sec^{-1}; the band (0–0) is the principal band of this system.

(iii) The first Herzberg system, A $^3\Sigma_u^+$–$^3\Sigma_g^-$, is a forbidden system with an emission probability of the order of 10^{-2} sec^{-1} and is observed in the airglow [4, 5]. Absorption bands lead to the observed continuum at 2424 Å.

(iv) The second Herzberg system $^1\Sigma_u^-$–$^3\Sigma_g^-$ should have a transition probability of the order of 10^{-3} sec^{-1}.

(v) The new Herzberg system $^3\Delta_u$–$^3\Sigma_g^-$ does not follow the rule $\Delta\Lambda = 0, \pm 1$ and its transition probability cannot be more than 10^{-5} sec^{-1}.

(vi) The 12 other states $^1\Pi$, $^3\Pi$, $^5\Sigma$, $^5\Pi$, and $^5\Delta$ leading to two normal atoms must correspond to highly excited electron configurations and are therefore probably unstable.

The contribution from the Herzberg continuum to the photodissociation rate is essentially constant in the mesosphere since the absorption cross section varies between 10^{-24} and 10^{-23} cm^2 while the O_2 total content is of the order of 4×10^{21} cm^{-2} above the stratopause. Therefore, with the solar fluxes given in Chapter 7, the photodissociation coefficient is

$$J_{O_2}(\text{Herzberg}) = 1.3 \times 10^{-9} \quad \text{sec}^{-1}. \tag{8.2}$$

In the Schumann–Runge bands, where predissociation occurs ($v' > 3$) the absorption is extremely variable and allowance must be made for the temperature variation [6]. It has been shown that a wave number resolution of 0.5 cm^{-1} is necessary to analyze the absorption in the Schumann–Runge bands. For example, *Fig. 8.1* shows the structure of the photodissociation coefficient between 1923 and 1942 Å at the stratopause [7]. For the Schumann–Runge bands average values given in *Table 8.1* can be adopted. These values can also be obtained with a mean absorption cross section [8] corresponding physically to an optical depth of one. This procedure is, however, not valid for the photodissociation of minor constituents [7, 8]

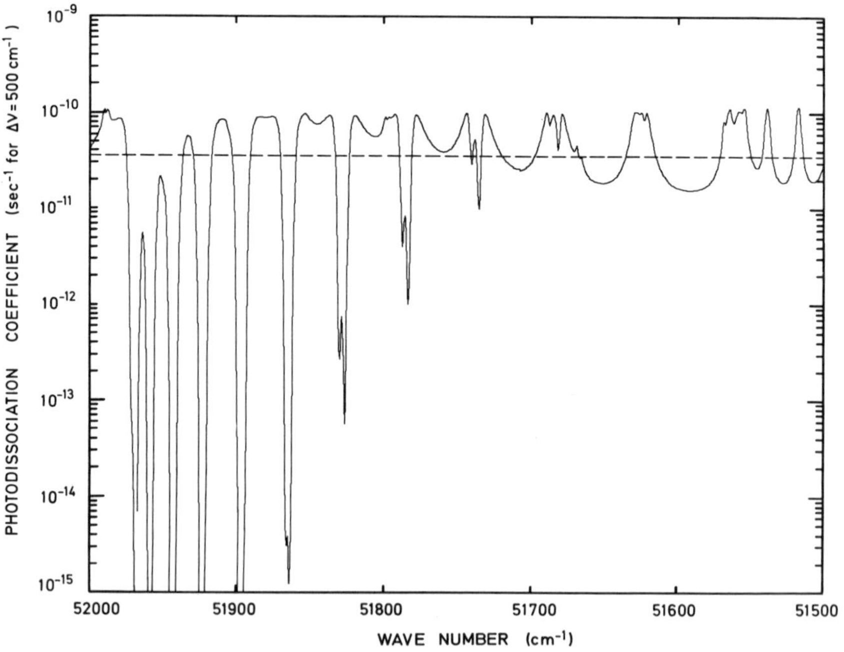

Fig. 8.1 Example of the structure of the molecular oxygen photodissociation coefficient at 50 km altitude between 1923 and 1942 Å.

TABLE 8.1

PHOTODISSOCIATION COEFFICIENT[a] OF O_2 IN THE SCHUMANN–RUNGE BANDS FOR AN OVERHEAD SUN

Altitude (km)	J_{O_2}(S–R)	Altitude (km)	J_{O_2}(S–R)
100	9.14×10^{-8}	60	7.47×10^{-10}
95	6.32 ·	55	4.52
90	3.62	50	2.53
85	1.73	45	1.51
80	8.80×10^{-9}	40	8.88×10^{-11}
75	4.40	35	4.45
70	2.35	30	1.79
65	1.29		

[a] Values in $\sec^{-1}$.

such as H_2O. Between 90 and 70 km predissociation in the Schumann–Runge bands represents the major contribution to the total O_2 photodissociation into two $O(^3P)$ atoms.

The second limit of dissociation corresponds to electronic transitions for which one of the two oxygen atoms is in the excited state ^1D. These various triplets $^3\Sigma^+$, 2X $^3\Sigma^-$, 3X $^3\Pi$, 2X $^3\Delta$, and $^3\Phi$ and singlets $^1\Sigma_u^+$ and $^1\Delta_u$ correspond to the following process:

$$O_2 + h\nu(\lambda \leq 1750 \text{ Å}) \rightarrow O(^3P) + O(^1D). \tag{8.3}$$

The Schumann–Runge continuum is actually the major cause of the O_2 dissociation.

A computation made for a blackbody at 5000°K gives the following value for the dissociation coefficient J (see *Table 8.2*):

$$J_\infty(5000°\text{K}) = 1.3 \times 10^{-5} \quad \sec^{-1}. \tag{8.4}$$

A lower temperature such as 4500°K will lead to a much smaller coefficient, i.e.,

$$J_\infty(4500°\text{K}) = 1.6 \times 10^{-6} \quad \sec^{-1}. \tag{8.5}$$

Since the number of photons available at the top of the earth's atmosphere

TABLE 8.2

PHOTODISSOCIATION RATE COEFFICIENTS OF O_2 AS A FUNCTION OF THE TOTAL NUMBER OF MOLECULES PER CM^{-2}

Molecules (cm^{-2})	$T = 4500°K$	$T = 5000°K$
10^{15}	1.57×10^{-6}	1.26×10^{-5}
2×10^{15}	1.55	1.22
5×10^{15}	1.52	1.19
10^{16}	1.47	1.15
2×10^{16}	1.38	1.07
5×10^{16}	1.14	8.69×10^{-6}
10^{17}	8.63×10^{-7}	6.43
2×10^{17}	5.56	3.98
5×10^{17}	2.58	1.74
10^{18}	1.33	8.62×10^{-7}
2×10^{18}	5.91×10^{-8}	3.74×10^{-7}

varies by about 1.5 for a temperature range of 100°K, it is necessary to consider that

$$J_\infty(O_2) = (4 \pm 2) \times 10^{-6} \quad \text{sec}^{-1} \qquad \text{[blackbody]} \tag{8.6}$$

for the Schumann–Runge continuum.

The value given by (8.6) corresponds to photodissociation of oxygen in the absence of screening absorption by any constituent. It shows that the time required to reduce an initial concentration of O_2 by 50% is more than 33 hr and less than 100 hr, i.e., between 2 and 6 days. Several hundreds of hours (100 to 300 hr) of permanent solar radiation are needed to reduce the O_2 concentration to one-tenth of its initial value, even if the optical depth is small; $n(O_2)H(O_2) \leq 10^{15}$ molecules cm^{-2}. It is known that molecules which are dissociated at sufficiently great height can be replaced by an upward vertical transport due to diffusion. At lower heights, for which there is absorption of solar radiation, the dissociation coefficient decreases (see *Table 8.3*) and the lifetime of O_2 in the sun's radiation field increases. For example, at 100 km the total number of O_2 molecules is of the order of 10^{18} cm^{-2} and the time necessary to reduce the concentration to 50% of its initial value is more than 20 days of permanent solar radiation. Therefore, O_2 cannot be in photoequilibrium in the thermosphere [9].

TABLE 8.3

ADOPTED PHOTODISSOCIATION RATE COEFFICIENT OF O_2 IN THE TERRESTRIAL ATMOSPHERE FOR AN OVERHEAD SUN

O_2 molecules (cm^{-2})	$J(O_2)$ (sec^{-1})	O_2 molecules (cm^{-2})	$J(O_2)$ (sec^{-1})	O_2 molecules (cm^{-2})	$J(O_2)$ (sec^{-1})
1.0×10^{14}	3.7×10^{-6}	7.5×10^{17}	4.6×10^{-7}	1.0×10^{20}	7.3×10^{-9}
1.0×10^{15}	3.7×10^{-6}	1.0×10^{18}	3.6×10^{-7}	2.5×10^{20}	3.9×10^{-9}
1.0×10^{16}	3.4×10^{-6}	2.5×10^{18}	1.6×10^{-7}	5.0×10^{20}	2.6×10^{-9}
2.5×10^{16}	3.1×10^{-6}	5.0×10^{18}	6.2×10^{-8}	7.5×10^{20}	2.2×10^{-9}
5.0×10^{16}	2.6×10^{-6}	7.5×10^{18}	4.6×10^{-8}	1.0×10^{21}	2.0×10^{-9}
7.5×10^{16}	2.2×10^{-6}	1.0×10^{19}	3.4×10^{-8}	2.5×10^{21}	1.5×10^{-9}
1.0×10^{17}	1.9×10^{-6}	2.5×10^{19}	1.7×10^{-8}	5.0×10^{21}	1.2×10^{-9}
2.5×10^{17}	1.0×10^{-6}	5.0×10^{19}	1.0×10^{-8}	7.5×10^{21}	8.4×10^{-10}
5.0×10^{17}	6.3×10^{-7}	7.5×10^{19}	8.6×10^{-9}	1.0×10^{22}	6.4×10^{-10}

With regard to Lyman α, a flux of 3×10^{11} photons cm^{-2} sec^{-1} and an absorption cross section of 10^{-20} cm^2 (see Chapter 6) lead to a photodissociation coefficient of

$$J_{O_2}(\text{Lyman } \alpha) = 3 \times 10^{-9} \quad sec^{-1}. \tag{8.7}$$

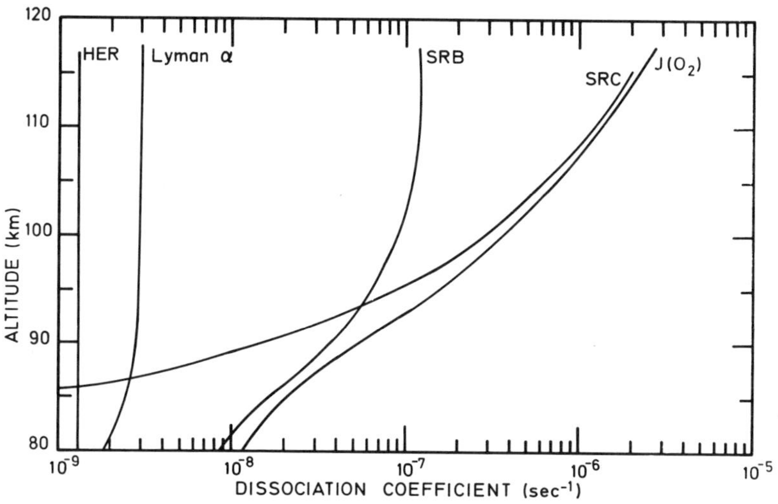

Fig. 8.2 Photodissociation coefficient of molecular oxygen in the lower thermosphere with an overhead sun. $J(O_2)$ is the total photodissociation coefficient resulting from the Schumann–Runge continuum (SRC), from Schumann–Runge bands (SRB), from Lyman α (Lyman α), and from the Herzberg continuum (HER).

In the thermosphere [8] (see *Fig. 8.2*), the principal process in the O_2 photodissociation is due to the absorption in the Schumann–Runge continuum. The predissociation process begins to play a role near the mesopause and becomes predominant in the mesosphere [7, 8, 10] (see *Fig. 8.3*). At the stratopause, the absorption in the Herzberg continuum leads to the major contribution for $J(O_2)$.

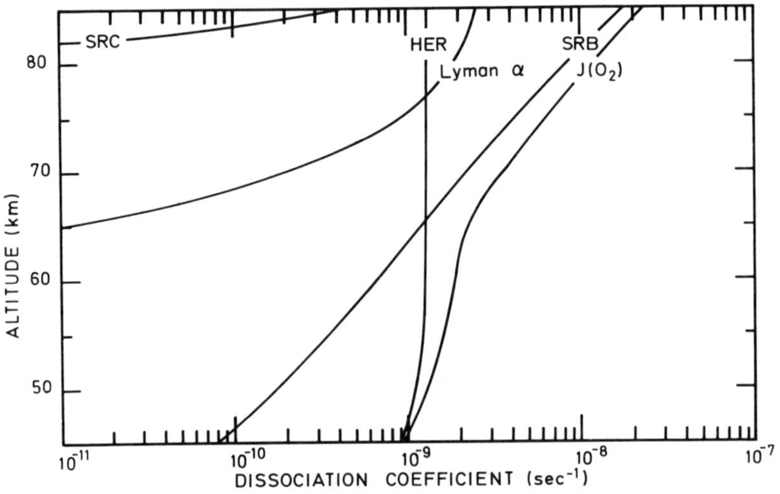

Fig. 8.3 Photodissociation coefficient of molecular oxygen in the mesosphere with an overhead sun. See Fig. 8.2 for the spectral ranges.

When the total number of oxygen molecules in a vertical column is less than 10^{19} cm^{-2}, i.e., the number above 90 km (see Table 3.2) for an overhead sun, the photodissociation of oxygen comes from the absorption in the Schumann–Runge continuum (see *Fig. 8.4*). The photodissociation coefficient decreases from about 2×10^{-6} sec^{-1} for 10^{17} molecules cm^{-2} to less than 10^{-8} sec^{-1} for a vertical column of 10^{19} O_2 molecules cm^{-2}. The $O(^1D)$ production is due to the absorption in the Schumann–Runge continuum and it is essentially (see *Fig. 8.5*) a thermospheric process. Below the mesopause, the photodissociation due to Lyman α is more important than that due to the Schumann–Runge continuum (see Fig. 8.3). However, the dissociation due to predissociation in the Schumann–Runge band system is the principal process in the mesosphere since the coefficient

$$J_{O_2}(\text{S–R bands}) = 1.7\times10^{-8} \quad \text{sec}^{-1} \tag{8.8}$$

at 85 km for an overhead sun is to be compared with the value of 1.3×10^{-9}

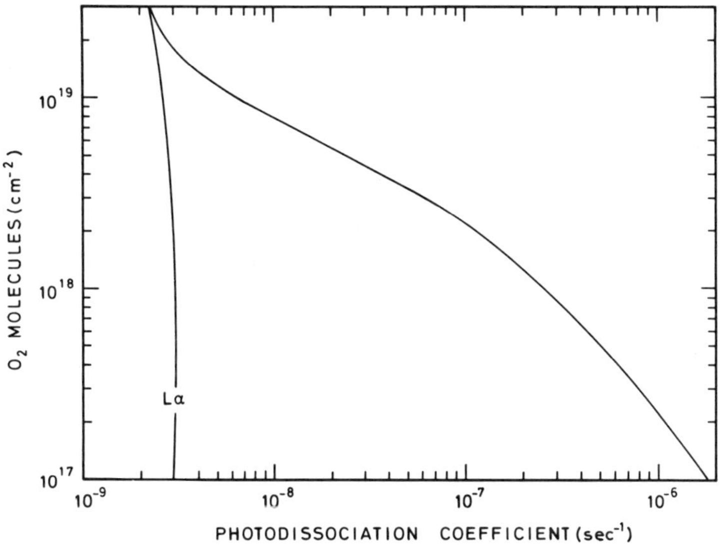

Fig. 8.4 Photodissociation coefficient of molecular oxygen for the Schumann–Runge continuum ($\lambda < 1754$ Å) and for Lyman α (1215.7 Å) in the lower thermosphere where the total content is greater than 10^{17} O_2 molecules cm^{-2}.

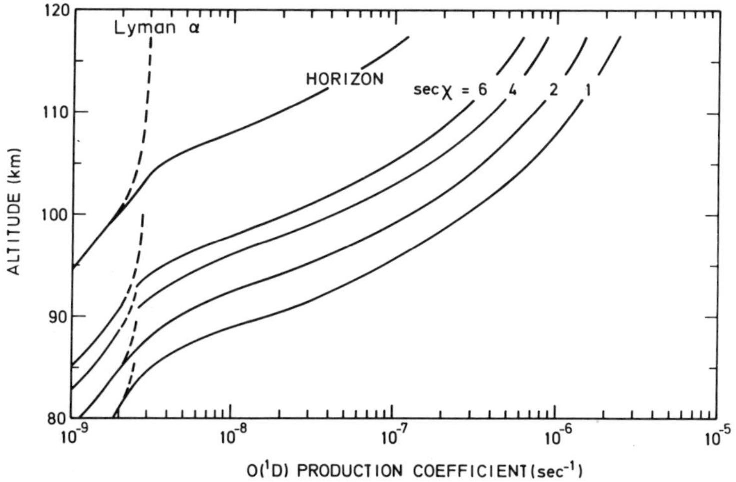

Fig. 8.5 Variation of the $O(^1D)$ production with solar zenith angle in the lower thermosphere. $\lambda < 1754$ Å.

sec^{-1} for the Herzberg continuum. Actually, the total production (see *Fig. 8.6*) of $O(^3P)$ atoms [8] is due to the predissociation in the Schumann–Runge bands in the mesosphere and to the dissociation in the Herzberg continuum in the stratosphere. Figure 8.6 also shows the importance of the variation of the O_2 dissociation in the stratosphere as a function of the solar zenith distance.

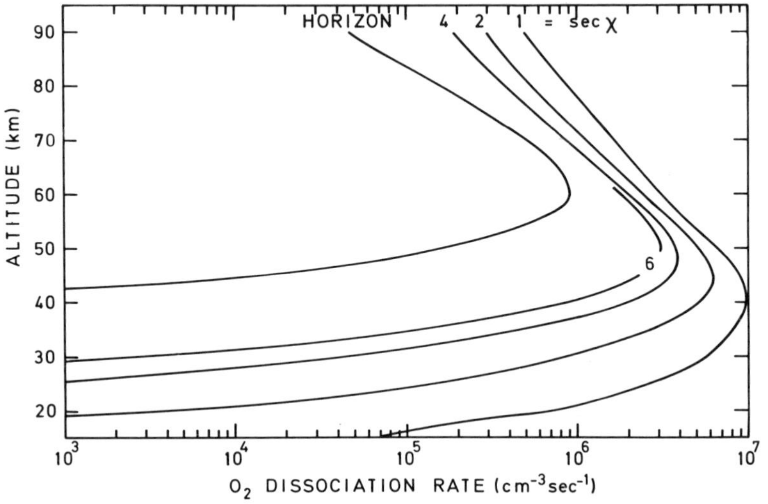

Fig. 8.6 Photodissociation rate of molecular oxygen for various solar zenith angles in the mesosphere and stratosphere.

8.2 Ozone

Since the photochemical equilibrium of ozone has already been examined, our purpose is not to carry through another analysis. However, since ozone plays a role in various reactions in the stratosphere and in the mesosphere, it is necessary to give some numerical values. Although the absorption cross sections have been measured, it is difficult to know what fraction of the absorption leads directly to dissociation; the variation of absorption coefficients with temperature in certain spectral ranges shows clearly that the absorption efficiency for direct photodissociation has not yet been determined. Nevertheless, if one compares the spectral distribution of solar radiation and ozone absorption, one can say that dissociation must result from the continuum between 2900 and 2400 Å, covering about 80% of the total possible photodissociation corresponding to a continuum not subject to variation with temperature.

If Chappuis bands in the visible spectrum and Huggins bands at the beginning of the ultraviolet spectrum ($\lambda < 3100$ Å) lead to a complete dissociation of O_3, the coefficient J_{O_3}(Chappuis) is [8]

$$J_{O_3}(\text{Chappuis}) = 3\times10^{-4} \quad \text{sec}^{-1} \tag{8.9}$$

and the coefficient J_{O_3}(Huggins) is

$$J_{O_3}(\text{Huggins}) = 1\times10^{-4} \quad \text{sec}^{-1}. \tag{8.10}$$

However, since the absorption cross section increases rapidly with decreasing wavelength, the photodissociation coefficient for the spectral range λ 3100–3000 Å is of the order of 2×10^{-4} sec^{-1}.

Calculation for the whole spectrum yields a photodissociation coefficient

$$J_{\infty}(O_3) = 1.0\times10^{-2} \quad \text{sec}^{-1} \tag{8.11}$$

corresponding to the absorption of solar radiation of $\lambda > 2300$ Å for which the absorption cross section in the Hartley bands is greater than 5×10^{-18} cm^{-2}. Since the continuous absorption of O_2 begins at $\lambda < 2424$ Å, it is clear that the ozone photodissociation is almost independent of the absorption by molecular oxygen. In aeronomy, ozone absorption is important

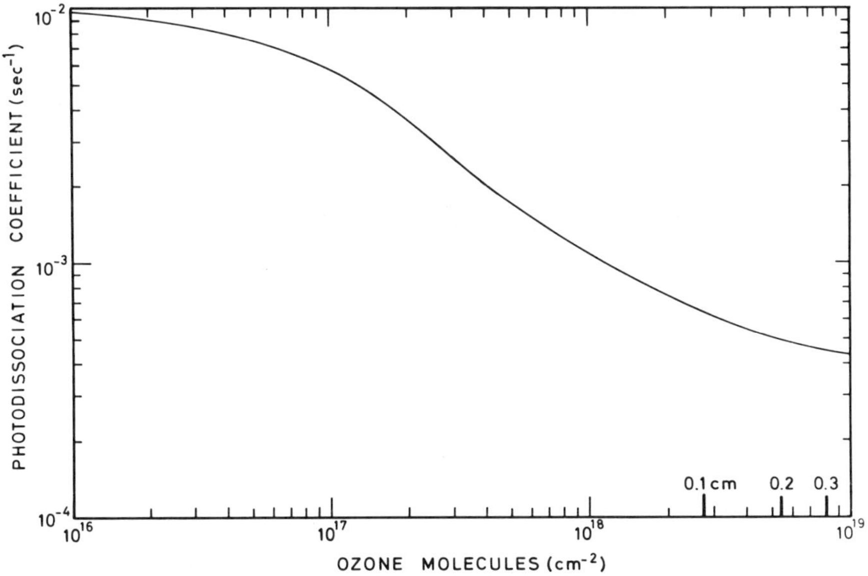

Fig. 8.7 Ozone photodissociation coefficient as a function of the total content.

because it limits the penetration of solar radiation of $\lambda < 2400$ Å, which dissociates O_2. The ozone photodissociation coefficient is given in *Fig. 8.7* as a function of the O_3 total content. It is clear that at least 10^{18} ozone molecules cm^{-2} are required to decrease the photodissociation coefficient J_{O_3} by a factor of 10. In the mesosphere, the ozone content is less than 10^{17} molecules cm^{-2} and oxygen dissociation is therefore not affected by ozone absorption above the stratopause. The various chemical mechanisms resulting from the ozone photodissociation [8] will be discussed in Chapter 10.

8.3 Molecular Nitrogen

The direct photodissociation of N_2 is not important. The predissociation mechanism [11]

$$N_2(X\,^1\Sigma_g) + h\nu(\lambda\lambda\ 1200\text{–}1250\ \text{Å}) \rightarrow N_2(a\,^1\Pi_g) \rightarrow N + N \quad (8.12)$$

is a very slow process, at least 10^6 times slower than the parallel processes for O_2.

Considering the Herzberg–Herzberg mechanism associated with the Lyman–Birge–Hopfield band (8–0), one finds $P(2) = 1227.02$ Å, $P(10) = 1228.20$ Å, $Q(1) = 1226.91$ Å, $Q(10) = 1226.15$ Å, $R(0) = 1226.84$ Å, and $R(10) = 1227.17$ Å. The N_2 dissociation rate is taken as [12, 13]

$$J_{N_2} \simeq 5 \times 10^{-12} \quad \text{sec}^{-1} \quad (8.13)$$

for zero optical depth if the radiative lifetime of the a $^1\Pi_g$ state is $(1.7 \pm 0.3) \times 10^{-4}$ sec^{-1} and the intensity distribution of the band is normal.

8.4 Water Vapor

Water vapor undergoes photodissociation in accordance with the following process:

$$H_2O + h\nu \rightarrow OH(X\,^2\Pi) + H(^2S) \quad (8.14)$$

which begins near 2400 Å. In addition to this generally accepted primary process, molecular hydrogen and atomic oxygen may be produced [14]:

$$H_2O + h\nu \rightarrow H_2 + O. \quad (8.15)$$

At 1236 Å, the probability of process (8.14) is three times that of process (8.15).

The absorption cross section is only of the order of 10^{-20} cm^2 at 1880 Å, but rises to 10^{-19} cm^2 near 1850 Å and to about 10^{-18} cm^2 near 1795 Å. Its value equals 2.7×10^{-18} cm^2 at the beginning of the Schumann–Runge continuum [15–17]. The overall dissociation coefficient at zero optical depth is [8]

$$J_\infty(H_2O) = 1 \times 10^{-5} \quad \text{sec}^{-1} \tag{8.16}$$

and is mainly due to the Lyman α radiation and to the Schumann–Runge continuum. The photodissociation of H_2O in the mesosphere is affected by the absorption in the spectral region of the O_2 bands of the Schumann–Runge system.

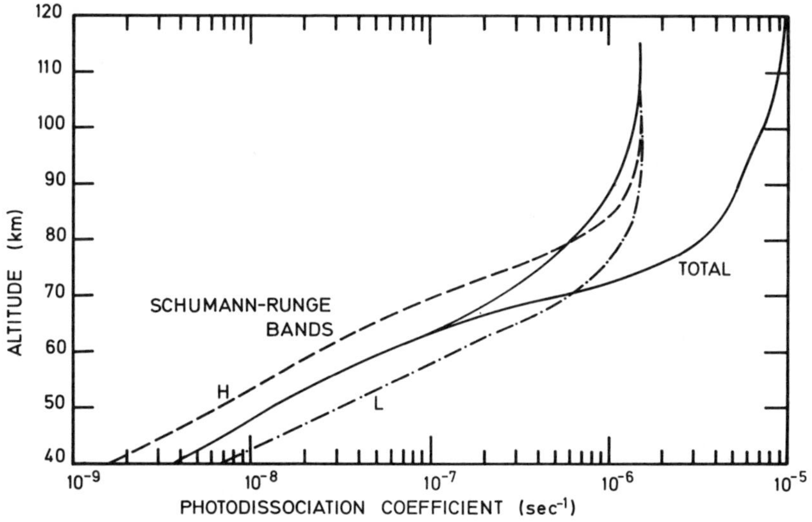

Fig. 8.8 Water vapor photodissociation coefficient.

In *Fig. 8.8*, the water vapor photodissociation coefficient is shown in the mesosphere. The detailed structure of the Schumann–Runge bands has been taken into account [7]. At the mesopause photodissociation by Lyman α is the most important process leading to

$$H_2O + h\nu(\lambda = 1216 \text{ Å}) \rightarrow H(^2S) + OH(^2\Sigma^+)_{v' \leq 2} \tag{8.17}$$

where the excited OH leads to a fluorescence process corresponding to several percent of the total dissociation [18]. Direct production of H_2 is also possible [14, 19] by

$$H_2O + h\nu(\lambda = 1216 \text{ Å}) \rightarrow H_2 + O(^1D). \tag{8.18}$$

The process (8.18) is, however, a small fraction of the general process (8.14) which is the principal mechanism for H_2O dissociation in the lower thermosphere and upper mesosphere [8].

8.5 Other Constituents

Methane, which undergoes dissociation for radiation with $\lambda < 1450$ Å, is not subject to general photodissociation in the mesosphere due to O_2 absorption. However, its dissociation cross section at Lyman α is about [20] 1.4×10^{-17} cm^2, so that the dissociation coefficient J_{CH_4}(Lyman α) at zero optical depth equals [21]

$$J_{CH_4}(\text{Lyman } \alpha) = 5 \times 10^{-6} \quad \text{sec}^{-1}. \tag{8.19}$$

The concentration of methane above 75 km should be reduced by Lyman α to 50% of its initial value in about 36 hr of sunlight. Collision processes involving excited atoms of oxygen are also effective [22] in producing dissociation of methane in the mesosphere.

Molecules such as CH_3 and CH_2 may be dissociated at 2160 Å and $\lambda < 1950$ Å, respectively, but oxidation processes are also effective in producing dissociation of such constituents.

The photodissociation rate for NO corresponds to the process

$$NO + h\nu(\lambda \leq 1900 \text{ Å}) \rightarrow N + O + 150 \quad \text{kcal mole}^{-1} \tag{8.20}$$

and involves predissociation observed in the γ, β, δ, and ε bands [23]. No emission is observed from levels above the dissociation energy of the ground electronic state. With a radiative lifetime [24] for NO* of about 2×10^{-7} sec, the dissociation coefficient is about [25]

$$J_{NO} = 5 \times 10^{-6} \quad \text{sec}^{-1} \tag{8.21}$$

at the mesopause where there is no effective absorption by O_2.

In a sunlit atmosphere, NO_2 can be photodissociated by the process

$$NO_2 + h\nu(\lambda < 3975 \text{ Å}) \rightarrow NO + O. \tag{8.22}$$

The photodissociation rate can be determined with precision since the absorption cross section at $\lambda < 3979$ Å, where the last sharp rotational lines appear [26], depends on the vibronic structure. Adopting absorption cross sections (*Figs. 8.9* and *8.10*) which have been measured over [27, 28]

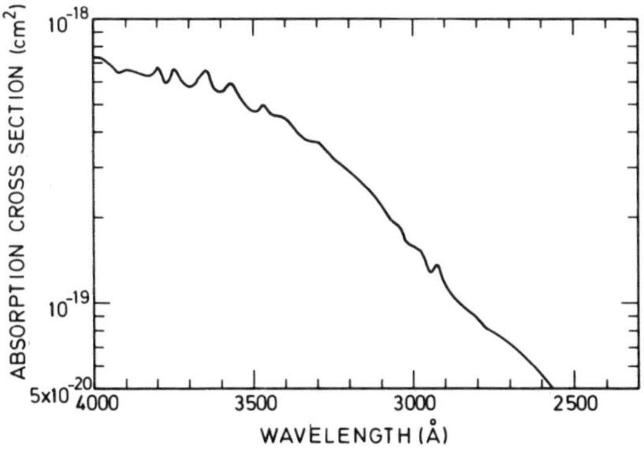

Fig. 8.9 Absorption cross section of NO_2 for wavelengths greater than 3000 Å. From Hall and Blacet [27].

the range 4000–1000 Å, the following average photodissociation coefficient is adopted:

$$J_{NO_2} = 8.1 \times 10^{-3} \quad sec^{-1} \tag{8.23}$$

with

$$J_{NO_2}(\lambda > 3300 \text{ Å}) = 7.3 \times 10^{-3} \quad sec^{-1} \tag{8.24}$$

and

$$J_{NO_2}(\lambda > 3000 \text{ Å}) = 8.0 \times 10^{-3} \quad sec^{-1}. \tag{8.25}$$

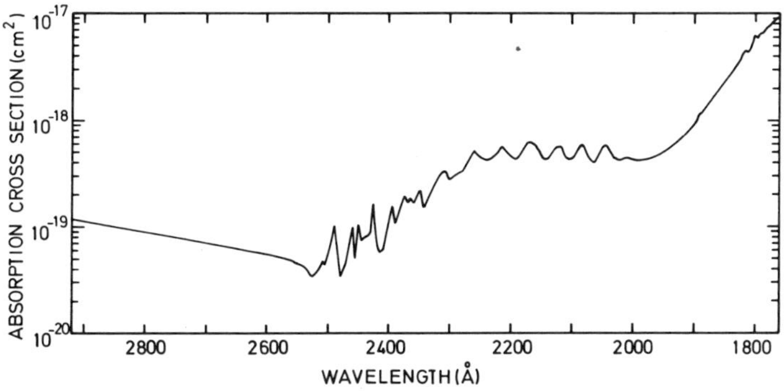

Fig. 8.10 Absorption cross section of NO_2 in the ultraviolet. From Nakayama *et al.* [28].

Carbon dioxide is completely transparent in the visible and near ultraviolet. A small absorption cross section of 2×10^{-24} cm^2 has been measured [29] at 2160 Å and the cross section increases slowly toward 2000 Å. Below 2000 Å an exponential increase of the cross section has been observed [29, 30] down to 1780 Å. According to the energy diagram for the $CO_2 \rightleftarrows CO + O$ system [31], the transition from a $^3B_2$ state of CO_2 to

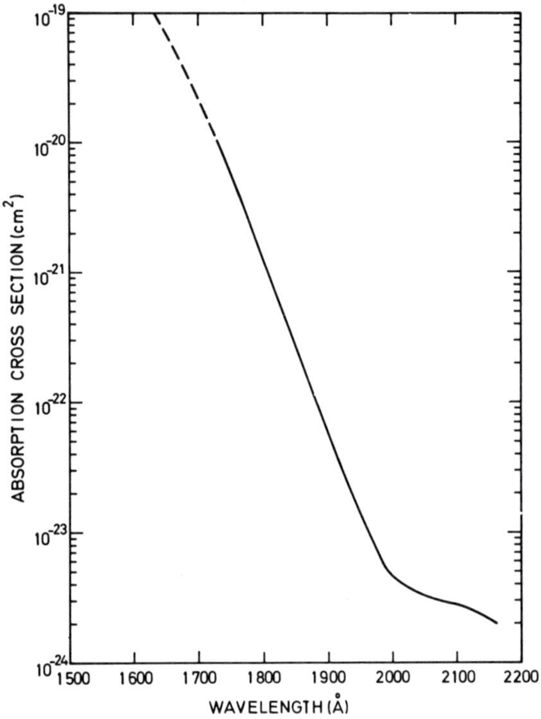

Fig. 8.11 Absorption cross section of CO_2 for wavelengths greater than 1700 Å. Solid line, from Ogawa [29]; dashed line, from Heimerl [32].

a $^1\Sigma_g{}^+$ state of CO is forbidden, but a predissociation process can occur through a $^1B_2$ state. A band structure is observed below 1980 Å and it overlaps a weak continuum [29] shown on *Fig. 8.11*. This corresponds to the dissociation process

$$CO_2 + h\nu(\lambda < 2274 \text{ Å}) \rightarrow CO(^1\Sigma_g{}^+) + O(^3P) \qquad (8.26)$$

which is not the principal photodissociation mechanism since the absorption cross section is less than 10^{-20} cm^2 (see Fig. 8.11).

For wavelengths between 1700 and 1200 Å, absorption measurements [32, 33] reveal the presence of diffuse bands superposed on a continuum. Two maxima appear in the absorption cross section: at 1475 Å, $\sigma = 6\times 10^{-19}$ cm², and at 1330 Å, $\sigma = 9\times 10^{-19}$ cm² (see *Fig. 8.12*). This spectral region corresponds to the dissociation process

$$CO_2 + h\nu(\lambda < 1671 \text{ Å}) \rightarrow CO(^1\Sigma_g{}^+) + O(^1D). \quad (8.27)$$

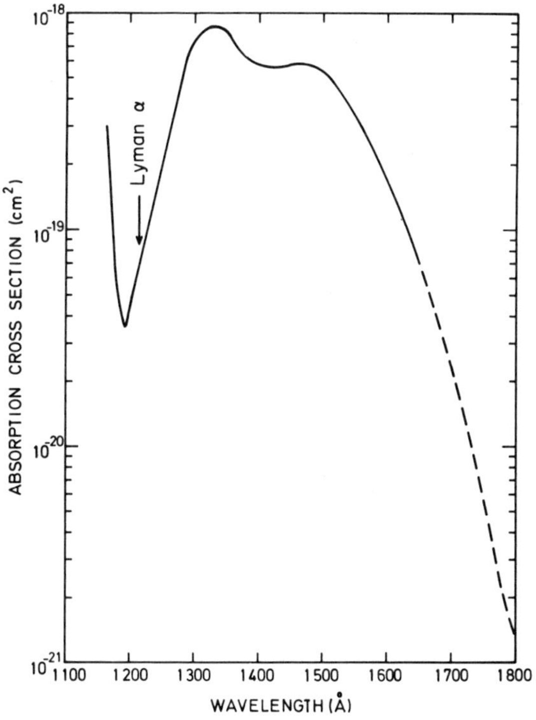

Fig. 8.12 Absorption cross section of CO_2 for wavelengths less than 1700 Å. Solid line, from Inn *et al.* [33], dashed line, from Heimerl [32].

The absorption cross section [33] of 7.3×10^{-20} cm² at Lyman α indicates that this radiation is important for the CO_2 photodissociation around the mesopause, since the absorption by molecular oxygen only becomes predominant below 80 km.

Around 1190 Å the CO_2 absorption cross section reaches a minimum of the order of 3.5×10^{-20} cm². Strong bands appear at shorter wavelengths [34] and converge to the first ionization potential of CO_2 at 900 Å. Below

1200 Å, the following dissociation processes are possible:

$$CO_2 + h\nu(\lambda < 1285\ \text{Å}) \rightarrow CO(^1\Sigma_g{}^+) + O(^1S) \tag{8.28}$$

and

$$CO_2 + h\nu(\lambda < 1082\ \text{Å}) \rightarrow CO(a\ ^3\Pi) + O(^3P). \tag{8.29}$$

A production of CO in an excited a $^1\Pi$ state is also possible below 900 Å, but must compete with the direct photoionization of CO_2. Absorption and photoionization cross sections have been measured [35–37] below 900 Å. The photoionization yield [35, 36] between 600 and 700 Å ranges from 90 to 100%.

With the absorption cross sections of Figs. 8.11 and 8.12 and with the solar fluxes given in Chapter 7, the CO_2 photodissociation coefficient can be computed as a function of altitude for an overhead sun (see *Fig. 8.13*). At the top of the atmosphere, the total photodissociation coefficient is

$$J_{CO_2} = 1.9 \times 10^{-7} \quad \text{sec}^{-1} \tag{8.30}$$

with a contribution by Lyman α radiation of

$$J_{CO_2}(\text{Lyman } \alpha) = 2.2 \times 10^{-8} \quad \text{sec}^{-1}. \tag{8.31}$$

Around the mesopause, Lyman α makes the major contribution to the total

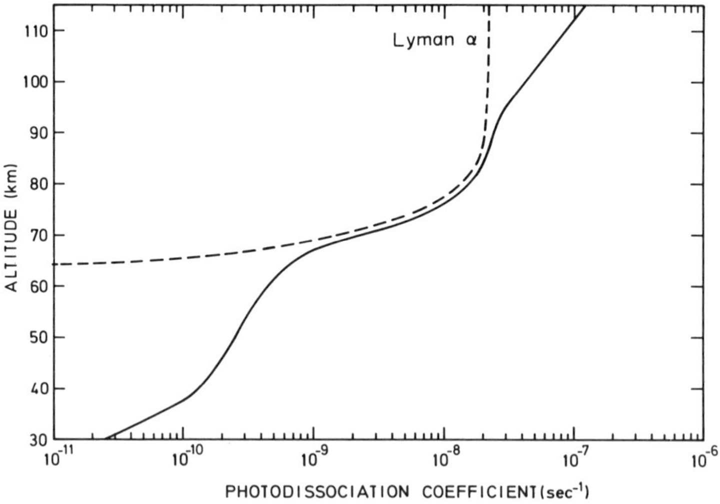

Fig. 8.13 CO_2 photodissociation rate coefficient for an overhead sun.

photodissociation coefficient. To make this calculation the effect of the detailed structure in the molecular oxygen Schumann–Runge bands [7] must be taken into account.

REFERENCES

1. G. Herzberg, *Can. J. Phys.* **30**, 185 (1952).
2. G. Herzberg, *Can. J. Phys.* **31**, 657 (1953).
3. R. M. Badger, A. C. Wright, and R. F. Whitlock, *J. Chem. Phys.* **43**, 4345 (1968).
4. J. W. Chamberlain, *Physics of the Aurora and Airglow*. Academic Press, New York, 1961.
5. V. I. Krassovsky, N. N. Shefov, and V. I. Yarin, *Planet. Space Sci.* **9**, 883 (1962).
6. M. Ackerman, F. Biaumé, and G. Kockarts, *Planet. Space Sci.* **18**, 1639 (1970).
7. G. Kockarts, in *Mesopheric Models and Related Experiments* (G. Fiocco, ed.), pp. 160–176. Reidel Publ., Dordrecht, The Netherlands, 1971.
8. M. Nicolet, in *Mesospheric Models and Related Experiments* (G. Fiocco, ed.), pp. 1–51. Reidel Publ., Dordrecht, The Netherlands, 1971.
9. M. Nicolet and P. Mange, *J. Geophys. Res.* **59**, 15 (1954).
10. R. D. Hudson, V. L. Carter, and E. L. Breig, *J. Geophys. Res.* **74**, 4079 (1969).
11. G. Herzberg and L. Herzberg, *Nature* (*London*) **161**, 283 (1948).
12. D. R. Bates, *Ann. Géophys.* **8**, 194 (1952).
13. D. R. Bates, in *The Solar System* (G. P. Kuiper, ed.), Vol. 2, *The Earth as a Planet*, pp. 576–643. Univ. of Chicago Press, Chicago, Illinois 1954.
14. J. R. McNesby, I. Tanaka, and H. Okabe, *J. Chem. Phys.* **36**, 605 (1962).
15. K. Watanabe and M. Zelikoff, *J. Opt. Soc. Amer.* **43**, 753 (1953).
16. A. H. Laufer and J. R. McNesby, *Can. J. Chem.* **43**, 3487 (1965).
17. K. Watanabe and A. S. Jursa, *J. Chem. Phys.* **41**, 1650 (1964).
18. T. Carrington, *J. Chem. Phys.* **41**, 2012 (1964).
19. L. J. Stieff, *J. Chem. Phys.* **44**, 277 (1966).
20. R. W. Ditchburn, *Proc. Roy. Soc. London* **A229**, 44 (1955).
21. M. Nicolet, *Ann. Géophys.* **26**, 531 (1970).
22. W. B. De More and O. F. Raper, *J. Chem. Phys.* **46**, 2500 (1967).
23. J. Heicklen and N. Cohen, *Advan. Photochem.* **5**, 157 (1968).
24. A. B. Callear and I. W. M. Smith, *Discuss. Faraday Soc.* **37**, 96 (1964).
25. M. Nicolet, *Planet. Space Sci.* **18**, 1111 (1970).
26. A. E. Douglas and K. P. Huber, *Can. J. Phys.* **43**, 74 (1965).
27. T. C. Hall, Jr. and F. F. Blacet, *J. Chem. Phys.* **20**, 1745 (1952).
28. T. Nakayama, M. Y. Kitamura, and K. Watanabe, *J. Chem. Phys.* **30**, 1180 (1959).
29. M. Ogawa, *J. Chem. Phys.* **54**, 2550 (1971).
30. B. A. Thompson, P. Harteck, and R. R. Reeves, Jr., *J. Geophys. Res.* **68**, 6431 (1963).
31. M. C. Lin and S. H. Bauer, *J. Chem. Phys.* **50**, 3377 (1969).
32. J. Heimerl, *J. Geophys. Res.* **75**, 5574 (1970).
33. E. C. Y. Inn, K. Watanabe, and M. Zelikoff, *J. Chem. Phys.* **21**, 1648 (1953).
34. Y. Tanaka, A. S. Jursa, and F. J. LeBlanc, *J. Chem. Phys.* **32**, 1199 (1960).
35. R. B. Cairns and J. A. R. Samson, *J. Geophys. Res.* **70**, 99 (1965).
36. R. S. Nakata, K. Watanabe, and F. M. Matsunaga, *Sci. Light* **14**, 54 (1965).
37. G. R. Cook, P. H. Metzger, and M. Ogawa, *J. Chem. Phys.* **44**, 2935 (1966).

CHAPTER 9

Collision Processes

9.1 Introduction

The study of collisions and reactions between atoms, molecules, and their ions lies at the heart of aeronomy. In this chapter the basic equations describing the interactions of these particles are given in a kinematic sense, avoiding chemical processes where the identity of the colliding particles is altered in the collision. The importance of such nonreactive collisions lies in studies of the electron and ion thermal balance, the diffusion of gases, heat conduction, radio wave propagation effects, ionization of gases, and photon emission.

9.2 Collision Processes

Consider a gas composed of two types of particles (subscripts 1 and 2) in thermal and dynamic equilibrium. The distribution of velocities of the particles composing this gas is given by the Maxwellian distribution

$$f(v_i) = n_i(m_i/2\pi kT_i)^{3/2} \exp[-m_i v_i^2/2kT_i] \tag{9.1}$$

where i identifies the type of particle, n is the concentration, m is the particle mass, T is the temperature, and k is Boltzmann's constant (1.38×10^{-16} erg deg^{-1}).

As usual, the normalization for $f(v_i)$ is chosen so that

$$\int f(v)\, d\mathbf{v} = 4\pi \int_0^\infty v^2 f(v)\, dv = n_i \tag{9.2}$$

taking the integral over all particle velocities in a spherical coordinate system.

The various collision properties of the gas follow from the velocity distribution function and the laws of interaction applying to the gas particles [1]. For example, the relative speed between particles of types 1 and 2 is

$$v_{12} = |\mathbf{v}_1 - \mathbf{v}_2| \triangleq g \tag{9.3}$$

so that the average relative speed can be written in terms of the individual velocity distribution functions as

$$\bar{v}_{12} = \int\int |\mathbf{v}_1 - \mathbf{v}_2| f_1(v_1) f_2(v_2)\, d\mathbf{v}_1\, d\mathbf{v}_2 . \tag{9.4}$$

Using Eq. (9.1), these integrals can be evaluated to give

$$\bar{v}_{12} = \left(\frac{8k}{\pi}\right)^{1/2} \left(\frac{T_1}{m_1} + \frac{T_2}{m_2}\right)^{1/2} \tag{9.5}$$

or, for thermal equilibrium,

$$\bar{v}_{12} = (8kT/\pi\mu)^{1/2} \tag{9.6}$$

where the reduced mass μ is given by

$$\mu = m_1 m_2/(m_1 + m_2). \tag{9.7}$$

The average kinetic energy, ε, of a particle in the gas is found by averaging $\frac{1}{2}mv_i^2$ over the velocity distribution:

$$\varepsilon_i = \int \tfrac{1}{2} m_i v_i^2 f(v_i)\, d\mathbf{v}_i \tag{9.8}$$

or, using (9.1),

$$\varepsilon_i = \tfrac{3}{2}kT_i . \tag{9.9}$$

The problem of collisions in a gas is actually very subtle since certain laws of interparticle interaction, such as the Coulomb force, do not lead to abrupt changes of particle momenta which are normally counted as collisions. For the present, suppose that the atomic properties of the gas particles are such that the cross section can be represented as $\sigma(g)$; i.e., a quantity which depends only upon the relative speed between the colliding particles. In this situation the number of collisions per unit time made by a particle of type 1 with type 2 particles is

$$\nu_{12}(g) = n_2 g\sigma(g) \tag{9.10}$$

where the velocity dependence of the collision frequency ν_{12} is emphasized.

To obtain the average collision frequency, Eq. (9.10) must be averaged over all relative velocities available from $f(v_1)$ and $f(v_2)$. Thus,

$$\bar{\nu}_{12} = (1/n_1) \int g\sigma(g) f(v_1) f(v_2)\, d\mathbf{v}_1\, d\mathbf{v}_2 \tag{9.11}$$

which is a function only of density, temperature, mass, and atomic parameters associated with the collision cross section.

The velocity-dependent time interval between successive collisions of a particular particle and other particles is

$$\tau_{12} = 1/\nu_{12} = 1/n_2 g\sigma \tag{9.12}$$

while the mean time can be obtained using the velocity distribution function as

$$\bar{\tau}_{12} = \int (n_2 g\sigma)^{-1} f(v_1) f(v_2)\, d\mathbf{v}_1\, d\mathbf{v}_2 . \tag{9.13}$$

Because the mean time between collisions is not a precise concept, it is usual to define the mean collision time as

$$\bar{\tau}_{12} = 1/\bar{\nu}_{12} . \tag{9.14}$$

In a similar fashion the mean distance or free path, λ_{12}, between collisions of a particle of type 1 and particles of type 2 is

$$\lambda_{12} = \bar{\tau}_{12}\bar{v}_{12} = \bar{v}_{12}/\bar{\nu}_{12} . \tag{9.15}$$

To obtain specific results it is necessary to use a particular form for the collision cross section $\sigma(g)$. The simplest model of an atomic interaction is that occurring between two rigid spheres of radii r_1 and r_2. Collisions between such particles are assumed to be elastic with no excitation of internal energy states. The collision cross section for this situation is

$$\sigma_0 = \pi(r_1 + r_2)^2 \tag{9.16}$$

while the velocity-dependent collision frequency becomes

$$\nu_{12} = n_2 \sigma_0 g \tag{9.17}$$

and the average collision frequency is

$$\bar{\nu}_{12} = n_2 \sigma_0 \bar{v}_{12} \tag{9.18}$$

with $\bar{v}_{12}$ being given by Eq. (9.5). Using Eq. (9.14) the mean collision time is

$$\bar{\tau}_{12} = (n_2 \sigma_0 \bar{v}_{12})^{-1} \tag{9.19}$$

while the mean free path is

$$\lambda_{12} = (n_2 \sigma_0)^{-1} \tag{9.20}$$

independent of relative velocity.

The concept of binary collisions considered above can be extended to include the possibility of nearly simultaneous collisions of three separate particles to form a temporary three-body collision complex. To compute the rate at which this occurs, consider a gas composed of three types of particles (subscripts 1, 2, and 3). Suppose that the binary collision frequency $\bar{\nu}_{12}$ is given by (9.11). A collision between particles 1 and 2 will take a certain length of time, t_{12}, corresponding roughly to the mean diameter of the two particle complex divided by the mean thermal speed, $\bar{v}_{12}$. During the time t_{12} there exists the opportunity for further binary collisions between types 1 and 3 and 2 and 3 particles. Thus, the three-body collision frequency is approximately given by

$$\bar{\nu}_{123} = \bar{\nu}_{12}(\bar{\nu}_{13} + \bar{\nu}_{23}) t_{12}. \tag{9.21}$$

The value for t_{12} is always uncertain, but a suitable approximation is

$$t_{12} = (\sigma_0/\pi)^{1/2}/\bar{v}_{12} \tag{9.22}$$

so that Eq. (9.21) becomes

$$\bar{\nu}_{123} = \bar{\nu}_{12}(\bar{\nu}_{13} + \bar{\nu}_{23})[(\sigma_0/\pi)^{1/2}/\bar{v}_{12}]. \tag{9.23}$$

To give an idea of the orders of magnitude for the collision frequencies defined by Eqs. (9.11) and (9.23) we can adopt rigid sphere approximations with typical molecular parameters: $\sigma_0 = 5 \times 10^{-15}$ cm², $m = 2.8 \times 10^{-23}$ gm (corresponding to atomic oxygen), and $T = 300$°K

$$\bar{\nu}_{12} = 1.1 \times 10^{-10} n_2 \quad \text{sec}^{-1} \tag{9.24}$$

$$\bar{\nu}_{123} = 4.4 \times 10^{-32} n_2 n_3 \quad \text{sec}^{-1}. \tag{9.25}$$

These values are typical of many aeronomic collisions between ions and neutral particles but not for electrons and neutrals, electrons and ions, or ions with ions, due to differing factors involving the particle masses and collision cross sections.

For electron collisions with neutral particles Eq. (9.24) gives values which are somewhat too small. Through consideration of Eqs. (9.5) and (9.18) with the condition $m_e \ll m_2$ one obtains

$$\bar{\nu}_{en} = n_n \sigma_{en} (8kT_e/\pi m_e)^{1/2}. \tag{9.26}$$

As a result of the mass factor, $\bar{\nu}_{en} > \bar{\nu}_{in}$ for equal collision cross sections. In addition, the electron collision frequency varies as $T_e^{1/2}$ for the case where the electron-neutral cross section is independent of relative velocity. This temperature dependence arises from the fact that most of the relative velocity between electrons and neutral particles derives from the much more rapid electron thermal motions. When the cross section depends on relative velocity, as is the actual case for most applications, the temperature dependence of the collision frequency is much more complex than indicated by Eq. (9.26). It is also usual to find that the electron-neutral cross section is substantially smaller than that for ion-neutral collisions.

In another application, it is possible to compute the fraction of collisions which occur with a relative collision kinetic energy greater than some value ε_0. Using Eq. (9.11) with suitable limits of integration,

$$\bar{\nu}_{12}(\varepsilon > \varepsilon_0)/\bar{\nu}_{12} = [(\varepsilon_0/kT) + 1] \exp[-\varepsilon_0/kT] \tag{9.27}$$

so that if $\varepsilon_0 \to 0$, the ratio approaches the value 1. This result is obtained only for rigid spheres.

The concept of a reaction rate for a chemical process is similar to the collision frequency. Given a two-body reaction of the form

$$A + B \xrightarrow{k} \text{some product}$$

the reaction rate k is given by

$$k = \overline{\sigma g} \tag{9.28}$$

where σ is the interaction cross section, g is the relative velocity, and the bar indicates an appropriate average over the velocity distribution functions of the particles, taking into account any energy thresholds which may exist.

9.3 Cross Sections Which Depend upon Velocity

The results of the last section are based on interparticle forces which yield cross sections independent of relative velocity. Furthermore, the cross section itself was undefined in its physical meaning.

In the study of gas transport phenomena it is found that there are three cross sections useful in the description of the behavior of a gas or mixture of gases perturbed from the equilibrium state [2]. These are the total cross section σ_T, the viscosity cross section σ_V, and the momentum transfer cross section σ_D. The differences between these cross sections arise from the ways in which the angular dependence of the differential cross section $q(\theta, g)$ are emphasized. The definitions are

$$\sigma_T(g) = 2\pi \int q(\theta, g) \sin\theta \, d\theta \tag{9.29}$$

$$\sigma_V(g) = 2\pi \int q(\theta, g) \sin^3\theta \, d\theta \tag{9.30}$$

$$\sigma_D(g) = 2\pi \int q(\theta, g)(1 - \cos\theta) \sin\theta \, d\theta \tag{9.31}$$

where θ is the center of mass scattering angle.

For most aeronomic problems involving collisions of electrons, ions, and neutral particles, the momentum transfer cross section is of greatest importance since it relates to both energy and momentum exchange.

Using Eq. (9.10), the velocity dependent momentum transfer collision frequency can be defined as

$$\nu_{12} = n_2 g \sigma_D(g) \tag{9.32}$$

where the relative speed g is given by $g = |\mathbf{v}_1 - \mathbf{v}_2|$.

To obtain the average momentum transfer cross section it is necessary to average Eq. (9.32) over all relative speeds consistent with two independent Maxwellian velocity distributions. Following the methods of the last section [3], one obtains

$$\bar{\nu}_{12} = \tfrac{4}{3} n_2 \bar{v}_{12} \bar{Q}_D \tag{9.33}$$

where the average relative velocity $\bar{v}_{12}$ is

$$\bar{v}_{12} = \left(\frac{8k}{\pi}\right)^{1/2} \left(\frac{T_1}{m_1} + \frac{T_2}{m_2}\right)^{1/2} \tag{9.34}$$

and the average momentum transfer cross section $\bar{Q}_D$ is given by

$$\bar{Q}_D = (1/c^6) \int_0^\infty g^5 \sigma_D(g) \exp[-g^2/c^2]\, dg \tag{9.35}$$

where the constant c is given by

$$c^2 = \left(\frac{2kT_1}{m_1} + \frac{2kT_2}{m_2} \right). \tag{9.36}$$

Although the factor $\frac{4}{3}$ appears in Eq. (9.33) it can be shown that for isotropic scattering there is no difference between Eqs. (9.33) and (9.18) where a total scattering cross section was used.

For electron collisions in a gas of real particles the momentum transfer frequency is, taking $T_e/m_e \gg T_2/m_2$

$$\bar{\nu}_{en} = \tfrac{4}{3} n_n (8kT_e/\pi m_e)^{1/2} \bar{Q}_D \tag{9.37}$$

where

$$\bar{Q}_D = (1/c_e^6) \int_0^\infty v_e^5 \sigma_D(v_e) \exp[-v_e^2/c_e^2]\, dv_e \tag{9.38}$$

and

$$c_e^2 = 2kT_e/m_e \tag{9.39}$$

with v_e being the electron speed.

The collision frequency defined by Eq. (9.33) is the momentum transfer collision frequency for the center of mass coordinate system [4]. The effective collision frequency in the laboratory system is

$$\bar{\nu}_L = [m_2/(m_1 + m_2)]\bar{\nu}_{12}. \tag{9.40}$$

This distinction is important in dealing with the application of the momentum transfer collision frequency to the collisional momentum exchange term appearing in the momentum balance equations for individual gas species. As shown later (see Chapter 19) the collision term has the form

$$m_1 n_1 \bar{\nu}_L (\mathbf{v}_2 - \mathbf{v}_1) \tag{9.41}$$

or, using Eq. (9.40),

$$n_1 [m_1 m_2/(m_1 + m_2)] \bar{\nu}_{12} (\mathbf{v}_2 - \mathbf{v}_1). \tag{9.42}$$

Another application of the momentum transfer cross section lies in the equation of energy exchange between two gases having different temperatures [3]. In this case the rate of energy exchange, du_1/dt is

$$du_1/dt = -3n_1[m_1m_2/(m_1 + m_2)^2]k(T_1 - T_2)\bar{\nu}_{12} \tag{9.43}$$

or, in terms of (9.39)

$$du_1/dt = -3n_1[m_1/(m_1 + m_2)]k(T_1 - T_2)\bar{\nu}_{\mathrm{L}}. \tag{9.44}$$

In the preceding discussion the importance of the momentum transfer collision frequency has been emphasized. It is clear, however, that the collision frequency is subject to arbitrary definitions and many different collision frequencies are obtained from experiments and theory. Thus, in comparing various data it is necessary to compare the methods whereby the velocity-dependent collision frequencies are converted into mean or average values.

The equations relating to the propagation of radio waves in the D region are particularly affected by such considerations. To show this, consider the velocity-dependent momentum transfer collision frequency given by (9.32) with $g \simeq v_{\mathrm{e}}$, the electron velocity:

$$\nu_{\mathrm{en}} = n_{\mathrm{n}} v_{\mathrm{e}} \sigma_{\mathrm{D}}(v_{\mathrm{e}}). \tag{9.45}$$

In the analysis of experimental data, reference is often made to the collision frequency for monoenergetic electrons, ν_{M}. Such a quantity is obtained from Eq. (9.45) through substitution of $v_{\mathrm{e}}{}^2 = 2kT_{\mathrm{e}}/m_{\mathrm{e}}$ to eliminate v_{e}. In general, the collision frequency for monoenergetic electrons differs from that for momentum transfer by some function which depends upon the actual velocity dependence of the collision cross section. For example, if $\sigma_{\mathrm{D}} = \alpha v_{\mathrm{e}}$, then $\nu_{\mathrm{M}} = \alpha(2kT_{\mathrm{e}}/m_{\mathrm{e}})$ while $\bar{\nu}_{\mathrm{en}} = \frac{5}{2}\alpha(2kT_{\mathrm{e}}/m_{\mathrm{e}})$ so that $\bar{\nu}_{\mathrm{en}}/\nu_{\mathrm{M}} = \frac{5}{2}$.

From the theory for the conductivity of a weakly ionized gas subjected to an alternating electric field of frequency ω, an effective collision frequency ν_{eff} if defined as [4]

$$\nu_{\mathrm{eff}} = \frac{\int_0^\infty v_{\mathrm{e}}{}^4[\nu_{\mathrm{en}}/(\nu_{\mathrm{en}}^2 + \omega^2)] \exp(-m_{\mathrm{e}} v_{\mathrm{e}}{}^2/2kT_{\mathrm{e}})\, dv_{\mathrm{e}}}{\int_0^\infty v_{\mathrm{e}}{}^4[1/(\nu_{\mathrm{en}}^2 + \omega^2)] \exp(-m_{\mathrm{e}} v_{\mathrm{e}}{}^2/2kT_{\mathrm{e}})\, dv_{\mathrm{e}}}. \tag{9.46}$$

When $\omega^2 \gg \nu_{\mathrm{en}}^2$ [note that ν_{en} is the velocity-dependent momentum transfer cross section given by Eq. (9.45)], Eq. (9.46) reduces to

$$\nu_{\mathrm{eff}} = (8/3\pi^{1/2} c_{\mathrm{e}}{}^5) \int_0^\infty v_{\mathrm{e}}{}^4 \nu_{\mathrm{en}} \exp[-v_{\mathrm{e}}{}^2/c_{\mathrm{e}}{}^2]\, dv_{\mathrm{e}}. \tag{9.47}$$

If, by chance, $\sigma_D = \alpha v_e$ then

$$\nu_{eff} = \bar{\nu}_{en} = \tfrac{5}{2}\nu_M \qquad (9.48)$$

but for any other velocity dependence of the cross section $\nu_{eff} \neq \bar{\nu}_{en}$ and the simple relation of Eq. (9.48) will not be true.

For the case $\omega^2 \ll \nu_{en}^2$ Eq. (9.46) reduces to

$$\nu_{eff} = \frac{\int_0^\infty v_e^4(1/\nu_{en}) \exp(-m_e v_e^2/2kT_e)\, dv_e}{\int_0^\infty v_e^4(1/\nu_{en}^2) \exp(-m_e v_e^2/2kT_e)\, dv_e}. \qquad (9.49)$$

Again, if $\sigma_{en} = \alpha v_e$, then solution of Eq. (9.49) shows that

$$\nu_{eff} = \tfrac{3}{5}\bar{\nu}_{en} = \tfrac{3}{2}\nu_M. \qquad (9.50)$$

As shown in the next section, the cross section for electron collisions with N_2 is proportional to electron velocity for energies below 0.1 eV. As a result, the relations (9.48) and (9.50) are approximately correct for the stated conditions.

9.4 Elastic Collisions of Electrons with Neutral Particles

Since theoretical methods usually do not give accurate values for the momentum transfer cross section it is necessary to depend upon experimental data. Unfortunately, the appropriate measurements are difficult to make and substantial uncertainty exists with certain cross sections. The results given here have been taken from detailed analyses [3, 5–7] of the available results.

Table 9.1 lists the velocity-dependent momentum transfer cross sections for the principal atmospheric gases. Using Eq. (9.38) the average momentum transfer cross sections can be obtained (see *Fig. 9.1*). The cross section results are listed in *Table 9.2*, while the final momentum transfer collision frequencies are presented in *Table 9.3* and *Fig. 9.2*.

With regard to the accuracy of these values complete discussions are given in the literature [8]. For N_2 and O_2 the cross sections are probably accurate to within 20% over the temperature range $100° < T_e < 4500°K$. The values for O are more uncertain, both in absolute magnitude and temperature dependence. Experimental values range from less than 10^{-16} cm^2 to 5×10^{-15} cm^2. The most recent experimental data [9], adopted here,

TABLE 9.1

ELECTRON MOMENTUM TRANSFER CROSS SECTIONS[a] $\sigma_D(\varepsilon)$

Element	Elastic cross section (cm^2)
N_2	$(18.3 - 7.3\varepsilon)\varepsilon^{1/2} \times 10^{-16}$
O_2	$(2.2 + 5.1\varepsilon^{1/2}) \times 10^{-16}$
O	1×10^{-15}
H	$(54.7 - 28.7\varepsilon) \times 10^{-16}$
He	5.6×10^{-16}

[a] ε is the electron energy in electron volts.

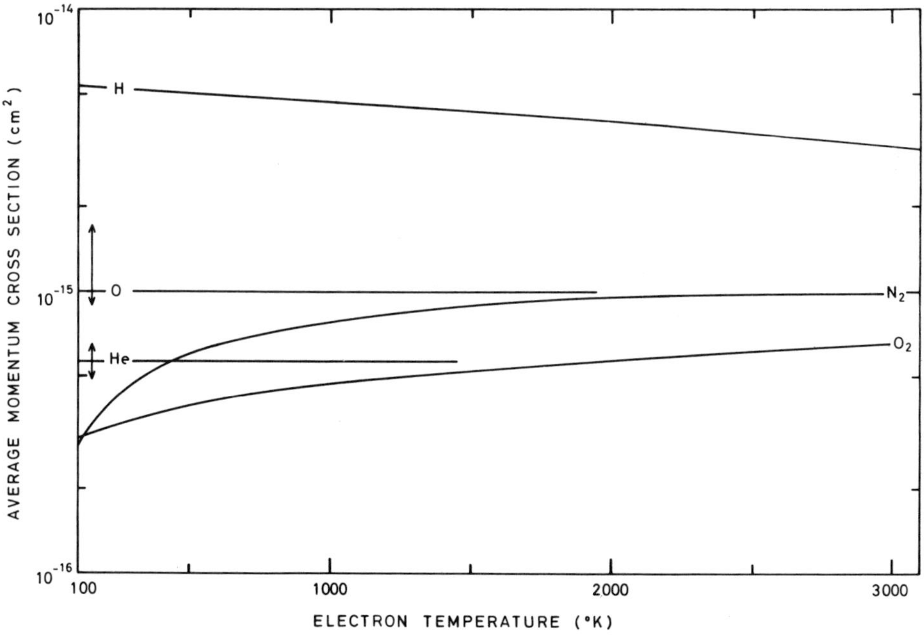

Fig. 9.1 Average momentum transfer cross section for electrons in the atmospheric gases.

TABLE 9.2

AVERAGE MOMENTUM TRANSFER CROSS SECTIONS[a] $Q_D(T_e)$

Element	Cross section (cm^2)
N_2	$(2.82 - 3.41 \times 10^{-4} T_e) T_e^{1/2} \times 10^{-17}$
O_2	$2.2 \times 10^{-16}(1 + 3.6 \times 10^{-2} T_e^{1/2})$
O	1×10^{-15}
H	$(54.7 - 7.45 \times 10^{-3} T_e) \times 10^{-16}$
He	5.6×10^{-16}

[a] T_e is measured in degrees Kelvin.

give a value near 10^{-15} cm^2. For H theoretical results have been used [3] and the error may easily be larger than $\pm 25\%$. The cross sections for He are well known with a total uncertainty less than 10% over the energy range 0–5 eV. The values of the different cross sections as a function of electron temperatures are shown in Fig. 9.1 while the momentum transfer collision frequencies are given in Fig. 9.2.

The collision frequency of electrons with neutral gases in the D region is of practical importance. At low temperatures the velocity-dependent

TABLE 9.3

ELECTRON MOMENTUM TRANSFER COLLISION FREQUENCIES ν_{en}

Element	Collision frequency (sec^{-1})
N_2	$2.33 \times 10^{-11} n(N_2)[1 - 1.2 \times 10^{-4} T_e] T_e$
O_2	$1.8 \times 10^{-10} n(O_2)[1 + 3.6 \times 10^{-2} T_e^{1/2}] T_e^{1/2}$
O	$8.2 \times 10^{-10} n(O) T_e^{1/2}$
H	$4.5 \times 10^{-9} n(H)[1 - 1.35 \times 10^{-4} T_e] T_e^{1/2}$
He	$4.6 \times 10^{-10} n(He) T_e^{1/2}$

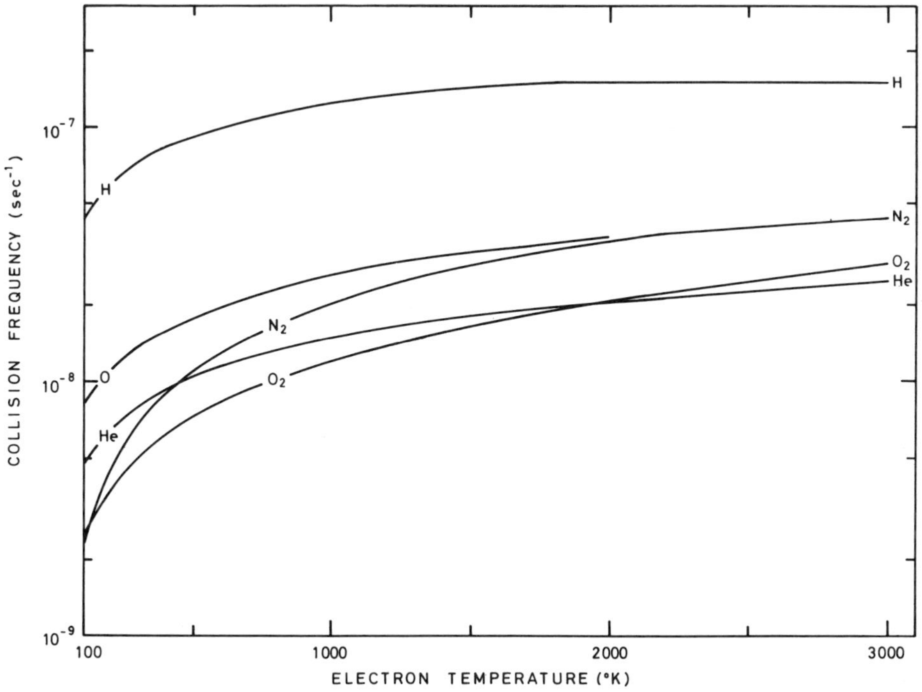

Fig. 9.2 Average momentum transfer collision frequencies for electrons in the atmospheric gases (for concentration $n = 1$).

cross sections for N_2 and O_2 can be written as [3]

$$\sigma_D(e, N_2) = 18.3 \times 10^{-16} \varepsilon^{1/2} \quad \text{cm}^2 \tag{9.51}$$

$$\sigma_D(e, O_2) = (2.2 + 5.1\varepsilon^{1/2}) \times 10^{-16} \quad \text{cm}^2 \tag{9.52}$$

with the electron energy ε measured in electron volts.

For an accuracy to within $\pm 15\%$ the contribution of O_2 to the total atmospheric collision frequency can be neglected and the average momentum transfer cross section taken as

$$\bar{\nu}_{e,\,air} = 1.8 \times 10^8 p \quad \text{sec}^{-1} \tag{9.53}$$

where p is the air pressure measured in Torr and thermal equilibrium between electrons and neutral particles is assumed.

Adopting the pressure distribution of Table 3.1 the collision frequency $\bar{\nu}_{e,\,air}$ is given between 20 and 100 km in *Table 9.4*.

TABLE 9.4

MOMENTUM TRANSFER COLLISION FREQUENCY FOR ELECTRONS IN THE HOMOSPHERE

z (km)	T (°K)	Pressure (Torr)	Collision frequency (sec^{-1})	z (km)	T (°K)	Pressure (Torr)	Collision frequency (sec^{-1})
20	219	3.8×10^{1}	7.0×10^{9}	47	274	9.5×10^{-1}	1.7×10^{8}
21	221	3.3	5.9	48	274	8.4	1.5
22	222	2.8	5.1	49	274	7.4	1.3
23	224	2.4	4.2	50	274	6.6	1.2
24	225	2.1	3.8	51	274	5.8	1.0
25	227	1.8	3.2	52	274	5.1	9.3×10^{7}
26	229	1.6	2.8	53	274	4.5	8.0
27	230	1.3	2.3	54	274	4.0	7.2
28	232	1.2	2.1	55	274	3.6	6.4
29	234	1.0	1.8	56	269	3.1	5.7
30	235	8.6×10^{0}	1.5	57	265	2.8	5.1
31	238	7.5	1.4	58	261	2.4	4.4
32	242	6.5	1.2	59	257	2.1	3.8
33	245	5.7	1.0	60	253	1.9	3.4
34	248	4.9	8.9×10^{8}	61	249	1.6	3.0
35	252	4.3	7.6	62	244	1.4	2.5
36	255	3.8	6.8	63	240	1.2	2.3
37	258	3.3	5.9	64	236	1.1	2.1
38	262	2.9	5.3	65	232	9.4×10^{-2}	1.7
39	265	2.6	4.7	66	228	8.1	1.5
40	268	2.2	4.0	67	224	7.0	1.3
41	269	2.0	3.6	68	220	6.0	1.1
42	271	1.8	3.2	69	215	5.2	9.3
43	272	1.6	2.8	70	211	4.4	7.8
44	273	1.4	2.5	71	210	3.8	6.8
45	274	1.2	2.1	72	208	3.2	5.7
46	274	1.1	1.9	73	207	2.7	4.9

TABLE 9.4 (*Continued*)

z (km)	*T* (°K)	Pressure (Torr)	Collision frequency (sec^{-1})	*z* (km)	*T* (°K)	Pressure (Torr)	Collision frequency (sec^{-1})
74	206	2.3×10^{-2}	4.2×10^{6}	88	194	2.2×10^{-3}	3.8×10^{5}
75	204	2.0	3.6		170	1.6	3.0
	194	1.9	3.4	89	195	1.8	3.4
76	203	1.7	3.0		173	1.3	2.3
	191	1.6	2.8	90	197	1.5	2.8
77	201	1.4	2.5		177	1.1	2.0
	187	1.3	2.3	91	198	1.3×10^{-3}	2.3
78	200	1.2	2.1		180	9.1×10^{-4}	1.6
	184	1.1	2.0	92	199	1.1×10^{-3}	2.0
79	199	1.0	1.8		183	7.6×10^{-4}	1.4
	181	9.5×10^{-3}	1.7	93	200	9.4×10^{-4}	1.7
80	197	8.6	1.5		186	6.3	1.1
	177	7.9	1.4	94	202	7.9	1.4
81	196	7.3	1.3		190	5.3	9.5×10^{4}
	174	6.5	1.2	95	203	6.7	1.2×10^{5}
82	194	6.1	1.1		193	4.5	8.0×10^{4}
	170	5.4	9.7×10^{5}	96	204	5.7	1.0×10^{5}
83	193	5.2	9.3		196	3.8	6.8×10^{4}
	167	4.4	7.8	97	205	4.8	8.7
84	192	4.3	7.8		200	3.2	5.7
	164	3.6	6.6	98	207	4.1	7.4
85	190	3.6	6.6		203	2.7	4.9
	160	2.9	5.3	99	208	3.5	6.4
86	192	3.1	5.5		206	2.3	4.0
	164	2.4	4.2	100	209	3.0	5.5
87	193	2.6	4.7		209	2.0	3.6
	167	2.0	3.6				

9.5 Inelastic Electron Collisions with Neutral Particles

The cross sections and collision frequencies considered in the last section refer exclusively to elastic processes where internal energy states of the target atom, molecule, or ion are not excited. For many aspects of aeronomical studies, however, the collisions of energetic electrons with neutral particles which lead to excited atomic and molecular states must be considered [10–13]. Important consequences of inelastic collisions between electrons and the atmospheric gases include the auroral emissions, the photoelectron dayglow, and the cooling of the ionospheric electron gas.

The electron impact excitation of rotational states in molecules possessing a permanent dipole moment can provide an important loss of electron kinetic energy. In the atmosphere, however, such molecules are minor constituents and can be important only with respect to their own characteristic spectrum of emitted radiation. It has been found, however, that the rotational excitation of N_2 and O_2 can take place through an electric quadrupole interaction [10].

The rate at which an electron of energy ε looses kinetic energy in the excitation of homonuclear molecular rotational states is

$$d\varepsilon/dt = \sum_J [(E_{J+2} - E_J)\nu_{J+2;J} - (E_J - E_{J-2})\nu_{J;J+2}] \tag{9.54}$$

where E is the molecular energy associated with the subscripted rotational energy level and the summation is taken over all accessible energy states. The collision frequencies are given by

$$\nu_{J+2;J} = n_{J+2} v_e q_{J+2;J} \tag{9.55}$$

$$\nu_{J;J+2} = n_J v_e q_{J;J+2} \tag{9.56}$$

where n is the population of the indicated level, v_e is the electron speed, and q is the speed dependent rotational cross section for the indicated transition. Cross sections for use in Eqs. (9.55) and (9.56) have been computed but are subject to substantial error for electron energies in the range $0.1 < \varepsilon < 1.0$ eV [12–16].

In a collision between an electron and a homonuclear molecule such as O_2 or N_2, excitation of vibrational states is possible. The general behavior of the vibrational cross section for electron impact has been investigated experimentally [17–19] and theoretically [20, 21]. The total vibration excitation cross section for N_2 is shown in *Fig. 9.3*. An important characteristic

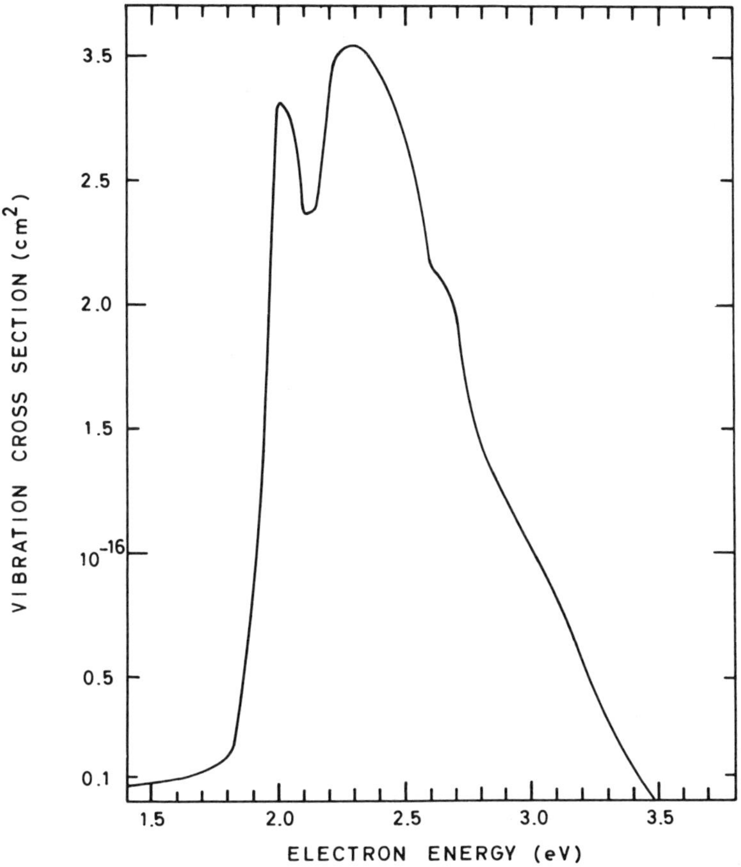

Fig. 9.3 Total N_2 vibration cross section for electron collisions $\sum_v = 1\text{–}8$. From Schulz [19].

of the N_2 vibrational cross section is its rapid decrease in magnitude for electron energies below 1.5 eV (cross section $< 10^{-17}$ cm^2). The peak value of 3×10^{-16} cm^2 at 2.3 eV assures a very rapid rate of N_2 excitation and consequent electron energy loss [13]. Thus, in the low altitude regions of the ionosphere where electron–N_2 collisions are frequent, an energetic electron which has undergone a number of inelastic collisions by excitation of atomic and molecular electronic energy states will continue to lose kinetic energy rapidly due to vibrational excitation of N_2 until its energy is less than about 1.5 eV. Below this energy the N_2 vibrational cross section is small and other energy transfer processes, such as rotational excitation, play a part.

The vibration cross sections for O_2 have been measured in the laboratory [22, 23]. In addition, analysis of measured electron transport coefficients by means of the Boltzmann equation also gives some indication of the cross sections [24]. Unlike those of N_2, the vibration cross sections of O_2 show a sharp resonance behavior starting from the excitation threshold of 0.19 eV as shown in *Fig. 9.4*. While the O_2 peak vibrational cross section is more than an order of magnitude smaller than the N_2 value, the presence of the resonances in the low electron energy ranges means that O_2 can be a non-negligible source of electron energy loss [13].

The excitation of atomic and molecular electronic energy states through electron impact leads to many important atmospheric emissions, both visual, such as in the aurorae, and subvisual, such as in the polar-glow aurorae and the dayglow. The cross sections for excitation are not well established for all possible transitions in the atmospheric gases. Those leading to intense radiative emission have been investigated most thoroughly [25–27].

The cross sections corresponding to excitation of electronic states from ground state N_2 and O_2 have been measured [28, 29] and calculated [30]. Analytic expressions for these and many other transitions which have recently been obtained are useful for detailed calculations of electron energy distributions and for determining the intensities of collision induced radiation [31–35].

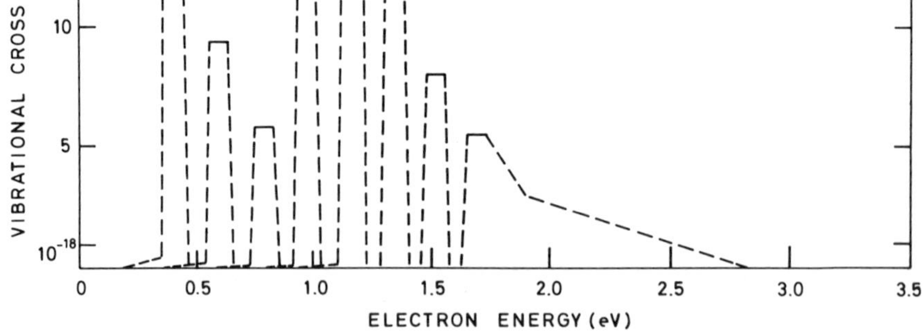

Fig. 9.4 Total O_2 vibration cross section for electron impact. From Hake and Phelps [24].

TABLE 9.5

ELECTRON IMPACT EXCITATION CROSS SECTION PARAMETERS[a]

Gas	State	W	$c_0 f_0$	Ω	ν	γ
N_2	A $^3\Sigma_u^+$	6.14	0.226	3.0	1.0	1.0
	B $^3\Pi_g$	7.30	0.178	3.0	1.0	3.0
	C $^3\Pi_u$	11.03	0.28	3.0	1.0	3.0
	a $^1\Pi_g$	9.10	0.136	1.0	1.0	1.0
	b $^1\Pi_u$	12.85	0.67	0.75	3	1.0
	b′ $^1\Sigma_u^+$	14.0	0.33	0.75	3	1.0
	Σ Rydberg	13.75	2.66	0.75	3	1.0
O_2	a $^1\Delta_g$	0.98	0.0005	3.0	1.0	3.0
	b $^1\Sigma_g$	1.64	0.0005	3.0	1.0	3.0
	A $^3\Sigma_u^+$	4.5	0.021	0.9	1.0	3.0
	B $^3\Sigma_u^-$	8.4	0.23	0.75	2.0	1.0
	9.9 eV allowed	9.9	0.08	0.75	3.0	1.0
	Σ Rydberg	13.5	2.77	0.75	3.0	1.0
O	^{1}D	1.96	0.01	1.0	2.0	1.0
	^{1}S	4.17	0.0042	1.0	1.0	0.5
	^{3}S	9.53	0.0465	0.75	3.0	1.0
	^{5}S	9.15	0.023	2.0	1.0	1.0
	$\Sigma\,(\Delta l = 1, \Delta s = 0)$	14.2	0.367	0.75	3.0	1.0
	$\Sigma\,(\Delta s = 1)$	14.7	0.694	2.0	1.0	1.0
	$\Sigma\,(\Delta l = 0, \Delta s = 0)$	13.5	0.043	0.75	1.0	2.0

[a] From Stolarski and Green [25].

For the semianalytical approach, electron impact cross sections for the atmospheric gases are given by the formula [31]

$$\sigma(\varepsilon) = (q_0 c_0 f_0 / W^2)[1 - (W/\varepsilon)^\gamma]^\nu (W/\varepsilon)^\Omega \quad \text{cm}^2 \tag{9.57}$$

which has been found to accurately represent a large number of experimental data. Here, $\sigma(\varepsilon)$ is the energy dependent cross section, $q_0 = 6.51 \times 10^{-14}$ eV^2 cm^2, the product $c_0 f_0$ is a constant, W is the excitation threshold energy measured in eV, and γ, ν, and Ω are tabulated constants. Values of these parameters are given in *Table 9.5* while the sum of all inelastic processes for N_2, O_2, and O are plotted as a function of energy in *Fig. 9.5*.

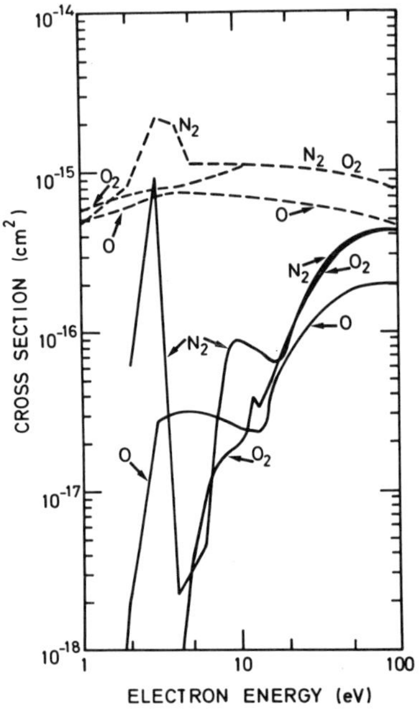

Fig. 9.5 Elastic and total excitation cross sections for electron impact on N_2, O_2, and O. Solid line, total inelastic cross section; dashed line, total elastic cross section.

Excitation of atomic oxygen by electron impact is not only an important means of electron energy loss, but also provides an opportunity to study atmospheric behavior through photon emission (see *Table 9.6*). An energy level diagram for O(I) is shown in *Fig. 9.6*.

Fine structure transitions of ground state oxygen O(I)(^3P_J) alter the energy of colliding electrons through inelastic collisions. Although the

TABLE 9.6

FORBIDDEN TRANSITIONS OF O(I)

Transition	Wavelength	Probability (sec^{-1})	Lifetime (sec)
$^3P_2$–$^1S_0$	2958.4 Å	3.7×10^{-4}	2.70×10^3
$^3P_1$–$^1S_0$	2972.3 Å	6.7×10^{-2}	14.9
$^1D_2$–$^1S_0$	5577.35 Å	1.34	0.75
$^3P_2$–$^1D_2$	6300.2 Å	5.10×10^{-3}	1.96×10^2
$^3P_1$–$^1D_1$	6363.9 Å	1.64×10^{-3}	6.10×10^2
$^3P_1$–$^3P_2$	63 μm	8.95×10^{-5}	1.1×10^4
$^3P_0$–$^3P_1$	147 μm	1.70×10^{-5}	5.9×10^4

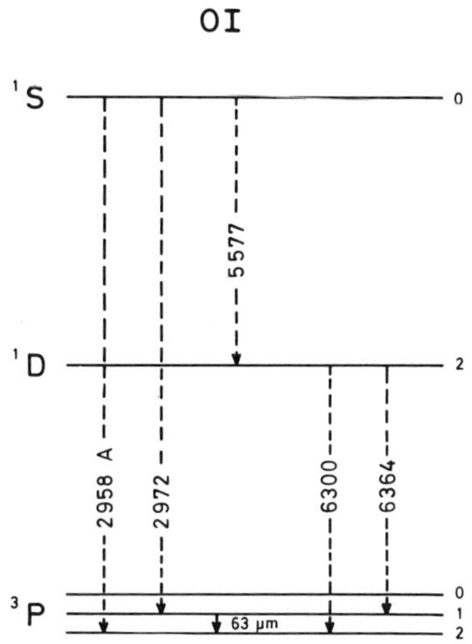

Fig. 9.6 Atomic oxygen energy levels.

energy changes for these inelastic collisions are small, the cross sections for electron interaction are large [36]. Values of the cross sections and electron cooling rates in the ionosphere [37] and interstellar medium [38] have been computed.

Atomic oxygen impact excitation cross sections $^3P \rightarrow {}^1S$ and $^3P \rightarrow {}^1D$ have been computed [39] and are shown in *Fig. 9.7* for electron energies

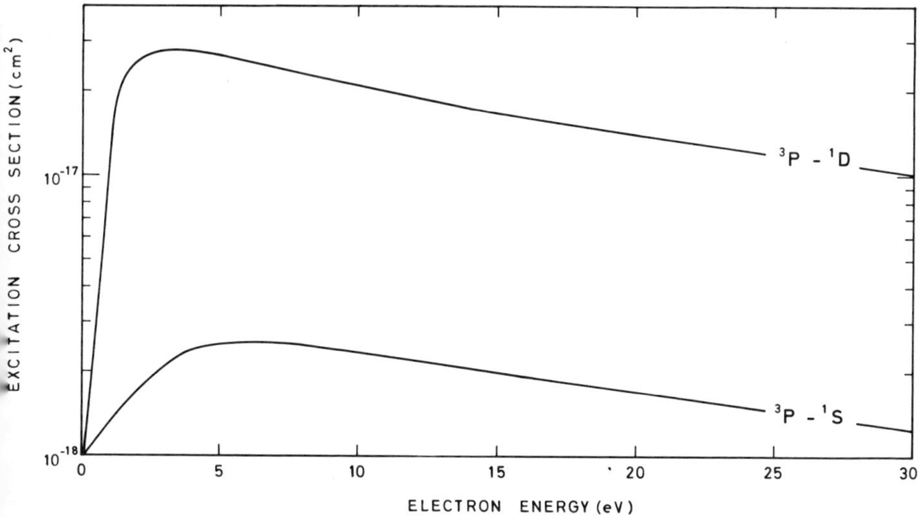

Fig. 9.7 Atomic oxygen excitation cross sections for electron impact.

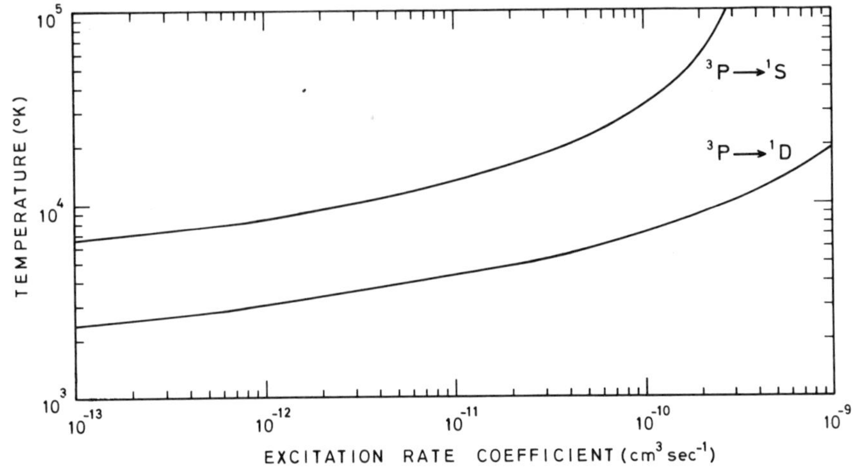

Fig. 9.8 Atomic oxygen excitation coefficients for electron impact.

ranging between 0 and 30 eV. Using these cross sections, it is possible to compute the rate at which the $^3P \rightarrow {}^1S$ and $^3P \rightarrow {}^1D$ transitions will be excited by thermal electrons having a Maxwellian distribution [40]. The resulting excitation coefficients, shown in *Fig. 9.8*, indicate the strong dependence upon electron temperature.

Similar data are available for atomic nitrogen. *Table 9.7* summarizes the important transitions, while *Fig. 9.9* shows the relevant energy level diagram. The electron impact cross sections for the transitions $N(^4S) \rightarrow N(^2D)$ and $N(^4S) \rightarrow N(^2P)$ are shown in *Fig. 9.10* while *Fig. 9.11* gives the temperature dependent excitation coefficients [39].

TABLE 9.7

FORBIDDEN TRANSITIONS OF N(I)

Transition	Wavelength	Probability (sec^{-1})	Lifetime (sec)
$^4S^0_{3/2}$–$^2D^0_{3/2}$	5200.7 Å	7.0×10^{-6}	1.42×10^5
$^4S^0_{3/2}$–$^2D^0_{3/2}$	5198.5 Å	1.6×10^{-5}	6.25×10^4
$^4S_{3/2}$–$2P^0_{1/2,\,3/2}$	3466.4 Å	4.96×10^{-3}	2.0×10^2
$^2D_{5/2}$–$^2P^0_{3/2,\,1/2}$	10395.4 Å	4.96×10^{-2}	20.2
$^2D_{3/2}$–$2P^0_{3/2,\,1/2}$	10404.1 Å	3.21×10^{-2}	31.2

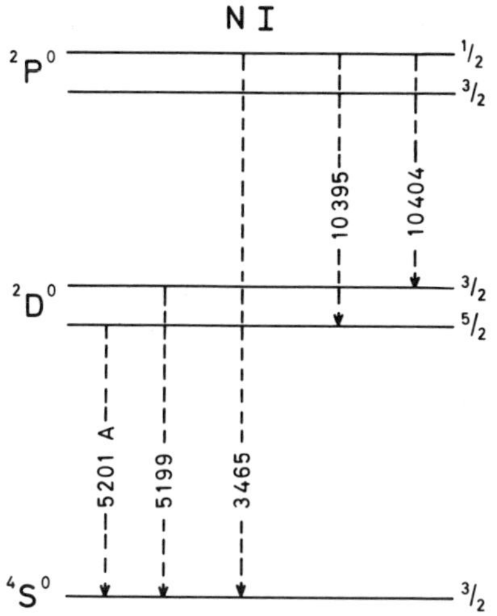

Fig. 9.9 Atomic nitrogen energy levels.

With regard to the electronic excitation of N_2 and O_2, the impact cross sections of Table 9.5 can be used to compute appropriate values.

The problem of the ionization of gases by electron impact has received considerable emphasis and fairly precise cross section data are available

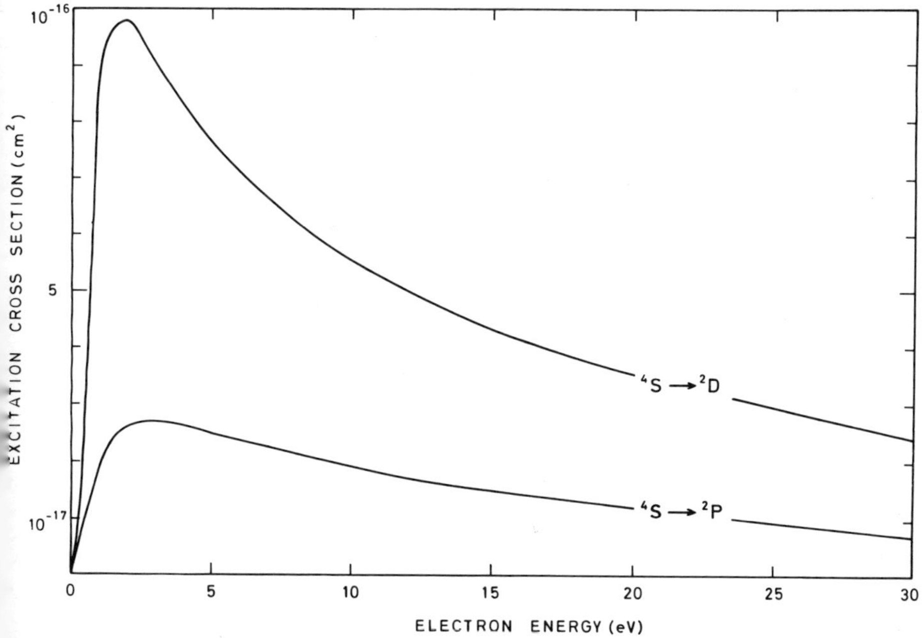

Fig. 9.10 Atomic nitrogen cross sections for electron impact.

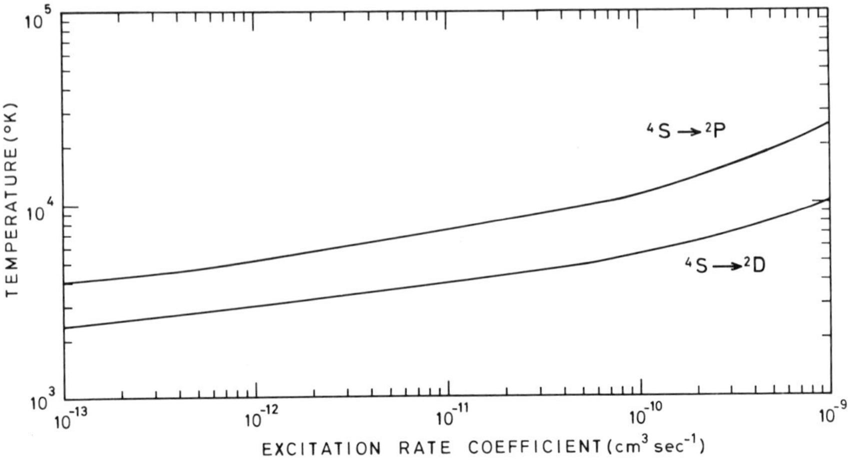

Fig. 9.11 Atomic nitrogen electron excitation coefficients.

giving the total ionization cross sections to all ion states as a function of electron energy. Adopted cross sections taken from laboratory data [25, 27] are shown in *Figs. 9.12–9.18*.

Analytical approximations to the ionization cross sections have been obtained [26, 31, 32]. Such expressions are able to predict not only the

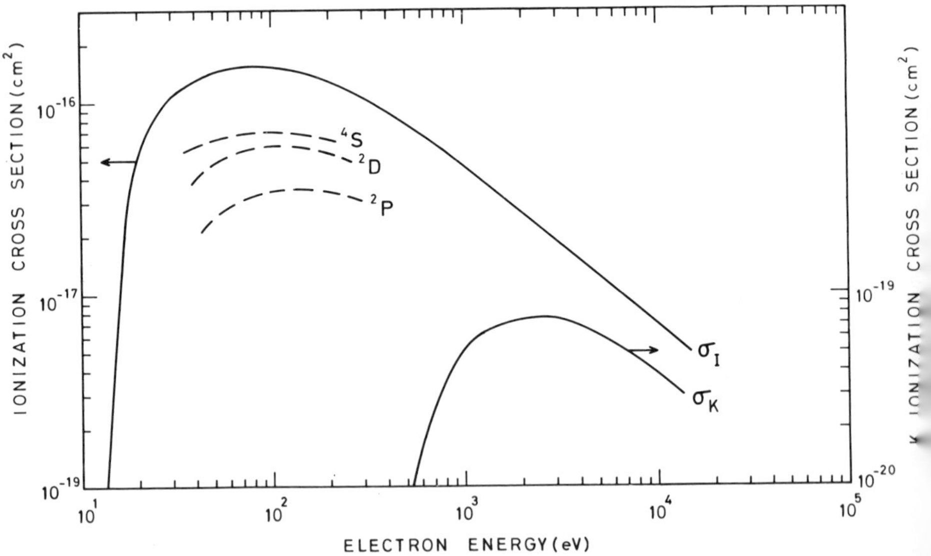

Fig. 9.12 Electron impact cross sections for ionization of atomic oxygen.

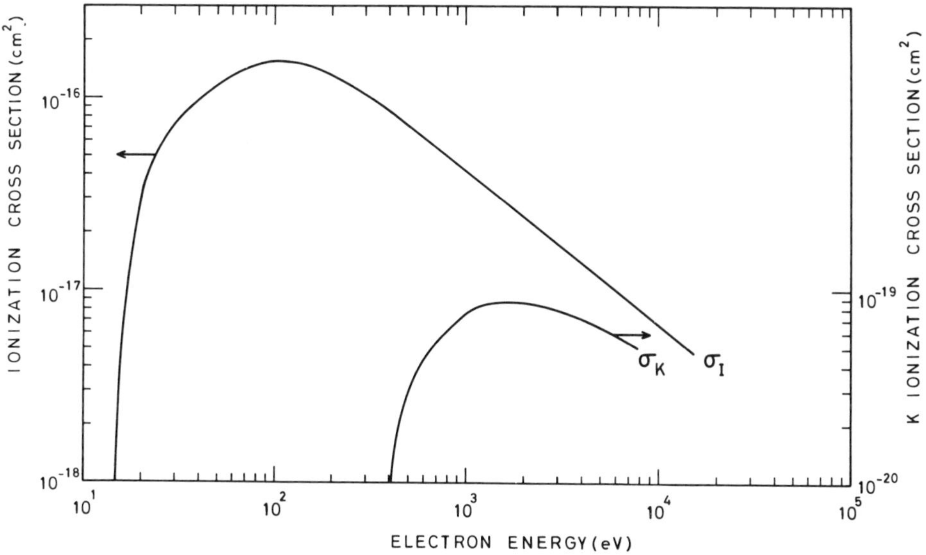

Fig. 9.13 Electron impact cross sections for ionization of atomic nitrogen [28].

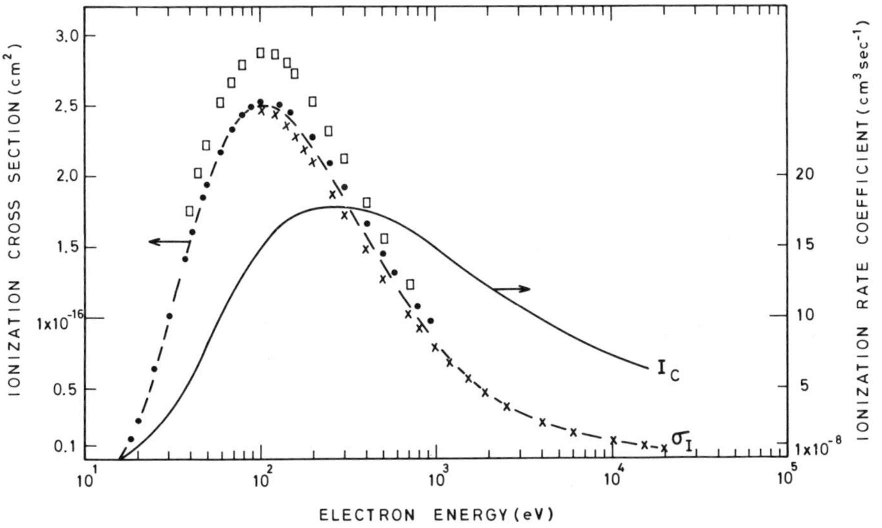

Fig. 9.14 Electron impact cross sections for ionization of molecular nitrogen. The experimental data are from Tate and Smith [55], squares; Rapp and Englander-Golden [56], dots; and Schram *et al.* [57, 58], crosses. The dashed curve is the one adopted.

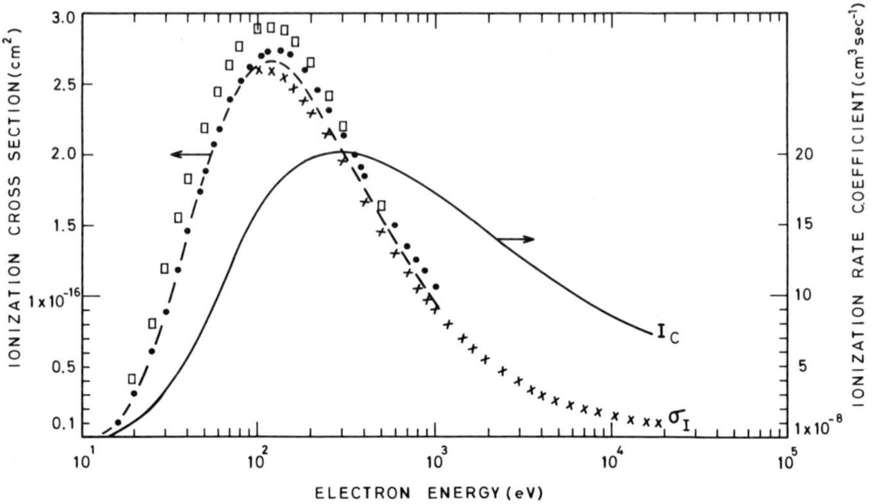

Fig. 9.15 Electron impact cross sections for ionization of molecular oxygen [28]. See Fig. 9.14 for explanation of figure key.

total impact ionization cross section, but also the cross sections for ions in various excited states. The energy-dependent cross sections, $\sigma(\varepsilon)$ are given by

$$\sigma(\varepsilon) = \int_I^{(\varepsilon+I)/2} S(\varepsilon, W)\, dW \tag{9.58}$$

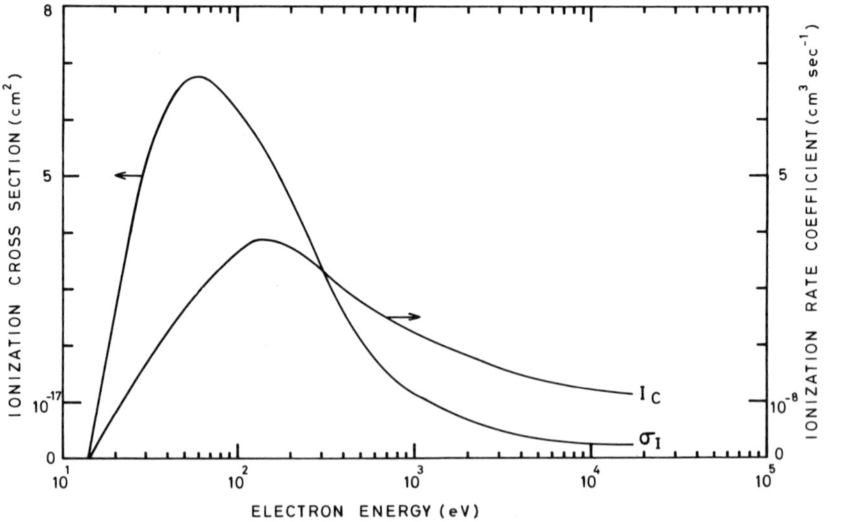

Fig. 9.16 Electron impact cross sections for ionization of atomic hydrogen.

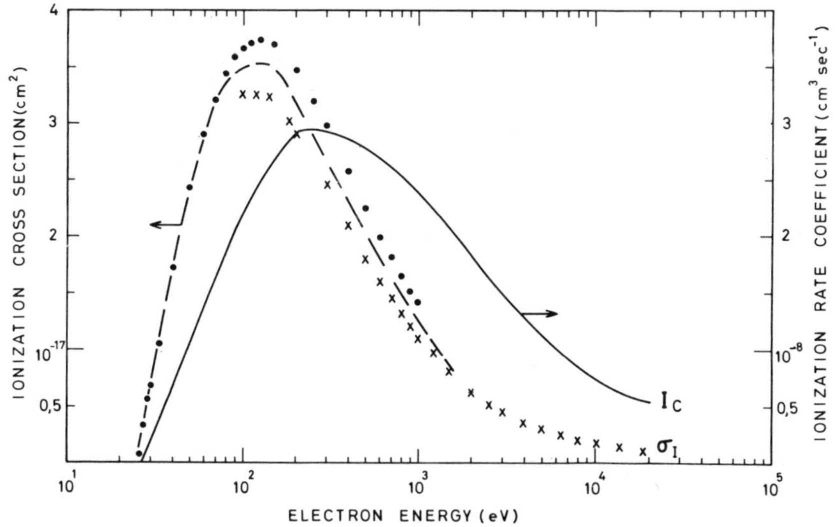

Fig. 9.17 Electron impact cross sections for ionization of helium [28]. The experimental data are from Rapp and Englander-Golden [56], dots; from Schramm *et al.* [57, 58], crosses. The adopted curve is shown as a broken line.

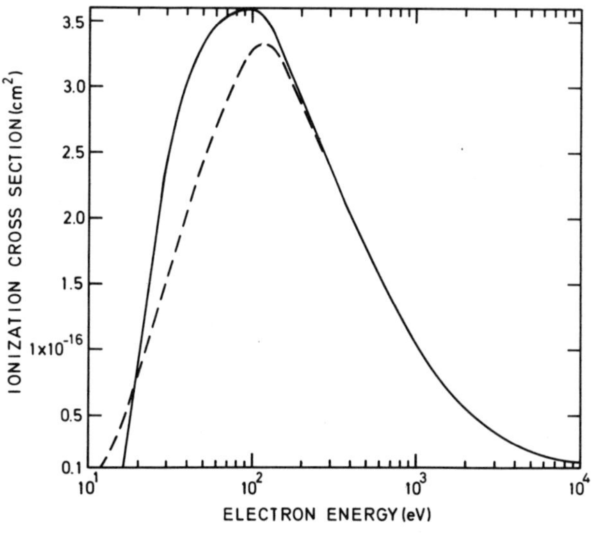

Fig. 9.18 Electron impact cross sections for ionization of argon (solid curve) and nitric oxide (dashed curve) [28].

where

$$S(\varepsilon, W) = (q_0 c_0 f_0 / W^2)(I/W)^P [1 - (W/\varepsilon)^\gamma]^\nu (W/\varepsilon)^\Omega. \tag{9.59}$$

Here I is the ionization threshold for the particular ion state, ε is the electron energy, and P, γ, ν, and Ω are parameters obtained from the best fit of Eq. (9.59) to experimental and theoretical results. *Table 9.8* lists values of the parameters for impact ionization of N_2, O_2, and O with numerous final ion states.

When electron impact occurs on molecular ions there is an opportunity for both dissociation and ionization. Energetic electrons moving through the atmosphere thus enhance the photodissociation effects of the solar ultraviolet radiation. Studies of such processes are especially important for auroras where intense streams of energetic electrons can dissociate appreciable amounts of N_2 and O_2 in regions of darkness [41]. In addition, photoelectrons released through the absorption of x rays can lead to important D region effects through creation of atomic ions and neutrals.

TABLE 9.8

ELECTRON IMPACT IONIZATION CROSS SECTION PARAMETERS[a]

	Final ion state		I	$c_0 f_0$	Ω	ν	γ	P
N_2	N_2^+	$X\,^2\Sigma_g^+$	15.58	0.370	0.80	3.0	1.0	1.2
		$A\,^2\Pi_u$	16.73	0.160	0.83	1.0	1.0	1.2
		$B\,^2\Sigma_u^+$	18.75	0.073	0.83	2.0	1.0	1.2
		$D\,^2\Pi_g$	22.0	0.056	0.83	2.0	1.0	1.2
		$C\,^2\Sigma_u^+$	23.6	0.060	0.83	2.0	1.0	1.2
		Dissociative	25.0	0.380	0.96	2.0	1.0	1.2
O_2	O_2^+	$X\,^2\Pi_g$	12.1	0.058	0.80	2.0	1.0	1.1
		$a\,^4\Pi_u$	16.1	0.150	0.80	2.0	1.0	1.1
		$A\,^2\Pi_u$	16.9	0.150	0.80	2.0	1.0	1.1
		$b\,^4\Sigma_g^-$	18.2	0.130	0.80	2.0	1.0	1.1
		B	23.0	0.064	0.80	2.0	1.0	1.1
		Diss. ($O^+\,^4S$)	18.0	0.400	0.93	3.0	1.0	1.1
		Diss. ($O^+\,^2D$)	22.0	0.250	0.93	3.0	1.0	1.1
O	O^+	^4S	13.6	0.290	0.85	1.0	0.3	1.2
		^2D	16.9	0.360	0.85	1.0	0.3	1.2
		^2P	18.5	0.190	0.85	1.0	0.3	1.2

[a] From Stolarski and Green [25].

Electron impact ionization of N_2 leads not only to N_2^+, but also to N_2^{2+} and the dissociation ionization products $N + N^+$ according to the reactions

$$e + N_2 \rightarrow N_2^+ + 2e \tag{9.60}$$

$$e + N_2 \rightarrow N_2^{2+} + 3e \tag{9.61}$$

$$e + N_2 \rightarrow N + N^+ + 2e. \tag{9.62}$$

The most recent laboratory data [28, 42–44] indicates that the values of these cross sections, shown in Fig. 9.14, are known to an accuracy better than 10%. The maximum cross section for ionization is about 2.5×10^{-16} cm^2 near 100 eV and a mean value of $(2 \pm 0.5) \times 10^{-16}$ cm^2 can be adopted over the range 40–400 eV. The ionization coefficient is $(1.4 \pm 0.4) \times 10^{-7}$ cm^3 sec^{-1} for electrons of energy 50–5000 eV.

Dissociative ionization [Eq. (9.62)] begins at an energy of 24.3 eV and accounts for about $20 \pm 5\%$ of the total ionization [44, 45] since the peak cross section for this process is about 6×10^{-17} cm^2 (see *Figs. 9.19* and *9.20*).

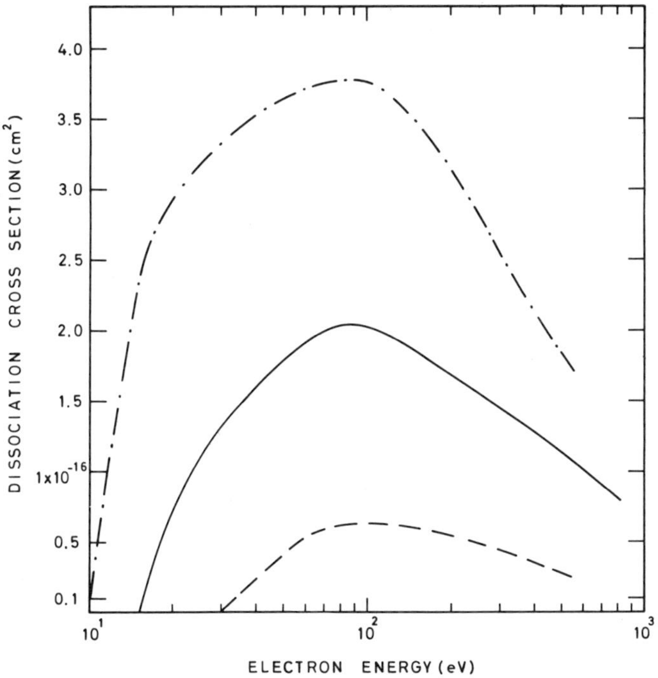

Fig. 9.19 Electron impact cross sections leading to N_2^+, $N + N^+$, and $N + N$. (–·–) Van Zyl and Dunn [43], $N_2^+ \rightarrow N + N^+$; (——) Winters [44], $N_2 \rightarrow N + N$; (- - -) Rapp *et al.* [42].

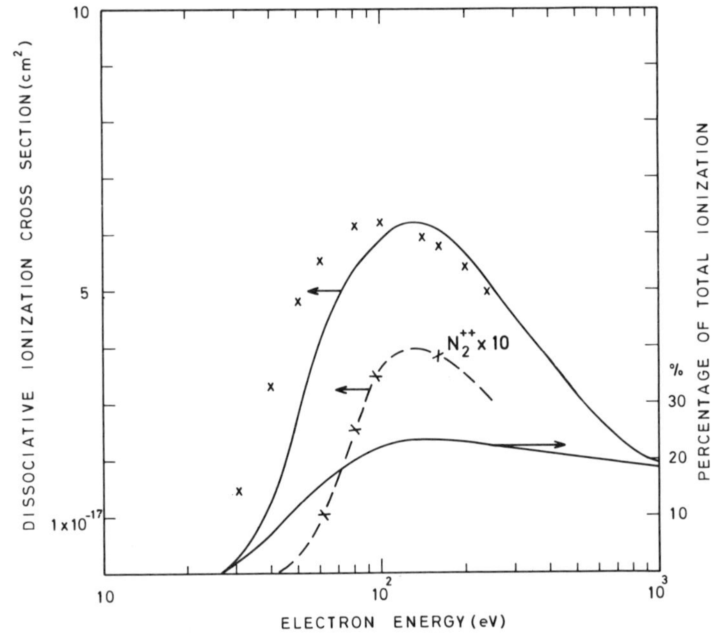

Fig. 9.20 Electron impact cross sections leading to N^{2+} and $N + N^+$. Values given in curve from Rapp *et al.* [42] and as crosses from Daly and Powell [45].

It is clear that the effect of dissociative ionization must not be neglected when electrons of energy greater than 50 eV produce appreciable direct ionization effects in the atmosphere.

One of the most important aspects of electron impact ionization of N_2 is the production of excited states of N_2^+ leading to photon emission. In particular the N_2^+ state $B\ ^2\Sigma_u^+$ is responsible for the first Negative System from the transition $N_2^+(B\ ^2\Sigma_u^+) \rightarrow N_2^+(X\ ^2\Sigma_g^+)$.

Electron excitation of $N_2^+(B\ ^2\Sigma_u^+)$ from the state $N_2^+(X\ ^2\Sigma_g^+)$ occurs very readily with relatively low energy particles ($\sigma \sim 1.8 \times 10^{-14}\ \text{cm}^2$ at 4 eV) [44]. As discussed in Section 6.4.7, the 0–0 transition of $(B\ ^2\Sigma_u^+ \rightarrow X\ ^2\Sigma_g^+)$ corresponds to photon emission at 3914 Å. The cross sections for electron impact excitation of this radiation from $N_2^+(X\ ^2\Sigma_g^+)$ as a function of energy are:

ε (eV):	4	6	9	14	16	22	60
σ ($\times 10^{-16}$ cm²):	80.	50.	20.	10.	8.0	6.0	0.13

The simultaneous ionization and excitation of ground state N_2 to $N_2^+(B\ ^2\Sigma_u^+)$ with a subsequent emission of photons in various vibrational bands (0–0: 3914 Å; 0–1: 4278 Å; 0–2: 4709 Å; 0–3: 5228 Å) is important since a direct relation exists between the photon emission rate and the electron energy flux. Numerous measurements of the cross sections have been made [46–54], the adopted results of which are shown in *Fig. 9.21.* Examination of the existing data indicates that the ratio of the total ionization to excitation of 3914 Å radiation is 17 ± 2, taking into account the variation of the ionization cross section. In addition, the ratio of ionization leading to N_2^+ to that for 3914 Å is 14. Thus, the 3914 Å production efficiency by electron impact on N_2 is 0.06 and about 10% of N_2 ionization leads to $N_2^+(B\ ^2\Sigma_u^+)$.

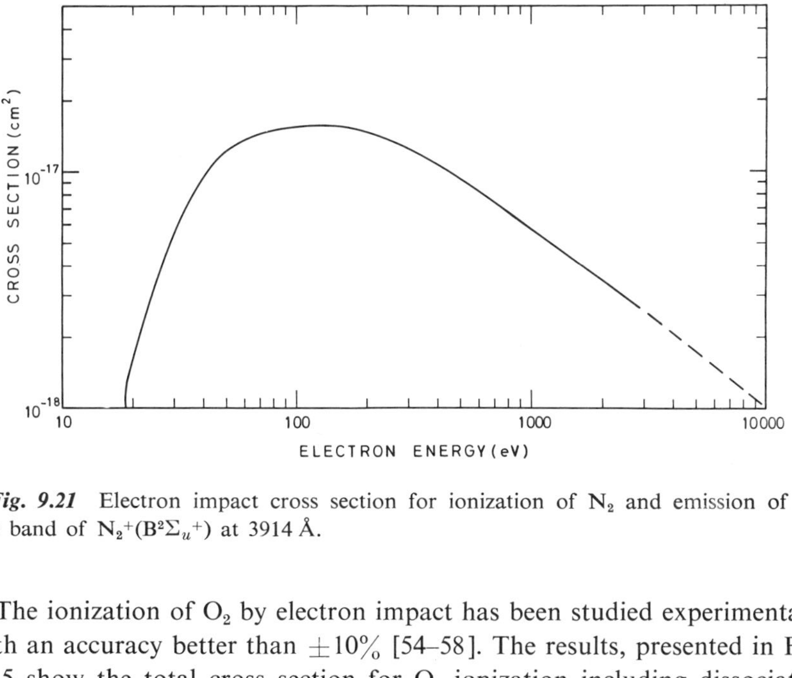

Fig. 9.21 Electron impact cross section for ionization of N_2 and emission of the 0–0 band of $N_2^+(B^2\Sigma_u^+)$ at 3914 Å.

The ionization of O_2 by electron impact has been studied experimentally with an accuracy better than ±10% [54–58]. The results, presented in Fig. 9.15 show the total cross section for O_2 ionization including dissociative ionization leading to O^+. The cross section for dissociative ionization, as shown in *Fig. 9.22,* reaches a peak of 10^{-16} cm² at an electron energy of 150 eV and accounts for about 35% of the total ionization.

For electrons with kinetic energy greater than a few keV, the rate of energy loss as a result of inelastic electron collision with neutral particles leading to excitation and ionization can be given by the Bethe equation [59–61]

$$d\varepsilon/dx = \{[1.3\times 10^{-12} n(n)]/\varepsilon\} \ln[1.17\varepsilon/I] \quad \text{eV cm} \tag{9.63}$$

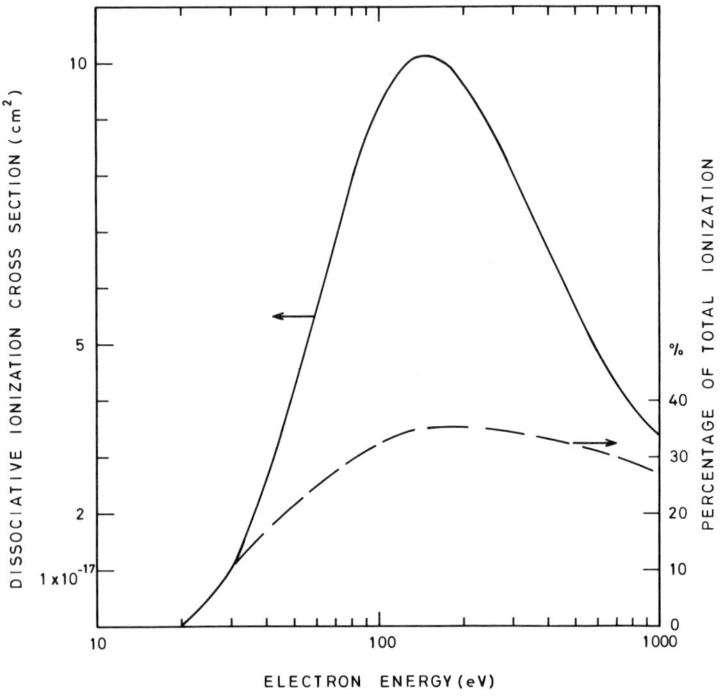

Fig. 9.22 Electron impact cross section for dissociative ionization of O_2. The experimental data are from Rapp *et al.* [42].

where ε is the incident electron energy in eV and I corresponds to the average excitation energy of the neutral gas, measured in eV. For air the value $I = 85$ eV can be used [62–64].

To relate (9.63) to the ionization rate, it is necessary to know the relative rates at which electron energy is expended in ionization and excitation. Thus, the ionization rate, q, due to a flux, ϕ, of energetic electrons is given by

$$q = (d\varepsilon/dx)(\phi/\Delta\varepsilon) \tag{9.64}$$

where $\Delta\varepsilon$ is the average energy needed to produce an electron–ion pair. For energies above 500 eV, the values listed in *Table 9.9* are appropriate; below 500 eV, $\Delta\varepsilon$ increases rapidly and particular assumptions are needed to produce the ionization rate from (6.93) [64].

For electron energies between 5 and 54 keV experimental data relating to the penetration and ionization by electrons in neutral gases are available. In terms of the electron range, R (in gm cm^{-2}) electrons are able to travel

TABLE 9.9

MEAN IONIZATION ENERGIES

Species	$\Delta\varepsilon$ (eV)
O_2	31
N_2	35
Air	34

downwards into the atmosphere to the depth given by the relation [65]

$$R = 4.57 \times 10^{-6} \varepsilon_0^{1.75} \quad \text{gm cm}^{-2} \tag{9.65}$$

where

$$R = \int \varrho(z)\, dz \tag{9.66}$$

is the range, and ε_0 is the initial electron energy in keV.

The local ionization rate, q, at a range r in the atmosphere ($r < R$) resulting from an incident flux, Φ, of monoenergetic electrons is given by the relation [65]

$$q(r) = (\varepsilon_0/R\,\Delta\varepsilon)\varrho(r)f(r/R) \tag{9.67}$$

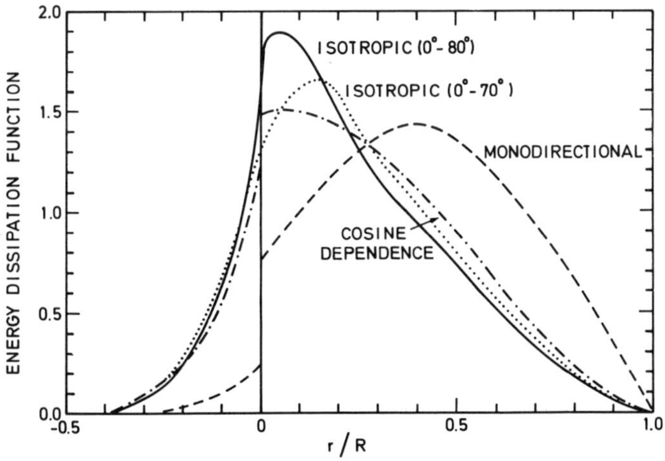

Fig. 9.23 Normalized electron energy dissipation distribution function for angular dispersions of an incident electron stream in air [65].

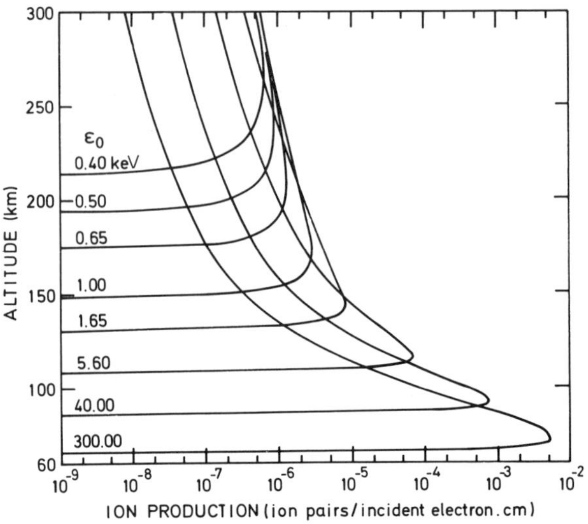

Fig. 9.24 Ionization production by isotropic stream (0°–80°) of monoenergetic electrons [65].

where $\varrho(r)$ is the mass density at the gas depth, r, defined by

$$r = \int_z^\infty \varrho(z)\, dz, \tag{9.68}$$

$\Delta\varepsilon$ is the mean energy loss per electron–ion pair, and $f(r/R)$ is a normalized energy dissipation function shown in *Fig. 9.23*.

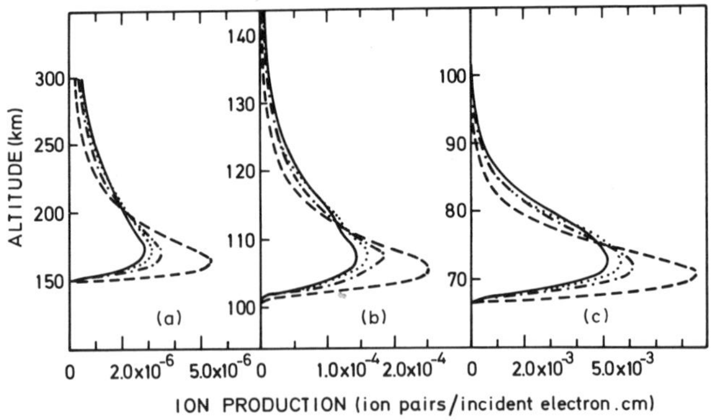

Fig. 9.25 Ionization production by monoenergetic electron streams having various angular distributions: (a) $\varepsilon_0 = 1.0$ keV; (b) $\varepsilon_0 = 10$ keV; (c) $\varepsilon_0 = 300$ keV [65]. (——) Isotropic (0°–80°); (···) Isotropic (0°–70°); (–·–) Cosine dependence; (– – –) Monodirectional.

Because electrons having different initial pitch angles travel different path lengths in reaching a given altitude in the atmosphere, an assumption regarding the pitch angle distribution of electron velocities must be made in order to evaluate $f(r/R)$. Four specific distributions have been shown in Fig. 9.23 to indicate the differences in the final dissipation rate.

The production of ionization by electrons in the earth's atmosphere having initial isotropic angular distribution (0–80°) and a variety of initial energies are illustrated in *Fig. 9.24* [65]. The effect of different initial pitch angle distributions upon the ionization production rate for several energies is shown in *Fig. 9.25* [65].

9.6 Collisions between Ions and Neutral Particles

At low temperatures the most important ion–neutral interaction arises from an induced dipole attraction which has a potential [2]

$$\phi = -\alpha e^2/2r^4. \tag{9.69}$$

Here, α is the neutral atom polarizability, e is the electric charge, and r is the radial separation. This potential corresponds to the force law acting between Maxwellian molecules and the details of the transport coefficients under conditions of thermal equilibrium are well known. In particular, it is found that the collision cross section is proportional to $T^{-1/2}$ and that the collision frequency is independent of the temperature [66, 67].

For temperatures greater than 300°K, the induced dipole force of attraction is countered by a short range quantum mechanical repulsion. For sufficiently high temperatures, the polarization contribution is negligible and the collision cross section becomes nearly constant, yielding a collision frequency which varies as $T^{1/2}$. Unlike the polarization force, which is independent of the chemical nature of a given ion and depends only upon the atomic polarizability of the neutral gas, the nature of the repulsive force is linked directly to the details of the ion and neutral particle orbital electron structure. Thus, variations in cross section are to be expected for different ions in the same neutral gas at elevated temperatures.

Because there exist virtually no data for either collision frequencies or cross section at high temperatures, the study of ion–neutral collisions must be based upon various assumptions about the interaction forces [68]. A detailed evaluation of the ion–neutral interaction potential has been made [69] in which the polarization potential terminates at a small atomic radius with an elastic sphere repulsive potential to simulate the quantum repulsion effects. This theoretical cross section was then compared with reported measurements of ionic mobilities. For equal ion and neutral temperatures below 300°K, the theoretical results appear to be accurate. Above 300°K, the detailed variation of the momentum transfer cross section depends upon the specific ion and neutral gas [68].

With there being only a few experimental data for temperatures above 300°K, it is necessary to accept the collision equations developed for the polarization force alone as representing the true interaction over the entire range of ion and neutral temperatures for several interactions. Accordingly, we take the ion-neutral velocity dependent momentum transfer cross sections for single charge ions to be [66]

$$q_D = 2.21\pi(\alpha e^2/\mu g^2)^{1/2}, \tag{9.70}$$

where α is the neutral gas atomic polarizability, μ is the ion-neutral reduced mass, e is the electron charge, and g is the ion-neutral particle relative velocity.

Using Eq. (9.35), the average momentum transfer cross section for ion–neutral collisions becomes [67]

$$\overline{Q_D} = (3\sqrt{2}/16)\pi^{3/2}(4.88\alpha e^2/\mu)^{1/2}[(kT_i/m_i) + (kT_n/m_n)]^{-1/2} \tag{9.71}$$

or numerically,

$$\overline{Q_D} = 13.3\times 10^{-14}(\alpha_0/\mu_A)^{1/2}[(T_i/A_i) + (T_n/A_n)]^{-1/2} \quad \text{cm}^2, \tag{9.72}$$

where A_i and A_n are the particle masses in atomic mass units (amu), μ_A is the reduced mass in amu, α_0 is the atomic polarizability (in units of 10^{-24} cm^3), and the subscripts apply to ions and neutrals, respectively.

Using Eqs. (9.72) and (9.33) the average ion-neutral collision frequency for momentum transfer is obtained as

$$\bar{\nu}_{in} = 2.6\times 10^{-9} n_n(\alpha_0/\mu_A)^{1/2} \quad \text{sec}^{-1} \tag{9.73}$$

independent of either gas temperature.

Table 9.10 gives values of the atomic and molecular polarizabilities for the atmospheric gases.

Values of the ion–neutral momentum transfer collision frequency at low temperatures for ions moving in various gases are given in *Table 9.11* taken from laboratory studies [68] and, where necessary, from Eq. (9.73). Uncertainties are indicated for collisions which have been recently computed [68]. The remaining cross sections are probably accurate to within ±30%.

TABLE 9.10

POLARIZABILITIES OF NEUTRAL GASES

Neutral gas	α_0 (10^{-24} cm^3)
N_2	1.76
O_2	1.59
H_2	0.82
O	0.79 ± 0.02
H	0.667
He	0.21 ± 0.01
N	1.1 ± 0.1
NO	1.74
CO	1.97
CO_2	2.63
N_2O	3.00
NH_3	2.22
SO_2	3.89

Charge exchange exerts a significant influence upon the momentum transfer cross section and is important for both the problems of ion energy transfer and diffusion. For this process, it is found that although there is a transfer of an electric charge between two particles, each tends to retain its original kinetic energy. Thus, even though the identity of the ion has changed, the reaction is still fundamentally elastic in preserving the total kinetic energy. Charge exchange consequently becomes of particular importance to the ion energy balance since it can provide a rapid means for energetic ions to be transformed into energetic neutral particles. For many of the possible atmospheric ion-neutral collision pairs it is found that the probability for charge exchange is much larger than that for the normal kinetic collisions. Therefore, while an originally energetic ion could cool to the ambient gas temperature by means of multiple elastic collisions, it can lose all of its excess energy in a single charge exchange reaction.

Charge exchange also plays an important part in enhancing the magnitude of the momentum transfer cross section above those values normally associated with high temperature ion–neutral reactions. This action arises from the conversion of very slightly glancing collisions into what appear to be nearly direct impacts with a consequent backscattering. Under these circumstances there exists a relation between the charge exchange, σ_E, and momentum transfer, σ_D, cross sections given as $\sigma_D = 2\sigma_E$. The relation-

TABLE 9.11

ION–NEUTRAL MOMENTUM TRANSFER COLLISION FREQUENCIES

Ion	Neutral	Collision frequency / $n(n)$ (10^{-10} cm^3 sec^{-1})		
		300°K	1000°K	1500°K
NO^+	Air	7.5 ± 0.1	6.2 ± 0.3	6.1 ± 0.5
	O	6.0	4.7	4.5
O_2^+	N_2	7.6	5.9	5.7
	O_2	9.9	*a*	*a*
	O	5.9	4.6	4.4
N_2^+	N_2	14.1	*a*	*a*
	O_2	7.2	5.7	5.4
	O	7.2	*b*	*a*
O^+	N_2	9.2	9.5 ± 2.0	9.5 ± 3.0
	O_2	10.2	10.8	11.2
	O	9.1	*a*	*a*
	He	4.8	4.6	5.1
N^+	N_2	11.4	12.1	12.5
H^+	H	30.0	*a*	*a*
	He	13.3	*b*	*b*
	N_2	35.1	*a*	*b*
He^+	He	8.3	*a*	*a*
	O	12.8	*b*	*b*
	N_2	18.4	*b*	*b*
	H	23.8	*b*	*b*

[a] See Table 9.13 and 9.14.
[b] High temperature data not available.

ship is valid only for high temperatures where the ion–neutral polarization effects can be neglected. In general, an exact solution to the problem of determining the charge exchange cross section is not possible because of the uncertainties in the form of the energy wave functions for all possible energy states of two colliding particles. Therefore, it is necessary to resort to experimental results to determine specific resonance charge exchange cross sections.

It has been shown [69] that the velocity dependent charge exchange cross section, σ_E, can be expressed as

$$\sigma_E = (A - B \log \mathscr{E})^2, \tag{9.74}$$

for particle energies below 500 eV. Here A and B are constants characteristic of a given gas and $\mathscr{E}$ is the kinetic energy of relative motion, measured in electron volts. Using $\sigma_D = 2\sigma_E$, the high temperature velocity dependent momentum transfer cross section can be given as [69]

$$\sigma_D = 2(A - B \log \mathscr{E})^2. \tag{9.75}$$

It is important to note that the total cross section for momentum transfer is not equal to the sum of the charge exchange and polarization contributions. Because there exist mutual interaction effects, the polarization potential and charge exchange action combine to form a total cross section whose variation with temperature follows closely the individual cross sections in their own regions of dominance [70]. In the transition region there is a slight enhancement of the cross section which arises from curvature in the individual particle orbits due to the polarization interaction. This effect increases the probability for charge exchange to values greater than predicted by Eq. (9.74). For temperatures above the transition region, σ_D is determined exclusively by charge exchange, while below this point the polarization forces are predominant. This effect is illustrated in *Fig. 9.26* for the specific example of N_2^+–N_2 collisions [67]. Here the transition region extends, in terms of $\Gamma = T_i + T_n$, from 200° to 700°K and is cen-

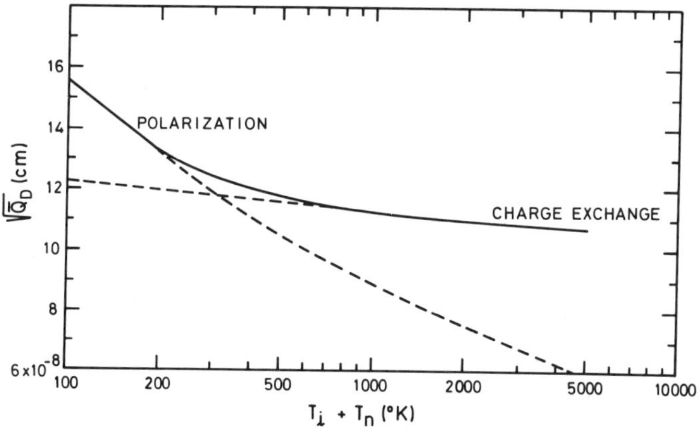

Fig. 9.26 Momentum transfer cross sections for N_2^+ moving through N_2.

tered about the point $\Gamma = 340°\mathrm{K}$. It is seen that for $\Gamma > 750°\mathrm{K}$ there is essentially no contribution to $\overline{Q_{\mathrm{D}}}$ from polarization.

In order to determine suitable temperature dependent charge exchange and momentum transfer cross sections, it is necessary to average Eqs. (9.74) and (9.75) over the ion and neutral velocity distributions. Thus,

$$\overline{Q_{\mathrm{D}}} = 2\overline{Q_{\mathrm{E}}} = \iint f_{\mathrm{i}} f_{\mathrm{n}} \sigma_{\mathrm{D}}\, d\mathbf{v}_{\mathrm{i}}\, d\mathbf{v}_{\mathrm{n}}, \tag{9.76}$$

where σ_{D} is given by Eq. (9.75), f is the particle velocity distribution function, $d\mathbf{v}$ is a velocity space volume element, and the subscripts refer to ions and neutrals, respectively. The solution to Eq. (9.76) for conditions where $T_{\mathrm{i}} = T_{\mathrm{n}}$ is known [71, 72]. When the ions and neutrals have separate Maxwellian distributions with no mean mass motion ($T_{\mathrm{i}} \neq T_{\mathrm{n}}$) it is possible to obtain [61]

$$\overline{Q_{\mathrm{E}}} = [(A + 3.96B) - B\log(T_{\mathrm{i}} + T_{\mathrm{n}})]^2 \times 10^{-16} \quad \mathrm{cm}^2, \tag{9.77}$$

where A and B, measured in units of 10^{-8} cm, are the coefficients appearing in Eq. (9.74).

It is possible to define a resonance ion–neutral collision frequency, $\overline{\nu_{\mathrm{R}}}$, as

$$\overline{\nu_{\mathrm{R}}} = \tfrac{4}{3} n (8k/\pi m)^{1/2} (T_{\mathrm{i}} + T_{\mathrm{n}})^{1/2} \overline{Q_{\mathrm{D}}}, \tag{9.78}$$

with m the ion or neutral particle mass. In a similar manner the average collision frequency for charge exchange, ν_{E}, is defined as

$$\overline{\nu_{\mathrm{E}}} = \tfrac{4}{3} n (8k/\pi m)^{1/2} (T_{\mathrm{i}} + T_{\mathrm{n}})^{1/2} \overline{Q_{\mathrm{E}}}, \tag{9.79}$$

the factor of $\frac{4}{3}$ appearing since we desire to use this equation in an energy transfer equation. From Eqs. (9.74) and (9.75), we have

$$\overline{\nu_{\mathrm{E}}} = \tfrac{1}{2}\overline{\nu_{\mathrm{R}}}, \tag{9.80}$$

for the temperature region where polarization effects are negligible.

The resonance ion–neutral collision frequencies for charge exchange and momentum transfer are immediately obtained from the previous formulas [67] (see *Fig. 9.27* and *Tables 9.12–9.14*) of the charge exchange rate coefficient for values of $T_{\mathrm{i}} + T_{\mathrm{n}}$ above the transition temperature. At low temperature only the polarization effect is present, and the formulas of Table 9.11 can be used.

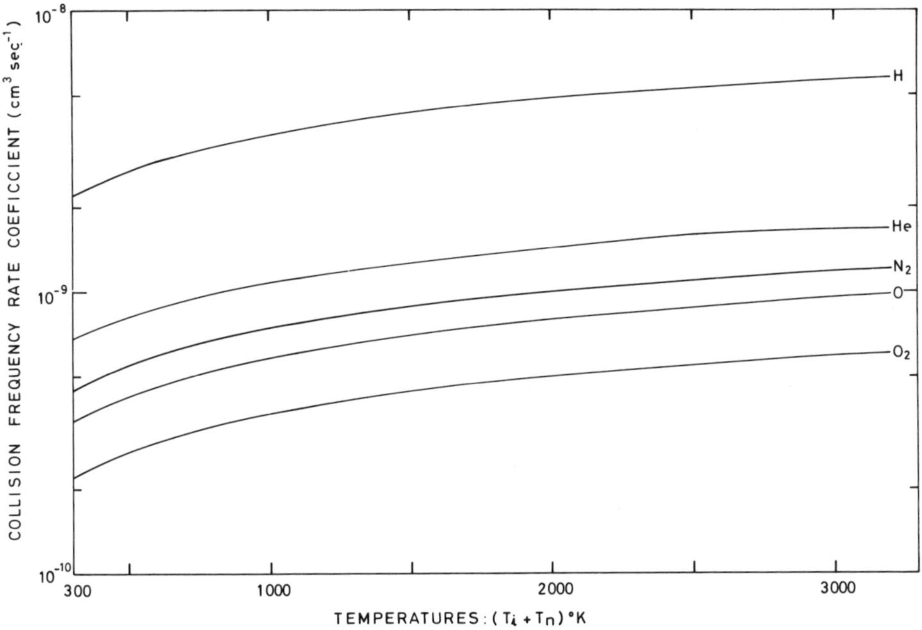

Fig. 9.27 Charge exchange collision frequencies for the atmospheric gases.

The process of resonance charge exchange occurs most readily between an ion and its parent atom where the energy defect is zero. For certain ion–neutral pairs, however, there occurs an accidental resonance in ionization energies, permitting charge exchange to proceed rapidly at thermal energies. The most important accidentally resonant charge exchange process in the

TABLE 9.12

Reduced Charge Exchange Collision Frequencies
$\log_{10}\Gamma = \log_{10}(3000°\text{K})$

Species	$\bar{\nu}_E$ (sec^{-1})
H^+–H	$1.0 \times 10^{-10} n(\text{H})\Gamma^{1/2}$
O^+–O	$1.6 \times 10^{-11} n(\text{O})\Gamma^{1/2}$
N^+–N	$1.7 \times 10^{-11} n(\text{N})\Gamma^{1/2}$
He^+–He	$3.0 \times 10^{-11} n(\text{He})\Gamma^{1/2}$
O_2^+–O_2	$1.1 \times 10^{-11} n(\text{O}_2)\Gamma^{1/2}$
N_2^+–N_2	$2.2 \times 10^{-11} n(\text{N}_2)\Gamma^{1/2}$

TABLE 9.13

AVERAGE ION–NEUTRAL RESONANCE MOMENTUM TRANSFER COLLISION FREQUENCIES

Species	Γ (°K)	$\bar{\nu}_R$ (sec^{-1})
H^+, H	>100	$1.9 \times 10^{-12} n(H) \Gamma^{1/2} (14.4 - 1.17 \log_{10} \Gamma)^2$
O^+, O	>470	$4.8 \times 10^{-13} n(O) \Gamma^{1/2} (10.6 - 0.67 \log_{10} \Gamma)^2$
N^+, N	>550	$5.2 \times 10^{-13} n(N) \Gamma^{1/2} (10.4 - 0.64 \log_{10} \Gamma)^2$
He^+, He	>100	$9.7 \times 10^{-13} n(He) \Gamma^{1/2} (11.6 - 1.05 \log_{10} \Gamma)^2$
O_2^+, O_2	>1600	$3.4 \times 10^{-13} n(O_2) \Gamma^{1/2} (10.6 - 0.76 \log_{10} \Gamma)^2$
N_2^+, N_2	>340	$3.7 \times 10^{-13} n(N_2) \Gamma^{1/2} (14.3 - 0.96 \log_{10} \Gamma)^2$

TABLE 9.14

CHARGE EXCHANGE COLLISION FREQUENCIES, $\bar{\nu}_E$ BETWEEN IONS AND NEUTRAL ATOMS OR MOLECULES[a]

Temperatures $T_i + T_n$	N_2^+–N_2	O_2^+–O_2	O^+–O	He^+–He	H^+–H
300				6.8×10^{-10}	2.2×10^{-9}
500	5.6×10^{-10}			8.3	2.7
1000	7.5		5.6×10^{-10}	1.1×10^{-9}	3.6
1500	8.9		6.6	1.3	4.3
2000	1.0×10^{-9}	4.9×10^{-10}	7.6	1.4	4.8
2500	1.1	5.4	8.3	1.6	5.3
3000	1.2	5.9	8.9	1.7	5.7
3500	1.3	6.2	9.6	1.8	6.0
4000	1.4	6.6	1.0×10^{-9}	1.9	6.3
5000	1.5	7.2	1.1	2.1	6.9
7500	1.8	8.6	1.3	2.4	8.2
10,000	2.0	9.7	1.5	2.7	9.1

[a] Values in $cm^3\ sec^{-1}$.

upper atmosphere involves atomic oxygen and hydrogen atoms and ions according to the relation [73–75]

$$O^+(^4S) + H(^2S) \rightleftharpoons O(^3P_J) + H^+(1p) + \Delta\varepsilon \tag{9.81}$$

the energy defect for each value of J being:

$\Delta\varepsilon$ (eV):	−0.00833	+0.00000	+0.01965
J:	0	1	2

The reaction rates for the forward and reverse reactions of Eq. (9.81) are related by the partition functions, Z, for each reactant [76]. Thus, for thermal equilibrium

$$\frac{k_E(O, H^+)}{k_E(H^+, O)} = \frac{Z(O^+)Z(H)}{Z(O)Z(H^+)}\, e^{-\Delta E/kT}. \tag{9.82}$$

For O^+, H, and H^+ the partition functions are given by the ground state degeneracies $2S + 1$; i.e., 4, 2, and 1, respectively. The partition function for $O(^3P)$ at high temperatures is $(2S + 1)(2L + 1) = 9$. However, the spin–orbit splitting of the ground state levels of $O(^3P)$ is not necessarily large compared with kT if one works with temperatures below 1000°K. The general expression for $Z(O)$ is [76]

$$Z(O) = \sum_{J=1}^{3} (2J + 1)e^{-\varepsilon_J/kT} \tag{9.83}$$

where ε_J are the energies of the J levels. Using numerical values, (9.83) becomes

$$Z(O) = 5 + 3e^{-227.8/T} + e^{-325.6/T} \tag{9.84}$$

with values given below:

T (°K):	300	600	1000	2000	∞
$Z(O)$:	6.7	7.6	8.1	8.5	9

Thus, the relation [77]

$$k_E(O^+\text{–}H) = \tfrac{9}{8}k_E(H^+\text{–}O) \tag{9.85}$$

is a convenient approximation valid at high temperatures.

Since electron excitation cross sections are large, the temperature referred to in (9.83) and (9.84) may actually be that of the electron gas.

Cross-section measurements of reaction (9.81) and its reverse are in essential accord. From beam experiments the value [78, 79]

$$\sigma_E(H^+\text{–}O) = (3.47 - 0.20 \log_{10} \varepsilon)^2 \times 10^{-16} \quad cm^2 \tag{9.86}$$

has been obtained over the energy range 50–10,000 eV. Recent measurements [76] show that (9.86) is valid down to 0.04 eV; i.e., even below thermospheric mean energies.

Using (9.77) in an extended form the average charge exchange cross section is

$$\bar{Q}_E(H^+\text{–}O) = (4.26 - 0.2 \log_{10} T_i)^2 \times 10^{-16} \quad cm^2. \tag{9.87}$$

Adopting an average value $T_i = 1500°K$, one finds $\bar{Q}_E(H^+\text{–}O) = 13.1 \times 10^{-16}$ cm²; i.e., large enough to be the principal sink [and by (9.81), the source] of H^+ in the upper atmosphere. For aeronomic calculations, the mean value

$$\bar{Q}_E(H^+\text{–}O) = (1.5 \pm 0.5) \times 10^{-15} \quad cm^2 \tag{9.88}$$

can be adopted.

From experimental data [76] and (9.79) the charge exchange rate can be taken as

$$\nu_E(H^+\text{–}O) = 2.2 \times 10^{-11} T_i^{1/2} n(O) \quad sec^{-1}. \tag{9.89}$$

Adopting relation (9.85) the direct charge exchange rate for (9.81) becomes

$$\nu_E(O^+\text{–}H) = 2.5 \times 10^{-11} T_n^{1/2} n(H) \quad sec^{-1} \tag{9.90}$$

in agreement with measurements [80].

Such a result depends, however, on the atomic oxygen partition function and at sufficiently low temperatures (9.90) will be significantly smaller. The dependence upon $T_i^{1/2}$ in (9.89) and $T_n^{1/2}$ in (9.90) arises from the mean thermal velocities between the colliding ions and neutrals [81, 82].

The rate of energy loss by high velocity protons and alpha particles in air has been discussed extensively [62, 63, 83]. Experimental curves of proton range, R, are shown in *Fig. 9.28* as a function of proton kinetic energy for a standard atmosphere [63]. The conversion of proton ranges to proton energy loss rates involves differentiating the range curves, the result of which is shown [63] in *Fig. 9.29*. It is seen that the proton energy loss rate increases for low energies, progressively shortening the penetration depth of the particle.

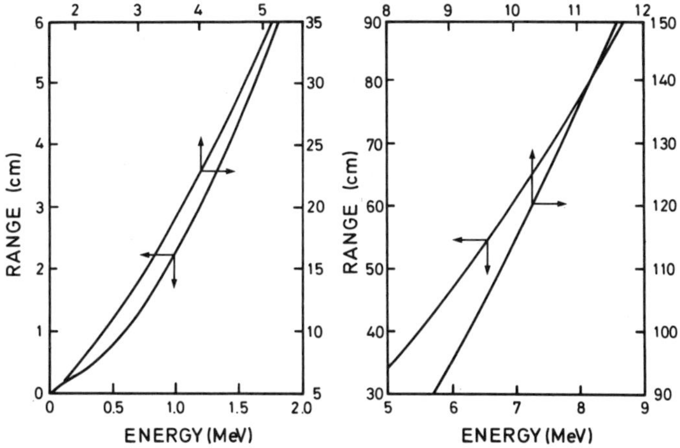

Fig. 9.28 Energy–range curves for protons in air (15°C, 760 Torr) up to 10,000 MeV [63].

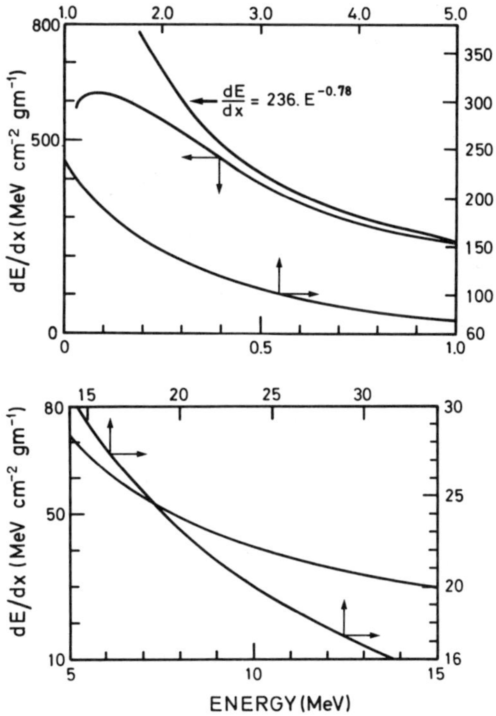

Fig. 9.29 Proton energy loss rate in air. The approximate form given by Eq. (9.91) is also shown [63].

For the case of protons in the energy range $0.2 \leq \varepsilon \leq 400$ MeV, the energy loss rate in air can be approximated by the expression [41]

$$d\varepsilon/dR = -236\varepsilon^{-0.78} \tag{9.91}$$

where ε is the proton energy in MeV and R is the gas depth defined by (9.66).

Integrating (9.91) the proton energy varies with gas depth (in gm cm²) as

$$\varepsilon = \varepsilon_0[1 - (420R/\varepsilon_0^{1.78})]^{0.56} \quad \text{MeV} \tag{9.92}$$

where ε_0 is the initial proton energy at $R = 0$.

The range R_0 of a proton in air is found from (9.92) by taking $\varepsilon = 0$, giving

$$R_0 = (\varepsilon_0/30)^{1.78}. \tag{9.93}$$

With regard to the earth's atmosphere, the penetration depth of energetic protons can be found from *Fig. 9.30*. The effect of pitch angle is also indicated with alignment along the magnetic field, assumed to be vertical, corresponding to $\alpha = 0$.

Using the energy loss rate given by (9.91), the fast proton ionization rate can be found by adopting a value of 35 eV for the mean ionization loss. The results for single protons of different energies and zero pitch angle are shown in *Fig. 9.31* while the ionization rates appropriate to 2π isotropic fluxes of protons are shown in *Fig. 9.32*.

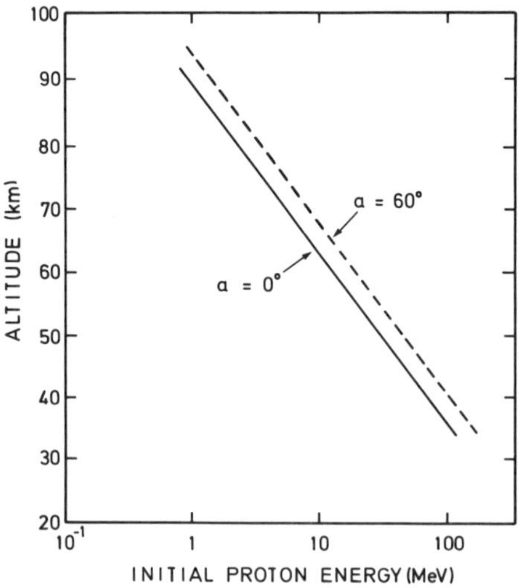

Fig. 9.30 Proton penetration depths for various initial energies and two pitch angles.

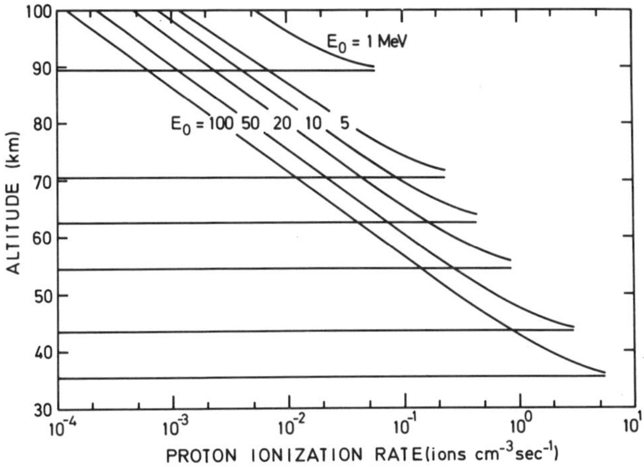

Fig. 9.31 Proton ionization rates for various initial energies and vertical incidence.

The cross sections for proton ionization in atmospheric gases are shown in *Figs. 9.33–9.35* [2]. Also shown are the electron ionization cross sections. From theory, it is predicted and observed that electron and proton ionization cross sections are the same at equal velocities. This equivalence breaks down, however, for electron energies below about 300 eV.

Since the proton energy loss rate depends upon the proton energy, it follows that the energy spectrum of protons incident upon the upper atmosphere depends upon altitude [83]. To illustrate this consider a flux of 2π

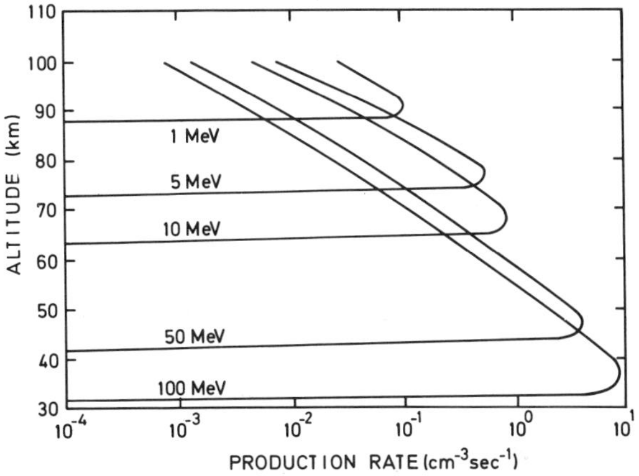

Fig. 9.32 Proton ionization production rates (the production rates of electron-ion pairs for monoenergetic 2π-isotropic intensities of 1 proton cm^{-2} sec^{-1} $ster^{-1}$) as a function of altitude for five different proton energies.

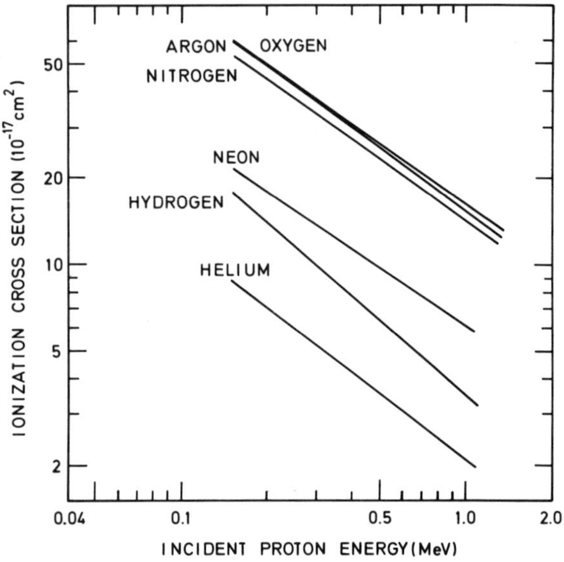

Fig. 9.33 Apparent ionization cross sections for protons on He, Ar, Ne, H_2, and N_2 [2].

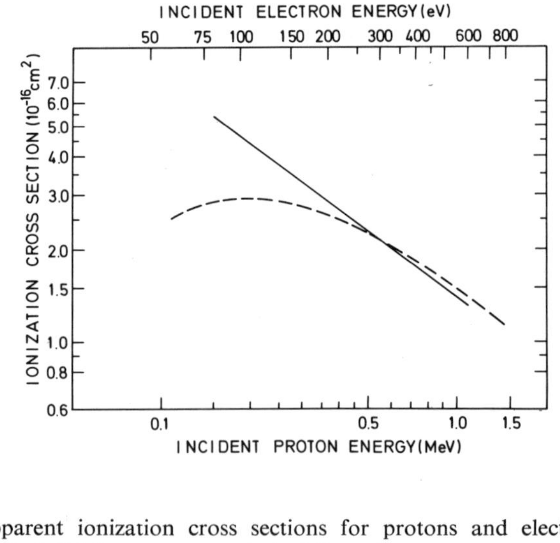

Fig. 9.34 Apparent ionization cross sections for protons and electrons of equal velocity in N_2 [2]. Solid line, protons; dashed line, electrons.

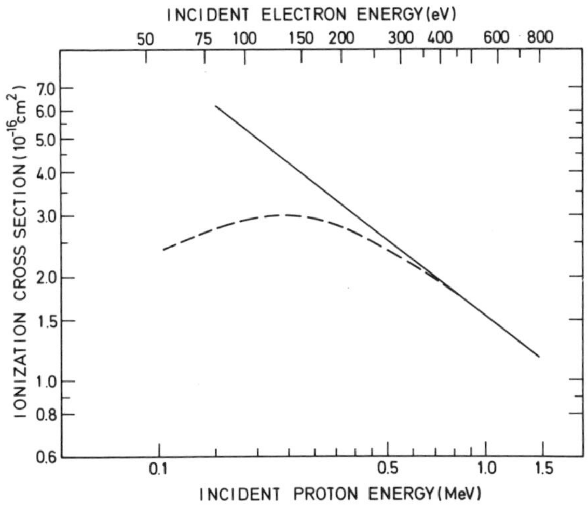

Fig. 9.35 Apparent ionization cross sections for protons and electrons of equal velocity in O_2 [2]. Solid line, protons; dashed line, electrons.

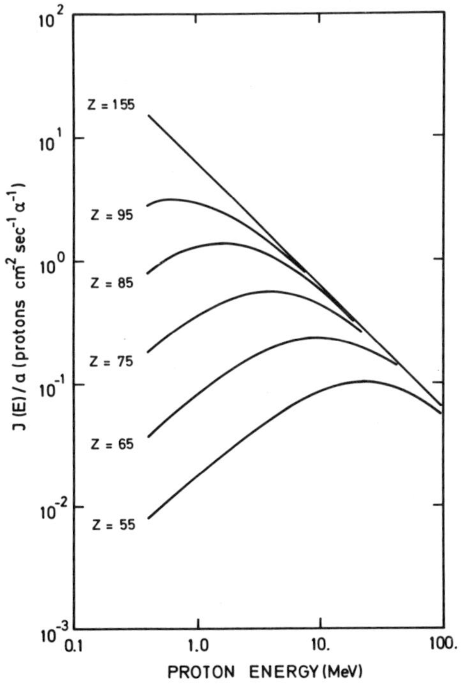

Fig. 9.36 Proton energy spectra for $J(E) = 2\pi\alpha E^{-1}$ where $J_0(E)\,dE = 2\pi\alpha E^{-1}\,dE$. Z is in km.

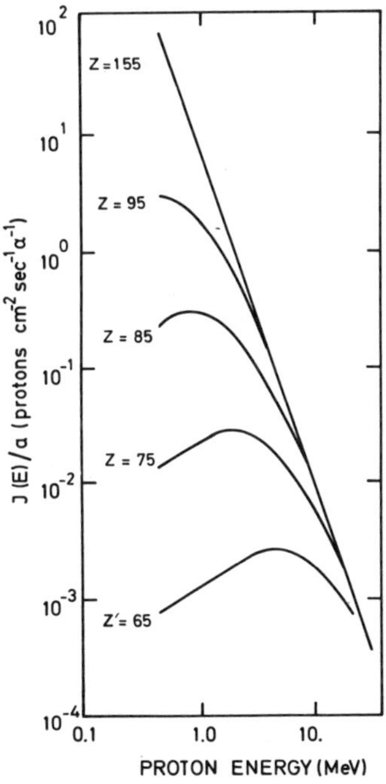

Fig. 9.37 Proton energy spectra for $J(E) = 2\pi\alpha E^{-2.76}$ where $J_0(E)\,dE = 2\pi\alpha E^{-2.76}\,dE$. Z is in km.

isotropic protons incident upon the upper atmosphere which have the initial spectrum

$$J_0(E) = 2\pi\alpha E^{-n} \quad \text{protons cm}^{-2}\ \text{sec}^{-1}\ \text{MeV}^{-1}.$$

The typical hardening of the proton spectrum is shown in *Figs. 9.36* and *9.37* for $n = 1$ and $n = 2.78$ for various altitudes. This behavior is explained by the more rapid rate of proton energy loss at lower energies.

9.7 Charged Particle Collisions

The force between charged particles obeys the Coulomb law

$$F_{12} = Z_1 Z_2 e^2/r^2 \tag{9.94}$$

where Z is the charge state, e the magnitude of the electric charge (4.8×10^{-10} esu), and r is the radial separation.

The differential scattering cross section appropriate to binary collisions between charged particles was derived from (9.94) by Rutherford [2]:

$$q(\theta, g) = (Z_1 Z_2 e^2/2\mu g^2)^2 \sin^{-4}(\theta/2). \tag{9.95}$$

Here μ is the two particle reduced mass, g is the relative velocity, and θ is the center of mass scattering angle. Using the cross section equations given by Eqs. (9.29)–(9.31), the angular dependence of (9.95) can be removed. Integrations of these equations over the limits $(\theta_{\min}, \pi/2)$ gives the following results:

$$\sigma_{\mathrm{T}}(g) = 2\pi(Z_1 Z_2 e^2/\mu g^2)^2[\cos\theta_{\mathrm{m}}/(1-\cos\theta_{\mathrm{m}})] \tag{9.96}$$

$$\sigma_{\mathrm{V}}(g) = 2\pi(Z_1 Z_2 e^2/\mu g^2)^2[2\ln(1-\cos\theta_{\mathrm{m}})^{-1} - \cos\theta_{\mathrm{m}}] \tag{9.97}$$

$$\sigma_{\mathrm{D}}(g) = 2\pi(Z_1 Z_2 e^2/\mu g^2)^2[\ln(1-\cos\theta_{\mathrm{m}})^{-1}] \tag{9.98}$$

where θ_{m} is some minimum scattering angle.

It is noted that all three cross sections diverge at small scattering angles as a result of the infinite range of the Coulomb force. This divergence is especially important for the total scattering cross section. The viscosity and momentum transfer cross sections, however, diverge logarithmically and, using the results of plasma kinetic theory [4], a minimum scattering angle θ_{m} can be chosen consistent with the concept of the plasma Debye length. The result is that

$$1 - \cos\theta_{\mathrm{m}} \simeq (Z_1 Z_2 e^2/\mu g^2 \lambda_{\mathrm{D}}) \triangleq \Lambda \tag{9.99}$$

where λ_{D}, the Debye length, is given as

$$\lambda_{\mathrm{D}} = \left[4\pi \sum_j (n_j e^2 Z_j/kT_j)\right]^{-1} \tag{9.100}$$

with the summation taken over all charged particle species present.

Using Eq. (9.99), the basic collision cross sections can be written

$$\sigma_{\mathrm{T}} = 2\pi(Z_1 Z_2 e^2/\mu g^2)^2[(1-\Lambda)/\Lambda] \tag{9.101}$$

$$\sigma_{\mathrm{V}} = 2\pi(Z_1 Z_2 e^2/\mu g^2)^2[2\ln\Lambda + \Lambda - 1] \tag{9.102}$$

$$\sigma_{\mathrm{D}} = 2\pi(Z_1 Z_2 e^2/\mu g^2)^2 \ln\Lambda. \tag{9.103}$$

Before continuing, it is worthwhile to give numerical values of the Debye distance for the ionosphere (see the accompanying tabulation). From Eq. (9.100) the following value can be used for calculations:

$$\lambda_{\mathrm{D}} = 6.9(T/n_{\mathrm{e}})^{1/2} \quad \mathrm{cm}. \tag{9.104}$$

T (°K):	100	400	900	1600	2500	2600	4900
For $n_{\mathrm{e}} = 10^2$ cm^{-3}							
λ_{D} (cm):	7	14	21	28	35	41	48
For $n_{\mathrm{e}} = 10^4$ cm^{-3}							
λ_{D} (cm):	0.7	1	2	3	3.5	4	5
For $n_{\mathrm{e}} = 10^6$ cm^{-3}							
λ_{D} (cm):	0.07	0.1	0.2	0.3	0.35	0.4	0.5

The Debye length varies from several decimeters in the D region to less than 1 cm in the F region.

Values of $\ln\Lambda$ are given in *Table 9.15* as a function of Debye length and kinetic energy of relative motion, ε.

The term $\ln\Lambda$ is common to all calculation of ionized gases as indicated by Eqs. (9.101)–(9.103); it can be taken as a constant in the ionosphere if

TABLE 9.15

VALUES OF $\ln \Lambda$

ε (eV)	λ_D (cm)								
	0.1	0.5	1.0	2.0	5.0	10.0	20.0	50.0	100.0
1×10^{-2}	9.5	11.1	11.8	12.5	13.5	14.1	14.8	15.8	16.4
5×10^{-2}	11.1	12.8	13.5	14.1	15.1	15.8	16.4	17.4	18.1
1×10^{-1}	11.8	13.5	14.1	14.8	15.8	16.4	17.1	18.1	18.8
5×10^{-1}	13.5	15.1	15.8	16.4	17.4	18.1	18.8	19.7	20.4
1×10^{0}	14.1	15.8	16.4	17.1	18.1	18.8	19.4	20.4	21.1
5×10^{0}	15.8	17.4	18.1	18.8	19.7	20.4	21.1	22.0	22.7
1×10^{1}	16.4	18.1	18.8	19.4	20.4	21.1	21.7	22.7	23.4
5×10^{1}	18.1	19.7	20.4	21.1	22.0	22.7	23.4	24.3	25.0
1×10^{2}	18.8	20.4	21.0	21.7	22.6	23.4	24.0	24.9	25.7

a certain degree of approximation is acceptable. Using Eq. (9.99), a constant value $\ln \Lambda = 15 \pm 1.5$ is possible, provided that

with n_e (cm^{-3}):	10^2	10^4	10^6
we have T_e (°K) between:	100	500	2500
and:	900	4000	10,000

With Eqs. (9.35) and (9.103), the average momentum transfer cross section for two charged gases, each having its own Maxwellian velocity distribution, is [3]

$$\overline{Q_D} = \frac{\pi}{2} \left[\frac{(m_1 + m_2) Z_1 Z_2 e^2}{m_1 m_2} \right]^2 \left[\frac{kT_1}{m_1} + \frac{kT_2}{m_2} \right]^{-2} \ln \Lambda \tag{9.105}$$

where the parameter Λ is given by Eq. (9.99).

From Eqs. (9.33) and (9.105) the average collision frequency for momentum transfer of charged particles is

$$\overline{\nu_{12}} = \frac{4n_2}{3} (2\pi)^{1/2} \left[\frac{(m_1 + m_2) Z_1 Z_2 e^2}{m_1 m_2} \right]^2 \left[\frac{kT_1}{m_1} + \frac{kT_2}{m_2} \right]^{-3/2} \ln \Lambda . \tag{9.106}$$

The corresponding collision interval is defined by

$$\overline{\tau_{12}} = \frac{3 \ln \Lambda}{4(2\pi)^{1/2}} \left[\frac{m_1 m_2}{(m_1 + m_2) Z_1 Z_2 e^2}\right]^2 \left[\frac{kT_1}{m_1} + \frac{kT_2}{m_2}\right]^{3/2} \frac{1}{n_2} \tag{9.107}$$

while the free path between successive collisions is

$$\overline{\lambda_{12}} = \frac{3 \ln \Lambda}{2\pi} \left[\frac{m_1 m_2}{(m_1 + m_2) Z_1 Z_2 e^2}\right]^2 \left[\frac{kT_1}{m_1} + \frac{kT_2}{m_2}\right]^{2} \frac{1}{n_2}. \tag{9.108}$$

To apply these results to electron–electron, electron–ion, and ion–ion collisions it is necessary to specify the various parameters.

The average momentum transfer cross section for a thermal electron moving in a gas of thermal ions follows from Eq. (9.105) as

$$\overline{Q_{\mathrm{D}}} = (\pi/2)(e^2/kT_{\mathrm{e}})^2 \ln \Lambda \tag{9.109}$$

where it is assumed that $T_{\mathrm{e}}/m_{\mathrm{e}} \gg T_{\mathrm{i}}/m_{\mathrm{i}}$. With $\ln \Lambda = 15$, the numerical value for (9.109) is

$$\overline{Q_{\mathrm{D}}} = 6.6 \times 10^{-5}/T_{\mathrm{e}}^2 \quad \mathrm{cm}^2. \tag{9.110}$$

This cross section is much larger than that found for electron collisions with neutral particles. At $T_{\mathrm{e}} = 1000$°K, for example, the cross section of 6.6×10^{-11} cm² corresponds to a collision diameter of about 900 Å.

The average electron–ion collision frequency is obtained from Eqs. (9.37) and (9.109) as

$$\overline{\nu_{\mathrm{ei}}} = \frac{4(2\pi)^{1/2}}{3} \frac{Z_{\mathrm{i}}^2 e^4 \ln \Lambda}{m_{\mathrm{e}}^{1/2}(kT_{\mathrm{e}})^{3/2}} n_{\mathrm{i}} \tag{9.111}$$

or

$$\overline{\nu_{\mathrm{ei}}} = \frac{(54 \pm 5)}{T_{\mathrm{e}}^{3/2}} n_{\mathrm{i}} \quad \mathrm{sec}^{-1}, \tag{9.112}$$

indicating the $\pm 10\%$ uncertainty associated with the choice of a minimum scattering angle in Eq. (9.102). *Table 9.16* gives values of the electron–ion momentum transfer collision frequency as a function of electron density and electron temperature.

For collisions between ions of different species it is possible to assume $T_1 = T_2 \triangleq T_{\mathrm{i}}$ and $m_1 \neq m_2$. Thus, from Eq. (9.105) we have

$$\overline{Q_{\mathrm{D}}} = \frac{\pi}{2} \frac{(Z_1 Z_2 e^2)^2 \ln \Lambda}{(kT_{\mathrm{i}})^2} \tag{9.113}$$

TABLE 9.16

ELECTRON COLLISION FREQUENCIES[a] WITH POSITIVE IONS

n_e (cm^{-3})	600°K	800°K	1000°K	1200°K	1400°K	1600°K	1800°K	2000°K
10^2	4.1×10^{-1}	2.7×10^{-1}	2.0×10^{-1}	1.5×10^{-1}	1.2×10^{-1}	1.0×10^{-1}	8.7×10^{-2}	7.5×10^{-2}
2.5×10^2	1.0	6.7×10^{-1}	4.9×10^{-1}	3.8×10^{-1}	3.0×10^{-1}	2.5×10^{-1}	2.1×10^{-1}	1.8×10^{-1}
5.0×10^2	2.0	1.3	9.5×10^{-1}	7.4×10^{-1}	5.9×10^{-1}	4.9×10^{-1}	4.2×10^{-1}	3.6×10^{-1}
7.5×10^2	2.9	1.9	1.4	1.1	8.8×10^{-1}	7.3×10^{-1}	6.2×10^{-1}	5.3×10^{-1}
10^3	3.8	2.6	1.9	1.4	1.2	9.6×10^{-1}	8.2×10^{-1}	7.0×10^{-1}
2.5×10^3	1.2×10^1	6.2	4.5	3.5	2.9	2.3	2.0	1.7
5.0×10^3	1.8×10^1	1.2×10^1	8.9	6.9	5.5	4.6	3.9	3.3
7.5×10^3	2.7×10^1	1.8×10^1	1.3×10^1	1.0×10^1	8.2	6.8	5.8	5.0
10^4	3.5×10^1	2.4×10^1	1.7×10^1	1.3×10^1	1.1×10^1	9.0	7.6	6.6
2.5×10^4	8.6×10^1	5.7×10^1	4.2×10^1	3.3×10^1	2.6×10^1	2.2×10^1	1.8×10^1	1.6×10^1
5.0×10^4	1.7×10^2	1.1×10^2	8.2×10^1	6.4×10^1	5.1×10^1	4.3×10^1	3.6×10^1	3.1×10^1
7.5×10^4	2.5×10^2	1.7×10^2	1.2×10^2	9.4×10^1	7.6×10^1	6.3×10^1	5.3×10^1	4.6×10^1
10^5	3.3×10^2	2.2×10^2	1.6×10^2	1.2×10^2	1.0×10^2	8.3×10^1	7.1×10^1	6.1×10^1
2.5×10^5	7.9×10^2	5.3×10^2	3.9×10^2	3.0×10^2	2.4×10^2	2.0×10^2	1.7×10^2	1.5×10^2
5.0×10^5	1.5×10^3	1.0×10^3	7.6×10^2	5.9×10^2	4.7×10^2	3.9×10^2	3.3×10^2	2.9×10^2
7.5×10^5	2.3×10^3	1.5×10^3	1.1×10^3	8.7×10^2	7.0×10^2	5.8×10^2	4.9×10^2	4.3×10^2
10^6	3.0×10^3	2.0×10^3	1.5×10^3	1.1×10^3	9.2×10^2	7.7×10^2	6.5×10^2	5.6×10^2
2.5×10^6	7.2×10^3	4.8×10^3	3.5×10^3	2.8×10^3	2.2×10^3	1.9×10^3	1.6×10^3	1.4×10^3
5.0×10^6	1.4×10^4	9.4×10^3	6.9×10^3	5.4×10^3	4.3×10^3	3.6×10^3	3.4×10^3	2.6×10^3

[a] Values in sec^{-1}

which, except for the ion temperature and ion charges, is exactly the same as that given by Eq. (9.109) for electron–ion collisions.

Using (9.33) the average ion–ion momentum transfer collision frequency is

$$\overline{\nu_{12}} = \tfrac{4}{3}(2\pi)^{1/2} n_2 \frac{(Z_1 Z_2 e^2)^2}{\mu^{1/2}(kT_\mathrm{i})^{3/2}} \ln \Lambda \qquad (9.114)$$

or

$$\overline{\nu_{12}} = 1.3 n_2 \frac{(Z_1 Z_2)^2}{\mu_\mathrm{A}^{1/2} T_\mathrm{i}^{3/2}} \quad \mathrm{sec}^{-1} \qquad (9.115)$$

where μ_A is the ion–ion reduced mass measured in atomic mass units.

REFERENCES

1. J. B. Hasted, *Physics of Atomic Collisions*. Butterworths, London and Washington, D.C., 1964.
2. E. W. McDaniel, *Collision Phenomena in Ionized Gases*. Wiley, New York, 1964.
3. P. Banks, *Planet. Space Sci.* **14**, 1085 (1966).
4. E. H. Holt and R. E. Haskell, *Foundations of Plasma Dynamics*. Macmillan, New York, 1965.
5. A. Dalgarno, *Ann. Géophys.* **17**, 16 (1961).
6. M. Nicolet, *Phys. Fluids* **2**, 95 (1959).
7. E. V. Thrane and W. R. Piggott, *J. Atmos. Terr. Phys.* **28**, 721 (1966).
8. Y. Itikawa, *Planet. Space Sci.* **19**, 993 (1971).
9. J. W. Daiber and H. F. Waldron, *Phys. Rev.* **151**, 51 (1966).
10. A. V. Phelps, *Rev. Mod. Phys.* **40**, 399 (1968).
11. E. W. McDaniel, V. Cermak, A. Dalgarno, E. E. Ferguson, and L. Friedman, *Ion-Molecule Reactions*. Wiley, New York, 1970.
12. K. Takayanagi and Y. Itikawa, *Advan. At. Mol. Phys.* **6**, 105–153 (1970).
13. A. Dalgarno, *Can. J. Chem.* **47**, 1723 (1969).
14. E. Gerjuoy and S. Stein, *Phys. Rev.* **97**, 1671 (1955).
15. K. Takayanagi and S. Geltman, *Phys. Rev.* **138**, A1003 (1965).
16. D. H. Sampson and R. C. Mjolsness, *Phys. Rev.* **140**, A1466 (1965).
17. G. J. Schulz, *Phys. Rev.* **116**, 1141 (1959).
18. G. J. Schulz, *Phys. Rev.* **125**, 229 (1962).
19. G. J. Schulz, *Phys. Rev.* **135**, A988 (1964).
20. A. G. Engelhardt, A. V. Phelps, and C. G. Risk, *Phys. Rev.* **135**, A1566 (1964).
21. J. C. Y. Chen, *J. Chem. Phys.* **45**, 2710 (1966).
22. G. J. Schulz and J. T. Dowell, *Phys. Rev.* **128**, 174 (1962).
23. D. Spence and G. J. Schulz, *Phys. Rev. A* **2**, 1802 (1970).
24. R. D. Hake, Jr. and A. V. Phelps, *Phys. Rev.* **158**, 70 (1967).
25. R. S. Stolarski and A. E. S. Green, *J. Geophys. Res.* **72**, 3967 (1967).
26. A. E. S. Green and C. A. Barth, *J. Geophys. Res.* **72**, 3975 (1967).
27. A. Dalgarno, *Ann. Géophys.* **20**, 65 (1964).
28. L. J. Kieffer and G. H. Dunn, *Rev. Mod. Phys.* **38**, 1 (1966).

29. B. L. Moisewitsch and S. V. Smith, *Rev. Mod. Phys.* **40**, 238 (1968).
30. W. Lotz, *Astrophys. J. Suppl. Ser.* **14**, 207 (1967).
31. A. E. S. Green and C. A. Barth, *J. Geophys. Res.* **70**, 1083 (1965).
32. A. E. S. Green and S. K. Dutta, *J. Geophys. Res.* **72**, 3933 (1967).
33. A. T. Jusick, C. E. Watson, L. R. Peterson, and A. E. S. Green, *J. Geophys. Res.* **72**, 3943 (1967).
34. R. S. Stolarski, V. A. Dulock, Jr., C. E. Watson, and A. E. S. Green, *J. Geophys. Res.* **72**, 3953 (1967).
35. C. E. Watson, V. A. Dulock, Jr., R. S. Stolarski, and A. E. S. Green, *J. Geophys. Res.* **72**, 3961 (1967).
36. E. L. Breig and C. C. Lin, *Phys. Rev.* **151**, 67 (1966).
37. A. Dalgarno and T. C. Degges, *Planet. Space Sci.* **16**, 125 (1968).
38. R. E. Gershberg, *Izv. Krymsk. Astrofiz. Observ.* **26**, 324 (1961).
39. R. J. W. Henry, P. G. Burke, and A.-L. Sinfailam, *Phys. Rev.* **178**, 218 (1969).
40. R. J. W. Henry and R. E. Williams, *Publ. Astron. Soc. Pac.* **80**, 669 (1969).
41. K. Maeda, *J. Geophys. Res.* **68**, 185 (1963).
42. D. Rapp, P. Englander-Golden, and D. D. Briglia, *J. Chem. Phys.* **42**, 4081 (1965).
43. B. Van Zyl and G. H. Dunn, *Phys. Rev.* **163**, 43 (1967).
44. H. F. Winters, *J. Chem. Phys.* **44**, 1472 (1966).
45. N. R. Daly and R. E. Powell, *Proc. Phys. Soc.* **89**, 273 (1966).
46. A. R. Lee and N. P. Carleton, *Phys. Lett.* **27A**, 195 (1968).
47. D. T. Stewart, *Proc. Phys. Soc.* **69A**, 437 (1956).
48. S. Hayakawa and H. Nishimura, *J. Geomagn. Geoelec.* **16**, 72 (1964).
49. J. W. McConkey and I. D. Latimer, *Proc. Phys. Soc.* (*London*) **86**, 463 (1965).
50. R. F. Holland, Los Alamos Sci. Lab. Tech. Rep. LA-3793 (1967).
51. B. N. Srivastava and I. M. Mirza, *Phys. Rev.* **176**, 137 (1968).
52. B. N. Srivastava and I. M. Mirza, *Can. J. Phys.* **47**, 475 (1969).
53. J. F. M. Aarts, F. J. de Heer, and D. A. Vroom, *Physica* **40**, 197 (1968).
54. P. N. Stanton and R. M. St. John, *J. Opt. Soc. Amer.* **59**, 252 (1969).
55. S. H. Tate and P. T. Smith, *Phys. Rev.* **39**, 270 (1932).
56. D. Rapp and P. Englander-Golden, *J. Chem. Phys.* **43**, 1464 (1965).
57. B. L. Schram, F. J. De Heer, M. J. van der Wiel, and J. Kistemaker, *Physica* **31**, 94 (1965).
58. B. L. Schram, H. R. Moustafa, J. Schutten, and F. J. De Heer, *Physica* **32**, 734 (1966).
59. E. N. Lassettre, *Can. J. Chem.* **47**, 1733 (1969).
60. L. J. Kieffer, *At. Data* **1**, 19 (1969).
61. L. J. Kieffer, *At. Data* **1**, 120 (1969).
62. B. Rossi, *High Energy Particles.* Prentice-Hall, Englewood Cliffs, New Jersey, 1956.
63. E. Segré, *Experimental Nuclear Physics.* Wiley, New York, 1953.
64. A. Dalgarno, *Adv. At. Mol. Phys.* **4**, 381 (1968).
65. M. H. Rees, *Planet. Space Sci.* **11**, 1209 (1963).
66. A. Dalgarno, M.R.C. McDowell, and A. Williams, *Proc. Roy. Soc.* **A250**, 411 (1958).
67. P. Banks, *Planet. Space Sci.* **14**, 1105 (1966).
68. E. A. Mason, *Planet. Space Sci.* **18**, 137 (1970).
69. A. Dalgarno, *Phil. Trans. Roy. Soc. London* **A250**, 426 (1958).
70. H. Knof, E. A. Mason, and J. T. Vanderslice, *J. Chem. Phys.* **40**, 3548 (1964).
71. J. W. Sheldon, NASA Tech. Note D-2408, Washington, D.C. (August 1964).
72. E. A. Mason and J. T. Vanderslice, *Phys. Rev.* **114**, 497 (1959).

73. J. W. Dungey, *Physics of the Ionosphere*. Inst. Phys. Phys. Soc., London, 1955.
74. F. S. Johnson, *J. Geophys. Res.* **65,** 577 (1960).
75. J. W. Chamberlain, *Astrophys. J.* **124**, 390 (1956).
76. F. C. Fehsenfeld and E. E. Ferguson, *J. Chem. Phys.* **55**, 2120 (1971).
77. W. B. Hanson and I. B. Ortenburger, *J. Geophys. Res.* **66**, 1425 (1961).
78. R. F. Stebbings, A. C. H. Smith, and H. Ehrhardt, *J. Geophys. Res.* **69**, 2349 (1964).
79. R. F. Stebbings, B. R. Turner, and A. C. H. Smith, *J. Chem. Phys.* **38**, 2277 (1963).
80. R. F. Stebbings and J. A. Rutherford, *J. Geophys. Res.* **73**, 1035 (1968).
81. P. M. Banks and T. E. Holzer, *Planet. Space Sci.* **16**, 1019 (1968).
82. T. E. Holzer, and P. M. Banks, *Planet. Space Sci.* **17**, 1074 (1969).
83. G. W. Adams and A. J. Masley, *Planet. Space Sci.* **14**, 277 (1966).

CHAPTER 10

Aeronomic Reactions

10.1 Introduction

The last chapter discussed in detail the kinematic aspects of particle collisions where the chemical identity of the particles did not change. In the present chapter these basic ideas are extended to consider chemical reactions which are of fundamental importance to the physical state of the upper atmosphere and ionosphere. To a large extent aeronomic studies require a knowledge of reaction rates between different neutral and ion species over a large range of temperatures and densities including the possibility of excited internal energy states. Above 130 km only binary collisions are important, but at lower altitudes three-body collisions and the possibility of ion clusters must be considered.

It is generally found that experimental values are not sufficiently well known for precise analysis in aeronomy and that theoretical determinations, no matter how detailed, cannot provide exact values. Thus, the values that are adopted here may be subject to systematic errors when experimental results vary widely.

10.2 Radiative Recombination of Atoms

During a collision of two atoms, there is a possibility of molecular formation if the kinetic energy of relative motion is transformed into radiation. In this case photon emission must occur during the collision time interval t_{12}. Thus, the number of recombinations $N_{12}(r)$ per unit volume and time is related to the total number of collisions N_{12} per unit volume and time by the relation

$$N_{12}(r) = N_{12}t_{12}GA_{ji} \quad (10.1)$$

in which A_{ji} is the probability of spontaneous emission and G, the statistical factor, is the ratio of the statistical weights of the recombination state to all other electronic states. If a collision time $t_{12} = (\sigma/\pi)^{1/2}/v$ is used, a very rough approximation is obtained which gives some indication of the magnitude of $N_{12}(r)$.

With this approximation for the collision time, Eq. (10.1) leads to

$$N_{12}(r) = (n_1 n_2 \sigma^{3/2}/\pi^{1/2})GA_{ji}. \quad (10.2)$$

Taking $\sigma = 10^{-15}$ cm^2, an approximate numerical result is given by

$$N_{12}(r) = 10^{-23} n_1 n_2 GA_{ij}. \quad (10.3)$$

If oxygen atoms are considered, one obtains:

(i) The Schumann–Runge system with $G = 1.8$ for the three states of $^3\Sigma$ and the two states of $^3\Pi$ and with an oscillator strength of the order of 0.16,

$$N_{0-0}(\text{S–R}) = 4\times 10^{-16} n(^3\text{P}) n(^1\text{D}) \quad \text{cm}^{-3}\,\text{sec}^{-1}. \quad (10.4)$$

(ii) Herzberg system with $G = 1/43$ and with $GA_{ij} = 0.4$,

$$N_{0-0}(\text{H}) = 4\times 10^{-24} n^2(^3\text{P}) \quad \text{cm}^{-3}\,\text{sec}^{-1}. \quad (10.5)$$

Numerical values such as those of (10.4) and (10.5) are order of magnitude estimates of radiative recombination for allowed and forbidden transitions. Since recombination of two normal atoms generally leads to forbidden transitions, the recombination coefficient is low. Moreover, the coefficient for allowed transitions will not be greater than 5×10^{-16} cm^3 sec^{-1}. As a result, radiative recombination is unimportant when compared with other atmospheric reactions, but it must be considered as a possibility when specific emissions in the airglow are studied.

Consideration can be given to molecules which have an excited state arising from the combination of ground state atoms. In the terrestrial atmosphere the most important molecular products are [1]:

$$\text{O}(^3\text{P}) + \text{H}^+(^1\text{S}) \rightarrow \text{OH}^+(^3\Pi) \rightarrow \text{OH}^+(^3\Sigma) + h\nu \quad (10.6)$$

$$\text{O}(^3\text{P}) + \text{H}(^2\text{S}) \rightarrow \text{OH}(^2\Sigma) \rightarrow \text{OH}(^2\Pi) + h\nu \quad (10.7)$$

$$N(^4S) + O(^3P) \rightarrow NO(^4\Pi) \rightarrow NO(^2\Pi) + h\nu \tag{10.8}$$

$$N^+(^3P) + O(^3P) \rightarrow NO^+(^1\Pi) \rightarrow NO^+(^1\Sigma) + h\nu \tag{10.9}$$

$$N^+(^3P) + N(^4S) \rightarrow N_2{}^+(^2\Pi) \rightarrow N_2{}^+(^2\Sigma) + h\nu \tag{10.10}$$

$$O^+(^4S) + O(^2P) \rightarrow O_2{}^+(^2\Pi) \rightarrow O_2{}^+(^2\Pi) + h\nu. \tag{10.11}$$

A complete theoretical study [2] has been made of the rate at which free atoms form molecules by radiative association. The result, expressed as (10.2), has been used to obtain the rate coefficient of Eq. (10.10). Over a wide temperature range up to at least 2000°K the rate coefficient $r(N^+, N)$ is of the order of

$$r(N^+, N) = 3 \times 10^{-17} \quad cm^3\ sec^{-1}. \tag{10.12}$$

Experimental data are available [3, 4] for reaction (10.8). The process corresponds to a preassociation (which is the inverse of predissociation) with δ band emission from the C $^2\Pi$ state of NO (see *Fig. 10.1*):

$$N(^4S) + O(^3P) \rightleftharpoons NO(a\ ^4\Pi) \rightarrow NO(C\ ^2\Pi) \rightarrow NO(X\ ^2\Pi) + h\nu. \tag{10.13}$$

This association of two atoms results in nitric oxide formation with a rate coefficient

$$k(N, O) = 1 \times 10^{-17} \quad cm^3\ sec^{-1}. \tag{10.14}$$

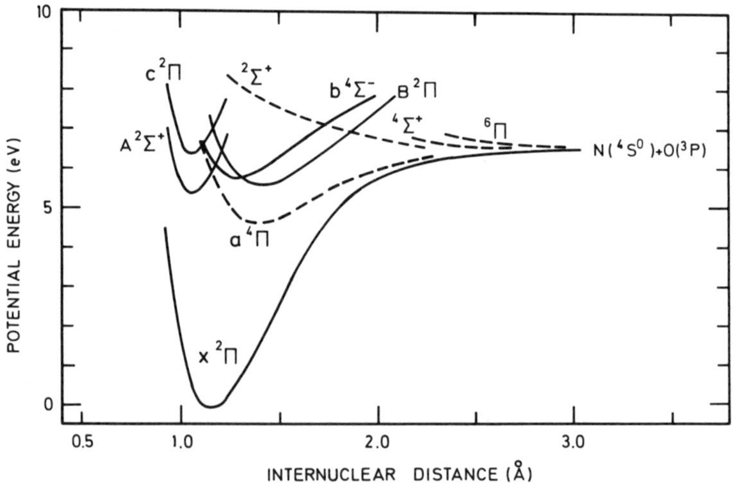

Fig. 10.1 Potential energy curves for NO.

The chemiluminescent combination [5] of O and NO can be a two-body process [6]

$$O + NO \rightarrow NO_2 + h\nu \tag{10.15}$$

with a rate coefficient of the order [7, 8]

$$k(O, NO) = 6\times 10^{-17}(300/T)^{3/2} \quad cm^3 \, sec^{-1}. \tag{10.16}$$

The spectral distribution of emitted radiation (3875 to 14,000 Å) is given [9] in *Fig. 10.2*.

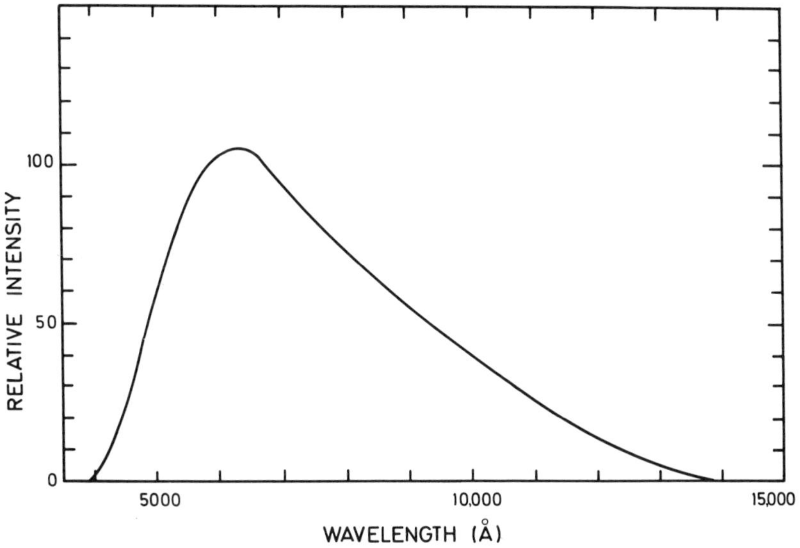

Fig. 10.2 Relative intensity of photon emission from the reaction $NO + O + M \rightarrow NO_2 + M + h\nu$. From Fontijn *et al.* [9].

It is generally found that reactions such as (10.13) and (10.15) are more readily accomplished through a three-body association rather than by two-body recombination.

10.3 Recombination of Atoms by Three-Body Collisions

Recombination by a three-body collision is an effective process as long as the concentration of the third body is sufficiently large. Following an analysis similar to that of the last section it can be shown that values of the recombination rate $N_{123}(r)$ such as

$$N_{123}(r) \sim 2 \times 10^{-32} n(\mathrm{M}) n^2 \quad \mathrm{cm^{-3}\ sec^{-1}} \tag{10.17}$$

should be considered as typical. For example, the reaction [10]

$$\mathrm{H + H + Ar \rightarrow H_2 + Ar + 103.2 \quad kcal\ mole^{-1}} \tag{10.18}$$

has a rate coefficient

$$a_{34} = 2 \times 10^{-32} \quad \mathrm{cm^6\ sec^{-1}} \tag{10.19}$$

when an argon atom is the third body and when the temperature is of the order of 300°K.

The association of oxygen atoms has been the subject of several studies in the laboratory [11, 12] and has a rate coefficient of the order of 3×10^{-33} cm^6 sec^{-1}. Recent experimental analysis [13] indicates that the molecular formation in the presence of molecular nitrogen

$$\mathrm{O + O + N_2 \rightarrow O_2 + N_2 + 118 \quad kcal\ mole^{-1}} \tag{10.20}$$

has a rate coefficient varying with temperature. It can be expressed as

$$k(\mathrm{O_2, N_2}) = 3 \times 10^{-33} (300/T)^3 \quad \mathrm{cm^6\ sec^{-1}}. \tag{10.21}$$

An important aspect of the association of oxygen atoms is the emission of molecular bands (*Fig. 10.3*) which are observed in the airglow and for which measurements have been made in laboratory [14, 15]. The Herzberg system

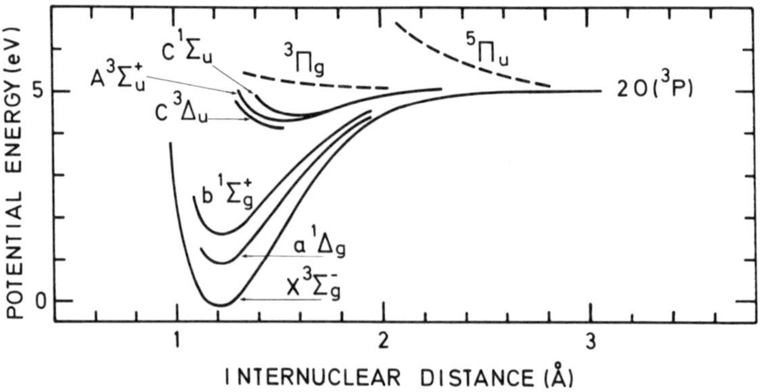

Fig. 10.3 Potential energy curves for O_2.

is due to the three-body process

$$O(^3P) + O(^3P) + N_2 \rightarrow O_2(A\ ^3\Sigma_u^+) + N_2 \tag{10.22}$$

which has a rate coefficient [14]

$$k(O_2, A\ ^3\Sigma_u^+) = 2\times 10^{-37} \quad cm^6\ sec^{-1}. \tag{10.23}$$

The red Atmospheric band system can be produced by

$$O(^3P) + O(^3P) + N_2 \rightarrow O_2(b\ ^1\Sigma_g^+) + N_2 \tag{10.24}$$

with a rate coefficient [14]

$$k(O_2, b\ ^1\Sigma_g^+) = 2\times 10^{-37} \quad cm^6\ sec^{-1}. \tag{10.25}$$

The kinetics of three-body chemiluminescent processes can be represented by a simplified scheme of the following type:

(k_{XY}): $X + Y + M \rightarrow XY^* + M$ (association)

(k_M): $XY^* + M \rightarrow XY + M$ (quenching)

(k_E): $XY^* \rightarrow XY + h\nu$ (emission).

This sequence implies an emission rate I given by

$$I = \frac{k_E k_{XY} n(M) n(X) n(Y)}{k_E + k_M n(M)} \tag{10.26}$$

so at low pressure when $k_E \gg k_M n(M)$

$$I = k_{XY} n(M) n(X) n(Y) \tag{10.27}$$

while at high pressure when $k_E \ll k_M n(M)$

$$I = (k_{XY} k_E / k_M) n(X) n(Y). \tag{10.28}$$

Unfortunately, it does not appear possible to explain the observed nightglow emissions of the Herzberg ($A\ ^3\Sigma_u^+ \rightarrow X\ ^3\Sigma_g^-$) and Atmospheric ($b\ ^1\Sigma_g^+ \rightarrow X\ ^3\Sigma_g^-$) bands by atomic association since the rate coefficients (10.23) and (10.25) are too small.

Laboratory investigation [14,15] of the rate coefficient for the Chapman mechanism

$$O + O + O \rightarrow O_2 + O(^1S) \tag{10.29}$$

support the identification of this process as the source of the 5577 Å green line in the airglow. The reaction rate for (10.29) is measured as [157]

$$k(O_2, O\ ^1S) = 4.8\times10^{-33} \quad cm^6\ sec^{-1}. \tag{10.30}$$

When atomic nitrogen is involved the reaction, processes leading to O(I)(^{1}S) are

$$N + N + O \rightarrow N_2 + O(^1S) \tag{10.31}$$

with

$$k(N_2, O\ ^1S) = 1\times10^{-33} \quad cm^6\ sec^{-1} \tag{10.32}$$

and

$$N + O + O \rightarrow NO + O(^1S) \tag{10.33}$$

with

$$k(NO, O\ ^1S) = 3\times10^{-33} \quad cm^6\ sec^{-1}. \tag{10.34}$$

If the reaction rates for (10.29), (10.31), and (10.33) are correct, at least 10^{12} cm^{-3} oxygen atoms are needed in order to explain the observed peak of the 5577 Å nightglow near 95 km. If the concentration peak of oxygen atoms is less than 10^{12} cm^{-3}, a proportionally larger rate coefficient for (10.29) is needed. Since experimental rates [14, 15] for the processes given by Eqs. (10.29), (10.31), and (10.33) lead to values between 1.5×10^{-34} and 5×10^{-33} cm^6 sec^{-1}, it is reasonable to assume that the three-body association process of Eq. (10.29) is the principal excitation mechanism for O(I)(^{1}S).

Direct nitrogen and oxygen atom recombination with molecular nitrogen as a third body leads to

$$N + O + N_2 \rightarrow NO + N_2 + 151 \quad kcal\ mole^{-1} \tag{10.35}$$

which has a rate coefficient on the order of [13]

$$k(NO, N_2) = 1.1\times10^{-32}(300/T)^{1/2} \quad cm^6\ sec^{-1}. \tag{10.36}$$

The molecular excitation of NO(B $^2\Pi$) leading to the emissions of the β bands has a smaller rate coefficient given by

$$k(NO, \beta) = 1\times10^{-34} \quad cm^6\ sec^{-1}. \tag{10.37}$$

A wide diversity of experimental results are available [11, 16, 17] for the association of nitrogen atoms according to the process (see *Fig. 10.4*)

$$N + N + N_2 \rightarrow N_2 + N_2 + 225 \quad kcal\ mole^{-1}. \tag{10.38}$$

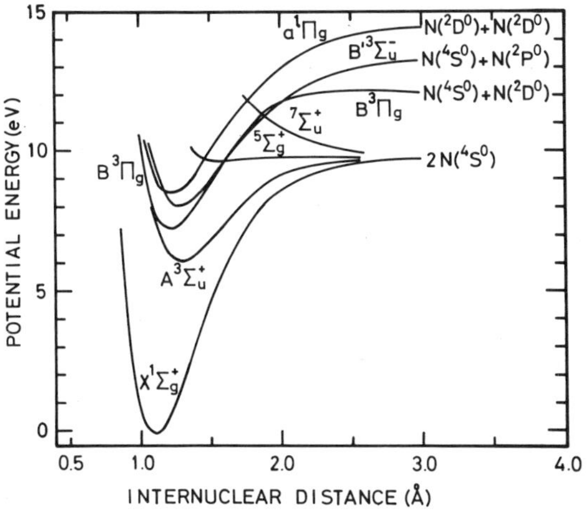

Fig. 10.4 Potential energy curves for N_2.

The rate coefficient may be

$$k(N_2, N_2) = 4\times 10^{-33}(300/T)^2 \quad cm^6 \ sec^{-1}. \tag{10.39}$$

Ground state (^4S) nitrogen atoms in (10.38) yield the $^5\Sigma_g{}^+$ electronic state which crosses the B $^3\Pi_g$ state and other upper electronic states of N_2. These lead to the emission of the N_2 First Positive system and other infrared (B $^3\Sigma_u{}^- \rightarrow$ B $^3\Pi_g$) and ultraviolet (a $^1\Pi_g \rightarrow$ X $^1\Sigma_g{}^+$) bands.

The rate coefficient of the chemiluminescent process

$$N(^4S) + N(^4S) + M \rightleftharpoons N_2(^5\Sigma_g{}^+) + M \tag{10.40}$$

with

$$N_2(^5\Sigma_g{}^+) \rightarrow N_2(B\ ^3\Pi_g) \rightarrow N_2(A\ ^3\Sigma_u{}^+) + h\nu \tag{10.41}$$

is

$$k(N_2, {}^5\Sigma_g{}^+) = 2\times 10^{-34} \quad cm^6 \ sec^{-1}. \tag{10.42}$$

Since the nitrogen nightglow from (10.41) is at least a thousand times less intense than that from the A $^3\Sigma_u{}^+$ and b $^1\Sigma_g{}^+$ states of O_2 arising from oxygen atom recombination, it follows that atomic nitrogen is not an important constituent of the lower thermosphere.

Reactions in which two atoms and a molecule are involved are important in the lower thermosphere where the concentration of atoms reaches a maximum. But association reactions involving one atom and two molecules are usually more important in the mesosphere and stratosphere. The most important three-body process of this type is the association

$$O + O_2 + N_2 \rightarrow O_3 + N_2 + 24 \quad \text{kcal mole}^{-1} \tag{10.43}$$

whose rate coefficient is uncertain. A possible value is [18, 19]

$$k(O, O_2) = (5 \pm 2.5)\times 10^{-34}(300/T)^3 \quad \text{cm}^6 \text{ sec}^{-1} \tag{10.44}$$

indicating that the variation due to temperature is almost a factor of 10 from the stratosphere to the mesopause. If a negative activation energy is introduced in the formula of the rate coefficient, Eq. (10.44) can be written as

$$k(O, O_2) = (3.3 \pm 1)\times 10^{-35}e^{950/T} \quad \text{cm}^6 \text{ sec}^{-1}. \tag{10.45}$$

In any case, the rates indicated by Eqs. (10.44) and (10.45) have been chosen to follow experimental values normalized to 300°K.

The three-body association of O with NO leads to

$$O + NO + N_2 \rightarrow NO_2 + N_2 + 72 \quad \text{kcal mole}^{-1} \tag{10.46}$$

and has a rate coefficient of 1.1×10^{-31} cm^6 sec^{-1} at 300°K. Its variation with temperature suggests a negative activation energy such that [7, 12, 20]

$$k(O, NO) = 3\times 10^{-33}e^{900/T} \quad \text{cm}^6 \text{ sec}^{-1}. \tag{10.47}$$

Such an expression corresponds to

$$k(O, NO) = 6\times 10^{-32}(300/T)^{3/2} \quad \text{cm}^6 \text{ sec}^{-1}. \tag{10.48}$$

The chemiluminescent association of O and NO can be considered as both a three-body process after Eq. (10.46) and as a much slower two-body process after Eq. (10.15). This association also leads to a nightglow continuum in the low thermosphere.

Another three-body reaction which is important in the mesosphere is the association of hydrogen atoms with molecular oxygen

$$H + O_2 + M \rightarrow HO_2 + M + 46 \quad \text{kcal mole}^{-1} \tag{10.49}$$

Since the rate coefficient is $(2.2 \pm 0.2)\times 10^{-32}$ cm^6 sec^{-1} at 293°K with argon Ar as third particle M, the following value can be used for aeronomic purposes ($T < 300°K$) [21–24]

$$k(\mathrm{H, O_2}) = 5\times 10^{-32}(300/T)^2 \quad \mathrm{cm^6\ sec^-} \tag{10.50}$$

while a general formula can be adopted as

$$k(\mathrm{H, O_2}) = 7.5\times 10^{-33} e^{600/T} \quad \mathrm{cm^6\ sec^{-1}}. \tag{10.51}$$

10.4 Atom Interchange

Reaction such as

$$\mathrm{XY + Z \rightarrow XZ + Y} \tag{10.52}$$

in which a single bond is broken require an activation energy A which sometimes can derive from the kinetic energy of relative motion. Because A can be large, many reactions of the form (10.52) are very slow at atmospheric temperature. Taking the activation energy into account, the reaction rate for Eq. (10.52) can be written as

$$k \propto f e^{-A/RT} \quad \mathrm{cm^3\ sec^{-1}} \tag{10.53}$$

in which R is the gas constant (1.987 cal $\mathrm{deg^{-1}\ mole^{-1}}$) and f is the steric hindrance factor.

Activation energies are not generally well known. Using the results of the last chapter, however, an approximate form for (10.53) is

$$k = 1.5\times 10^{-11} T^{1/2} f \exp(-503.3A/T) \quad \mathrm{cm^3\ sec^{-1}} \tag{10.54}$$

in which A is expressed in kcal $\mathrm{mole^{-1}}$.

Various interchange reactions will be considered in more detail in the analysis of chemical processes in Chapters 11–13. Some of the most important atom interchange reactions are

$$\mathrm{O_3 + H \rightarrow OH + O_2} \tag{10.55}$$

$$\mathrm{OH + O \rightarrow O_2 + H} \tag{10.56}$$

$$\mathrm{O_2 + N \rightarrow NO + O} \tag{10.57}$$

$$\mathrm{NO + N \rightarrow N_2 + O} \tag{10.58}$$

$$\mathrm{O_3 + O \rightarrow O_2 + O_2}. \tag{10.59}$$

Our experimental knowledge concerning the rate coefficients for these processes is not yet complete, but has increased rapidly in recent years. Systematic reviews of the existing data can be found in the literature for reactions involving hydrogen [24, 25], nitrogen, and oxygen [26].

The important reaction (10.55) leads to excited OH molecules with $v'' \leq 9$, according to the process [26]

$$(a_2):\quad \mathrm{H} + \mathrm{O_3} \rightarrow \mathrm{O_2} + \mathrm{OH}(\mathrm{X}\ ^2\Pi)_{v''\leq 9} + 77 \quad \text{kcal mole}^{-1}. \tag{10.60}$$

The level $v'' = 9$ of X $^2\Pi$ has an energy of 75 kcal and is the highest energetically accessible level which can be vibrationally excited in the exchange reaction Eq. (10.60). This reaction was postulated [27] to explain the observed airglow emission of OH up to the vibrational level $v'' = 9$ but not $v'' = 10$ (81 kcal). Subsequent laboratory investigations have confirmed that Eq. (10.60) gives rise to OH* excited up to $v'' = 9$.

The rate coefficient for (10.60) has practically no activation energy and is approximately [26, 28]

$$a_2 = 1.5\times 10^{-12} T^{1/2} \quad \text{cm}^3\ \text{sec}^{-1}. \tag{10.61}$$

A reaction similar to (10.60) of particular interest in the stratosphere and lower mesosphere is

$$(b_4):\quad \mathrm{NO} + \mathrm{O_3} \rightarrow \mathrm{O_2} + \mathrm{NO_2}^* + 48 \quad \text{kcal mole}^{-1} \tag{10.62}$$

with a chemiluminescence in the red and infrared [29]. The overall rate for the formation of ground state and electronically excited NO_2 is [30–32]

$$b_4 = 5\times 10^{-14} T^{1/2} e^{-1200/T} \quad \text{cm}^3\ \text{sec}^{-1}; \tag{10.63}$$

while for the chemiluminescent reaction alone [33],

$$b_4{}^* = 5\times 10^{-14} T^{1/2} e^{-2000/T} \quad \text{cm}^3\ \text{sec}^{-1}. \tag{10.64}$$

The reaction

$$(k_3):\quad \mathrm{O}(^3\mathrm{P}) + \mathrm{O_3} \rightarrow 2\mathrm{O_2} + 94 \quad \text{kcal mole}^{-1} \tag{10.65}$$

is of particular importance in the stratosphere and mesosphere since it is the final loss process of atomic oxygen. Its rate coefficient, which is of great importance in aeronomy, is not well known [12]. Recent information suggests that at 300°K its value should be [33]

$$k_3(\mathrm{O}, \mathrm{O_3}) = 9\times 10^{-15} \quad \text{cm}^3\ \text{sec}^{-1}. \tag{10.66}$$

The product molecules of reaction (10.65) are not necessarily in their ground states and the reaction may proceed more rapidly to give

$$O(^3P) + O_3 \rightarrow O_2^*(^3\Sigma_g^-) + O_2(^1\Delta_g) + 71 \quad \text{kcal mole}^{-1} \qquad (10.67)$$

which makes somewhat less than 71 kcal available for the vibrational excitation of $O_2(^3\Sigma_g^-)$. The exact distribution of energy between translation and vibration is not known, but it is not possible to use this energy through the reaction

$$O_2^*(^3\Sigma_g^-) + O_3 \rightarrow 2O_2(^3\Sigma_g^-) + O(^1D) \qquad (10.68)$$

since at least 69 kcal are required, the energies being:

$$67 \qquad 70 \qquad 73 \ \text{kcal mole}^{-1}$$

for

$$v = 17 \qquad 18 \qquad 19$$

The destruction of the vibrationally excited molecules may proceed through near resonance collisions [34] such as

$$O_2^*(v_1) + O_2^*(v_2) \rightarrow O_2^*(v_3) + O_2^*(v_4), \qquad (10.69)$$

where $v_1 + v_2 \simeq v_3 + v_4$, with a possibility of an exchange of internal vibrational energies more frequently than one in every hundred collisions. Furthermore, the process of atom–atom interchange

$$O + O_2^*(v_1) \rightarrow O_2^*(v < v_1) + O \qquad (10.70)$$

with no activation energy can be an efficient deactivating mechanism [35].

The atom–atom interchange reaction (10.56) leads to the formation of O_2^* with vibrational energy less than 13 kcal mole^{-1} ($v \leq 3$) since

$$O(^3P) + OH(X\,^2\Pi) \rightarrow H + O_2^*(^3\Sigma_g^-) + 16.6 \quad \text{kcal mole}^{-1} \qquad (10.71)$$

Since this reaction has a negligible activation energy, it is an important production mechanism of hydrogen atoms in the mesosphere. The reaction rate for (10.71) appears to be $(5 \pm 2) \times 10^{-11}$ cm^3 sec^{-1} at 300°K [24].

Atomic nitrogen undergoes two important reactions via (10.57) and (10.58). In the thermosphere the atom–atom interchange reaction

$$N + O_2 \rightarrow O + NO^* + 32 \quad \text{kcal mole}^{-1} \qquad (10.72)$$

leads to an excited nitric oxide molecule. Since the vibrational energy

$v'' = 6$ is about 31 kcal mole^{-1}, the NO* molecule can be populated up to the sixth vibrational level. The activation energy of Eq. (10.72) is relatively high, being not less than 6 kcal mole^{-1}. Various laboratory determinations of the rate coefficients of reaction (10.72) can be fitted by different expressions. Considering all the experimental values [24], the following rate coefficient is adopted:

$$b(\mathrm{N, O_2}) = 5\times 10^{-13} T^{1/2} e^{-3500/T} \quad \mathrm{cm^3\ sec^{-1}}. \qquad (10.73)$$

Values such as [36]

$$b(\mathrm{N, O_2}) = 2\times 10^{-13} T^{1/2} e^{-3000/T} \quad \mathrm{cm^3\ sec^{-1}} \qquad (10.74)$$

have been used previously and give different values at very low temperatures such as found at the mesopause. To emphasize the temperature dependence of (10.73), it is found that there is a variation of a factor of more than 10^6 between the mesopause ($T < 200$°K with $b(\mathrm{N, O_2}) = 2\times 10^{-19}$ cm^3 sec^{-1}) and the thermosphere [$T > 1000$°K with $b(\mathrm{N, O_2}) = 4\times 10^{-13}$ cm^3 sec^{-1}].

As far as reaction (10.58) is concerned, consideration of spin indicates that only O(^{3}P) atoms are produced as [28, 37]

$$\mathrm{N + NO \rightarrow O(^3P) + N_2^* + 75} \quad \mathrm{kcal\ mole^{-1}} \qquad (10.75)$$

with negligible activation energy. The N_2^* is formed with sufficient energy in vibrational states to dissociate O_3 [12]. The rate coefficient may be taken as [12]

$$b(\mathrm{N, NO}) = 1.5\times 10^{-12} T^{1/2} \quad \mathrm{cm^3\ sec^{-1}}. \qquad (10.76)$$

A discussion of aeronomic effects of the interchange reactions will be given in Chapters 11–13.

10.5 Radiative Recombination of Atomic Ions

When a collision occurs between an electron and an ion, a recombination is possible whereby the ionization energy of the neutral atom is released as a photon and internal excitation. This process, represented as

$$(\alpha_{\mathrm{R}}):\quad \mathrm{e + X^+ \rightarrow X + } h\nu \qquad (10.77)$$

has a recombination coefficient of about 10^{-12} cm^3 sec^{-1} which is relatively insensitive to the identity of the ion. Little laboratory information is available about the temperature dependence of radiative recombination but theoretical studies [38, 39] indicate that the recombination coefficient varies as $T_e^{-0.5}$ for capture into low lying electron energy states. The overall radiative recombination coefficient is proportional to $T_e^{-0.7}$ normalized to about 5×10^{-12} cm^3 sec^{-1} at $T_e = 250$°K; i.e.,

$$\alpha_R \simeq 5\times10^{-12}(250/T_e)^{0.7} \quad \text{cm}^3\ \text{sec}^{-1}. \tag{10.78}$$

Theoretical computations made for atomic oxygen [40] give the following values:

T_e (°K):	250	500	1000	2000
α_R (10^{-12} cm^3 sec^{-1}):	3.7	2.4	1.5	0.9

Since the dissociative recombination of electrons and molecular ions, discussed in the next section, gives recombination coefficients on the order of 10^{-7} cm^3 sec^{-1}, the radiative recombination of atomic ions is seldom of importance except in the equatorial regions where tropical ultraviolet airglow emissions are observed.

10.6 Dissociative Recombination of Ions

The process

$$(\alpha_D):\quad XY^+ + e \rightarrow X' + X'' \tag{10.79}$$

is rapid ($\alpha_D \sim 10^{-6}$ cm^3 sec^{-1}) and the principal means of ion recombination in the ionosphere [41, 42].

Dissociative recombination occurs as a result of a transition to some state of the molecule in which the constituents move apart and gain kinetic energy under the action of their mutual repulsion so that the neutralization is rendered permanent by virtue of the Frank–Condon principle. It is clear from the theoretical analysis that it is practically impossible to predict the temperature variations of the dissociative recombination characterized by α_D. Since the dissociation depends on the presence of a repulsive curve, it may be assumed that the absolute value of the recombinance should be different for different molecules. The problem is quite complicated since there is no simple law establishing the variation of the repulsive curves of molecules and the absolute value of α_D and its dependence on temperature must be found experimentally [43]. In any event, it must be remembered that the electron–ion collision frequency decreases with temperature ($\propto T_e^{-3/2}$) and the normal tendency of the temperature dependence of α_D, therefore, should be, in a complicated way, a decrease with increasing temperatures. In other words, if values of α_D between 10^{-6} and 10^{-8} cm^3 sec^{-1} and variation such as $T_e^{(-1\pm0.5)}$ are assumed, they do not contradict theoretical considerations.

Since dissociative recombination is the final loss process of electrons in the thermosphere it is essential to know the exact values of the dissociative recombination processes. These are given in the following sections for the most important atmospheric molecular ions.

10.6.1 Nitric Oxide Ions

The recombination

$$NO^+ + e \rightarrow N + O \tag{10.80}$$

is one of the only few processes of this ion. It may give nitrogen atoms in ^4S and ^2D states and oxygen atoms in ^3P and ^1D states (see Figs. 9.6 and 9.9) according to the following processes:

$$NO^+(^1\Sigma^+) + e \rightarrow N(^4S) + O(^3P) + 2.76 \text{ eV} \tag{10.81}$$

$$NO^+(^1\Sigma^+) + e \rightarrow N(^4S) + O(^1D) + 0.79 \text{ eV} \tag{10.82}$$

$$NO^+(^1\Sigma^+) + e \rightarrow N(^2D) + O(^3P) + 0.38 \text{ eV.} \tag{10.83}$$

The recombination process Eq. (10.82) does not conserve spin and this reaction is generally neglected in aeronomic problems. In any case, no experimental analysis has been made on the nature of the excited atoms, and the recombination coefficients refer to the total recombinance which is given by [44–51]

$$\alpha_D(NO^+) = (5 \pm 2) \times 10^{-7} \quad \text{cm}^3 \text{ sec}^{-1} \tag{10.84}$$

which is greater than the value which was suggested a few years ago for aeronomic problems. A variation with temperature suggested by laboratory measurements [49] gives values such as:

T_e (°K):	200	300	600	1500
$\alpha_D(NO^+)$ (10^{-7} cm^3 sec^{-1}):	7.4	4.1	2.1	0.85

An adequate approximation to the temperature dependent recombination for NO^+ can be adopted as

$$\alpha_D(NO^+) = 4.1 \times 10^{-7}(300/T_e) \quad \text{cm}^3 \text{ sec}^{-1} \tag{10.85}$$

It is noted that a maximum value of about 10^{-6} cm^3 sec^{-1} can be reached at the mesopause for $T_e \simeq 150$°K.

10.6.2 Molecular Nitrogen Ions

Dissociative recombination of N_2^+ by the process

$$N_2^+ + e \rightarrow N^* + N^{**} \tag{10.86}$$

can lead to nitrogen atoms in ^4S, ^2D, and ^3S states according to

$$N_2^+(^2\Sigma_g^+) + e \rightarrow N(^4S) + N(^4S) + 5.82 \quad eV \tag{10.87}$$

$$N_2^+(^2\Sigma_g^+) + e \rightarrow N(^4S) + N(^2D) + 3.44 \quad eV \tag{10.88}$$

$$N_2^+(^2\Sigma_g^+) + e \rightarrow N(^4S) + N(^2P) + 2.24 \quad eV \tag{10.89}$$

$$N_2^+(^2\Sigma_g^+) + e \rightarrow N(^2D) + N(^2D) + 1.06 \; eV. \tag{10.90}$$

The total recombination coefficient was measured at 300°K as [38, 52, 53]

$$\alpha_D(N_2^+) = (2.9 \pm 0.3) \times 10^{-7} \text{ cm}^3 \text{ sec}^{-1}. \tag{10.91}$$

If a temperature dependence is considered such as [54]

$$\alpha_D(N_2^+) = 3 \times 10^{-7} (300/T_e)^{1/3} \quad \text{cm}^3 \text{ sec}^{-1} \tag{10.92}$$

the variation with electron temperature leads to $\alpha_D(N_2^+) = 3.8 \times 10^{-7}$ cm^3 sec^{-1} at 150°K and 1.4×10^{-7} cm^3 sec^{-1} at 3000°K. The variation of the ionic recombinations leading to atoms in the excited states ^2D and ^2P is not yet known.

10.6.3 Molecular Oxygen Ions

Dissociative recombination of O_2^+

$$O_2^+ + e \rightarrow O^* + O^* \tag{10.93}$$

leads to oxygen atoms in ^3P, ^1D, and ^1S states:

$$O_2^+(^2\Pi_g) + e \rightarrow O(^3P) + O(^3P) + 6.95 \quad eV \tag{10.94}$$

$$O_2^+(^2\Pi_g) + e \rightarrow O(^3P) + O(^1D) + 4.98 \quad eV \tag{10.95}$$

$$O_2^+(^2\Pi_g) + e \rightarrow O(^3P) + O(^1S) + 2.76 \quad eV \tag{10.96}$$

$$O_2^+(^2\Pi_g) + e \rightarrow O(^1D) + O(^1D) + 3.01 \quad eV \tag{10.97}$$

$$O_2^+(^2\Pi_g) + e \rightarrow O(^1D) + O(^1S) + 0.79 \quad eV. \tag{10.98}$$

According to recent laboratory measurements [43, 55–57] the recombination coefficient should vary as:

T_e (°K):	200	300	400	500	600	700
$\alpha_D(O_2^+)$ (10^{-7} cm³ sec⁻¹):	3.0	2.2	1.6	1.3	1.2	1.0

A suitable expression for the temperature dependent dissociative recombination rate is [43]

$$\alpha_D(O_2^+) = 2.2\times 10^{-7}(300/T_e)^{0.7 \text{ to } 1.0} \tag{10.99}$$

for temperatures above 300°K.

10.6.4 Other Ions

The dissociative recombination coefficient for CO_2^+ has been measured [49] as $(3.8 \pm 0.5)\times 10^{-7}$ cm^3 sec^{-1} at 300°K. At 210°K a value of $\sim 4\times 10^{-7}$ cm^3 sec^{-1} was obtained.

At the present time laboratory data for the dissociative recombination of H_3O^+ and its hydrated forms $H_3O^+ \cdot (H_2O)_n$ is becoming increasingly available.

The recombination process

$$H_3O^+ + e \rightarrow H + OH^* \tag{10.100}$$

has been measured in the laboratory. At 2300°K a value $\alpha_D(H_3O^+) = 2 \times 10^{-7}$ cm^3 sec^{-1} was found [58], while at 300°K [38],

$$\alpha_D(H_3O^+) = (1.3 \pm 0.6)\times 10^{-6} \quad \text{cm}^3\,\text{sec}^{-1}. \tag{10.101}$$

Dissociative recombination rates for the hydrated forms of H_3O^+ give [38]

$$\alpha_D(H_3O^+ \cdot H_2O) = (2.7 \pm 1)\times 10^{-6} \quad \text{cm}^3\,\text{sec}^{-1} \tag{10.102}$$

$$\alpha_D(H_3O^+ \cdot (H_2O)_2) = (4.6 \pm 1)\times 10^{-6} \quad \text{cm}^3\,\text{sec}^{-1}. \tag{10.103}$$

For hydrated ions of larger mass it has been found [38] that the dissociative recombination coefficient approaches 10^{-5} cm^3 sec^{-1}.

10.7 Ion–Neutral Reactions

Thermal ion–atom interchange or charge transfer reactions have until recently presented a great problem to aeronomy. A few years ago laboratory measurements of such reactions were not sufficiently accurate [59] to permit an understanding of the behavior in the ionosphere. Today many of the more important reaction rates have been obtained from experimental measurements [60–63] and it is now possible to obtain a consistent picture of ionospheric reactions without assuming values based on questionable theoretical arguments.

10.7.1 Atomic Oxygen Ions

The predominant atomic ion in the ionosphere, O^+, is subject to the following principal ion–atom interchange reactions (*Fig. 10.5*):

$$(\gamma_1):\quad O^+(^4S) + N_2(^1\Sigma) \rightarrow N(^4S) + NO^+(^1\Sigma) + 1.10\quad \text{eV} \tag{10.104}$$

and

$$(\gamma_2):\quad O^+(^4S) + O_2(^3\Sigma) \rightarrow O(^3P) + O_2^+(^2\Pi) + 1.55\quad \text{eV}. \tag{10.105}$$

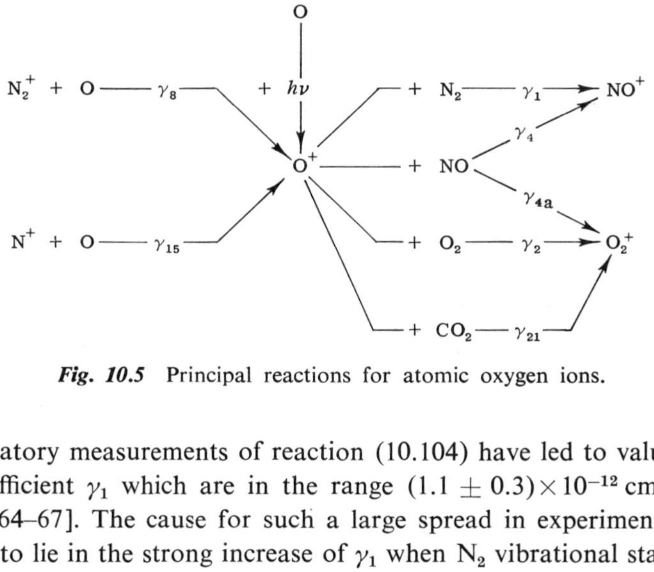

Fig. 10.5 Principal reactions for atomic oxygen ions.

Laboratory measurements of reaction (10.104) have led to values of the rate coefficient γ_1 which are in the range $(1.1 \pm 0.3)\times 10^{-12}$ cm³ sec⁻¹ at 300°K [64–67]. The cause for such a large spread in experimental values appears to lie in the strong increase of γ_1 when N_2 vibrational states $v \geq 2$ are present [68]. The variation of γ_1 with neutral gas temperature is shown in *Fig. 10.6*.

For translation temperatures in the range 80–600°K an approximate expression for γ_1 as a function of the N_2 translational temperature, $T(N_2)$ is

$$\gamma_1 = 1.2\times 10^{-12}(300/T(N_2))\quad \text{cm}^3\,\text{sec}^{-1} \tag{10.106}$$

with an accuracy $\pm 15\%$.

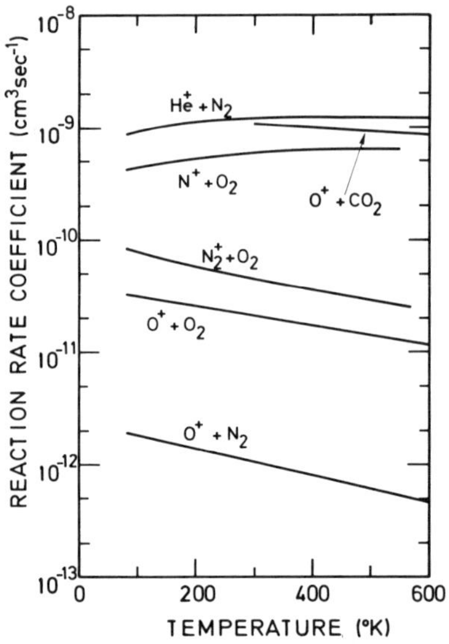

Fig. 10.6 Rate coefficients as a function of translational temperature [62].

Above 600°K Eq. (10.104) has a strong dependence upon the degree oJ vibrational excitation. Measurements for the reaction

$$O^+ + N_2(v) \rightarrow NO^+ + N \tag{10.107}$$

are shown in *Fig. 10.7*. Since reaction (10.107) dominates at temperatures

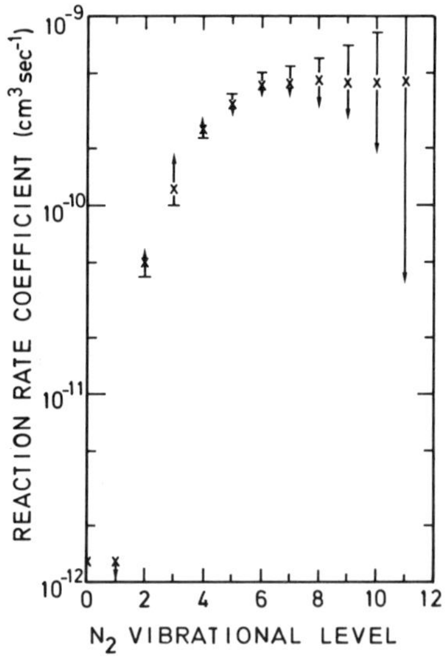

Fig. 10.7 $O^+ + N_2(v)$ reaction rate as a function of vibrational level [65]. $O^+ + N_2 \rightarrow NO^+ + N$.

above 1200 °K, it is important to examine the vibrational temperature dependence of the process. Experimental data given in *Fig. 10.8* show that from 300° to 1200°K there is little variation in γ_1 (the translational temperature is constant at 300°K), showing that the $v = 1$ state contributes little to the reaction. Above 1200°K, however, a rapid increase occurs which changes γ_1 by an order of magnitude between 1000° and 3500°K.

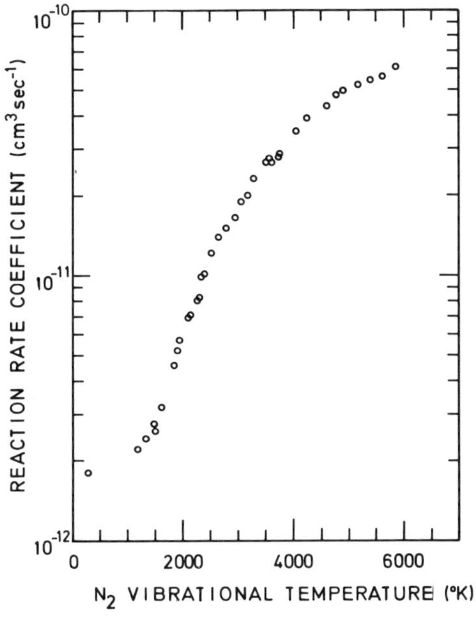

Fig. 10.8 $O^+ + N_2(v)$ reaction rate as a function of vibrational temperature [102]. $O^+ + N_2 \rightarrow NO^+ + N$.

Measurements of the overall reaction rate in the aeronomically important translational range 600°–2000°K have been made [69, 70] and computations normalized to available data [71, 72] can be used to indicate the expected variation. The results, shown in *Fig. 10.9*, imply an extremely variable behavior for γ_1 even when the vibrational levels of N_2 are in thermal equilibrium with the gas temperature. For disturbed conditions it is possible for T_v to be greater than T_n. In this situation large values of γ_1 may exist even in a relatively cool neutral atmosphere. In fact, if $T_v > 1200$°K γ_1 is determined almost completely by the vibrationally excited component of N_2 for the complete range of thermospheric temperatures.

A reaction related to (10.104) is

$$(\gamma_1^*): \quad O^+(^2D) + N_2(X\,^1\Sigma_g, v' = 0) \rightarrow N_2^+(A\,^2\Pi_u, v' = 1) + O(^3P). \tag{10.108}$$

Measurements [73] show that this charge transfer occurs rapidly for O^+ kinetic energies near 5 eV. Extrapolation of this data to thermal energies would indicate a reaction rate $\sim 10^{-9}$ cm^3 sec^{-1}. Since about half the O^+ produced through photoionization is present as ^2D, it is clear that reaction (10.108) is of importance to the F region (see Chapter 20).

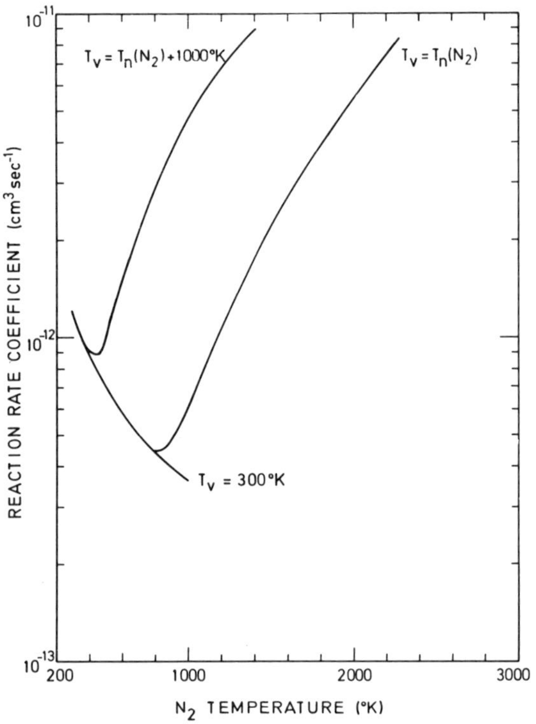

Fig. 10.9 $O^+ + N_2(v)$ reaction rate as a function of neutral gas and vibrational temperature.

The reaction of O^+ with O_2 has been extensively measured [64, 66, 67, 74, 75]. A rate coefficient

$$\gamma_2 = (2.0 \pm 0.2) \times 10^{-11} \quad \text{cm}^3\ \text{sec}^{-1} \tag{10.109}$$

provides an adequate mean value at 300°K. The temperature dependence of the reaction, shown in Fig. 10.6, can be taken into account by an equation such as

$$\gamma_2 = (2.0 \pm 0.2) \times 10^{-11} (300/T)^{1/2}. \tag{10.110}$$

In order to consider all ion–atom interchange processes in which atomic oxygen ions are involved, the following unimportant reactions may be added to Eqs. (10.104) and (10.105):

$$(\gamma_4): \quad O^+(^4S) + NO(^2\Pi) \rightarrow O(^3P) + NO^+(^1\Sigma) + 4.35 \quad \text{eV} \tag{10.111}$$

and

$$(\gamma_4{}^*): \quad O^+(^4S) + NO(^2\Pi) \rightarrow N(^4S) + O_2{}^+(^2\Pi) + 0.16 \quad \text{eV}. \tag{10.112}$$

The rate coefficient for Eq. (10.111) has been measured with the result that [76]

$$\gamma_4 < 1.3 \times 10^{-12} \quad \text{cm}^3 \text{ sec}^{-1} \tag{10.113}$$

at 278°K. In addition $\gamma_4{}^* \ll 10^{-12}$ cm^3 sec^{-1}. As a result, neither (10.111) nor (10.112) can be important for the loss of O^+ in the lower thermosphere.

Recent laboratory measurements [77, 78] have shown the importance of O^+ reaction with CO_2:

$$(\gamma_{21}): \quad O^+(^4S) + CO_2(^1\Sigma) \rightarrow CO(^1\Sigma) + O_2{}^+(^2\Pi) + 1.22 \quad \text{eV}. \tag{10.114}$$

This reaction has a large cross section. From laboratory measurements

$$\gamma_{21} = (1.2 \pm 0.4) \times 10^{-9} \quad \text{cm}^3 \text{ sec}^{-1} \tag{10.115}$$

with a temperature dependence tending to decrease γ_{21} for increasing temperatures.

Reactions of O^+ with NO_2 and N_2O have been measured [79]:

$$(\gamma_{47}): \quad O^+ + NO_2 \rightarrow NO_2{}^+ + O \tag{10.116}$$

$$(\gamma_{48}): \quad O^+ + N_2O \rightarrow N_2O^+ + O \tag{10.117}$$

with the result that

$$\gamma_{47} = 1.6 \times 10^{-9} \quad \text{cm}^3 \text{ sec}^{-1} \tag{10.118}$$

$$\gamma_{48} = 6.3 \times 10^{-10} \quad \text{cm}^3 \text{ sec}^{-1} \tag{10.119}$$

with an accuracy of $\pm 30\%$ at 278°K.

10.7.2 Molecular Oxygen Ions

The reactions for $O_2{}^+$ are somewhat simpler to describe than those for O^+. *Figure 10.10* shows the aeronomically important processes.

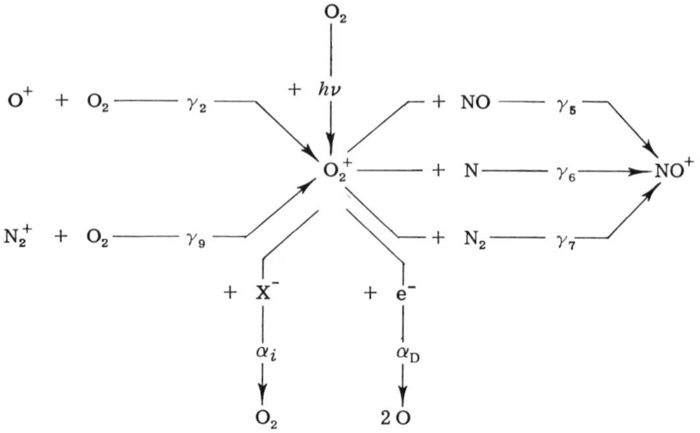

Fig. 10.10 Principal reactions for molecular oxygen ions.

The reaction

$$(\gamma_5): \quad O_2^+(^2\Pi) + NO(^2\Pi) \rightarrow O_2(^3\Sigma) + NO^+(^1\Sigma) + 2.81 \quad eV \qquad (10.120)$$

has been observed in the laboratory [80–82]. Its rate coefficient appears to be

$$\gamma_5 = 6.3 \times 10^{-10} \quad cm^3\, sec^{-1} \qquad (10.121)$$

with an accuracy $\pm 30\%$ and little temperature dependence.

Two ion–atom interchange reactions with nitrogen are

$$(\gamma_6): \quad O_2^+(^2\Pi) + N(^4S) \rightarrow O(^3P) + NO^+(^1\Sigma) + 4.19 \quad eV \qquad (10.122)$$

with [81]

$$\gamma_6 = (1.5 \pm 0.5) \times 10^{-10} \quad cm^3\, sec^{-1} \qquad (10.123)$$

at room temperature and

$$(\gamma_7): \quad O_2^+(^2\Pi) + N_2(^1\Sigma) \rightarrow NO(^2\Pi) + NO^+(^1\Sigma) + 0.94 \quad eV. \qquad (10.124)$$

This last reaction is probably very slow since a double-bond breakage is required and it has not been observed in laboratory [83]; γ_7 is certainly less than 2×10^{-16} cm^3 sec^{-1}. Neverthless, such very small values could still permit reaction (10.124) to be important in the lower ionosphere [84].

With regard to CO_2, a charge transfer reaction with O_2^+ does not occur since such a reaction would be endothermic.

10.7.3 Molecular Nitrogen Ions

The principal processes affecting N_2^+ are shown in *Fig. 10.11*. In the thermosphere the most important reaction is

$$(\gamma_3): \quad N_2^+(^2\Sigma) + O(^3P) \rightarrow N(^4S) + NO^+(^1\Sigma) + 3.06 \quad \text{eV} \tag{10.125}$$

for which the measured rate coefficient at 295°K is [80, 85]

$$\gamma_3 = (1.4 \pm 0.7) \times 10^{-10} \quad \text{cm}^3\,\text{sec}^{-1}. \tag{10.126}$$

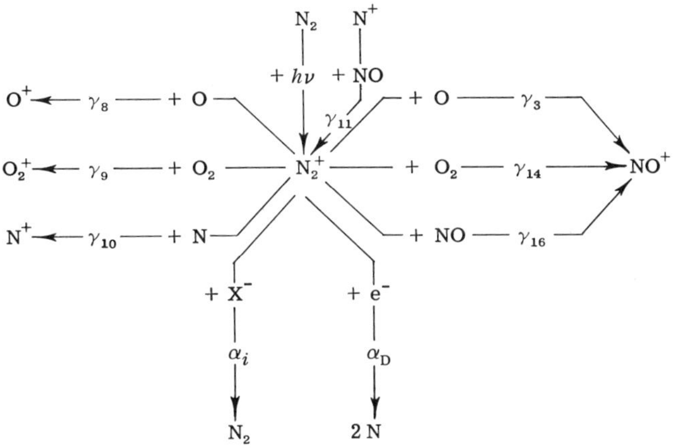

Fig. 10.11 Principal reactions for molecular nitrogen ions.

Charge transfer and ion–atom interchange reactions must also be considered. The process

$$(\gamma_8): \quad N_2^+(^2\Sigma) + O(^3P) \rightarrow N_2(^1\Sigma) + O^+(^4S) + 1.96 \quad \text{eV} \tag{10.127}$$

appears to have a rate coefficient which is negligible compared with reaction (10.125); i.e. [85],

$$\gamma_8 < 10^{-11} \quad \text{cm}^3\,\text{sec}^{-1}.$$

The charge transfer with molecular oxygen

$$(\gamma_9): \quad N_2^+(^2\Sigma) + O_2(^3\Sigma) \rightarrow N_2(^1\Sigma) + O_2^+(^2\Pi) + 3.5 \quad \text{eV} \tag{10.128}$$

is a rapid process since [75, 86, 87]

$$\gamma_9 = (5 \pm 2) \times 10^{-11} \quad \text{cm}^3\,\text{sec}^{-1}. \tag{10.129}$$

The charge transfer

$$(\gamma_{10}):\quad N_2^+(^2\Sigma) + N(^4S) \rightarrow N_2(^1\Sigma) + N^+(^3P) + 1.03 \quad \text{eV} \tag{10.130}$$

which has not been measured, is not important compared with Eq. (10.128).

An exothermic reaction such as

$$(\gamma_{14}):\quad N_2^+(^2\Sigma) + O_2(^3\Sigma) \rightarrow NO(^2\Pi) + NO^+(^1\Sigma) + 4.45 \quad \text{eV} \tag{10.131}$$

is not a competing reaction with the charge transfer process Eq. (10.128) since it requires a double-bond breakage. Thus, it can be neglected even for the formation of NO and NO^+.

The charge transfer of N_2^+ with NO has been measured [80]:

$$(\gamma_{16}):\quad N_2^+(^2\Sigma) + NO(^2\Pi) \rightarrow N_2(^1\Sigma) + NO^+ + 6.31 \quad \text{eV} \tag{10.132}$$

$$\gamma_{16} = 3.3 \times 10^{-10} \quad \text{cm}^3\,\text{sec}^{-1} \tag{10.133}$$

with an accuracy of $\pm 30\%$. However, since $n(O_2) \gg n(NO)$ such a reaction is not important for the loss of N_2^+ except perhaps in auroras.

Several reactions occur with carbon oxides. Since the ionization potential of N_2 is greater than that of CO_2, the change transfer

$$(\gamma_{22}):\quad N_2^+(^2\Sigma) + CO_2(^1\Sigma) \rightarrow N_2(^1\Sigma) + CO_2^+(^2\Pi) + 1.81 \text{ eV} \tag{10.134}$$

is rapid with [88–90]

$$\gamma_{22} = 9 \times 10^{-10} \quad \text{cm}^3\,\text{sec}^{-1}. \tag{10.135}$$

The parallel charge transfer in CO is [97]

$$(\gamma_{23}):\quad N_2^+(^2\Sigma) + CO(^1\Sigma) \rightarrow N_2(^1\Sigma) + CO^+(^2\Sigma) + 1.57 \text{ eV} \tag{10.136}$$

with

$$\gamma_{23} = 7 \times 10^{-11} \quad \text{cm}^3\,\text{sec}^{-1}. \tag{10.137}$$

10.7.4 Atomic Nitrogen Ions

The most important reactions involving N^+ are shown in *Fig. 10.12.* Two reactions are possible in O_2:

$$(\gamma_{12}):\quad N^+(^3P) + O_2(^3\Sigma) \rightarrow N(^4S) + O_2^+(^2\Pi) + 2.49 \quad \text{eV} \tag{10.138}$$

$$(\gamma_{12}^*):\quad N^+(^3P) + O_2(^3\Sigma) \rightarrow O(^3P) + NO^+(^1\Sigma) + 6.67 \quad \text{eV}. \tag{10.139}$$

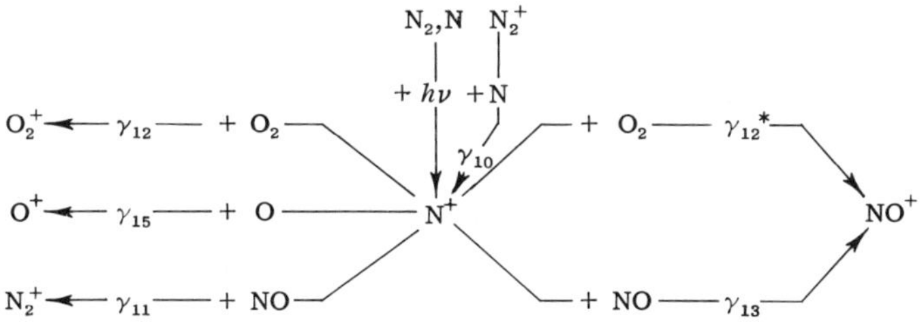

Fig. 10.12 Principal reactions for atomic nitrogen ions.

Reactions (10.138) and (10.139) involve ion–atom interchange as well as charge transfer. Laboratory measurements [64, 67] give

$$\gamma_{12} + \gamma_{12}^* = (7 \pm 3) \times 10^{-10} \quad \text{cm}^3\,\text{sec}^{-1} \tag{10.140}$$

with

$$\gamma_{12} \leq 5 \times 10^{-10} \quad \text{cm}^3\,\text{sec}^{-1} \tag{10.141}$$

and

$$\gamma_{12}^* \geq 2 \times 10^{-10} \quad \text{cm}^3\,\text{sec}^{-1}. \tag{10.142}$$

Reactions with NO involve similar processes:

$$(\gamma_{11}): \quad N^+(^3P) + NO(^2\Pi) \rightarrow O(^3P) + N_2{}^+(^2\Sigma) + 2.2 \quad \text{eV}, \tag{10.143}$$

$$(\gamma_{13}): \quad N^+(^3P) + NO(^2\Pi) \rightarrow N(^4S) + NO^+(^1\Sigma) + 5.28 \quad \text{eV}. \tag{10.144}$$

The last reaction is more important than the former. Measurements give [81]

$$\gamma_{13} = (7 \pm 2) \times 10^{-10} \quad \text{cm}^3\,\text{sec}^{-1}. \tag{10.145}$$

The charge transfer process

$$(\gamma_{15}): \quad N^+(^3P) + O(^3P) \rightarrow N(^4S) + O^+(^4S) + 0.93 \quad \text{eV} \tag{10.146}$$

could possibly be a loss of N^+, but its rate has not been measured. Other similar reactions such as

$$(\gamma_{24}): \quad N^+(^3P) + CO_2(^1\Sigma) \rightarrow N(^4S) + CO_2{}^+(^2\Pi) + 0.78 \quad \text{eV} \tag{10.147}$$

$$(\gamma_{25}): \quad N^+(^3P) + CO(^1\Sigma) \rightarrow N(^4S) + CO^+(^2\Sigma) + 0.54 \quad \text{eV} \tag{10.148}$$

yield reaction rates [77]

$$\gamma_{24} = 1.3 \times 10^{-9} \quad \text{cm}^3\,\text{sec}^{-1}, \tag{10.149}$$

$$\gamma_{25} = 5 \times 10^{-10} \quad \text{cm}^3\,\text{sec}^{-1}. \tag{10.150}$$

10.7.5 Helium Ions

The most important reactions of helium ions in the upper ionosphere can be summarized as

$$(\gamma_{17}):\quad He^+(^2S) + O_2(^3\Sigma) \rightarrow He(^1S) + O(^3P) + O^+(^4S) + 5.85 \quad eV, \tag{10.151}$$

$$(\gamma_{17}^*):\quad He^+(^2S) + O_2(^3\Sigma) \rightarrow He(^1S) + O_2^+(^2\Pi) + 12.25 \quad eV, \tag{10.152}$$

$$(\gamma_{18}):\quad He^+(^2S) + N_2 \rightarrow He(^1S) + N(^4S) + N^+(^3P) + 0.28 \quad eV, \tag{10.153}$$

$$(\gamma_{18}^*):\quad He^+(^2S) + N_2 \rightarrow He(^1S) + N_2^+(C\,^2\Sigma)_{v=3,4}, \tag{10.154}$$

$$(\gamma_{19}):\quad He^+(^2S) + NO(^2\Pi) \rightarrow He(^1S) + O(^3P) + N^+(^3P) + 3.54 \quad eV. \tag{10.155}$$

A special discussion on processes (10.151) to (10.155) is required since such reactions can have at times an important effect upon the ionospheric ion distribution to above the F_2 layer.

Reactions (10.151) to (10.155) have been measured [92] and rate coefficients not less than 10^{-9} cm^3 sec^{-1} have been obtained. When He^+ reacts with O_2 there is a possibility of charge transfer or a dissociation with O and O^+ products. However, the first mechanism which must be considered is

$$He^+(^2S) + O_2(^3\Sigma) \rightarrow He(^1S) + O_2^+(c\,^4\Sigma)_{v=0}. \tag{10.156}$$

This corresponds to an accidentally resonant charge transfer process since the c $^4\Sigma_u$ state of the O_2^+ ion is the upper state of the Hopfield emission band of oxygen which lies at $(198{,}125 \pm 30)$ cm^{-1} (for $v' = 0$) and at $(199{,}665 \pm 30)$ cm^{-1} (for $v' = 1$) above the ground state of O_2. The ionization potential of helium is 198,315 cm^{-1}, and the transition to O_2^+ (c $^4\Sigma$) follows the Franck–Condon principle.

Experimental results show that [92–94]

$$\gamma_{17} = 1.5 \times 10^{-9} \quad cm^3\, sec^{-1} \tag{10.157}$$

and perhaps

$$\gamma_{17}^* < 2 \times 10^{-10} \quad cm^3\, sec^{-1}. \tag{10.158}$$

It appears that (10.156) is followed by predissociation which leads to an overall reaction (10.151) unless photon emission occurs, in which case (10.152) is obtained.

In any case, the emission band system from c $^4\Sigma_u^-$ is obtained at 2132, 2170, and 2218 Å corresponding to vibration assignments (0–4), (0–5), and (0–6), respectively. However, if most of the O^+ ions resulting from He^+–O_2 interaction have approximately 1 eV kinetic energy, the resonant charge transfer mechanism Eq. (10.151) is not the only possible process. In such circumstances experimental data would be consistent with the possible formation of a short-lived molecule HeO^+ postulated by Bates and Patterson [95].

The most important loss processes for He^+ in the upper ionosphere are reactions (10.153) and (10.154). The latter reaction involves a resonant charge transfer followed by the emission of the second negative system of N_2^+

$$N_2^+(C\ ^2\Sigma_u^+)_{v=3,4} \rightarrow N_2^+(X\ ^2\Sigma_g^+) + h\nu \qquad (10.159)$$

i.e., the emission of (3–9) and (4–10) bands at 1914 and 1910 Å, respectively, and also by a forbidden predissociation via a $^4\Pi$ state

$$N_2^+(C\ ^2\Sigma_u^+) \rightarrow N(^4S) + N^+(^3P). \qquad (10.160)$$

The He^+ to N_2 loss rate has been measured a number of times with all recent results falling in the range [96]

$$\gamma_{18} + \gamma_{18}^* = (1.4 \pm 0.3)\times 10^{-9} \quad \text{cm}^3\ \text{sec}^{-1}. \qquad (10.161)$$

The branching ratio between γ_{18} and γ_{18}^* is not so well known but seems to be in the range 1–2 [94, 97]. The total rate constant is independent of temperature (or relative kinetic energy) over the range 300°–1400°K and is insensitive to N_2 vibrational temperature changes in the range 300°–6000°K [96].

At one time it was thought that reaction (10.154) might involve ground state $N_2^+(^2\Sigma)$ and an excess kinetic energy of 9.01 eV. Such an energy, when shared between He and N_2^+, would be sufficient to permit the He atom to escape the earth's gravitational field (see Chapter 16). However, measurements show that the exothermicity of (10.154) is placed almost entirely in the internal energy states of N_2^+ and is not available as a source of kinetic energy. (For a review of this topic, see Patterson [98].)

Reaction (10.155) of He^+ with NO is rapid with a dissociative charge transfer rate coefficient [99, 100],

$$\gamma_{19} = 1.5\times 10^{-9} \quad \text{cm}^3\ \text{sec}^{-1}. \qquad (10.162)$$

Measurements have also been made of the reaction of He^+ with CO_2 [99]:

$$(\gamma_{49}):\quad He^+(^2S) + CO_2(^1\Sigma) \rightarrow He(^1S) + CO_2^+(^2\Pi) + 10.9 \quad eV$$
$$\rightarrow He + CO + O^+$$
$$\rightarrow He + O + CO^+ \qquad (10.163)$$

with an overall reaction rate

$$\gamma_{49} = 1.2\times 10^{-9} \quad cm^3\ sec^{-1} \qquad (10.164)$$

at 300°K. The branching ratios of (10.163) are such that 80% is associated with CO^+, 15% with O^+, and 5% with CO_2^+.

For He^+ in CO the reaction

$$(\gamma_{50}):\quad He^+ + CO \rightarrow He + O + C^+ + 2.2 \quad eV \qquad (10.165)$$

has been observed with a reaction rate [102]

$$\gamma_{50} = 1.7\times 10^{-9} \quad cm^3\ sec^{-1}. \qquad (10.166)$$

10.7.6 Carbon Oxide Ions

Reactions of CO_2^+ and CO^+ preceed rapidly with most atmospheric gases.

Charge transfer of CO_2^+ with O_2 leads to

$$(\gamma_{26}):\quad CO_2^+ + O_2 \rightarrow CO_2 + O_2^+ + 1.71 \quad eV \qquad (10.167)$$

with [101–103]

$$\gamma_{26} = 5\times 10^{-11} \quad cm^3\ sec^{-1}. \qquad (10.168)$$

In atomic oxygen two reactions follow [80]:

$$(\gamma_{26a}):\quad CO_2^+ + O \rightarrow O_2^+ + CO \qquad (10.169)$$

$$(\gamma_{26b}):\quad CO_2^+ + O \rightarrow O^+ + CO_2 \qquad (10.170)$$

with an overall reaction rate at 295°K of

$$\gamma_{26a} + \gamma_{26b} = 2.6\times 10^{-10} \quad cm^3\ sec^{-1}. \qquad (10.171)$$

The ratio $\gamma_{26a}/\gamma_{26b}$ is difficult to measure but values of $\frac{5}{3}$ have been suggested [80].

The reaction of CO_2^+ with H_2 gives

$$(\gamma_{40}): \quad CO_2^+ + H_2 \rightarrow CO_2H^+ + H \tag{10.172}$$

at a rate [104]

$$\gamma_{40} = 1.4 \times 10^{-9} \quad \text{cm}^3\,\text{sec}^{-1}. \tag{10.173}$$

The reaction of CO_2^+ with H was suspected to lead to the charge transfer [105]:

$$(\gamma_{51a}): \quad CO_2^+ + H \rightarrow CO_2 + H^+ + 0.2 \quad \text{eV}. \tag{10.174}$$

Laboratory measurements [106] have shown that the principal channel for (10.174) is actually

$$(\gamma_{51b}): \quad CO_2^+ + H \rightarrow COH^+ + O \tag{10.175}$$

with a reaction rate

$$\gamma_{51a} + \gamma_{51b} = 6 \times 10^{-10} \quad \text{cm}^3\,\text{sec}^{-1}. \tag{10.176}$$

Separation of γ_{51a} and γ_{51b} is difficult, but initial computations indicate the ratio $\gamma_{51a}/\gamma_{51b} = 0.2$ [106].

The reverse reaction,

$$(\gamma_{52}): \quad H^+ + CO_2 \rightarrow COH^+ + O \tag{10.177}$$

has a large cross section with a reaction rate of [106]

$$\gamma_{52} = 3 \times 10^{-9} \quad \text{cm}^3\,\text{sec}^{-1} \tag{10.178}$$

at 300°K with an accuracy of $\pm 30\%$. Such a process provides a rapid destruction of H^+ in CO_2 rich atmospheres such as found on Mars and Venus.

Reactions of CO^+ in similar gases leads to the following reactions [77, 102]:

$$(\gamma_{27}): \quad CO^+ + O_2 \rightarrow CO + O_2^+ + 1.95 \quad \text{eV} \tag{10.179}$$

$$\gamma_{27} = 2.0 \times 10^{-10} \quad \text{cm}^3\,\text{sec}^{-1} \tag{10.180}$$

$$(\gamma_{28}): \quad CO^+ + CO_2 \rightarrow CO + CO_2^+ + 0.24 \quad \text{eV} \tag{10.181}$$

$$\gamma_{28} = 1.1 \times 10^{-9} \quad \text{cm}^3\,\text{sec}^{-1} \tag{10.182}$$

$$(\gamma_{39}):\quad CO^+ + H_2 \rightarrow COH^+ + H \tag{10.183}$$

$$\gamma_{39} = 2\times 10^{-9} \quad cm^3\ sec^{-1} \tag{10.184}$$

$$(\gamma_{53}):\quad CO^+ + N \rightarrow NO^+ + C + 0.26 \quad eV \tag{10.185}$$

$$\gamma_{53} < 2\times 10^{-11} \quad cm^3\ sec^{-1} \tag{10.186}$$

$$(\gamma_{54}):\quad CO^+(^2\Sigma) + O(^3P) \rightarrow CO(^1\Sigma) + O^+(^4S) + 0.33 \quad eV \tag{10.187}$$

$$\gamma_{54} = 1.4\times 10^{-10} \quad cm^3\ sec^{-1} \tag{10.188}$$

$$(\gamma_{55}):\quad CO^+(^2\Sigma) + NO(^2\Pi) \rightarrow CO(^1\Sigma) + NO^+(^1\Sigma) + 4.75\ eV \tag{10.189}$$

$$\gamma_{55} = 3.3\times 10^{-10} \quad cm^3\ sec^{-1}. \tag{10.190}$$

Reactions of carbon ions with O_2 and CO_2 lead to the following ion–atom interchanges:

$$(\gamma_{29}):\quad C^+(^2P) + O_2(^3\Sigma) \rightarrow O(^3P) + CO^+(^2\Sigma) + 3.25 \quad eV \tag{10.191}$$

with [77]

$$\gamma_{29} = 1.1\times 10^{-9} \quad cm^3\ sec^{-1} \tag{10.192}$$

and

$$(\gamma_{30}):\quad C^+(^2P) + CO_2(^1\Sigma) \rightarrow CO(^1\Sigma) + CO^+(^2\Sigma) + 2.91\ eV \tag{10.193}$$

with [77]

$$\gamma_{30} = 1.9\times 10^{-9} \quad cm^3\ sec^{-1}. \tag{10.194}$$

10.7.7 Argon Ions

Argon, which is ionized in the D region by x rays, has a very short lifetime as an ion, as can be seen from the following exothermic reactions:

$$(\gamma_{31}):\quad Ar^+(^2P) + N_2(^1\Sigma) \rightarrow Ar(^1S) + N_2^+(^2\Pi) + 0.18 \quad eV \tag{10.195}$$

with

$$\gamma_{31} = 6\times 10^{-12} \quad cm^3\ sec^{-1}\ [107] \quad or \quad 6.6\times 10^{-11} \quad cm^3\ sec^{-1}\ [108]. \tag{10.196}$$

$$(\gamma_{32}):\quad Ar^+(^2P) + O_2(^3\Sigma) \rightarrow Ar(^1S) + O_2^+(^2\Pi) + 3.70 \quad eV \tag{10.197}$$

with [107, 108]

$$\gamma_{32} = 1.1\times 10^{-10} \quad \text{cm}^3\ \text{sec}^{-1}. \tag{10.198}$$

$$(\gamma_{33}):\quad \text{Ar}^+(^2\text{P}) + \text{CO}_2(^1\Sigma) \rightarrow \text{Ar}(^1\text{S}) + \text{CO}_2{}^+(^2\Pi) + 1.99\ \text{eV} \tag{10.199}$$

with [107, 108]

$$\gamma_{33} = 7\times 10^{-10} \quad \text{cm}^3\ \text{sec}^{-1}. \tag{10.200}$$

$$(\gamma_{34}):\quad \text{Ar}^+(^2\text{P}) + \text{CO}(^1\Sigma) \rightarrow \text{Ar}(^1\text{S}) + \text{CO}^+(^2\Sigma) + 1.75\quad \text{eV} \tag{10.201}$$

with [107, 108]

$$\gamma_{34} = 1\times 10^{-10} \quad \text{cm}^3\ \text{sec}^{-1}. \tag{10.202}$$

$$(\gamma_{35}):\quad \text{Ar}^+(^2\text{P}) + \text{NO}(^2\Pi) \rightarrow \text{Ar}(^1\text{S}) + \text{NO}^+(^1\Sigma) + 5.49\quad \text{eV} \tag{10.203}$$

with [108]

$$\gamma_{35} = 3.9\times 10^{-10} \quad \text{cm}^3\ \text{sec}^{-1}. \tag{10.204}$$

Reactions of argon ions with molecular hydrogen lead to the following sequence:

$$(\gamma_{36}):\quad \text{Ar}^+ + \text{H}_2 \rightarrow \text{H} + \text{ArH}^+ \tag{10.205}$$

$$(\gamma_{37}):\quad \text{ArH}^+ + \text{H}_2 \rightarrow \text{Ar} + \text{H}_3{}^+ \tag{10.206}$$

and

$$(\gamma_{38}):\quad \text{H}_3{}^+ + \text{Ar} + \text{Ar} \rightarrow \text{Ar} + \text{ArH}_3{}^+ \tag{10.207}$$

for which [104]

$$\gamma_{36} \simeq \gamma_{37} \simeq 10^{-9} \quad \text{cm}^3\ \text{sec}^{-1} \tag{10.208}$$

and

$$\gamma_{38} \simeq 10^{-31} \quad \text{cm}^6\ \text{sec}^{-1}. \tag{10.209}$$

10.7.8 Water Vapor and Hydroxyl Ions

Laboratory measurements of reactions in which hydroxyl radicals or water vapor are involved are of potential importance in the analysis of the lower ionosphere.

$$(\gamma_{41}):\quad \text{OH}^+ + \text{H}_2 \rightarrow \text{H} + \text{H}_2\text{O}^+ + 1.19\quad \text{eV} \tag{10.210}$$

with [104]

$$\gamma_{41} = 1.5\times 10^{-9} \quad \text{cm}^3\ \text{sec}^{-1}; \tag{10.211}$$

$$(\gamma_{42}): \quad H_2O^+ + H_2 \rightarrow H + H_3O^+ + 1.82 \quad \text{eV} \tag{10.212}$$

with [104]

$$\gamma_{42} = 1.4\times 10^{-9} \quad \text{cm}^3\ \text{sec}^{-1}; \tag{10.213}$$

$$(\gamma_{43}): \quad OH^+ + O_2 \rightarrow OH + O_2^+ + 1.12 \quad \text{eV} \tag{10.214}$$

with [104]

$$\gamma_{43} = 2\times 10^{-10} \quad \text{cm}^3\ \text{sec}^{-1}; \tag{10.215}$$

$$(\gamma_{44}): \quad H_2O^+ + O_2 \rightarrow H_2O + O_2^+ + 0.56 \quad \text{eV} \tag{10.216}$$

with [104]

$$\gamma_{44} = 2\times 10^{-10} \quad \text{cm}^3\ \text{sec}^{-1}; \tag{10.217}$$

$$(\gamma_{45}): \quad H_2O^+ + H_2O \rightarrow H_3O^+ + OH + 0.52 \quad \text{eV} \tag{10.218}$$

with [109]

$$\gamma_{45} = 4.9\times 10^{-10} \quad \text{cm}^3\ \text{sec}^{-1}; \tag{10.219}$$

$$(\gamma_{46}): \quad OH^+ + H_2O \rightarrow H_3O^+ + O + 1.82 \quad \text{eV} \tag{10.220}$$

with [109]

$$\gamma_{46} = 4.7\times 10^{-10} \quad \text{cm}^3\ \text{sec}^{-1}. \tag{10.221}$$

Thus OH^+ and H_2O^+ should have a short lifetime in an oxygen atmosphere. The associated problem of water vapor cluster ions is discussed in the next section.

10.8 Cluster Reactions of Positive Ions

One of the major advances in our understanding of the ionic structure of the D region revolves about the importance of three-body collisions and the influence of water vapor. For many years it was assumed from theoretical studies and a limited number of experiments that NO^+ and O_2^+ were the major positive ions between 70 and 85 km. Direct observations by rocket-borne mass spectrometers have shown, however, that ions such as H_3O^+ and $H_5O_2^+$ are actually the dominant species. Attempts to explain these ions on the basis of a direct series of water vapor ion reactions failed, principally because of the rapid charge transfer between H_2O^+ and O_2 [reaction (10.216)]. As a consequence, it has become necessary to devise more complicated reaction schemes which involve ion clusters formed in three-body collisions.

Because many of the relevant reactions rates are not known, a detailed study of all possible cluster reactions is difficult to make. For ionospheric purposes, there seem to be three major sequences separately involving molecular oxygen ions, nitric oxide ions, and water vapor cluster ions. Each reaction chain is discussed below.

10.8.1 Molecular Oxygen Ions

The formation of the dimer ion O_4^+ follows from the three-body reaction [110, 111]

$$(\varkappa_1):\quad O_2^+ + O_2 + M \rightarrow O_4^+ + M \tag{10.222}$$

for which the reaction rates shown below have been obtained [112–115]:

T (°K):	82	200	307
$\varkappa_1$ ($cm^6\ sec^{-1}$):	3.1×10^{-29}	2.4×10^{-30}	2.4×10^{-30}
M:	He	He	O_2

The loss of O_4^+ has been measured through two channels:

$$(\varkappa_2):\quad O_4^+ + O \rightarrow O_2^+ + O_3 \tag{10.223}$$

$$(\varkappa_3):\quad O_4^+ + H_2O \rightarrow O_2^+ \cdot H_2O + O_2. \tag{10.224}$$

Very recent experiments give [115]

$$\varkappa_2 = (3 \pm 2)\times 10^{-10}\quad cm^3\ sec^{-1} \tag{10.225}$$

at 295°K.

Since the atomic oxygen density increases rapidly toward the mesopause, such a rapid reaction quickly reduces the density of O_4^+ above 85 km and, as discussed later, prevents the formation of water vapor ions from the oxygen ion cycle [116].

Reaction (10.224) is the key process in the production chain for H_3O^+ and higher-order ions. Its rate has been measured in the range [117, 118]

$$\varkappa_3 = (2.2 \pm 1)\times 10^{-10}\quad cm^3\ sec^{-1}. \tag{10.226}$$

Three-body reactions involving O_2 and N_2 also lead to $O_2^+ \cdot H_2O$:

$$(\varkappa_4): \quad O_2^+ + H_2O + N_2 \rightarrow O_2^+ \cdot H_2O + N_2 \tag{10.227}$$

$$(\varkappa_5): \quad O_2^+ + H_2O + O_2 \rightarrow O_2^+ \cdot H_2O + O_2 \tag{10.228}$$

with measured rates

$$\varkappa_4 = 2.5 \times 10^{-28} \quad \text{cm}^6\ \text{sec}^{-1} \tag{10.229}$$

$$\varkappa_5 = 2.3 \times 10^{-28} \quad \text{cm}^6\ \text{sec}^{-1} \tag{10.230}$$

at 295°K [117, 119].

The cluster ion $O_2^+ \cdot H_2O$ is subject to several reactions involving water vapor. The most important of these are

$$(\varkappa_6): \quad O_2^+ \cdot H_2O + H_2O \rightarrow H_3O^+ + OH + O_2 \tag{10.231}$$

and

$$(\varkappa_7): \quad O_2^+ \cdot H_2O + H_2O \rightarrow H_3O^+ \cdot OH + O_2 \tag{10.232}$$

with [117, 119]

$$\varkappa_6 = (2 \pm 1) \times 10^{-10} \quad \text{cm}^3\ \text{sec}^{-1} \tag{10.233}$$

$$\varkappa_7 = (1.2 \pm 0.3) \times 10^{-9} \quad \text{cm}^3\ \text{sec}^{-1}. \tag{10.234}$$

As indicated, the ratio $\varkappa_6/\varkappa_7$ is subject to experimental uncertainty.

Another reaction involving a third collision partner is

$$(\varkappa_8): \quad O_2^+ \cdot H_2O + H_2O + O_2 \rightarrow O_2^+ \cdot (H_2O)_2 + O_2 \tag{10.235}$$

for which [120]

$$\varkappa_8 \sim 10^{-27} \quad \text{cm}^6\ \text{sec}^{-1} \tag{10.236}$$

has been given.

Examination of reactions (10.222)–(10.235) shows the oxygen ion cycle to proceed from O_2^+ to the hydrated ion $O_2^+ \cdot H_2O$ via the dimer O_4^+. The next step involves reactions with water vapor which deionize O_2 and leave H_3O^+ or $H_3O^+ \cdot OH$ or possibly $O_2^+ \cdot (H_2O)_2$.

10.8.2 Nitric Oxide Ions

Ions of nitric oxide are readily formed through several processes involving atoms and molecules of progressively lower ionization potential. At one time it was thought that NO^+ should be the major positive ion of the D region [121]. Rocket measurements [120], however, have shown that below a certain altitude, (in the range 75–85 km) H_3O^+ and $H_5O_2{}^+$ are the dominant ions, i.e., some reactions are either effective in converting NO^+ into water cluster ions or the ionization is being destroyed through recombination.

Laboratory measurements have shown that binary reactions involving NO^+ and the major atmospheric constituents are very slow. Three-body associative reactions of NO^+ have been observed, however.

The reactions

$$(\varkappa_9):\quad NO^+ + N_2 + He \rightarrow NO^+ \cdot N_2 + He \tag{10.237}$$

$$(\varkappa_{10}):\quad NO^+ + O_2 + He \rightarrow NO^+ \cdot O_2 + He \tag{10.238}$$

illustrate slow three-body associations with rate constants [122]

$$\varkappa_9 < 5\times 10^{-33} \quad cm^6\ sec^{-1} \tag{10.239}$$

$$\varkappa_{10} < 6\times 10^{-34} \quad cm^6\ sec^{-1}. \tag{10.240}$$

Unpublished data of Heimerl and Vanderhoff, however, indicate a faster rate for Eq. (10.237), i.e., $\varkappa_9 = 3.4 \pm 2.4\times 10^{-31}$ cm^6 sec^{-1}.

With CO_2, however, much more rapid reactions occur. Measurements of the rapid associations

$$(\varkappa_{11}):\quad NO^+ + CO_2 + He \rightarrow NO^+ \cdot CO_2 + He \tag{10.241}$$

$$(\varkappa_{12}):\quad NO^+ + CO_2 + N_2 \rightarrow NO^+ \cdot CO_2 + N_2 \tag{10.242}$$

$$(\varkappa_{13}):\quad NO^+ + CO_2 + Ar \rightarrow NO^+ \cdot CO_2 + Ar \tag{10.243}$$

show that CO_2 must be very important in the chemistry of NO^+ in the D region. Values for the rate coefficients have been measured as [122]

$$\begin{aligned} \varkappa_{11} &= (1 \pm 0.3)\times 10^{-29} \quad cm^6\ sec^{-1} && (T = 197°K) \\ &= (7.2 \pm 3.5)\times 10^{-30} \quad cm^6\ sec^{-1} && (T = 235°K) \\ &= (4.0 \pm 2.0)\times 10^{-30} \quad cm^6\ sec^{-1} && (T = 290°K) \end{aligned} \tag{10.244}$$

$$\varkappa_{12} = (2.5 \pm 1.5)\times 10^{-29} \quad cm^6\ sec^{-1} \qquad (T = 200°K) \tag{10.245}$$

$$\varkappa_{13} = (3.1 \pm 1.0)\times 10^{-29} \quad cm^6\ sec^{-1} \qquad (T = 196°K). \tag{10.246}$$

The ultimate fate of the association product is not known in detail. The reaction

$$(\varkappa_{14}):\quad NO^+ \cdot CO_2 + H_2O \rightarrow NO^+ \cdot H_2O + CO_2 \tag{10.247}$$

is rapid, however, and is a means for forming cluster ions of hydrated NO^+. The rate for (10.247) has been measured as [122]

$$\varkappa_{14} \simeq 10^{-9} \quad \text{cm}^3\ \text{sec}^{-1} \tag{10.248}$$

at 200°K.

Reactions between NO^+ and water vapor ions have been observed under laboratory conditions. The reaction

$$(\varkappa_{15}):\quad NO^+ + H_2O + N_2 \rightarrow NO^+ \cdot H_2O + N_2 \tag{10.249}$$

$$\varkappa_{15} = (1.5 \pm 0.1)\times 10^{-28} \quad \text{cm}^6\ \text{sec}^{-1} \quad (T = 295°\text{K}) \tag{10.250}$$

provides a somewhat slower hydration of NO^+ using the most abundant atmospheric gas as a buffer.

The subsequent reactions of $NO^+ \cdot H_2O$ lead to the production of $H_3O^+ \cdot (H_2O)_2$ and HNO_2. The reaction chains involved in such a conversion are

$$(\varkappa_{16}):\quad NO^+ \cdot H_2O + H_2O + M \rightleftharpoons NO^+ \cdot (H_2O)_2 + M \tag{10.251}$$

$$(\varkappa_{17}):\quad NO^+ \cdot (H_2O)_2 + H_2O + M \rightleftharpoons NO^+ \cdot (H_2O)_3 + M \tag{10.252}$$

$$(\varkappa_{18}):\quad NO^+ \cdot (H_2O)_3 + H_2O \rightarrow H_3O^+ \cdot (H_2O)_2 + HNO_2 \tag{10.253}$$

with reaction rates [117]

$$\varkappa_{16} = 1.0\times 10^{-27} \quad \text{cm}^6\ \text{sec}^{-1} \tag{10.254}$$

$$\varkappa_{-16} < 1\times 10^{-13} \quad \text{cm}^3\ \text{sec}^{-1} \tag{10.255}$$

$$\varkappa_{17} = 2.0\times 10^{-27} \quad \text{cm}^6\ \text{sec}^{-1} \tag{10.256}$$

$$\varkappa_{-17} = 1.3\times 10^{-12} \quad \text{cm}^3\ \text{sec}^{-1} \tag{10.257}$$

$$\varkappa_{18} = 8\times 10^{-11} \quad \text{cm}^3\ \text{sec}^{-1} \tag{10.258}$$

at 295°K for N_2 as the third body (reverse reaction rates are indicated with a minus sign). Thus, in a moist atmosphere there will be a rapid conversion of both $O_2{}^+$ and NO^+ to water cluster ions.

The production of $NO^+ \cdot H_2O$ follows from the two separate reactions (10.242) and (10.249). Because CO_2 is more abundant than H_2O (3×10^{-4} compared with 3×10^{-6} of the total atmospheric concentration), reaction (10.242) producing $NO^+ \cdot CO_2$ is very rapid. The next step, involving the replacement of CO_2 by H_2O is also rapid and provides an important step in the elimination of NO^+ in the lower ionosphere.

To avoid the difficulties involved with the slowness of reactions (10.241)–(10.253), the two-body reaction

$$NO^+ \cdot H_2O + H \rightarrow NO + H_3O^+ \tag{10.259}$$

was proposed [123]. Measurements [115] indicate that while this reaction is exothermic, it has a rate less than 10^{-11} cm³ sec⁻¹ at 300°K.

10.8.3 Water Vapor Ions

The reactions among water vapor ions act to determine the degree of clustering found in the lower ionosphere. The basic reactions involve:

$$(\varkappa_{19}):\quad H_3O^+ + H_2O + M \rightarrow H_3O^+ \cdot H_2O + M \tag{10.260}$$

$$(\varkappa_{20}):\quad H_3O^+ \cdot OH + H_2O \rightarrow H_3O^+ \cdot H_2O + OH \tag{10.261}$$

$$(\varkappa_{21}):\quad H_3O^+ \cdot H_2O + H_2O + M \rightleftharpoons H_3O^+ \cdot (H_2O)_2 + M \tag{10.262}$$

$$(\varkappa_{22}):\quad H_3O^+ \cdot (H_2O)_2 + H_2O + M \rightleftharpoons H_3O^+ \cdot (H_2O)_3 + M \tag{10.263}$$

with rate constants at 300°K of [124]

$$\varkappa_{19} = (2 \pm 1)\times10^{-27} \quad \text{cm}^6\ \text{sec}^{-1} \tag{10.264}$$

$$\varkappa_{20} = 3.2\times10^{-9} \quad \text{cm}^3\ \text{sec}^{-1} \tag{10.265}$$

$$\varkappa_{21} = 3\times10^{-27} \quad \text{cm}^6\ \text{sec}^{-1} \tag{10.266}$$

$$\varkappa_{-21} = 1.4\times10^{-17} \quad \text{cm}^3\ \text{sec}^{-1} \tag{10.267}$$

$$\varkappa_{22} = 2\times10^{-27} \quad \text{cm}^6\ \text{sec}^{-1} \tag{10.268}$$

$$\varkappa_{-22} = 4\times10^{-14} \quad \text{cm}^3\ \text{sec}^{-1}. \tag{10.269}$$

Equilibrium ratios K_i for $H^+ \cdot (H_2O)_{n-1}/H^+ \cdot (H_2O)_n$ have been measured [124] in the form $\ln K_i = A/T - B$.

The rate constants for the association reactions tend to increase significantly for decreasing temperatures, while the binary reverse reactions become slower. Thus, there may be appreciable changes from the laboratory values over the range 150°–250°K typical of the lower ionosphere.

10.9 Negative Ions

The explanation of the formation and loss of negative ions is an important aeronomical problem of the lower ionosphere [125–127]. In recent years there has been a rapid increase in laboratory data concerning the reactions of negative ions at low energies. However, only a few successful measurements [128, 129] have been made of the actual negative ion composition of the lower ionosphere so it is not clear whether current reaction models are complete or correct.

The first theoretical studies of negative ions in the D region considered O^- and O_2^- as the principal negative ionic constituents. Later, as laboratory data became available [130–133], it became clear that the molecular ions O_3^-, CO_3^-, NO_2^-, NO_3^-, and cluster ions O_4^-, CO_4^- are produced as a result of O^- and O_2^- collisions with O, O_3, CO_2, and oxides of nitrogen. Thus, it became evident that minor atmospheric constituents are important in the negative ion reaction chains

The first direct measurements of negative ions in the D region showed that such reaction schemes [128, 129] were substantially correct, even though discrepancies appeared, especially with respect to the presence of hydrated ions such as $CO_4^- \cdot H_2O$, $O_2^- \cdot (H_2O)_2$, and $CO_3^- \cdot H_2O$.

The formation of negative ions in the lower ionosphere starts from the attachment of electrons to oxygen molecules and, to a much less extent, atoms. The most important process involves a three-body collision between electrons and two oxygen molecules:

$$(\phi_1): \quad e + O_2 + O_2 \rightarrow O_2^- + O_2 + 0.45 \quad \text{eV} \tag{10.270}$$

for which a rate coefficient can be adopted as [134]

$$\phi_1 = 1.5 \times 10^{-29}(300/T)e^{-600/T} \quad \text{cm}^6 \text{ sec}^{-1} \tag{10.271}$$

or

T (°K):	150	200
ϕ_1 (cm^6 sec^{-1}):	5×10^{-31}	1.0×10^{-30}

With respect to the formation of O^-, the radiative process

$$(\phi_2): \quad e + O \rightarrow O^- + h\nu \tag{10.272}$$

has a rate coefficient

$$\phi_2 = 1.3 \times 10^{-15} \quad \text{cm}^3 \text{ sec}^{-1} \tag{10.273}$$

which is too low for reaction (10.272) to be of any significance.

In the stratosphere and mesosphere where ozone is present, the dissociative attachment

$$(\phi_3): \quad e + O_3 \rightarrow O^- + O_2 + 0.4 \quad \text{eV} \qquad (10.274)$$

can occur with a rate in the range $200° < T < 300°$K [135]

$$\phi_3 = 9 \times 10^{-12}(T/300)^{3/2} \quad \text{cm}^3\ \text{sec}^{-1}. \qquad (10.275)$$

In practice, reaction (10.270) is the only important source of negative ions in the D region.

Once created, O_2^- is transformed into other negative ions by means of charge exchange, interchange, clustering, and cluster interchange reactions. In the following subsections these reactions and the ultimate losses of negative ions are discussed for each negative ion.

10.9.1 Molecular Oxygen Negative Ions

A schematic representation of O_2^- reactions is given in *Fig. 10.13*. The reactions which lead to a loss of negative ions may be summarized as the associative detachments

$$(\lambda_1): \quad O_2^- + O \rightarrow O_3 + e + 0.6 \quad \text{eV} \qquad (10.276)$$

with [136]

$$\lambda_1 = 3.2 \times 10^{-10} \quad \text{cm}^3\ \text{sec}^{-1}; \qquad (10.277)$$

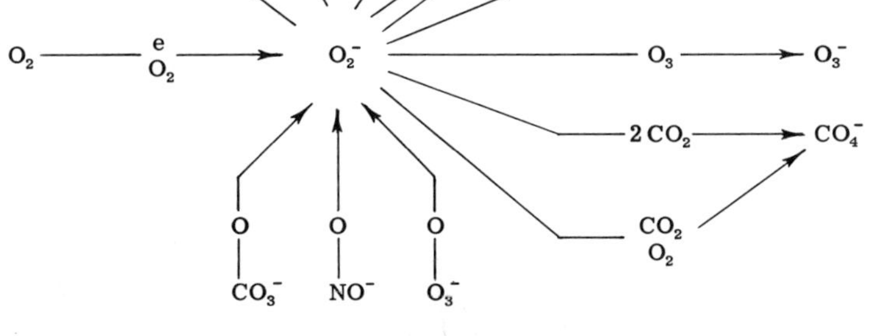

Fig. 10.13 Principal reactions for O_2^-.

$$(\lambda_2):\quad O_2^- + N \rightarrow NO_2 + e + 4.1 \quad eV \tag{10.278}$$

with [136]

$$\lambda_2 = 5.0 \times 10^{-10} \quad cm^3\,sec^{-1}; \tag{10.279}$$

and the collisional detachment involving excited oxygen molecules:

$$(\lambda_3):\quad O_2^- + O_2(^1\Delta_g) \rightarrow O_2 + O_2 + e + 0.6 \quad eV \tag{10.280}$$

with [137]

$$\lambda_3 \simeq 2.0 \times 10^{-10} \quad cm^3\,sec^{-1}. \tag{10.281}$$

Exposure of O_2^- to sunlight also leads to photodetachment by the process

$$(\lambda_4):\quad O_2^- + h\nu\ (\lambda < 2.8\ \mu m) \rightarrow O_2 + e \tag{10.282}$$

with a detachment rate of [138]

$$\lambda_4 = 0.3 \quad sec^{-1}. \tag{10.283}$$

Reactions leading to other negative molecular ion species involve minor atmospheric constituents:

$$(\delta_1):\quad O_2^- + O_3 \rightarrow O_3^- + O_2 \tag{10.284}$$

$$\delta_1 = (3.5 \pm 0.5) \times 10^{-10} \quad cm^3\,sec^{-1}\ [130] \tag{10.285}$$

$$(\delta_2):\quad O_2^- + NO_2 \rightarrow NO_2^- + O_2 \tag{10.286}$$

$$\delta_2 = 8 \times 10^{-10} \quad cm^3\,sec^{-1}\ [132]. \tag{10.287}$$

Cluster ions are formed through three-body reactions. For O_2^- the most important reactions are:

$$(\delta_3):\quad O_2^- + O_2 + O_2 \rightleftharpoons O_4^- + O_2 \tag{10.288}$$

$$\delta_3 = 4 \times 10^{-31} \quad cm^6\,sec^{-1}\ [139, 140] \tag{10.289}$$

at 300°K and

$$\delta_{-3} = 2.4 \times 10^{-14} \quad cm^3\,sec^{-1}.$$

$$(\delta_4):\quad O_2^- + CO_2 + CO_2 \rightarrow CO_4^- + CO_2 \tag{10.290}$$

$$\delta_4 = 9.0 \times 10^{-30} \quad cm^6\,sec^{-1}\ [141] \tag{10.291}$$

$$(\delta_5):\quad O_2^- + O_2 + CO_2 \rightarrow CO_4^- + O_2 \tag{10.292}$$

$$\delta_5 = 2.0 \times 10^{-29} \quad cm^6\ sec^{-1}\ [141]. \tag{10.293}$$

Recent measurements have given a rapid rate for the three-body association of H_2O with O_2^-:

$$(\delta_6):\quad O_2^- + H_2O + O_2 \rightarrow O_2^- \cdot H_2O + O_2 \tag{10.294}$$

$$\delta_6 = 3 \times 10^{-28} \quad cm^3\ sec^{-1} \tag{10.295}$$

at 300°K [139].

10.9.2 Atomic Oxygen Negative Ions

The various direct production and losses of O^- ions are shown in *Fig. 10.14*. Reactions which result in a loss of electrons occur rapidly with O^-. The photodetachment reaction

$$(\lambda_5):\quad O^- + h\nu\ (\lambda < 8460\ \text{Å}) \rightarrow O + e \tag{10.296}$$

has a detachment coefficient of [142]

$$\lambda_5 = 1.4 \quad sec^{-1} \tag{10.297}$$

for an overhead sun.

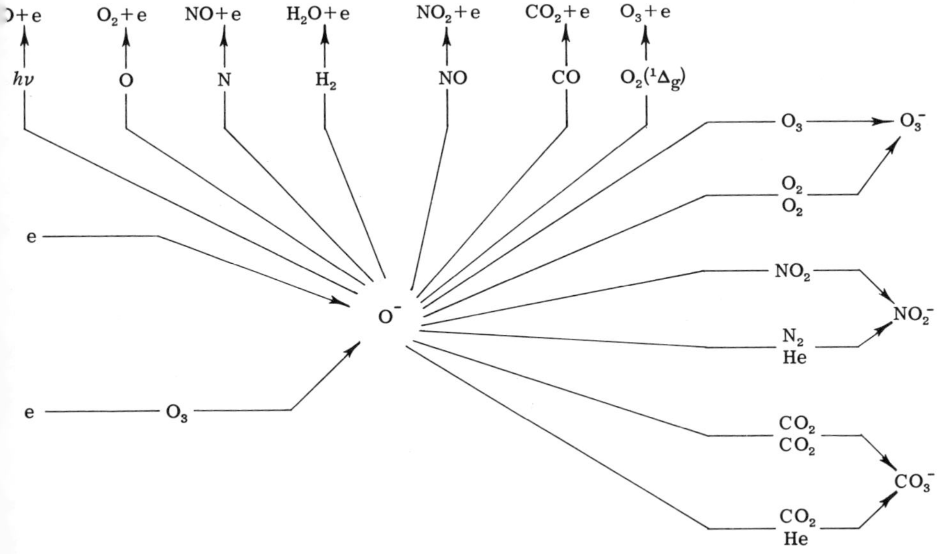

Fig. 10.14 Principal reactions for O^-.

Associative detachment reactions also lead to destruction of negative ions. The following reactions appear to occur readily in the D region:

$$(\lambda_6): \quad O^- + O \rightarrow O_2 + e + 3.6 \quad \text{eV} \tag{10.298}$$

$$\lambda_6 = 1.7 \times 10^{-10} \quad \text{cm}^3\,\text{sec}^{-1}\ [130] \tag{10.299}$$

$$(\lambda_7): \quad O^- + O_2(^1\Delta_g) \rightarrow O_3 + e + 0.6 \quad \text{eV} \tag{10.300}$$

$$\lambda_7 = 3.0 \times 10^{-10} \quad \text{cm}^3\,\text{sec}^{-1}\ [137] \tag{10.301}$$

$$(\lambda_8): \quad O^- + H_2 \rightarrow H_2O + e + 3.6 \quad \text{eV}\ [136, 143, 144] \tag{10.302}$$

$$\lambda_8 = 6 \times 10^{-10} \quad \text{cm}^3\,\text{sec}^{-1} \tag{10.303}$$

$$(\lambda_9): \quad O^- + NO \rightarrow NO_2 + e + 1.4 \quad \text{eV} \tag{10.304}$$

$$\lambda_9 = 1.6 \times 10^{-10} \quad \text{cm}^3\,\text{sec}^{-1}\ [136, 143, 144] \tag{10.305}$$

$$(\lambda_{10}): \quad O^- + CO \rightarrow CO_2 + e + 4.0 \quad \text{eV} \tag{10.306}$$

$$\lambda_{10} = 4.4 \times 10^{-10} \quad \text{cm}^3\,\text{sec}^{-1}\ [136, 143, 144] \tag{10.307}$$

$$(\lambda_{11}): \quad O^- + N \rightarrow NO + e + 5.1 \quad \text{eV} \tag{10.308}$$

$$\lambda_{11} = 2.0 \times 10^{-10} \quad \text{cm}^3\,\text{sec}^{-1}\ [136, 144]. \tag{10.309}$$

Associative detachment reactions of O^- with the dominant atmospheric gases are unimportant in the negative ion balance.

The transfer of negative charge to form other ions is accomplished through charge exchange and interchange reactions.

$$(\delta_7): \quad O^- + O_3 \rightarrow O_3^- + O \tag{10.310}$$

$$\delta_7 = (6.0 \pm 1.0) \times 10^{-10} \quad \text{cm}^3\,\text{sec}^{-1}\ [130, 144] \tag{10.311}$$

$$(\delta_8): \quad O^- + NO_2 \rightarrow NO_2^- + O \tag{10.312}$$

$$\delta_8 = (1.1 \pm 0.1) \times 10^{-9} \quad \text{cm}^3\,\text{sec}^{-1}\ [144, 145] \tag{10.313}$$

$$(\delta_9): \quad O^- + O_2 + O_2 \rightarrow O_3^- + O_2 \tag{10.314}$$

$$\delta_9 = (9 \pm 3) \times 10^{-31} \quad \text{cm}^6\,\text{sec}^{-1}\ [146\text{–}148] \tag{10.315}$$

$$(\delta_{10}): \quad O^- + CO_2 + CO_2 \rightarrow CO_3^- + CO_2 \tag{10.316}$$

$$\delta_{10} = 9.0 \times 10^{-30} \quad \text{cm}^6\,\text{sec}^{-1}\ [141] \tag{10.317}$$

$$(\delta_{11}): \quad O^- + CO_2 + He \rightarrow CO_3^- + He\ [133] \tag{10.318}$$

$$\delta_{11} = 1.5 \times 10^{-28} \quad \text{cm}^6\,\text{sec}^{-1}. \tag{10.319}$$

10.9.3 Ozone Negative Ions

The reactions for ozone are shown in *Fig. 10.15*. There is no direct formation of O_3^- involving electrons and O_3 but reactions involving O^-, O_2^-, and O_4^- are important.

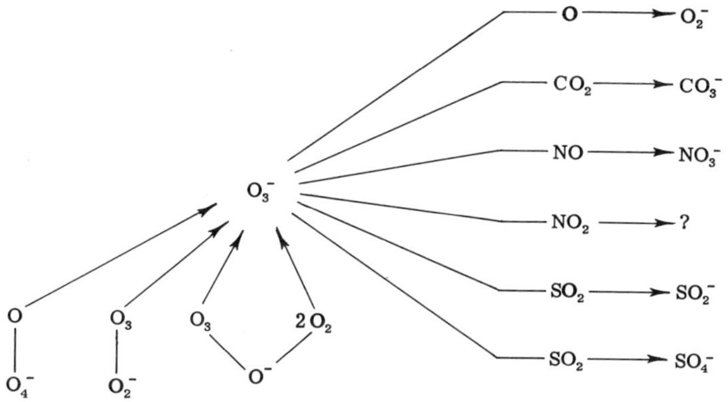

Fig. 10.15 Principal reactions for O_3^-.

The principal reactions are

$$(\delta_{12}): \quad O_3^- + O \rightarrow O_2^- + O_2. \tag{10.320}$$

The reaction rate for this process has not been measured but is estimated to be $\sim 10^{-10}$ cm^3 sec^{-1} [149].

$$(\delta_{13}): \quad O_3^- + CO_2 \rightarrow CO_3^- + O_2 \tag{10.321}$$

$$\delta_{13} = 4.0 \times 10^{-10} \quad \text{cm}^3\,\text{sec}^{-1}\ [130] \tag{10.322}$$

$$(\delta_{14}): \quad O_3^- + NO \rightarrow NO_3^- + O \tag{10.323}$$

$$\delta_{14} = 1.0 \times 10^{-11} \quad \text{cm}^3\,\text{sec}^{-1}\ [130] \tag{10.324}$$

$$(\delta_{15}): \quad O_3^- + NO_2 \rightarrow \text{products } (NO_2^- \text{ and/or } NO_3^-) \tag{10.325}$$

$$\delta_{15} = 1.9 \times 10^{-10} \quad \text{cm}^3\,\text{sec}^{-1}\ [133] \tag{10.326}$$

$$(\delta_{16}): \quad O_3^- + SO_2 \rightarrow SO_3^- + O_2 \tag{10.327}$$

$$\delta_{16} + \delta_{17} = 5.1 \times 10^{-10} \quad \text{cm}^3\,\text{sec}^{-1}\ [133] \tag{10.328}$$

$$(\delta_{17}): \quad O_3^- + SO_2 \rightarrow SO_4^- + O \tag{10.329}$$

$$\delta_{16} + \delta_{17} = 5.1 \times 10^{-10} \quad \text{cm}^3\,\text{sec}^{-1}\ [133] \tag{10.330}$$

Attempts have been made to observe reactions between O_3^- and N_2. Dissociative detachment rates appear to be much less than 10^{-15} cm^3 sec^{-1}, while interchange reactions are slower than 10^{-15} cm^3 sec^{-1} and 1.5×10^{-31} cm^6 sec^{-1} for two- and three-body collisions, respectively. In addition, dissociative detachment with CO_2 must have an extremely small rate.

A clustering reaction with water has been observed:

$$(\delta_{18}):\quad O_3^- + H_2O + O_2 \rightarrow O_3^- \cdot H_2O + O_2 \tag{10.331}$$

$$\delta_{18} = 2.1 \times 10^{-28} \quad \text{cm}^6\ \text{sec}^{-1} \tag{10.332}$$

at 300°K.

10.9.4 O_4^-

The ion O_4^- is produced principally through clustering reactions involving O_2^- [see reaction (10.288)]. The complete set of reactions for this ion has not yet been determined, but it is clear that O_4^- plays an important part in the negative ion balance.

The basic reactions are:

$$(\delta_{19}):\quad O_4^- + CO_2 \rightarrow CO_4^- + O_2 \tag{10.333}$$

$$\delta_1 = 4.3 \times 10^{-10} \quad \text{cm}^3\ \text{sec}^{-1}\ [132] \tag{10.334}$$

$$(\delta_{20}):\quad O_4^- + NO \rightarrow NO_3^- + O_2 \tag{10.335}$$

$$\delta_{20} = 2.5 \times 10^{-10} \quad \text{cm}^3\ \text{sec}^{-1}\ [132] \tag{10.336}$$

$$(\delta_{21}):\quad O_4^- + O \rightarrow O_3^- + O_2 \tag{10.337}$$

$$\rightarrow O^- + O_2 + O_2 \tag{10.338}$$

$$\delta_{21} = 4 \times 10^{-10} \quad \text{cm}^3\ \text{sec}^{-1}\ [132]. \tag{10.339}$$

Reaction (10.337) is the principal channel of the O_4^- reaction with O but no precise measurements have been made to separate the two effects.

Interchange reactions with water have also been observed. The reaction

$$(\delta_{22}):\quad O_4^- + H_2O \rightarrow O_2^- \cdot H_2O + O_2 \tag{10.340}$$

could be a source of O_2^- hydration in the D region. The value for δ_{22} is uncertain but is probably greater than 10^{-10} cm^3 sec^{-1}.

The interchange reaction of O_4^- and N_2 leads to

$$(\delta_{23}):\quad O_4^- + N_2 \rightarrow O_2^- \cdot N_2 + O_2 \tag{10.341}$$

with a slow rate of [113]

$$\delta_{23} \ll 10^{-11} \quad cm^3\ sec^{-1}. \tag{10.342}$$

Likewise, the reactions

$$(\delta_{24}): \quad O_4^- + CO \rightarrow CO_3^- + O_2 \tag{10.343}$$

$$(\delta_{25}): \quad O_4^- + N_2O \rightarrow O_2^- \cdot N_2O + O_2 \tag{10.344}$$

have rates [113]

$$\delta_{24} < 2\times 10^{-11} \quad cm^3\ sec^{-1} \tag{10.345}$$

$$\delta_{25} < 1\times 10^{-12} \quad cm^3\ sec^{-1}. \tag{10.346}$$

10.9.5 CO_3^- *and* CO_4^-

The ions CO_3^- and CO_4^- are destroyed through reactions with O, NO, and NO_2 (see *Fig. 10.16*):

$$(\delta_{26}): \quad CO_3^- + O \rightarrow O_2^- + CO_2 \tag{10.347}$$

$$\delta_{26} = 8.0\times 10^{-11} \quad cm^3\ sec^{-1}\ [130] \tag{10.348}$$

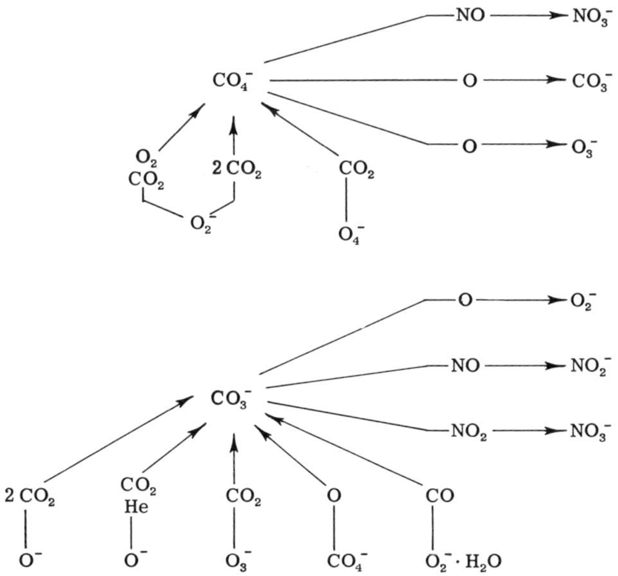

Fig. 10.16 Principal reactions for CO_3^- and CO_4^-.

$$(\delta_{27}):\quad CO_3^- + NO \rightarrow NO_2^- + CO_2 \tag{10.349}$$

$$\delta_{27} = 9.0\times 10^{-12} \quad \text{cm}^3\,\text{sec}^{-1}\ [132] \tag{10.350}$$

$$(\delta_{28}):\quad CO_3^- + NO_2 \rightarrow NO_3^- + CO_2 \tag{10.351}$$

$$\delta_{28} = 8.0\times 10^{-11} \quad \text{cm}^3\,\text{sec}^{-1}\ [133] \tag{10.352}$$

$$(\delta_{29}):\quad CO_4^- + O \rightarrow CO_3^- + O_2 \tag{10.353}$$

$$\rightarrow O_3^- + CO_2 \tag{10.354}$$

$$\delta_{29} = 1.5\times 10^{-10} \quad \text{cm}^3\,\text{sec}^{-1}\ [132]. \tag{10.355}$$

The branching ratio between reactions (10.353) and (10.354) is not known, but a suitable approximation is to ignore (10.354) [132].

$$(\delta_{30}):\quad CO_4^- + NO \rightarrow NO_3^- + CO_2 \tag{10.356}$$

$$\delta_{30} = 4.8\times 10^{-11} \quad \text{cm}^3\,\text{sec}^{-1}\ [132]. \tag{10.357}$$

As pointed out by Ferguson [127], the NO_3^- ions produced by (10.335) and (10.356) have quite different energies and configurations from those arising from (10.358) and (10.360) [O—O—N—O$^-$ and O—N(—O)—O$^-$, respectively]. Since both sources are rapid, such a difference introduces an additional complexity in the problem of D-region negative ion composition.

The photodetachment rates for CO_3^- and CO_4^- are not known. However, it is known that an energy greater than 0.8 ± 0.08 eV is sufficient to dissociate CO_4^- into $O_2^- + CO_2$.

10.9.6 NO_2^- *and* NO_3^-

The ions NO_2^- and NO_3^- originate from reactions involving negative ions of oxygen and oxides of carbon. NO_3^- is important because it presently seems to be a terminal ion, i.e., its only loss occurs via water vapor clustering and ion–ion neutralization (see *Fig. 10.17*).

The reactions leading to a loss of these ions are:

$$(\delta_{31}):\quad NO_2^- + O_3 \rightarrow NO_3^- + O_2 \tag{10.358}$$

$$\delta_{31} = 1.8\times 10^{-11} \quad \text{cm}^3\,\text{sec}^{-1}\ [131] \tag{10.359}$$

$$(\delta_{32}):\quad NO_2^- + NO_2 \rightarrow NO_3^- + NO \tag{10.360}$$

$$\delta_{32} = 4\times 10^{-12} \quad \text{cm}^3\,\text{sec}^{-1}\ [132] \tag{10.361}$$

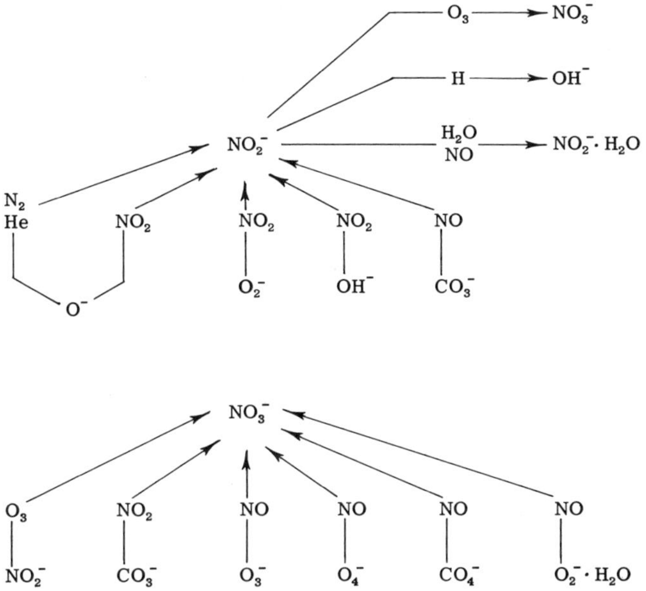

Fig. 10.17 Principal reactions for NO_2^- and NO_3^-.

$$(\delta_{33}): \quad NO_2^- + H \to OH^- + NO \tag{10.362}$$

$$\to HNO_2 + e \tag{10.363}$$

$$\delta_{33} = 3(+3, -2) \times 10^{-10} \quad \text{cm}^3\,\text{sec}^{-1}\ [155]. \tag{10.364}$$

This last reaction provides two channels although no direct evidence exists for (10.363). Reaction (10.362) is the major loss process for NO_2^- at altitudes above 75 to 80 km in the daytime D region [155]. The OH^- produced is rapidly destroyed by associative detachment reactions with O and H, i.e., (10.362) is the first step in an effective electron detachment chain.

NO_2^- can also react to produce water vapor clusters. The reaction

$$(\delta_{34}): \quad NO_2^- + H_2O + NO \to NO_2^- \cdot H_2O + NO \tag{10.365}$$

may be typical of cluster reactions involving N_2 and O_2 in place of NO. The rate coefficient [150]

$$\delta_{34} = 1.2 \times 10^{-28} \quad \text{cm}^6\,\text{sec}^{-1} \tag{10.366}$$

indicates that such a clustering would be rapid in the lower ionosphere.

The ion NO_3^- appears to be nonreactive with N_2 and O_2. Measurements of processes involving NO and N indicate rate coefficients less than 10^{-12} $cm^3 sec^{-1}$ and 10^{-11} $cm^3 sec^{-1}$, respectively. Cluster reactions involving NO_3^- and H_2O may proceed rapidly, however. The direct clustering

$$(\delta_{35}): \quad NO_3^- + H_2O + M \rightarrow NO_3^- \cdot H_2O + M \tag{10.367}$$

could provide a hydrated form of NO_3^-. An alternative sequence involves

$$(\delta_{36}): \quad NO_3^- + O_2 + M \rightarrow NO_3^- \cdot O_2 + M \tag{10.368}$$

$$(\delta_{37}): \quad NO_3^- \cdot O_2 + H_2O \rightarrow NO_3^- \cdot H_2O + O_2. \tag{10.369}$$

The reaction rates δ_{35}, δ_{36}, and δ_{37} have not been measured.

10.9.7 Other Negative Ions

The ion OH^- is produced through reaction of NO_2^- and H (10.362) and undergoes rapid transformations.

$$(\delta_{38}): \quad OH^- + NO_2 \rightarrow NO_2^- + OH \tag{10.370}$$

$$\delta_{38} = (1.2 \pm 0.2) \times 10^{-9} \quad cm^3\, sec^{-1} \text{ [144]} \tag{10.371}$$

$$(\delta_{39}): \quad OH^- + CO_2 + M \rightarrow HCO_3^- + M \tag{10.372}$$

whose reaction rate has been measured as [152]

$$\delta_{39} = 8.6 \times 10^{-28} \quad cm^6\, sec^{-1} \tag{10.373}$$

with M = Ar at 233°K [152].

In addition, OH^- leads to a number of associative detachment reactions

$$(\lambda_{12}): \quad OH^- + O \rightarrow HO_2 + e \tag{10.374}$$

$$\lambda_{12} = 2 \times 10^{-10} \quad cm^3\, sec^{-1} \text{ [144]} \tag{10.375}$$

$$(\lambda_{13}): \quad OH^- + H \rightarrow H_2O + e \tag{10.376}$$

$$\lambda_{13} = 1.0 \times 10^{-9} \quad cm^3\, sec^{-1} \text{ [144]} \tag{10.377}$$

which destroy negative ions. Due to the abundance of O and H above 80 km, reactions (10.362), (10.374), and (10.376) act as a rapid loss process for negative ions.

The ion NO^- charge exchanges with O_2:

$$(\delta_{40}):\quad NO^- + O_2 \rightarrow O_2^- + NO \tag{10.378}$$

$$\delta_{40} = 6.0\times 10^{-10} \quad cm^3\,sec^{-1}\ [156]. \tag{10.379}$$

Due to the rapidity of collisional detachment, NO^- cannot be an important negative ion in the D region.

The ion $O_2^- \cdot H_2O$ is formed from the reaction of O_4^- with water vapor. Two reactions involving this ion are known:

$$(\delta_{41}):\quad O_2^- \cdot H_2O + CO \rightarrow CO_4^- + H_2O \tag{10.380}$$

$$\delta_{41} = 5.8\times 10^{-10} \quad cm^3\,sec^{-1}\ [113] \tag{10.381}$$

$$(\delta_{42}):\quad O_2^- \cdot H_2O + NO \rightarrow NO_3^- + H_2O \tag{10.382}$$

$$\delta_{42} = 3.1\times 10^{-10} \quad cm^3\,sec^{-1}\ [113]. \tag{10.383}$$

Most of the ion reactions discussed in this section have been measured at a temperature of 300°K. In many cases there will be significant changes in the reaction rates for the temperatures characteristic of the mesosphere and upper stratosphere. Thus, a precise theory is not possible even if it were possible to choose the proper reactions and minor constituent densities.

In addition to loss of negative ions by means of associative detachment reactions, a mutual neutralization mechanism involving positive ions is always possible. Experimental information relating to the rate at which mutual neutralization can proceed is difficult to obtain. Recent measurements for $NO^+ + NO_2^-$ and $NO^+ + NO_3^-$ give [153]

$$\alpha(NO^+ + NO_2^-) = (1.75 \pm 0.6)\times 10^{-7} \quad cm^3\,sec^{-1} \tag{10.384}$$

$$\alpha(NO^+ + NO_3^-) = (3.4 \pm 1.2)\times 10^{-8} \quad cm^3\,sec^{-1} \tag{10.385}$$

at a temperature of 300°K. Such values agree with theory but disagree with other results [154] obtained under conditions where excited ions might exist.

REFERENCES

1. G. Herzberg, *Mém. Soc. Roy. Sci. Liège* **15**, 291 (1955).
2. D. R. Bates, *Month. Notices Roy. Astronom. Soc.* **111**, 303 (1951).
3. R. A. Young and R. L. Sharpless, *J. Chem. Phys.* **39**, 1071 (1963).
4. A. B. Callear and I. W. H. Smith, *Discuss. Faraday Soc.* **37**, 96 (1964).
5. F. Kaufman, *Proc. Roy. Soc. London* **A247**, 123 (1958).

6. R. R. Reeves, P. Harteck, and W. H. Chace, *J. Chem. Phys.* **41**, 764 (1964).
7. M. A. A. Clyne and B. A. Thrush, *Proc. Roy. Soc.* **269A**, 404 (1962).
8. R. A. Hartunian, W. P. Thompson, and E. W. Hewitt, *J. Chem. Phys.* **44**, 1765 (1966).
9. A. Fontijn, C. B. Meyer, and H. I. Schiff, *J. Chem. Phys.* **40**, 64 (1964).
10. F. S. Larkin and B. A. Thrush, *Discuss. Faraday Soc.* **37**, 112 (1964).
11. C. A. Barth, *Ann. Géophys.* **20**, 182 (1964).
12. H. I. Schiff, *Can. J. Chem.* **47**, 1903 (1969).
13. I. M. Campbell and B. A. Trush, *Proc. Roy. Soc.* **296A**, 222 (1967).
14. R. A. Young and G. Black, *J. Chem. Phys.* **44**, 3741 (1966).
15. R. A. Young, *Can. J. Chem.* **47**, 1927 (1969).
16. I. M. Campbell and B. A. Thrush, *Proc. Roy. Soc.* **296A**, 201 (1967).
17. K. M. Evenson and D. S. Burch, *J. Chem. Phys.* **45**, 2450 (1966).
18. M. A. A. Clyne, D. J. McKenney, and B. A. Thrush, *Trans. Faraday Soc.* **61**, 2701 (1965).
19. F. Kaufman and J. R. Kelso, *Discuss. Faraday Soc.* **37**, 26 (1964).
20. F. S. Klein and J. T. Herron, *J. Chem. Phys.* **41**, 1285 (1964).
21. M. A. A. Clyne and B. A. Thrush, *Proc. Roy. Soc.* **275A**, 559 (1963).
22. F. S. Larkin and B. A. Thrush, *Discuss. Faraday Soc.* **37**, 112 (1964).
23. R. W. Getzinger and G. L. Schott, *J. Chem. Phys.* **43**, 3237 (1965).
24. F. Kaufman, *Can. J. Chem.* **47**, 1917 (1969).
25. H. I. Schiff, *Ann. Géophys.* **20**, 115 (1964).
26. F. Kaufman, *Ann. Géophys.* **20**, 106 (1964).
27. D. R. Bates and M. Nicolet, *J. Geophys. Res.* **55**, 301 (1950).
28. L. F. Phillips and H. I. Schiff, *J. Chem. Phys.* **37**, 1233 (1962).
29. J. C. Greaves and D. Garvin, *J. Chem. Phys.* **30**, 348 (1959).
30. H. S. Johnston and H. J. Crosby, *J. Chem. Phys.* **22**, 689 (1954).
31. J. E. Marte, E. Tschuikow-Roux, and H. W. Ford, *J. Chem. Phys.* **39**, 3277 (1963).
32. M. A. A. Clyne, B. A. Thrush, and R. P. Wayne, *Trans. Faraday Soc.* **60**, 359 (1964).
33. A. Mathias and H. I. Schiff, *Discuss. Faraday Soc.* **37**, 38 (1964).
34. A. Dalgarno, *Planet. Space Sci.* **10**, 19 (1963).
35. D. R. Bates, *J. Atmos. Terr. Phys.* **6**, 171 (1955).
36. M. Nicolet, *J. Geophys. Res.* **70**, 679 (1965).
37. M. A. A. Clyne and B. A. Thrush, *Proc. Roy. Soc. London* **216A**, 259 (1961).
38. M. A. Biondi, M. T. Leu, and R. Johnson, *COSPAR Symp. Urbana, Illinois* (July 1971).
39. M. J. Seaton, *Mon. Notices Roy. Astronom. Soc.* **119**, 81 (1959).
40. H. S. W. Massey and D. R. Bates, *Rep. Progr. Phys.* **9**, 62 (1943).
41. D. R. Bates, *Phys. Rev.* **77**, 718 (1950).
42. D. R. Bates, *Phys. Rev.* **78**, 492 (1950).
43. M. A. Biondi, *Ann. Géophys.* **20**, 34 (1964).
44. S.-C. Lin and J. D. Teare, *Phys. Fluids* **6**, 355 (1963).
45. R. P. Stein, M. Scheibe, M. W. Syverson, T. M. Shaw, and R. C. Gunton, *Phys. Fluids* **7**, 1641 (1964).
46. R. C. Gunton and T. M. Shaw, *Phys. Rev.* **140**, A748 (1965).
47. R. A. Young and G. A. St. John, *Phys. Rev.* **152**, 25 (1966).
48. M. H. Mentzoni and J. Donohoe, *Can. J. Phys.* **45**, 1565 (1967).
49. C. S. Weller and M. A. Biondi, *Phys. Rev.* **172**, 198 (1968).

50. C. F. Hansen, *Phys. Fluids* **11**, 904 (1968).
51. J. Bardley and M. A. Biondi, *Advan. At. Mol. Phys.* **6**, 1–57 (1970).
52. W. H. Kasner and M. A. Biondi, *Phys. Rev.* **137**, A317 (1965).
53. W. H. Kasner, *Phys. Rev.* **164**, 194 (1967).
54. L. Frommhold and M. A. Biondi, *Bull. Amer. Phys. Soc.* **12**, 217 (1967).
55. M. H. Mentzoni, *J. Appl. Phys.* **36**, 57 (1965).
56. W. H. Kasner and M. A. Biondi, *Bull. Amer. Phys. Soc.* **12**, 218 (1967).
57. D. Smith and C. V. Goodall, *Planet. Space Sci.* **16**, 1177 (1968).
58. J. A. Green and T. M. Sugden, in *Ninth Symposium on Combustion* (The Combustion Inst., ed.), pp. 607–618. Academic Press, New York, 1963.
59. J. F. Paulson, *Ann. Géophys.* **20**, 75 (1964).
60. E. E. Ferguson, *Rev. Geophys.* **5**, 305 (1967).
61. E. E. Ferguson, *Ann. Géophys.* **26**, 589 (1970).
62. E. E. Ferguson, F. C. Fehsenfeld, and A. L. Schmeltekopf, *Advan. At. Mol. Phys.* **5**, 1–56 (1960).
63. W. L. Fite, *Can. J. Chem.* **47**, 1797 (1969).
64. P. Warneck, *Planet. Space Sci.* **15**, 1349 (1967).
65. A. L. Schmeltekopf, E. E. Ferguson, and F. C. Fehsenfeld, *J. Chem. Phys.* **48**, 2966 (1968).
66. D. Smith and R. A. Fouracre, *Planet. Space Sci.* **16**, 243 (1968).
67. M. M. Nakshbandi and J. B. Hasted, *Planet. Space Sci.* **15**, 1781 (1967).
68. A. L. Schmeltekopf, F. C. Fehsenfeld, G. I. Gilman, and E. E. Ferguson, *Planet. Space Sci.* **15**, 401 (1967).
69. Y. Kaneko, N. Kobayashi, and I. Kananatu, *Mass. Spectrosc.* **18**, 920 (1970).
70. C. F. Giese, *Advan. Chem. Ser.* **58**, 20 (1966).
71. T. F. O'Malley, *J. Chem. Phys.* **52**, 3269 (1970).
72. T. E. Van Zandt, (personal communication, 1971).
73. B. R. Turner, J. A. Rutherford, and D. M. J. Compton, *J. Chem. Phys.* **48**, 1602 (1968).
74. M. J. Copsey, D. Smith, and J. Sayers, *Planet. Space Sci.* **14**, 1047 (1966).
75. D. B. Dunkin, F. C. Fehsenfeld, A. L. Schmeltekopf, and E. E. Ferguson, *J. Chem. Phys.* **49**, 1365 (1968).
76. D. B. Dunkin, M. McFarland, F. C. Fehsenfeld, and E. E. Ferguson, *J. Geophys. Res.* **76**, 3820 (1971).
77. F. C. Fehsenfeld, E. E. Ferguson, and A. L. Schmeltekopf, *J. Chem. Phys.* **44**, 3022 (1966).
78. J. F. Paulson, R. L. Mosher, and F. Dale, *J. Chem. Phys.* **44**, 3025 (1966).
79. D. B. Dunkin, M. McFarland, F. C. Fehsenfeld, and E. E. Ferguson, *J. Geophys. Res.* **76**, 3820 (1971).
80. F. C. Fehsenfeld, D. B. Dunkin, and E. E. Ferguson, *Planet. Space Sci.* **18**, 1267 (1970).
81. P. D. Goldan, A. L. Schmeltekopf, F. C. Fehsenfeld, H. I. Schiff, and E. E. Ferguson, *J. Chem. Phys.* **44**, 4095 (1966).
82. P. Warneck, *J. Geophys. Res.* **72**, 1651 (1967).
83. M. M. Shahin, *J. Chem. Phys.* **45**, 2600 (1966).
84. M. Nicolet, *J. Geophys. Res.* **70**, 691 (1965).
85. E. E. Ferguson, F. C. Fehsenfeld, P. D. Goldan, A. L. Schmeltekopf, and H. I. Schiff, *Planet. Space Sci.* **13**, 823 (1965).

86. R. Johnsen, H. L. Brown, and M. A. Biondi, *Bull. Amer. Phys. Soc.* **14**, 260 (1969).
87. D. E. Golden, G. Sinnott, and R. N. Varney, *Phys. Rev. Lett.* **20**, 239 (1968).
88. F. C. Fehsenfeld, A. L. Schmeltekopf, and E. E. Ferguson, *J. Chem. Phys.* **44**, 4537 (1966).
89. F. C. Fehsenfeld, A. L. Schmeltekopf, and E. E. Ferguson, *J. Chem. Phys.* **45**, 23 (1966).
90. F. C. Fehsenfeld, A. L. Schmeltekopf, and E. E. Ferguson, *J. Chem. Phys.* **46**, 2019 (1967).
91. W. B. Maier, II, *Planet. Space Sci.* **16**, 477 (1968).
92. J. Sayers and D. Smith, *Discuss. Faraday Soc.* **37**, 167 (1964).
93. E. E. Ferguson, F. C. Fehsenfeld, D. B. Dunkin, A. L. Schmeltekopf, and H. I. Schiff, *Planet. Space Sci.* **12**, 1169 (1964).
94. P. Warneck, *J. Chem. Phys.* **47**, 4279 (1967).
95. D. R. Bates and T. N. L. Patterson, *Planet. Space Sci.* **9**, 599 (1962).
96. E. E. Ferguson, *Ann. Géophys.* **25**, 819 (1969).
97. E. E. Ferguson, *Rev. Geophys.* **5**, 305 (1967).
98. T. N. L. Patterson, *Rev. Geophys.* **6**, 553 (1968).
99. F. C. Fehsenfeld, A. L. Schmeltekopf, P. D. Goldan, H. I. Schiff, and E. E. Ferguson, *J. Chem. Phys.* **44**, 4087 (1966).
100. T. F. Moran and L. Friedman, *J. Chem. Phys.* **45**, 3837 (1966).
101. V. Aquilanti and G. G. Volpi, *Ric. Sci. Rend.* **36**, 359 (1966).
102. E. E. Ferguson, *Advan. Electron. Electron Phys.* **24**, 1–50 (1968).
103. R. B. Norton, E. E. Ferguson, F. C. Fehsenfeld, and A. L. Schmeltekopf, *Planet. Space Sci.* **14**, 969 (1966).
104. F. C. Fehsenfeld, A. L. Schmeltekopf, and E. E. Ferguson, *J. Chem. Phys.* **46**, 2802 (1967).
105. P. M. Banks and W. I. Axford, *Nature* (*London*) **225**, 924 (1970).
106. F. C. Fehsenfeld and E. E. Ferguson, *J. Geophys. Res.* **76**, 8453 (1971).
107. F. C. Fehsenfeld, E. E. Ferguson, and A. L. Schmeltekopf, *J. Chem. Phys.* **45**, 404 (1966).
108. P. Warneck, *J. Chem. Phys.* **46**, 513 (1967).
109. J. C. J. Thynne and A. G. Harrison, *Trans. Faraday Soc.* **62**, 2468 (1966).
110. D. C. Conway, *J. Geophys. Res.* **69**, 3304 (1964).
111. F. C. Fehsenfeld and E. E. Ferguson, *J. Geophys. Res.* **74**, 2217 (1969).
112. D. A. Durden, P. Kebarle, and A. Good, *J. Chem. Phys.* **50**, 805 (1969).
113. N. G. Adams, D. K. Bohme, D. B. Dunkin, F. C. Fehsenfeld, and E. E. Ferguson, *J. Chem. Phys.* **52**, 3133 (1970).
114. D. K. Bohme, D. B. Dunkin, F. C. Fehsenfeld, and E. E. Ferguson, *J. Chem. Phys.* **51**, 863 (1969).
115. F. C. Fehsenfeld and E. E. Ferguson, *Radio Sci.* **7**, 113 (1972).
116. E. E. Ferguson, in *Mesospheric Models and Related Experiments* (G. Fiocco, ed.), p. 188–197 Reidel Publ., Dordrecht, The Netherlands, 1971.
117. F. C. Fehsenfeld, M. Mosesman, and E. E. Ferguson, *J. Chem. Phys.* **55**, 2115 (1971).
118. A. Good, D. A. Durden, and P. Kebarle, *J. Chem. Phys.* **52**, 222 (1970).
119. C. J. Howard, V. M. Bierbaum, H. W. Rundle, and F. Kaufman, *J. Chem. Phys.* **57**, 3491 (1972).
120. R. S. Narcisi and W. Roth, *Advan. Electron. Electron Phys.* **29**, 79 (1970).
121. M. Nicolet and A. C. Aikin, *J. Geophys. Res.* **65**, 1469 (1960).

122. D. B. Dunkin, F. C. Fehsenfeld, A. L. Schmeltekopf, and E. E. Ferguson, *J. Chem. Phys.* **54**, 3817 (1971).
123. R. R. Burke, *J. Geophys. Res.* **75**, 1345 (1970).
124. P. Kebarle, S. K. Searles, A. Zolla, J. Scarborough, and M. Arshadi, *J. Amer. Chem. Soc.* **89**, 6393 (1967).
125. L. M. Branscomb, *Ann. Géophys.* **20**, 88 (1964).
126. J. F. Paulson, *Ann. Géophys.* **20**, 75 (1964).
127. E. E. Ferguson, D. B. Dunkin, and F. C. Fehsenfeld, *J. Chem. Phys.* **57**, 1459 (1972).
128. R. S. Narcisi, A. D. Bailey, L. Della Luca, C. Sherman, and D. M. Thomas, *J. Atmos. Terr. Phys.* **33**, 1147 (1971).
129. F. Arnold, J. Kissel, D. Krankowsky, H. Wieder, and J. Zähringer, *J. Atmos. Terr. Phys.* **33**, 1169 (1971).
130. F. C. Fehsenfeld, A. L. Schmeltekopf, H. I. Schiff, and E. E. Ferguson, *Planet. Space Sci.* **15**, 373 (1967).
131. F. C. Fehsenfeld and E. E. Ferguson, *Planet. Space Scí.* **16**, 701 (1968).
132. F. C. Fehsenfeld, E. E. Ferguson, and D. K. Bohme, *Planet. Space Sci.* **17**, 1759 (1969).
133. E. E. Ferguson, *Can. J. Chem.* **47**, 1815 (1969).
134. L. M. Chanin, A. V. Phelps, and M. A. Biondi, *Phys. Rev.* **128**, 219 (1962).
135. D. Stelman and A. V. Phelps, to appear *J. Chem. Phys.* (1973).
136. F. C. Fehsenfeld, E. E. Ferguson, and A. L. Schmeltekopf, *J. Chem. Phys.* **45**, 1844 (1966).
137. F. C. Fehsenfeld, D. L. Albritton, J. A. Burt, and H. I. Schiff, *Can. J. Chem.* **47**, 1793 (1969).
138. S. B. Woo, L. M. Branscomb, and E. C. Beaty, *J. Geophys. Res.* **74**, 2933 (1969).
139. J. L. Pack and A. V. Phelps, *Bull. Amer. Phys. Soc.* **16**, 214 (1971).
140. D. A. Parkes, *Trans. Faraday Soc.* **67**, 711 (1971).
141. J. L. Moruzzi and A. V. Phelps, *J. Chem. Phys.* **45**, 4617 (1966).
142. L. M. Branscomb, D. S. Burch, S. J. Smith, and S. Geltman, *Phys. Rev.* **111**, 504 (1958).
143. J. L. Moruzzi, J. W. Ekin, Jr., and A. V. Phelps, *J. Chem. Phys.* **48**, 3070 (1968).
144. E. E. Ferguson, F. C. Fehsenfeld, and A. L. Schmeltekopf, *Advan. Chem. Ser.* **80**, 83 (1969).
145. J. A. Rutherford and B. R. Turner, *J. Geophys. Res.* **72**, 3795 (1967).
146. D. S. Burch and R. Geballe, *Phys. Rev.* **106**, 183 (1957).
147. E. C. Beaty, L. M. Branscomb, and P. L. Patterson, *Bull. Amer. Phys. Soc.* **9**, 535 (1964).
148. L. M. Chanin, A. V. Phelps, and M. A. Biondi, *Phys. Rev. Lett.* **2**, 344 (1959).
149. R. E. LeLevier and L. M. Branscomb, *J. Geophys. Res.* **73**, 27 (1968).
150. L. J. Puckett and W. C. Lineberger, *Phys. Rev. A* **1**, 1635 (1970).
151. E. E. Ferguson, *Ann. Géophys.* **28**, 389 (1972).
152. E. E. Ferguson, personal communication (1972).
153. P. N. Eisner and M. N. Hirsh, *Phys. Rev. Lett.* **26**, 874 (1971).
154. J. R. Peterson, W. H. Aberth, J. T. Moseley, and J. R. Sheridan, *Phys. Rev. A* **3**, 1651 (1971).
155. F. C. Fehsenfeld and E. E. Ferguson, *Planet. Space Sci.* **20**, 295 (1972).
156. M. McFarland, D. B. Dunkin, F. C. Fehsenfeld, A. C. Schmeltekopf, and E. E. Ferguson, *J. Chem. Phys.* **56**, 2358 (1972).
157. W. Felder and R. A. Young, *J. Chem. Phys.* **56**, 6028 (1972).

CHAPTER 11

Oxygen and Ozone

11.1 Introduction

A study of photochemical reactions involving the principal constituents of the atmosphere such as nitrogen and oxygen is necessary before analyzing chemical reactions in which minor constituents may play a role. Nevertheless, important minor constituents such as ozone and OH are closely linked with the behavior of the principal constituents and must be studied very carefully in order to understand aeronomic processes completely.

An aeronomic study requires a knowledge of (i) the vertical distribution of constituents, (ii) absorption coefficients, (iii) activation energies, and (iv) the number of solar photons available at the top of the earth's atmosphere. In general, these parameters are not known simultaneously and it is necessary, therefore, to try to distinguish between principal and secondary reactions before reaching quantitative results. Furthermore, it is extremely important to determine the effects of solar radiations which lead to photodissociation and photoionization, in order to introduce reactions in which ions are involved. In other words, an aeronomic study of chemical reactions in the upper atmosphere is incomplete if the possibility of ionized constituents is not considered in the analysis.

11.2 Dissociation of Oxygen

It has been shown in Section 8.2 how oxygen is photodissociated and in Sections 10.2 and 10.3 how oxygen atoms may recombine in a nitrogen–oxygen atmosphere. However, the times of dissociation and recombination in the lower thermosphere are sufficient to lead to departures from photochemical and chemical equilibrium conditions. The effects of dissociation and recombination, therefore, must be written as follows [1]:

$$\int_z^\infty n(O_2)J_2\, dz = \int_z^\infty k_1 n(M) n^2(O)\, dz \tag{11.1}$$

in which the first term represents the total number of dissociations of molecular oxygen. The total number of solar photons Q_∞ ($\lambda \leq 1750$ Å) at the top of the earth's atmosphere is of the order of 4×10^{12} cm^{-2} sec^{-1} since the solar emission corresponds to a blackbody at 5000°K (see Section 7.2).

With the numerical values used earlier, a mean value is obtained for $n(O)$ from (11.1),

$$n(O) \leq 3 \times 10^{12} \quad \text{cm}^{-3} \tag{11.2}$$

which corresponds to the atmospheric region where a concentration peak must exist. Such a region corresponds to the lower thermosphere above 85 km. From the data of Table 3.2 the total number of molecules above 85 km is about 1×10^{20} cm^{-2} and

$$\overline{n^2(O)} = Q_\infty / k_1 n(M) H. \tag{11.3}$$

Since an atomic oxygen concentration of $1 \times 10^{12} < n(O) < 3 \times 10^{12}$ cm^{-3} near the production peak is approximately the concentration obtained in photochemical conditions, a conservative value of the order of $(2 \pm 1) \times 10^{12}$ oxygen atoms cm^{-3} may be adopted for this region.

A concentration greater than 10^{12} cm^{-3} in the lower thermosphere seems, at first glance, to lead to the confirmation of the Chapman theory, which attributed the emission of the green line in the airglow ($\lambda = 5577$ Å) to the three-body reaction

$$O + O + O \rightarrow O_2 + O(^1S). \tag{11.4}$$

The number of recombinations is certainly greater than 10^4 cm^{-3} sec^{-1} and it only requires an efficiency of the order of 10^{-2} to produce at least 100 cm^{-3} sec^{-1} atoms of $O(^1S)$. However, several experimental analyses [2] seem to indicate that the efficiency of (11.4) would be too low to provide sufficient numbers of $O(^1S)$. It is difficult, however, to find another acceptable process involving an oxygen atom which could replace (11.4). For an example, a reaction such as

$$N + O + O \rightarrow NO + O(^1S) \tag{11.5}$$

in which an oxygen atom is replaced by a nitrogen atom requires a large concentration of nitrogen atoms. It is noted, however, that the association

$$O(^3P) + O(^3P) + M \rightarrow O_2^* + M \tag{11.6}$$

seems to be the source of electronic systems observed in the airglow.

Since the recombination of oxygen atoms occurs in the lower thermosphere after a downward transport, temporal variations can occur which can lead to the patchiness and variable nature of the airglow observed in the green line.

Although the suggested peak values of atomic oxygen concentration such as given by (11.2) and (11.3) can be accepted as first estimates, more detailed studies which include an eddy diffusion effect indicate different values [3–5]. Instead of 3×10^{12} cm^{-3}, the atomic oxygen concentration can only reach a value of 3×10^{11} cm^{-3} at the peak below 100 km. Such a low value (see Section 10.3) cannot explain the oxygen airglow emissions if experimental values of the rate coefficients are accepted (*Fig. 11.1*). Thus, a round figure such as 10^{12} oxygen atoms cm^{-3} at 95 km instead of 3×10^{11} cm^{-3} cannot be rejected without a detailed analysis.

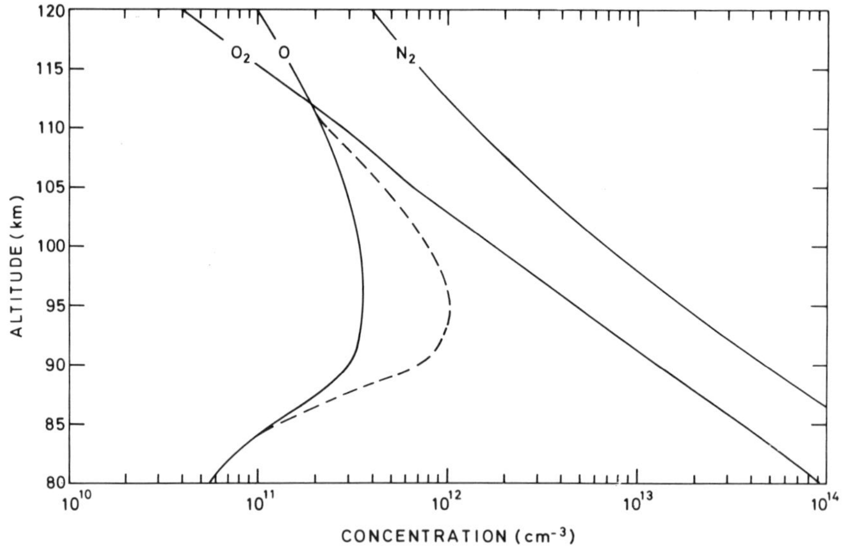

Fig. 11.1 Adopted vertical distribution of atomic oxygen in a nitrogen-oxygen atmosphere between 80 and 120 km. For numerical data see Table 4.3d. Such a model, which is needed for computation purposes, cannot represent all atmospheric conditions. The dashed line shows data from airglow.

In any case it is not yet possible to obtain an adequate view of the dissociation state of molecular oxygen in the turbopause zone at the beginning of the heterosphere. All measurements [6, 7] of the O/O_2 concentration ratio near 120 km indicate that most variations arise from various atmospheric conditions which are not necessarily related to normal diurnal variations. In such circumstances no model is able to represent all aeronomic conditions. A comparison at 120 km emphasizes the differences between the boundary values which are assumed for the calculation of general models and those found for atomic oxygen by mass spectrometry. For example, the atomic oxygen concentrations of Table 4.3d (see Fig. 11.1) are about a factor of 5 above the 1966 observational values near 120 km. As far as molecular oxygen and nitrogen are concerned no important differences exist.

Consequently, it is necessary to keep in mind that changes in the absolute values which are introduced here may alter the interpretation of observational results without modifying the general interpretation of aeronomic processes.

There is no simple way to resolve the problem of the vertical distribution of atomic oxygen in the lower thermosphere with that of the mesosphere and stratosphere.

Considering again the theoretical problem of oxygen dissociation (Section 4.2) and ozone formation (Section 4.3), we write (4.8) and (4.9) as follows:

$$\frac{dn(O)}{dt} = 2n(O_2)J_2 - 2k_1 n(M)n^2(O) - k_2 n(M)n(O_2)n(O) - k_3 n(O_3)n(O) + n(O_3)J_3 \tag{11.7}$$

$$\frac{dn(O_3)}{dt} = k_2 n(M)n(O_2)n(O) - k_3 n(O)n(O_3) - n(O_3)J_3 \tag{11.8}$$

which lead to the general form

$$\frac{dn(O)}{dt} + \frac{dn(O_3)}{dt} + 2k_1 n(M)n^2(O) + 2k_3 n(O_3)n(O) = 2n(O_2)J_2. \tag{11.9}$$

In the thermosphere and mesosphere where the photodissociation rate of O_3 is high ($10^{-3} \leq J_3 \leq 10^2$ sec^{-1}) there is a daytime photoequilibrium for ozone. Thus, from (11.8)

$$n(O_3)[\,J_3 + k_3 n(O)] = k_2 n(M)n(O_2)n(O). \tag{11.10}$$

With $k_3 < 10^{-15}$ cm^3 sec^{-1}, the following approximation can be made:

$$\frac{n(O_3)}{n(O)} = \frac{k_2 n(M) n(O_2)}{J_3}. \tag{11.11}$$

This shows that the ozone concentration depends on the unknown atomic oxygen concentration. At the mesopause level, Eq. (11.7) becomes, by using (11.10),

$$\frac{dn(O)}{dt} + 2\left[k_1 n(M) + \frac{k_3 k_2 n(M) n(O_2)}{J_3 + k_3 n(O)}\right] n^2(O) = 2n(O_2) J_2 \tag{11.12}$$

or, with (11.11),

$$\frac{dn(O)}{dt} + 2\left[k_1 n(M) + \frac{k_3 k_2 n(M) n(O_2)}{J_3}\right] n^2(O) = 2n(O_2) J_2. \tag{11.13}$$

Before discussing the numerical values of the various coefficients, it can be said that τ_0, the time which is required in a sunlit atmosphere to reach about 50% of the equilibrium value (or to reach 80% from 50%), is given by

$$\tau_0(50\%) = 0.275\left[\left(k_1 n(M) + \frac{k_3 k_2 n(M) n(O_2)}{J_3}\right) n(O_2) J_2\right]^{-1/2} \tag{11.14}$$

or

$$\tau_0(50\%) = 0.275 n^*(O)/n(O_2) J_2 \tag{11.15}$$

where $n^*(O)$ is the photoequilibrium concentration. It will be shown later that it is not possible to reach the photoequilibrium value of oxygen concentration at and above the mesopause.

A precise knowledge of k_1, k_2, and k_3, the rate coefficients for the formation of O_2 and O_3, and the loss of O and O_3, is required in order to determine the values which must be used in (11.14). The value of k_1 is given by (10.21) and is sufficiently precise for aeronomic purposes. As far as k_2 and k_3 are concerned the experimental values are not yet sufficiently precise.

The thermal decomposition of ozone has been studied in terms of the following mechanisms:

$$(k_2):\quad O_2 + O \rightleftharpoons O_3 (k_{-2}) \tag{11.16}$$

and

$$(k_3):\quad O + O_3 \rightarrow 2O_2. \tag{11.17}$$

From the usual Arrhenius plot for an *ozone* atmosphere between 343° and 383°K, the following values of k_{-2} and $k_{-2}k_3/k_2$ have been deduced [8]:

$$k_2 n(\mathrm{M})/k_3 = (1.4 \pm 0.14) \times 10^{-24} e^{(3300 \pm 300)/T} n(\mathrm{M}) \qquad (11.18)$$

in which $n(\mathrm{M})$ denotes the total concentration in a *nitrogen–oxygen* atmosphere. Adopting the result

$$k_2 n(\mathrm{M}) = 3.3 \times 10^{-35} e^{950/T} n(\mathrm{M}) \quad \mathrm{cm^3\ sec^{-1}} \qquad (11.19)$$

from recent measurements [9], the adopted expression for k_3 is, from (11.18),

$$k_3 = 2.4 \times 10^{-11} e^{-2350/T} \quad \mathrm{cm^3\ sec^{-1}}. \qquad (11.20)$$

The numerical values obtained from (11.18)–(11.20) are in agreement with certain recent experimental data [10–13] at 300°K. But there is still some uncertainty in the values of the above rate coefficients, particularly at low temperatures. Keeping the same activation energies, k_2 should be considered as a maximum value and k_3 as a minimum value. There is perhaps a possibility that $\frac{1}{2}k_2$ may correspond to a minimum value and $2k_3$ to a maximum value. An aeronomic calculation made with these two alternatives shows that the differences are not negligible. However, more important difficulties occur when a pure oxygen–nitrogen atmosphere is considered.

11.3 Ozone Distribution and Optical Depth

The vertical distribution of ozone and its absolute concentration cannot be explained without the introduction of other minor constituents. But, in order to show how the rate coefficients vary with height, numerical data are given in *Table 11.1* for the atmospheric model of Tables 3.1, 3.2, and 4.3d.

In the whole thermosphere and in the major part of the mesosphere, ozone molecules are always in photoequilibrium with oxygen atoms. It is clear from numerical values of $k_2 n(\mathrm{M}) n(\mathrm{O_2})$ [see Eq. (11.11) and column 5 in Table 11.1] that the ratio $n(\mathrm{O_3})/n(\mathrm{O}) < 1$ in the major part of the mesosphere. However, an absolute value of the $\mathrm{O_3}$ concentration cannot be determined since atomic oxygen is not in photochemical equilibrium at the mesopause level. Application of formula (11.14) shows that the following times are required to reach 50% of the photochemical equilibrium con-

TABLE 11.1

ADOPTED RATE COEFFICIENTS

Altitude (km)	$k_1n(M)$ ($cm^3\ sec^{-1}$)	$k_2n(M)$ ($cm^3\ sec^{-1}$)	k_3 ($cm^3\ sec^{-1}$)	$k_2n(M)n(O_2)$ (sec^{-1})	$k_2n(M)n(O_2)/k_3$ (cm^{-3})
15	1.1×10^{-14}	1.2×10^{-14}	3.5×10^{-16}	9.5×10^{3}	2.7×10^{19}
20	5.5×10^{-15}	5.2×10^{-15}	5.2×10^{-16}	1.6×10^{3}	2.9×10^{18}
25	2.1×10^{-15}	1.7×10^{-15}	7.7×10^{-16}	2.7×10^{2}	3.5×10^{17}
30	9.8×10^{-16}	6.7×10^{-15}	1.1×10^{-15}	5.0×10^{1}	4.5×10^{16}
35	4.6×10^{-16}	2.4×10^{-16}	2.1×10^{-15}	8.3×10^{0}	3.9×10^{15}
40	2.2×10^{-16}	9.3×10^{-17}	3.8×10^{-15}	1.6×10^{0}	4.2×10^{14}
45	1.2×10^{-16}	4.5×10^{-17}	4.6×10^{-15}	4.0×10^{-1}	8.7×10^{13}
50	6.4×10^{-17}	2.4×10^{-17}	4.5×10^{-15}	1.2×10^{-1}	2.6×10^{13}
55	3.5×10^{-17}	1.3×10^{-17}	4.5×10^{-15}	3.5×10^{-2}	7.8×10^{12}
60	2.0×10^{-17}	1.0×10^{-17}	2.2×10^{-15}	1.5×10^{-2}	6.9×10^{12}
65	1.1×10^{-17}	7.8×10^{-18}	9.6×10^{-16}	7.6×10^{-3}	6.7×10^{12}
70	5.6×10^{-18}	6.0×10^{-18}	3.5×10^{-16}	2.5×10^{-3}	7.2×10^{12}
75	2.7×10^{-18}	4.2×10^{-18}	1.3×10^{-16}	8.5×10^{-4}	6.4×10^{12}
80	1.2×10^{-18}	3.0×10^{-18}	4.2×10^{-17}	2.7×10^{-4}	6.5×10^{12}
85	4.9×10^{-19}	2.2×10^{-18}	1.0×10^{-17}	8.1×10^{-5}	7.9×10^{12}
90	1.7×10^{-19}	4.3×10^{-19}	4.0×10^{-17}	5.4×10^{-6}	1.3×10^{11}
95	6.2×10^{-20}	1.0×10^{-19}	1.2×10^{-16}	4.7×10^{-7}	3.8×10^{9}
100	2.5×10^{-20}	2.8×10^{-20}	3.2×10^{-16}	5.3×10^{-8}	1.7×10^{8}

centration of atomic oxygen:

Altitude (km):	90	85	80	75	70
Time (hr):	200	90	32	12	4

If photochemical values of atomic oxygen concentration are accepted, it is clear that the calculated values of O_3 concentrations are too large [14]. The introduction of ozone–hydrogen reactions [15] can change the situation and various studies [16–18] have been made. However, since at least superficial agreement is obtained by taking rate coefficients which fit observational data, it seems to us that an adopted model for the vertical distribution of ozone can more adequately describe the aeronomic processes which are involved. Such an analysis can show why the aeronomic behavior of minor constituents is so important. In *Fig. 11.2* the profile adopted for the distribution of the ozone concentration corresponds to conditions which

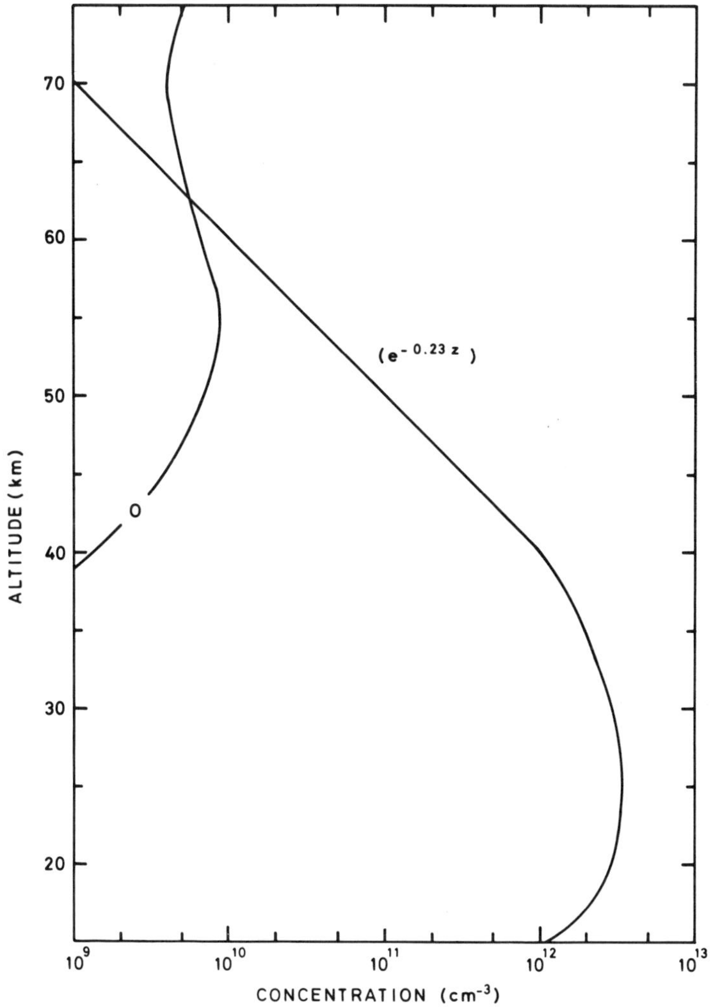

Fig. 11.2 Adopted vertical distribution of ozone. Atomic oxygen concentration is also shown above 40 km; it corresponds to photoequilibrium conditions which exist below 70 km. $n(O) = n(O)_3\, J_3/k_2 n(M) n(O_2)$.

can be compared with observational data [19–23] (*Fig. 11.3*). Such a vertical distribution of ozone leads to a simple density distribution above 40 km as follows:

Altitude (km):	40	50	60	70	80
Concentrations O_3 (cm^{-3}):	10^{12}	10^{11}	10^{10}	10^9	10^8

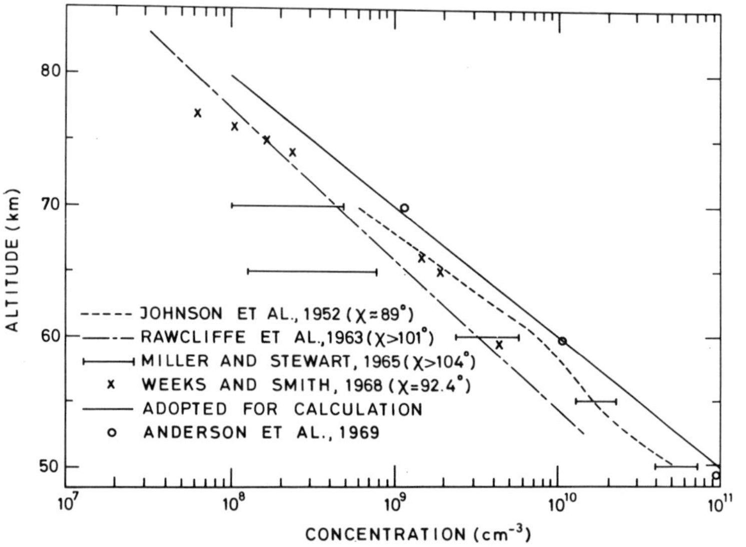

Fig. 11.3 Daytime observations ($\chi \geq 89°$) of the mesopheric ozone concentration compared with the adopted values (10^{11} cm^{-3} at 50 km, 10^{10} at 60 km, 10^9 at 70 km, and 10^8 at 90 km).

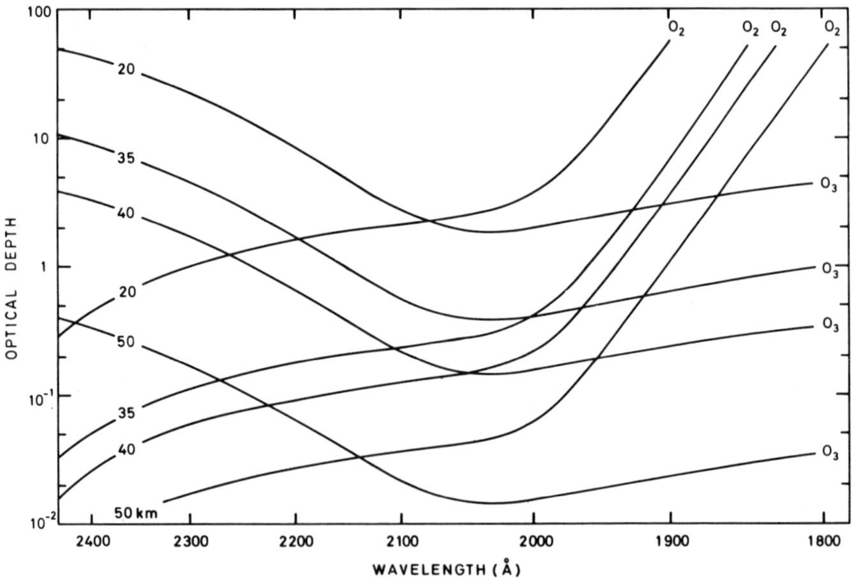

Fig. 11.4 Values of the optical depth between 2400 and 1800 Å for stratospheric ozone and molecular oxygen. Average values for $\Delta\nu = 500$ cm^{-1}.

The equilibrium concentration of atomic oxygen is also given in Fig. 11.2 for an overhead sun using Eq. (11.11): It is clear that $n(O) < 10^{10}\ cm^{-3}$ below 75 km is less than the concentrations obtained for photoequilibrium in a nitrogen–oxygen atmosphere.

With the vertical distribution of the O_3 concentration adopted in Fig. 11.2 it is possible to determine the variation of the optical depth in the stratosphere and mesosphere. Above the stratopause level the optical depth between 2400 Å, i.e., the first threshold for the O_2 dissociation, and 1754 Å, i.e., the second threshold, varies strongly. In *Fig. 11.4* the variation of the optical depth is shown for the stratospheric ozone and oxygen when mean values of the absorption cross sections are adopted for intervals of 500 cm^{-1}. Below 1900 Å molecular oxygen absorbs the solar radiation at the stratopause (50 km) and near 2400 Å the ozone absorption is not negligible. At 40 km and below, the optical depths of O_2 and O_3 are not too different so that the global absorption between 2100 and 2000 Å depends strongly on the ozone content. Special care must, however, be taken when the

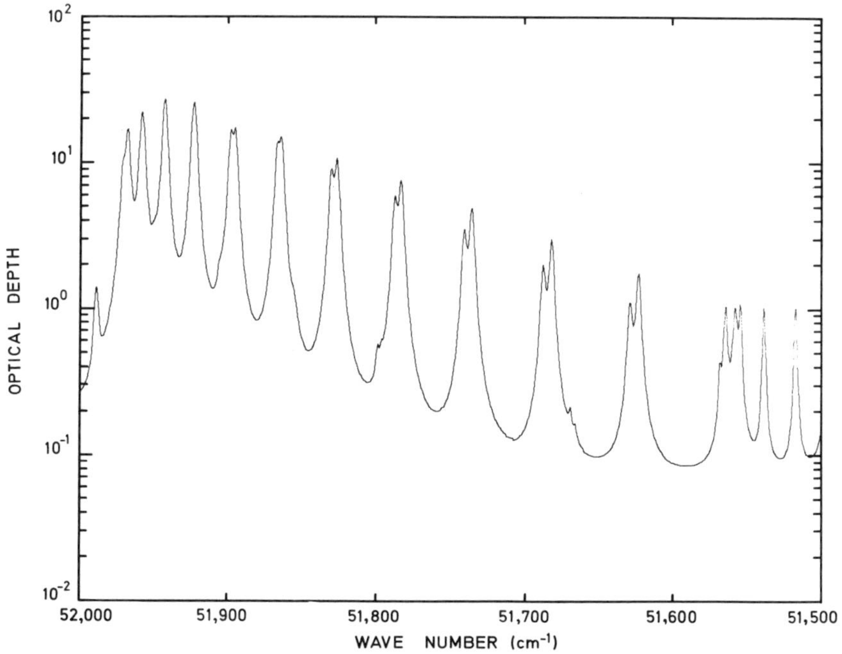

Fig. 11.5 Detailed structure ($\Delta\nu = 0.5\ cm^{-1}$) of the optical depth between 1923 and 1942 Å corresponding to the 4–0 band of the Schumann–Runge system of molecular oxygen. Optical depth values from 10^{-1} to more than 10 indicate that average values cannot be used for a complete analysis. The altitude used is 50 km.

Schumann–Runge bands are involved. From the example [24] shown in *Fig. 11.5*, it is clear that the optical depth at the stratopause varies by more than a factor of 100 in a spectral range of about 20 Å.

11.4 Products of the Ozone Photodissociation

When an analysis of the ozone photodissociation is made, it is necessary to distinguish between the various products in order to know the production of excited atoms or molecules. For $\lambda \leq 11{,}880$ Å

$$O_3 + h\nu\ (\lambda \leq 11{,}800\ \text{Å}) \rightarrow O_2(^3\Sigma_g{}^-) + O(^3P) \qquad (11.21)$$

with O_2 and O in their normal states.

In the spectral range of the Chappuis band (with predissociation) the *possible* products of the photodissociation [25] are, with spin changes,

$$O_3 + h\nu\ (\lambda \leq 6110\ \text{Å}) \rightarrow O_2(^1\Delta_g) + O(^3P) \qquad (11.22)$$

$$O_3 + h\nu\ (\lambda \leq 4630\ \text{Å}) \rightarrow O_2(^1\Sigma_g{}^+) + O(^3P). \qquad (11.23)$$

However (11.22) seems to be an important process in the Huggins bands [26], while

$$O_3 + h\nu\ (\lambda < 3100\ \text{Å}) \rightarrow O_2(^1\Delta_g) + O(^1D) \qquad (11.24)$$

leads to a complete production [25] of $O(^1D)$ for $\lambda < 3000$ Å and to 0.4 ± 0.15 of the total production at $\lambda = 3130$ Å.

Other processes leading to $O_2(^1\Sigma_g{}^+)$ and to $O(^1S)$ at shorter wavelengths should also be considered. They require, however, an experimental analysis in order to determine which fraction is produced by absorption in the Hartley bands. The spin forbidden photodissociation into $O_2(^1\Sigma_g{}^+)$ and $O(^1S)$ for $\lambda < 2360$ Å cannot play an important role, but the spin allowed transition

$$O_3 + h\nu\ (\lambda \leq 1995\ \text{Å}) \rightarrow O_2(a\ ^1\Delta_g) + O(^1S) \qquad (11.25)$$

may be considered below 2000 Å. The production of $O(^1S)$ atoms by (11.25) is, however, small compared to the production of $O(^1D)$ by (11.24) since there are fewer photons available at 2000 Å than at 3000 Å.

The total number of photodissociation processes of O_3 leading to O_2 and O is (see *Fig. 11.6*) about 10^6 cm^{-3} sec^{-1} at 80 km and reaches about 2×10^9 cm^{-3} sec^{-1} at a broad peak between 40 and 30 km. The production

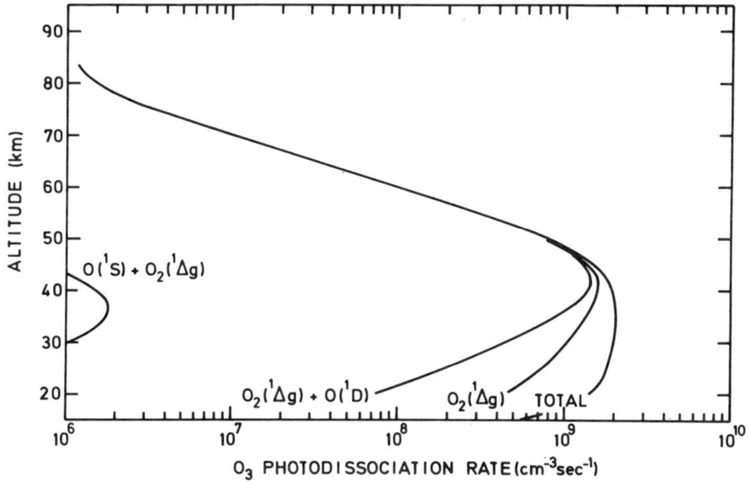

Fig. 11.6 Photodissociation rate of O_3 in various excited states of O_2 and O.

of $O(^1D)$ atoms and $O_2(^1\Delta_g)$ molecules is an important fraction (about 95%) of the total photodissociation in the mesosphere for an overhead sun. In the lower stratosphere the production of $O_2(^1\Delta_g)$ is about one-third of the total photodissociation rate while the production of $O(^1D)$ reaches only 1/30 for overhead sun conditions. The simultaneous production of $O(^1S)$ and $O_2(^1\Delta_g)$ cannot reach more than 1/1000 of the production of $O(^1D)$

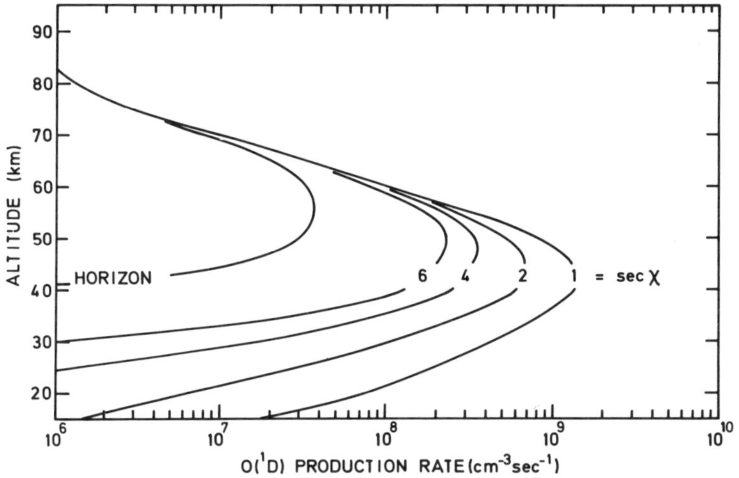

Fig. 11.7 Production rate of $O(^1D)$ atoms by O_3 photodissociation for various solar zenith angles χ.

atoms and $O_2(^1\Delta_g)$ molecules near 40 km for overhead sun conditions. The production rates of the excited atoms and molecules vary with solar distance. The $O(^1D)$ production rate is shown on *Fig. 11.7* for different values of the solar zenith angle. The most important variation occurs in the lower stratosphere. *Figure 11.8* indicates that molecular oxygen in the first excited state

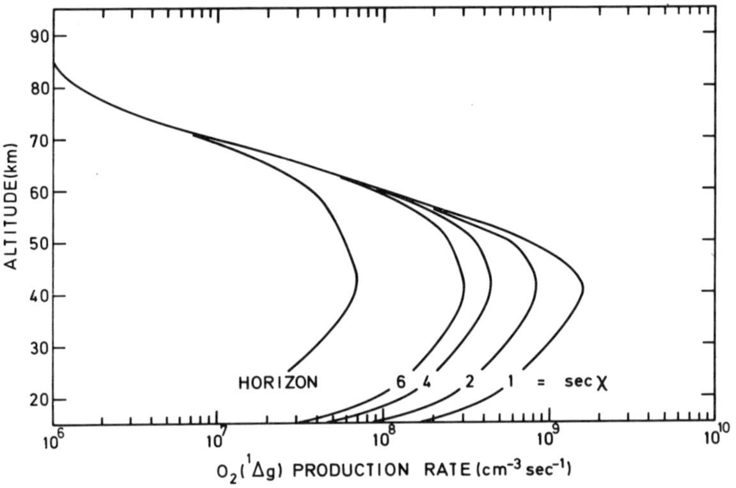

Fig. 11.8 Production rate of $O_2(^1\Delta_g)$ molecules for various solar zenith angles.

is produced in the stratosphere with a rate between 10^7 and 10^9 cm^{-3} sec^{-1} during the day. Actually the variation with solar zenith angle is more important for $O(^1D)$ than for $O_2(^1\Delta_g)$.

11.5 Concentration of Excited Atoms and Molecules

In order to determine the concentration of $O(^1D)$ atoms in a sunlit atmosphere it is necessary to know the principal loss processes. Experimental data [27] show that $O(^1D)$ is rapidly deactivated by N_2 and O_2, according to processes such as

$$O(^1D) + N_2(X\,^1\Sigma_g^+) \rightarrow N_2(X\,^1\Sigma_g^+) + O(^3P) \qquad (11.26)$$

$$O(^1D) + O_2(X\,^3\Sigma_g^-) \rightarrow O_2(b\,^1\Sigma_g^-) + O(^3P) \qquad (11.27)$$

$$\rightarrow O_2(a\,^1\Delta_g) + O(^3P) \qquad (11.28)$$

$$\rightarrow O_2(X\,^3\Sigma_g^-) + O(^3P). \qquad (11.29)$$

Thus,

$$dn[O(^1D)]/dt = n(O_3)J_3(^1D) - [k_Q(^1D, N_2)n(N_2) + k_Q(^1D, O_2)n(O_2)] \times n[O(^1D)] \quad (11.30)$$

where k_Q is the quenching rate coefficient. All experimental values [28–33] show that $k_Q(^1D, N_2, O_2)$ is greater than 10^{-11} cm^3 sec^{-1} and may reach 10^{-10} cm^3 sec^{-1}. Highest values are [32, 33] $k_Q(^1D, N_2) = 10^{-10}$ cm^3 sec^{-1} and $k_Q(^1D, O_2) = 6\times 10^{-11}$ cm^3 sec^{-1}. An aeronomic value of the same magnitude was deduced earlier [34] in order to explain the red line emitted in the airglow from $O(^1D)$. The following working value for the whole homosphere will be retained here:

$$k_Q(^1D, M) = 5\times 10^{-11} \quad \text{cm}^3 \text{ sec}^{-1}. \quad (11.31)$$

Equilibrium conditions, with (11.30) and (11.31), lead to

$$n[O(^1D)] = 2\times 10^{10} n(O_3)J_3(^1D)/n(M). \quad (11.32)$$

The photoequilibrium concentrations for various solar zenith angles are given [35] in *Fig. 11.9*. A concentration peak occurs near the stratopause level, and its maximum value is $n[O(^1D)] = 10^3$ cm^{-3} for an overhead sun. It may be pointed out that concentrations of $O(^1D)$ greater than 10^3 cm^{-3}

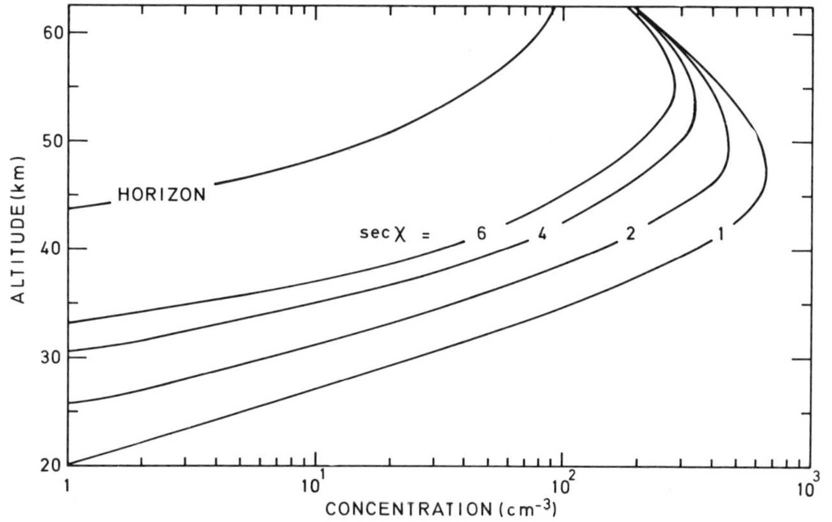

Fig. 11.9 Concentration of $O(^1D)$ atoms for various solar zenith angles (Nicolet [35]).

are not possible except when excessively low values of $k_Q(^1D, M)$, such as 10^{-12} cm^3 sec^{-1}, are taken. Thus, we may consider the values represented in Fig. 11.7 as maximum values, but the error involved should be less than 50%. In any case, these values (see *Table 11.2*) will be adopted when the reaction rates for oxidation of the hydrogen compounds are determined.

TABLE 11.2

CONCENTRATION OF $O(^1D)$ IN A SUNLIT ATMOSPHERE FOR EQUILIBRIUM CONDITIONS

Altitude (km)	sec $\chi = 1$ (cm^{-3})	sec $\chi = 2$ (cm^{-3})	sec $\chi = 4$ (cm^{-3})	sec $\chi = 6$ (cm^{-3})	Horizon (cm^{-3})
20	2.5	0.6			
25	1.1×10^1	3.2			
30	4.4×10^1	1.6×10^1			
35	1.5×10^2	6.6×10^1	2.4×10^1	1.1×10^1	4.0×10^{-4}
40	4.4×10^2	2.1×10^2	9.9×10^1	5.9×10^1	5.2×10^{-1}
45	7.8×10^2	4.3×10^2	2.1×10^2	1.4×10^2	8.0×10^0
50	7.9×10^2	6.0×10^2	3.8×10^2	2.6×10^2	2.7×10^1
55	5.6×10^2	5.1×10^2	4.3×10^2	3.6×10^2	5.7×10^1
60	3.3×10^2	3.2×10^2	3.0×10^2	2.9×10^2	1.0×10^2
65	2.0×10^2	1.9×10^2	1.9×10^2	1.9×10^2	1.3×10^2
70	1.2×10^2	1.2×10^2	1.2×10^2	1.2×10^2	1.0×10^2

At the mesopause level the concentration of $O(^1D)$ atoms begins to increase again due to the photodissociation of O_2 in the Schumann–Runge continuum. For an overhead sun the following concentrations are obtained:

Altitude (km):	95	100	105	110	115	120
Concentration $O(^1D)$ (cm^{-3}):	4×10^2	10^3	2×10^3	4×10^3	5×10^3	4×10^3

The importance of the production of the excited state of $O_2(^1\Delta_g)$ leading to the dayglow emission $^1\Delta_g \rightarrow {}^3\Sigma_g^-$ at 1.27 [34] and 1.58 μm [36] must be emphasized. The high values for the $O_2(^1\Delta_g)$ concentration (*Fig. 11.10*) compared with the $O(^1D)$ concentration are due to the $O_2(^1\Delta_g)$ low quenching rate coefficient.

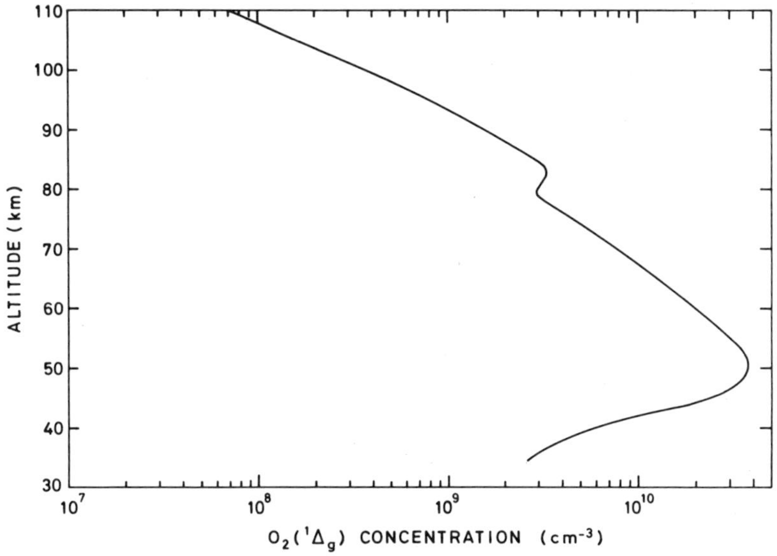

Fig. 11.10 Vertical distribution of excited oxygen molecules in the $^1\Delta_g$ state. Profile deduced from the first rocket flight White Sands, New Mexico, October 11, 1966 for a solar zenith angle $\chi = 75.5°$. Data are from dayglow emission (Evans *et al.* [34]).

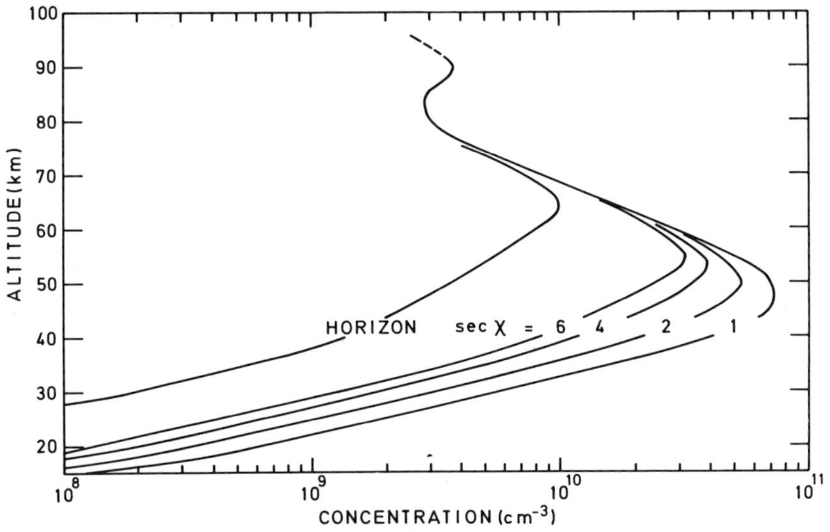

Fig. 11.11 Vertical distribution of $O_2(^1\Delta_g)$ molecules corresponding to photoequilibrium conditions for various solar zenith angles (Nicolet [35]).

With a quenching rate coefficient [37]

$$k_Q(^1\Delta_g) = 4.4 \times 10^{-19} \quad \text{cm}^3 \text{ sec}^{-1} \tag{11.33}$$

corresponding to quenching by molecular oxygen, the photoequilibrium concentration is very high [35] (*see Fig. 11.11*). For daytime conditions the concentration for $O_2(^1\Delta_g)$ is between 3×10^{10} and 7×10^{10} cm^{-3} near the stratopause. Such theoretical values, which correspond to the observations [38], show that the photolysis in the Hartley bands [39] (Eq. 11.24) is the primary process. It must be pointed out that the daytime $O_2(^1\Delta_g)$ concentration in the mesosphere is greater than the O_3 concentration.

As far as the O_2 b $^1\Sigma_g{}^+$–X $^3\Sigma_g{}^-$ atmospheric system is concerned, its day-

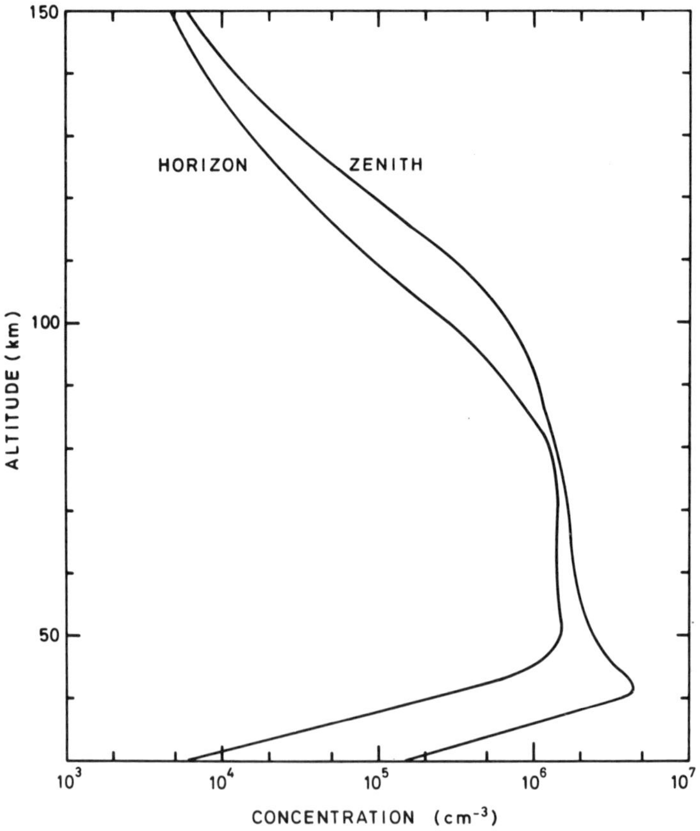

Fig. 11.12 Concentration of $O_2(^1\Sigma_g{}^+)$ molecules calculated from the analysis of rocket observations of the O_2 dayglow. From Wallace and Hunten [40].

glow emission at 7619 Å is also important. A quenching by atmospheric molecules corresponding to $k_Q(^1\Sigma_g^+, M) = 10^{-15}$ cm³ sec⁻¹ leads to daytime $O_2(^1\Sigma_g^+)$ concentrations such as those shown [40] in *Fig. 11.12*.

As far as the $O(^1S)$ atoms are concerned, their low collisional deactivation with N_2 [41],

$$k_Q(^1S, N_2) < 2\times 10^{-16} \quad \text{cm}^3\ \text{sec}^{-1} \tag{11.34}$$

and their moderate deactivation by O_2 [41, 42],

$$k_Q(^1S, O_2) = 3\text{–}5\times 10^{-13} \quad \text{cm}^3\ \text{sec}^{-1} \tag{11.35}$$

lead to equilibrium conditions written as follows:

$$n[O(^1S)] = 10^{13} n(O_3) J_3(^1S)/n(M). \tag{11.36}$$

Thus, Eqs. (11.32) and (11.36) lead to

$$\frac{n[O(^1S)]}{n[O(^1D)]} = \frac{500 J_3(^1S)}{J_3(^1D)}. \tag{11.37}$$

Numerical results for (11.37) are given in the accompanying tabulation.

Altitude (km):	100	50	30
$J_3(^1S)$ (sec⁻¹):	$\geq 3.5\times 10^{-6}$	2.7×10^{-6}	3.0×10^{-7}
$J_3(^1D)$ (sec⁻¹):	$= 8.6\times 10^{-3}$	7.4×10^{-3}	2.1×10^{-4}
$n[O(^1S)]/n[O(^1D)]$:	≤ 0.20	0.18	0.71

It is clear that the concentration of the ^1S oxygen atoms is smaller than that of the ^1D oxygen atoms produced by the ozone photodissociation in the mesosphere and stratosphere.

After sunset, $O(^1D)$ and $O(^1S)$ atoms and excited molecules disappear in the stratosphere and mesosphere. Nevertheless the night airglow emissions indicate that excitation processes occur in the lower thermosphere where atomic oxygen is a permanent constituent. The formation of ozone is the principal loss process of oxygen atoms in the stratosphere and mesosphere. When there is no photodissociation of O_3, atomic oxygen disappears, as indicated in *Fig. 11.13*. After 100 sec all oxygen atoms in the stratosphere

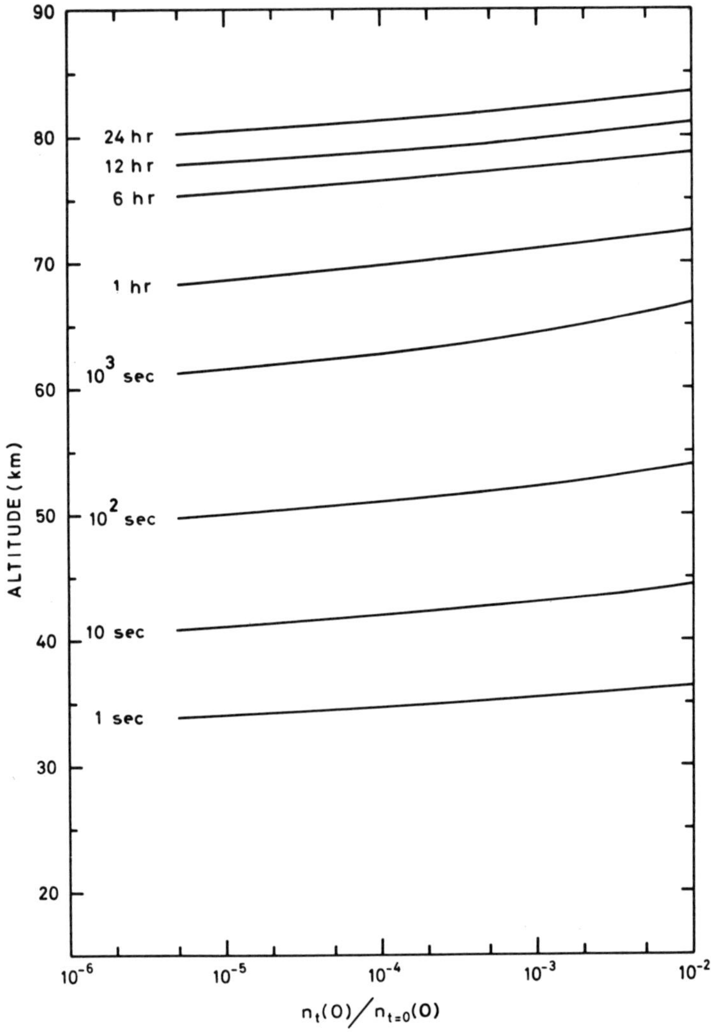

Fig. 11.13 Loss of oxygen atoms after sunset ($t = 0$). The ratio $n_t(O)/n_{t=0}(O)$ is given for various times from $t = 1$ sec to $t = 24$ hr, when only ozone is involved.

disappear. It takes about 1 hr near 70 km to remove oxygen atoms, but one night is needed to change the daytime concentration in the neighborhood of the mesopause. With the various numerical values adopted here it is possible to show (see *Fig. 11.14*) the vertical distribution of atomic oxygen for daytime and nighttime conditions. The maximum value of $O(^3P)$ concentrations in the stratosphere and mesosphere is 10^{10} cm^{-3}. At the end

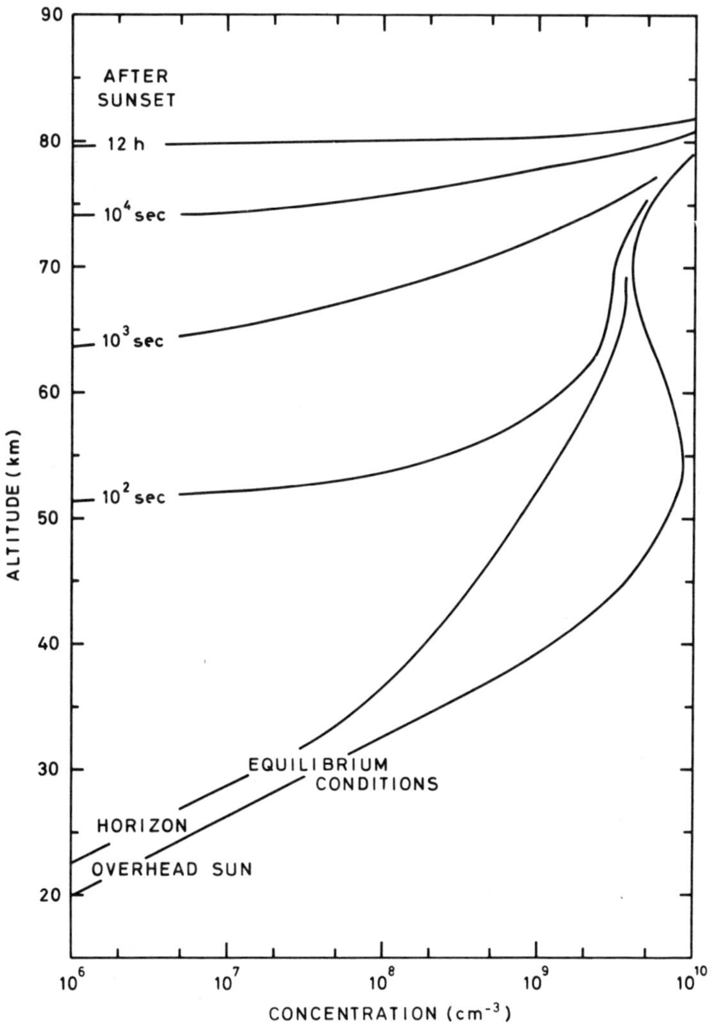

Fig. 11.14 Concentration of atomic oxygen $O(^3P)$ for daytime and nighttime conditions in an oxygen–nitrogen atmosphere.

of an equinoctal night, oxygen atoms are essentially gone at 80 km. Thus, the nighttime conditions should be studied in the mesosphere in order to determine the various changes resulting from the transition from atomic oxygen conditions to ozone conditions. However, the effect of other minor constituents, such as the hydrogen compounds, must still be introduced.

REFERENCES

1. M. Nicolet and P. Mange, *J. Geophys. Res.* **59**, 15 (1954).
2. R. A. Young and G. Black, *J. Chem. Phys.* **44**, 3741 (1966).
3. F. D. Colegrove, F. S. Johnson, and W. B. Hanson, *J. Geophys. Res.* **71**, 2227 (1966).
4. T. Shimazaki, *J. Atmos. Terr. Phys.* **30**, 1279 (1968).
5. T. J. Keneshea and S. P. Zimmerman, *J. Atmos. Sci.* **27**, 831 (1970).
6. W. T. Kasprzak, D. Krankowsky, and A. O. Nier, *J. Geophys. Res.* **73**, 6765 (1968).
7. U. von Zahn, *J. Geophys. Res.* **72**, 5933 (1967).
8. S. W. Benson and A. E. Axworthy, *J. Chem. Phys.* **26**, 1718 (1957).
9. M. A. A. Clyne and B. A. Thrush, *Proc. Roy. Soc.* **269A**, 404 (1962).
10. F. Kaufman and J. R. Kelso, *J. Chem. Phys.* **40**, 1162 (1964).
11. F. Kaufman and J. R. Kelso, *J. Chem. Phys.* **46**, 4541 (1967).
12. A. Mathias and H. I. Schiff, *Discuss. Faraday Soc.* **37**, 38 (1964).
13. C. J. Hochanadel, J. A. Ghormley, and J. W. Boyle, *J. Chem. Phys.* **48**, 2416 (1968).
14. M. Nicolet, *Discuss. Faraday Soc.* **37**, 7 (1964).
15. D. R. Bates and M. Nicolet, *J. Geophys. Res.* **55**, 301 (1950).
16. B. G. Hunt, *J. Geophys. Res.* **71**, 1385 (1966).
17. E. Hesstvedt, *Geofys. Publik. Norv.* **27** (4) (1968).
18. T. J. Shimazaki and A. R. Laird, *J. Geophys. Res.* **75**, 3221 (1970).
19. F. S. Johnson, J. D. Purcell, R. Tousey, and K. Watanabe, *J. Geophys. Res.* **57**, 157 (1952).
20. L. H. Weeks and L. G. Smith, *Planet. Space Sci.* **16**, 1189 (1968).
21. R. D. Rawcliffe, G. E. Meloy, R. M. Friedman, and E. H. Rogers, *J. Geophys. Res.* **68**, 6425 (1963).
22. D. E. Miller and K. H. Stewart, *Proc. Roy. Soc.* **288A**, 540 (1965).
23. G. P. Anderson, C. A. Barth, F. Cayla, and J. London, *Ann. Géophys.* **25**, 341 (1969).
24. G. Kockarts, in *Mesopheric Models and Related Experiments* (G. Fiocco, ed.), pp. 160–176. Reidel Publ., Dordrecht, The Netherlands, 1971.
25. W. B. De More and O. F. Raper, *J. Chem. Phys.* **44**, 1780 (1966).
26. I. T. N. Jones and R. P. Wayne, *J. Chem. Phys.* **51**, 3617 (1969).
27. W. B. De More and O. F. Raper, *Astrophys. J.* **139**, 1381 (1964).
28. D. R. Snelling and E. J. Bair, *J. Chem. Phys.* **47**, 228 (1967).
29. R. A. Young and G. Black, *J. Chem. Phys.* **47**, 2311 (1967).
30. R. A. Young, G. Black, and T. G. Slanger, *J. Chem. Phys.* **49**, 4758 (1968).
31. E. C. Zipf, *Can. J. Chem.* **47**, 1863 (1969).
32. J. F. Noxon, *Space Sci. Rev.* **8**, 92 (1968).
33. J. F. Noxon, *J. Chem. Phys.* **52**, 1852 (1970).
34. W. F. J. Evans, D. M. Hunten, E. J. Llewellyn, and A. Valance Jones, *J. Geophys. Res.* **73**, 2885 (1968).
35. M. Nicolet, in *Mesospheric Models and Related Experiments* (G. Fiocco, ed.), pp. 1–51. Reidel Publ. Dordrecht, The Netherlands, 1971.
36. R. L. Gattinger, *Can. J. Phys.* **46**, 1613 (1968).
37. I. D. Clark and R. P. Wayne, *Proc. Roy. Soc.* **314A**, 111 (1969).
38. W. F. J. Evans and E. J. Llewellyn, *Ann. Géophys.* **26**, 167 (1970).
39. A. Valance Jones and A. W. Harrison, *J. Atmos. Terr. Phys.* **13**, 45 (1958).
40. L. Wallace and D. M. Hunten, *J. Geophys. Res.* **73**, 4813 (1968).
41. S. V. Filseth, F. Stuhl, and K. H. Welge, *J. Chem. Phys.* **52**, 239 (1970).
42. G. Black, T. G. Slanger, G. A. St.John, and R. A. Young, *J. Chem. Phys.* **51**, 116 (1969).

CHAPTER 12

Nitrogen and Nitrogen Oxides

12.1 Introduction

Although molecular nitrogen is the most abundant atmospheric gas, the processes leading to the formation of atomic nitrogen and nitric oxide are not direct. In contrast with molecular oxygen, the photodissociation rate of molecular nitrogen and, consequently, the atomic nitrogen densities in the upper atmosphere are much lower than the atomic oxygen densities. In this chapter the chains of reactions leading to N and NO are described in some detail. To a large extent the problem of NO in the stratosphere and mesosphere is the subject of intense research and discussion among aeronomers and ionospheric physicists who are attempting to solve the relations existing between the minor constituents and the ionization of these regions.

12.2 Dissociation of Molecular Nitrogen

The direct photodissociation of molecular nitrogen, with a coefficient of 5×10^{-12} sec^{-1} (see Section 8.3) cannot be a major source of atomic nitrogen in the upper atmosphere. It appears that one of the most important processes leading to dissociation of molecular nitrogen is photoionization, followed by dissociative recombination:

$$(I_{N_2}): \quad N_2 + h\nu\ (\lambda < 796\ \text{Å}) \rightarrow N_2^+ + e \qquad (12.1)$$

with

$$\alpha_D(N_2):\quad N_2^+ + e \rightarrow N(^4S) + N(^4S) + 5.82 \quad eV \tag{12.2}$$

$$\rightarrow N(^4S) + N(^2D) + 3.44 \quad eV \tag{12.3}$$

$$\rightarrow N(^4S) + N(^2P) + 2.24 \quad eV \tag{12.4}$$

$$\rightarrow N(^2D) + N(^2D) + 1.06 \quad eV. \tag{12.5}$$

The ionization coefficient for N_2 is greater than 10^{-7} sec^{-1} in the F region (see Table 7.12). As far as x rays are concerned, the rate near 50 Å is about 10^{-10} sec^{-1} in the E layer. Since the total number of solar photons available with $\lambda < 800$ Å is about 5×10^{10} cm^{-2} sec^{-1}, the total production of nitrogen atoms resulting from processes (12.1)–(12.5) should be of the order of 10^{11} cm^{-2} sec^{-1}.

Another source of nitrogen atoms in the upper atmosphere is found from direct dissociation and dissociative ionization by electron impact. The ionization of neutral nitrogen molecules by electron impact leads not only to N_2^+ ions but also to doubly charged ions N_2^{2+} and to the dissociative ionization products $N^+ + N$. The total ionization cross section [1–3] has a maximum of about 2.5×10^{-16} cm^2 for an electron energy of the order of 100 eV (see Section 9.5) and an average value of $(1.4 \pm 0.4) \times 10^{-7}$ sec^{-1} for the ionization coefficient can be adopted for electron energies between 50 and 5000 eV. Dissociative ionization of N_2 is possible for electron energies greater than 24.3 eV. When solar x rays ($\lambda < 10$ Å) photoionize molecular nitrogen and oxygen, photoelectrons are produced with energies greater than 500 eV. The impact of these electrons on atmospheric N_2 leads to additional ionization (N^+) and a production of N atoms. Experimental results [4, 5] show that the fraction of atomic ions is about $20 \pm 5\%$ of the total ionization for electron energies greater than 70 eV. Therefore, about 40% of the total ionizations by x rays ($\lambda < 10$ Å) lead directly to nitrogen atoms.

The total dissociative cross section for N_2 by electron impact seems to be greater [6] than the cross section for dissociative ionization. If such cross sections are accepted it is clear that dissociation of N_2 by electron impact accounts for about one-half of the ionization and dissociative processes over the whole energy range. Finally, the dissociation of N_2^+ is not an important process, even if its cross section is very high [7], due to the small abundance of these ions.

12.3 Production of Atomic Nitrogen

The vertical distribution of the atomic nitrogen production coefficient for photodissociation with an overhead sun is as follows:

Altitude (km):	100	95	80	85	80
$2J_{N_2}$ (sec^{-1}):	5.4×10^{-12}	2.5×10^{-12}	4.3×10^{-13}	6.0×10^{-15}	2.5×10^{-17}

The corresponding atomic nitrogen production rate is

Altitude (km):	100	95	90	85	80
$P(N)$ ($cm^{-3}\ sec^{-1}$):	38	44	20	0.8	8×10^{-3}

In comparison, the production of nitrogen atoms by the predissociation process of Chapter 7 is very small in the mesosphere; less 1 atom $cm^{-3}\ sec^{-1}$ even for an overhead sun.

The production rates of N atoms through ionization by x rays are given in *Table 12.1* for solar conditions defined in Table 7.12 with the ionization effect of cosmic rays. The production of nitrogen atoms through the dissociative ionization of molecular nitrogen resulting from absorption of galactic

TABLE 12.1

PRODUCTION RATE[a] OF NITROGEN ATOMS IN THE MESOSPHERE[b]

Altitude (km)	Completely quiet sun	Low average	High average	Maximum	Disturbed	Flares
50	7.2×10^{-2}	7.2×10^{-2}	7.2×10^{-2}	7.2×10^{-2}	7.2×10^{-2}	7.2×10^{-2}
55	3.0×10^{-2}	3.0×10^{-2}	3.0×10^{-2}	3.0×10^{-2}	3.0×10^{-2}	3.0×10^{-2}
60	2.2×10^{-2}	2.2×10^{-2}	2.3×10^{-2}	2.5×10^{-2}	4.8×10^{-2}	2.8×10^{-1}
65	1.2×10^{-2}	1.2×10^{-2}	1.6×10^{-2}	2.6×10^{-2}	1.5×10^{-1}	1.4×10^{0}
70	6.3×10^{-3}	6.6×10^{-3}	9.0×10^{-3}	3.3×10^{-2}	2.8×10^{-1}	2.7×10^{0}
75	2.9×10^{-3}	3.8×10^{-3}	1.2×10^{-2}	9.1×10^{-2}	8.8×10^{-1}	8.8×10^{0}
80	1.2×10^{-2}	1.4×10^{-2}	5.3×10^{-2}	4.8×10^{-1}	4.8×10^{0}	4.8×10^{1}
85	8.0×10^{-1}	8.2×10^{-1}	1.0×10^{0}	3.2×10^{0}	2.5×10^{1}	2.4×10^{2}

[a] Values are in $cm^{-3}\ sec^{-1}$.

[b] The sources are solar x rays of $\lambda < 10$ Å for an overhead sun and galactic cosmic rays at geomagnetic latitude 50°–60° for average conditions.

cosmic rays can be considered using a production rate corresponding to about 40% of the ionization rate. The cosmic ray ionization rate coefficient adopted for Table 12.1 is 10^{-17} sec^{-1}, corresponding to average conditions at geomagnetic latitudes 50°–60°.

In the E region, the predissociation effect plays a role in the lower part of the layer since the ionization coefficient by x rays is not more than 1.2×10^{-11} sec^{-1} (see Table 7.13) for a quiet sun at 35 Å; it becomes less effective above 100 km where ionization of N_2 is due to x rays from 50 and 70 Å (see Table 7.13).

Thus, it follows that ionization processes are more important than dissociation processes in the production of nitrogen atoms even if there is a possibility for the predissociation processes to play a role in the lower thermosphere. The production rate of nitrogen atoms in the lower thermosphere is given in *Table 12.2* for various solar conditions which were defined

TABLE 12.2

PRODUCTION RATE[a] OF NITROGEN ATOMS IN THE LOWER THERMOSPHERE[b]

Altitude (km)	Completely quiet sun	Low average	High average	Maximum	Disturbed sun
90	1.0×10^1	1.2×10^1	1.4×10^1	1.6×10^1	2.4×10^1
95	6.3×10^1	9.1×10^1	1.2×10^2	1.6×10^2	2.4×10^2
100	1.2×10^2	2.5×10^2	3.7×10^2	5.3×10^2	9.6×10^2
105	2.1×10^2	4.1×10^2	8.0×10^2	1.2×10^3	2.0×10^3
110	3.1×10^2	5.6×10^2	1.0×10^3	1.5×10^3	2.4×10^3
115	6.2×10^2	9.6×10^2	1.5×10^3	2.4×10^3	3.0×10^3
120	1.3×10^3	1.8×10^3	2.8×10^3	4.0×10^3	4.5×10^3

[a] Values are in cm^{-3} sec^{-1}.

[b] The sources are solar x rays of >31 Å and ultraviolet radiation in addition to the predissociation process (50% of the total production).

in association with Table 7.13. Such values should be considered only as an indication of possible production rates of atomic nitrogen. In *Table 12.3* production rates for nitrogen atoms deduced from ionospheric conditions are given. The indirect production of atomic nitrogen through ionospheric reactions will be discussed in Section 12.5.

TABLE 12.3

PRODUCTION RATE[a] OF NITROGEN ATOMS IN THE THERMOSPHERE[b]

Altitude (km)	Completely quiet sun	Low average	High average	Maximum
130	2.2×10^3	2.6×10^3	3.3×10^3	4.6×10^3
140	2.6×10^3	2.9×10^3	3.5×10^3	4.7×10^3
150	2.6×10^3	3.0×10^3	3.5×10^3	4.6×10^3
160	2.2×10^3	2.7×10^3	3.3×10^3	4.2×10^3
170	1.8×10^3	2.3×10^3	2.9×10^3	3.8×10^3
180	1.3×10^3	1.9×10^3	2.6×10^3	3.4×10^3
190	1.0×10^3	1.6×10^3	2.2×10^3	2.9×10^3
200	7.5×10^2	1.3×10^3	1.9×10^3	2.5×10^3

[a] Values are in cm^{-3} sec^{-1}.
[b] The source is ultraviolet radiation (50% of the total production).

12.4 Formation of Nitric Oxide

Aeronomic studies have shown that nitric oxide is an essential constituent in the formation ionospheric D region [8]. The problem of nitrogen oxides was first studied in the chemosphere [9–11] and was developed in relation to the airglow. Calculations are difficult because the chemical aeronomy of nitrogen in an oxygen atmosphere cannot be completely resolved [12, 13] without observations in the stratosphere and mesosphere [14–19].

If nitrogen dissociation is operative at sufficiently low altitudes, the formation of nitric oxide may be due to a three-body process:

$$(b_{1a}): \quad N + O + M \rightarrow NO + M + 150 \text{ kcal mole}^{-1} \tag{12.6}$$

with

$$b_{1a} = 1 \times 10^{-32}(300/T)^{1/2} \quad \text{cm}^6 \text{ sec}^{-1}. \tag{12.7}$$

Three-body recombination of nitrogen atoms leads to the emission of various electronic bands of nitric oxide [20]. A preassociation process such as [21, 22]

$$N(^4S) + O(^3P) \rightarrow NO(^4\Pi) \rightarrow NO(C\ ^2\Pi) \tag{12.8}$$

followed by dissociation

$$NO(C\ ^2\Pi) \rightarrow N(^4S) + O(^3P); \quad A_{21} = 6.6\times 10^8\ \text{sec}^{-1} \tag{12.9}$$

and radiative process

$$NO(C\ ^2\Pi) \rightarrow NO(X\ ^2\Pi) + \delta\ \text{bands}; \quad A_{21} = 2.2\times 10^7\ \text{sec}^{-1} \tag{12.10}$$

results in nitric oxide formation with a rate coefficient [21, 22] of the order

$$b_{1\delta} = 10^{-17} \quad \text{cm}^3\,\text{sec}^{-1}. \tag{12.11}$$

Reaction (12.8) is more effective than reaction (12.6) when $n(\text{M})$ is less than 10^{15} cm^{-3}.

Measurements [23] of the rate coefficient for each of the vibrational transitions of the δ bands lead to the values of $b_{1\delta}$ given in *Table 12.4*.

If we adopt an emission coefficient $b_{1\delta}n(\text{O})$ of the order of 10^{-7} sec^{-1} in each band, corresponding to $n(\text{O}) \geq 10^{11}$ cm^{-3}, a concentration of 10^7 nitrogen atoms in a 10 km layer will lead to an airglow emission of about 10^6 photons cm^{-2} sec^{-1} for each band of the $C\ ^2\Pi \rightarrow X\ ^2\Pi$ system. Thus, if $n(\text{N}) > 10^7$ cm^{-3}, there would be a possibility to detect δ bands in the airglow.

The chemiluminescent production [21, 23] of the β and γ bands, the $B\ ^2\Pi \rightarrow X\ ^2\Pi$ and $A\ ^2\Sigma \rightarrow X\ ^2\Pi$ transitions, have been observed by the recombination of N and O atoms. From the $A\ ^2\Sigma^+$ state, responsible for the γ-band emission, the bands observed for $b_{1\gamma}$ are given in Table 12.4 [23].

The final overall result for the afterglow γ bands is [21]

$$b_{1\gamma} = 8.2\times 10^{-18} + 2.4\times 10^{-35} n(\text{M}) \quad \text{cm}^3\,\text{sec}^{-1} \tag{12.12}$$

or [23]

$$b_{1\gamma} = 1.2\times 10^{-17}(300/T)^{0.35} + 2.1\times 10^{-34}(300/T)^{1.24} n(N_2) \ \text{cm}^3\,\text{sec}^{-1} \tag{12.13}$$

using N_2.

Again, if the atomic nitrogen concentration is greater than 10^7 cm^{-3} in the lower thermosphere, where there is a peak for the atomic oxygen concentration, an airglow emission should be detected.

Although the mechanism responsible for the β-band emission is not yet completely understood, the emission bands $B\ ^2\Pi \rightarrow X\ ^2\Pi$ can be explained as the three-body process [21, 23]

$$(b_{1\beta}): \quad N + O + M \rightarrow NO^* + M \rightarrow NO(B\ ^2\Pi) + M \tag{12.14}$$

TABLE 12.4

RATE COEFFICIENT VALUES FOR VIBRATIONAL TRANSITIONS

λ (Å)	Transition	Rate coefficient
		$b_{1\delta}$ (cm^3 sec^{-1})
1914–1910	0–0	0.42×10^{-18}
1985–1981	0–1	0.94
2061–2056	0–2	1.64
2141–2136	0–3	1.60
2227–2221	0–4	1.37
2318–2311	0–5	0.53
2415–2408	0–6	0.25
		$b_{1\gamma}$ (cm^3 sec^{-1})
2269–2264	0–0	3.40×10^{-18}
2370–2263	0–1	4.35
2479–2471	0–2	3.55
2596–2587	0–3	2.56
2722–2713	0–4	1.33
2860–2850	0–5	0.57
3009–2998	0–6	0.27
		$b_{1\beta}$ (cm^6 sec^{-1})
2627–2621	0–4	2.63×10^{-35}
2754–2748	0–5	4.15
2893–2916	0–6	5.70
3043–3035	0–7	4.55
3207–3198	0–8	4.35
3386–3376	0–9	3.90
3583–3572	0–10	2.90
3801–3868	0–11	1.64
4042–4028	0–12	0.71
4310–4293	0–13	0.38

with a rate coefficient [23]

$$b_{1\beta} = 3 \times 10^{-34}(300/T)^{1.4} \quad \text{cm}^6\ \text{sec}^{-1}. \tag{12.15}$$

The rate coefficients $b_{1\beta}$ for each B $^2\Pi \rightarrow$ X $^2\Pi$ vibrational transition are shown in Table 12.4. Obviously, it is impossible to observe the β bands in the airglow if the emissions of γ and δ bands are not observed; the maximum emission coefficient for the β bands would be 10^{-10} sec^{-1}, requiring an excessively large number of nitrogen atoms in order to be an observable thermospheric airglow.

An average value for the radiative coefficient for the total β, γ, δ-band emissions is

$$b_1 = 2\times 10^{-17} \quad \text{cm}^3\ \text{sec}^{-1}. \tag{12.16}$$

Before considering other nitric oxide formation processes, it is necessary to continue the nitric oxide reaction chains through analysis of reactions with atomic oxygen and ozone, the principal chemically active constituents in the chemosphere.

First, the three-body process

$$(b_2):\quad \text{O} + \text{NO} + \text{M} \rightarrow \text{NO}_2{}^* + \text{M} + 72\ \text{kcal mole}^{-1} \tag{12.17}$$

has a relatively high rate coefficient

$$b_{2\text{a}_1} = 3\times 10^{-33} e^{900/T} \quad \text{cm}^6\ \text{sec}^{-1} \tag{12.18}$$

or, for N_2 alone,

$$b_{2\text{a}_2} = 6\times 10^{-32}(300/T)^{3/2} \quad \text{cm}^6\ \text{sec}^{-1}. \tag{12.19}$$

In order to explained the chemiluminescent reaction from excited nitrogen dioxide

$$(b_{2\text{b}}):\quad \text{NO}_2{}^* \rightarrow \text{NO}_2 + h\nu\ (\lambda > 3875\ \text{Å}) \tag{12.20}$$

with a rate coefficient

$$b_{2\text{b}} = 6\times 10^{-17}(300/T)^{3/2} \quad \text{cm}^3\ \text{sec}^{-1}, \tag{12.21}$$

the ratio of the emission probability, $A(NO_2{}^*)$, and of the quenching rate coefficient $a(NO_2{}^*)$ must be of the order of 10^{-15} cm^{-3}.

In fact,

$$\begin{aligned} n(\text{NO}_2{}^*)A(\text{NO}_2{}^*) &= n(\text{M})n(\text{O})n(\text{NO})b_2 \cdot [A/(an(\text{M}) + A)] \\ &\simeq n(\text{O})n(\text{NO})\times b_2(A/a) \end{aligned} \tag{12.22}$$

if there is no direct two-body process [24, 25]. However, for problems of chemical aeronomy it is important to know if (12.17) applies only to a three-body process. Above 75 km, where $n(M) < 10^{15}$ cm^{-3}, there must be a rapid decrease of the formation of NO_2, and the light emission would become pressure-dependent.

Two bimolecular processes involving NO_2 should be considered as important aeronomic reactions. The first is a reaction with oxygen atoms

$$(b_3):\quad O + NO_2 \rightarrow NO + O_2{}^* + 46 \text{ kcal mole}^{-1} \tag{12.23}$$

with O_2 molecules vibrationally excited to $v'' = 8$ (33.7 kcal) and perhaps electronically excited to $^1\Sigma_g$ ($v' \leq 2$) with energies between 37.5 and 45.5 kcal mole^{-1}. We adopt for (12.23) the rate coefficient [15, 16]

$$b_3 = 1.5 \times 10^{-12} T^{1/2} e^{-500/T} \quad \text{cm}^3 \text{ sec}^{-1}. \tag{12.24}$$

The second reaction involves nitric oxide with ozone

$$(b_4):\quad O_3 + NO \rightarrow NO_2 + O_2 + 48 \text{ kcal mole}^{-1} \tag{12.25}$$

and has the following rate coefficient: For ground state NO_2 molecules

$$b_4 = 5 \times 10^{-14} T^{1/2} e^{-1200/T} \quad \text{cm}^3 \text{ sec}^{-1} \tag{12.26}$$

and

$$b_4{}^* = 5 \times 10^{-14} T^{1/2} e^{-2100/T} \quad \text{cm}^3 \text{ sec}^{-1} \tag{12.27}$$

for electronically excited NO_2 molecules [26] leading to a red and infrared emission ($\lambda > 5900$ Å).

When photodissociation of NO_2 is included (see Section 8.5), the differential equation related to NO_2 can be written as

$$dn(NO_2)/dt + [b_3 n(O) + J_{NO_2}] n(NO_2) = [b_2 n(O) n(M) + b_4 n(O_3)] n(NO). \tag{12.28}$$

In a sunlit atmosphere, an equilibrium exists since the photodissociation $\tau_{NO_2} \leq 3 \times 10^2$ sec and, from (12.28), the ratio of nitrogen dioxide and nitrogen oxide concentration is

$$\frac{n(NO_2)}{n(NO)} = \frac{b_2 n(O) n(M) + b_4 n(O_3)}{J_{NO_2} + b_3 n(O)}. \tag{12.29}$$

From *Fig. 12.1* it is clear that $b_2 n(M) n(O)$ is negligible compared with $b_4 n(O_3)$ through the entire stratosphere and lower mesosphere. The reaction of NO with O_3 is a very important process in the stratosphere since it leads to a ratio $n(NO_2)/n(NO) \geq 1$. Numerical results are shown in *Table 12.5*

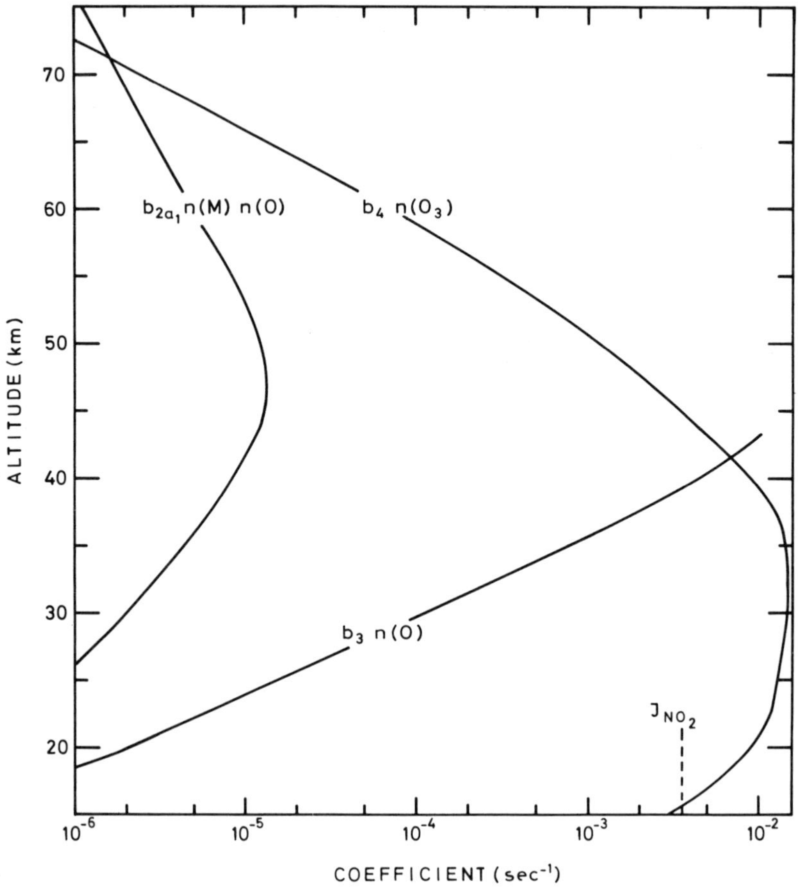

Fig. 12.1 Coefficients involved in the determination of the ratio $n(NO_2)/n(NO)$ for an overhead sun.

TABLE 12.5

Daytime Ratio $n(NO_2)/n(NO)$

Altitude	$n(NO_2)/n(NO)$	Altitude	$n(NO_2)/n(NO)$
15	0.8	35	3.2
20	2.6	40	1.2
25	3.5	45	0.2
30	3.8	50	0.04

for daytime conditions. Above the stratopause the ratio $n(NO_2)/n(NO)$ is so small that nitrogen dioxide can be neglected in the analysis of nitric oxide reactions in a sunlit atmosphere.

Adopting a relative value of $n(NO_2) + n(NO) = 1$ and mixing conditions, *Fig. 12.2* illustrates the possible vertical distribution of $n(NO)$ and $n(NO_2)$ in the stratosphere under mixing conditions. After twilight, when atomic

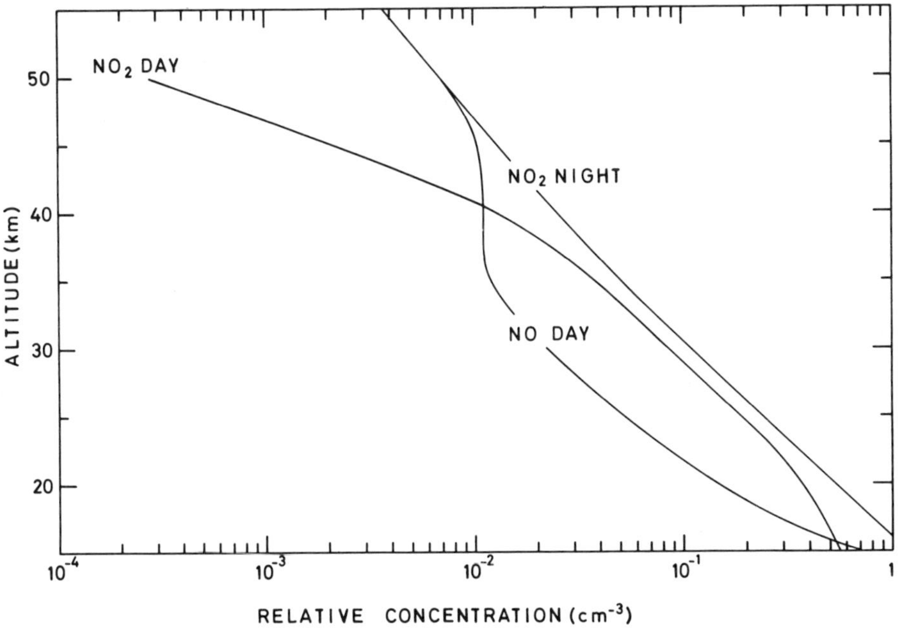

Fig. 12.2 Vertical distribution of nitrogen oxide and nitrogen dioxide in the stratosphere. Relative concentrations normalized to $n(NO) + n(NO_2) = 1$ at 16 km for mixing conditions.

oxygen is rapidly removed in the stratosphere by its association with molecular oxygen, nitric oxide disappears according to (12.25) as

$$dn(NO)/dt = -b_4 n(O_3) n(NO). \tag{12.30}$$

Figure 12.3 shows the transition from NO to NO_2 after sunset.

It will be shown later how NO and NO_2 may react with hydroxyl and perhydroxyl radicals to lead to nitrous acid and also to nitric acid, the latter substance having been recently observed [27, 28] in the stratosphere. To determine the amount of NO, NO_2, and HNO_3 in the stratosphere a detailed analysis of the infrared spectrum is required.

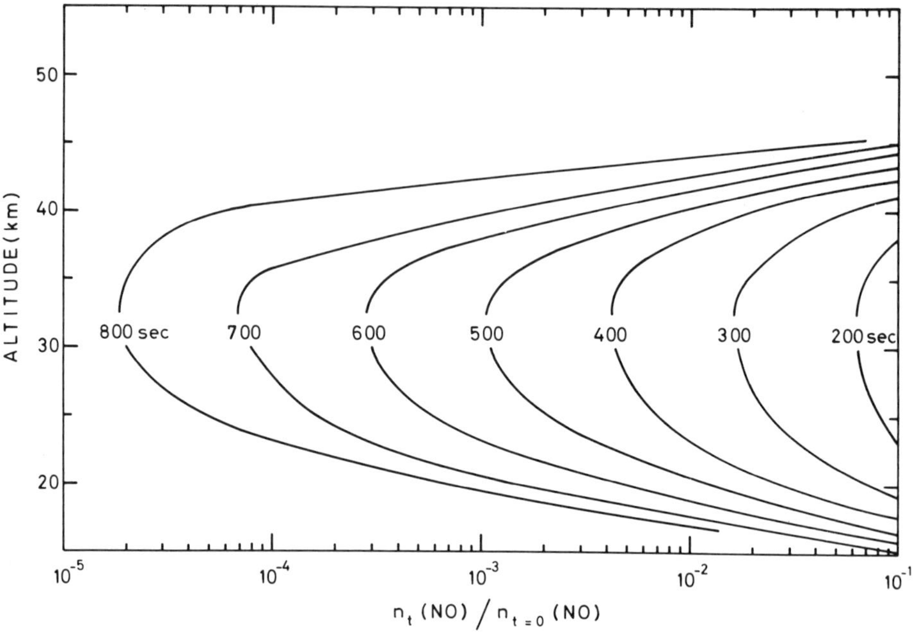

Fig. 12.3 Decrease in nitric oxide concentration in the stratosphere. Reaction of NO with O_3 from 200 to 800 sec after sunset.

To study the behavior of nitric oxide, it is necessary to examine aeronomic reactions in which atomic nitrogen is directly involved. In addition to (12.6) and (12.8), nitrogen atoms can react with nitrogen dioxide

$$(b_5):\quad N + NO_2 \rightarrow O + N_2O + 34 \quad \text{kcal mole}^{-1} \tag{12.31}$$

$$\rightarrow NO + NO + 78 \quad \text{kcal mole}^{-1} \tag{12.32}$$

$$\rightarrow N_2 + O_2 + 121 \quad \text{kcal mole}^{-1} \tag{12.33}$$

for which a rate coefficient, b_5, may be of the same magnitude as for other bimolecular reactions. The reaction with atomic oxygen is more important for the determination of the ratio $n(NO_2)/n(NO)$ than reactions such as (12.31)–(12.33). Reactions of atomic nitrogen with nitric oxide and molecular oxygen are also more important than (12.31)–(12.33) and are certainly the principal reactions in the chemosphere.

The following fast bimolecular reaction

$$(b_6):\quad N + NO \rightarrow N_2\,(v'' \leq 12) + O + 75 \quad \text{kcal mole}^{-1} \tag{12.34}$$

must have a very small activation energy and a rate coefficient such as

$$b_6 = (1.5 \pm 0.5)\times 10^{-12}T^{1/2} \quad \text{cm}^3\ \text{sec}^{-1} \tag{12.35}$$

which varies only by about a factor of 2 between 200 and 700°K.

The reaction of atomic nitrogen with molecular oxygen produces nitric oxide

$$(b_7):\quad \text{N} + \text{O}_2 \rightarrow \text{NO}^*\ (v'' \leq 6) + \text{O} + 32 \quad \text{kcal mole}^{-1} \tag{12.36}$$

and requires a relatively high activation energy. Using recent laboratory determinations [29–31] the expression

$$b_7 = 5\times 10^{-13}T^{1/2}e^{-3500/T} \quad \text{cm}^3\ \text{sec}^{-1} \tag{12.37}$$

was adopted. A value such as $b_7 = 2\times 10^{-13}T^{1/2}e^{-3000/T}$ cm^3 sec^{-1} could lead to higher values [14–16] at very low temperatures according to measurements below 300°K.

When an excited molecule such as $O_2(^1\Delta_g)$ reacts with atomic nitrogen, the following reaction [32]

$$(b_{7*}):\quad \text{N} + \text{O}_2(^1\Delta_g) \rightarrow \text{NO}\ (v'' \leq 11) + \text{O} + 54 \quad \text{kcal mole}^{-1} \tag{12.38}$$

can be important if it has a rate coefficient $b_7{}^*$ such that

$$b_7{}^*n(\text{O}_2{}^*) > b_7 n(\text{O}_2).$$

Experimental data [33, 34] show, however, that $b_7{}^*$ is of the order of 3×10^{-15} cm^3 sec^{-1} at 200°K.

In the same way, one may assume that the reaction

$$(b_{7**}):\quad \text{N} + \text{O}_2(^1\Sigma_g{}^+) \rightarrow \text{NO}\ (v'' \leq 14) + \text{O} + 69 \quad \text{kcal mole}^{-1} \tag{12.39}$$

is also possible with a rate coefficient b_7^{**}.

Other processes may produce nitric oxide in the chemosphere, such as the reaction of nitrogen atoms with ozone:

$$(b_8):\quad \text{N} + \text{O}_3 \rightarrow \text{NO} + \text{O}_2 + 126 \quad \text{kcal mole}^{-1} \tag{12.40}$$

which has a rate coefficient of about 5×10^{-12} cm^3 sec^{-1} at room temperature [35, 36]. The activation energy is not known. With an activation energy such as in (12.26), the following expression is obtained:

$$b_{8a} = 2\times 10^{-12}T^{1/2}e^{-1200/T} \quad \text{cm}^3\ \text{sec}^{-1} \tag{12.41}$$

and with a smaller activation energy such as in (12.23) the following form

$$b_{8b} = 2 \times 10^{-13} T^{1/2} e^{-500/T} \quad \text{cm}^3 \text{ sec}^{-1} \tag{12.42}$$

can be deduced.

Thus, the reaction of nitrogen atoms with ozone may be added to the reaction with molecular oxygen in the mesosphere [37].

Considering processes (12.14), (12.34), (12.36), (12.38)–(12.40) in which NO and N are involved, and including the production of atomic nitrogen P(N), the continuity equations for N and NO can be written as

$$\frac{dn(\text{N})}{dt} + [b_{1b}n(\text{O}) + b_6 n(\text{NO}) + b_7 n(\text{O}_2) + b_7{}^* n(\text{O}_2{}^*) + b_7^{**} n(\text{O}_2^{**}) + b_8 n(\text{O}_3)]n(\text{N}) = P(\text{N}) \tag{12.43}$$

and

$$\frac{dn(\text{NO})}{dt} + b_6 n(\text{N})n(\text{NO}) = [b_{1b}n(\text{O}) + b_7 n(\text{O}_2) + b_7{}^* n(\text{O}_2{}^*) + b_7^{**} n(\text{O}_2^{**}) + b_8 n(\text{O}_3)]n(\text{N}) \tag{12.44}$$

or

$$\frac{dn(\text{N})}{dt} + \frac{dn(\text{NO})}{dt} + 2b_6 n(\text{NO})n(\text{N}) = P(\text{N}). \tag{12.45}$$

When atomic nitrogen is in photochemical equilibrium (12.45) becomes

$$\frac{dn(\text{NO})}{dt} + 2b_6 n(\text{N})n(\text{NO}) = P(\text{N}). \tag{12.46}$$

Thus, the product $n(\text{N})n(\text{NO})$ always depends on the production of nitrogen atoms, and the ratio $n(\text{NO})/n(\text{N})$ depends on the loss of nitrogen atoms by various processes.

In the ozonosphere, the reaction

$$(b_9):\quad \text{O}_3 + \text{NO}_2 \rightarrow \text{NO}_3 + \text{O}_2 + 26 \quad \text{kcal mole}^{-1} \tag{12.47}$$

may occur. Its activation energy is relatively high, and a rate coefficient such as

$$b_9 = 5 \times 10^{-13} T^{1/2} e^{-3600/T} \quad \text{cm}^6 \text{ sec}^{-1} \tag{12.48}$$

leads to the maximum value of 1.5×10^{-17} cm^3 sec^{-1} at the stratopause.

However, nitrogen trioxide [38] may appear through a three-body reaction with atomic oxygen:

$$(b_{10}):\quad O + NO_2 + M \rightarrow NO_3 + M + 50 \quad \text{kcal mole}^{-1} \tag{12.49}$$

with a high rate coefficient

$$b_{10} = 5 \times 10^{-31} \quad \text{cm}^6\ \text{sec}^{-1}. \tag{12.50}$$

For daytime conditions such reactions are followed by the rapid reaction

$$(b_{11}):\quad NO + NO_3 \rightarrow 2NO_2 + 22 \quad \text{kcal mole}^{-1} \tag{12.51}$$

which has a small activation energy and may have a rate coefficient b_{11} of about

$$b_{11} = 5 \times 10^{-13} T^{1/2} e^{-1000/T} \quad \text{cm}^3\ \text{sec}^{-1}. \tag{12.52}$$

Therefore, the presence of NO_3 may be neglected in a sunlit atmosphere compared with NO and NO_2. However, the nighttime conditions are different in the stratosphere due to the absence of nitric oxide. The following processes are considered [35]:

$$(b_{12}):\quad NO_3 + NO_2 \rightarrow N_2O_5 + 21 \quad \text{kcal mole}^{-1} \tag{12.53}$$

with

$$b_{12} = 5 \times 10^{-12} T^{1/2} e^{-1000/T} \quad \text{cm}^3\ \text{sec}^{-1}; \tag{12.54}$$

$$(b_{13}):\quad NO_3 + NO_2 \rightarrow NO + NO_2 + O_2 - 4 \quad \text{kcal mole}^{-1} \tag{12.55}$$

with

$$b_{13} = 3 \times 10^{-14} T^{1/2} e^{-2000/T} \quad \text{cm}^3\ \text{sec}^{-1}; \tag{12.56}$$

and

$$(b_{14}):\quad NO_3 + NO_3 \rightarrow 2NO_2 + O_2 + 18 \quad \text{kcal mole}^{-1} \tag{12.57}$$

with

$$b_{14} = 1.5 \times 10^{-13} T^{1/2} e^{-3600/T} \quad \text{cm}^3\ \text{sec}^{-1}. \tag{12.58}$$

Since only (12.47) and (12.53) will result in an NO_2 loss process for nighttime conditions, the total loss will be small during a night of 12 hr. The coefficient $b_9 n(O_3)$ decreases from 1×10^{-5} sec^{-1} near 40 km to 1×10^{-6} sec^{-1} near 50 km.

A three-body association with two nitrogen dioxide molecules, such as

$$(b_{15}): \quad NO_2 + NO_2 + M \rightarrow N_2O_4 + M + 15 \quad \text{kcal mole}^{-1} \tag{12.59}$$

can produce nitrogen tetroxide, but it has a small rate coefficient

$$b_{15} = 8 \times 10^{-34} \quad \text{cm}^6 \text{ sec}^{-1}. \tag{12.60}$$

Thus, all reactions from (12.47) to (12.59) involving nitrogen dioxide (and ozone) have no practical effect in a sunlit atmosphere. During long nighttime conditions they may, however, contribute to the formation of various oxides of nitrogen.

In addition to the loss process (12.34) of nitric oxide occurring under day- and nighttime conditions, it is important to add photodissociation (see Section 8.5). *Table 12.6* shows that absorption becomes important in

TABLE 12.6

PHOTODISSOCIATION OF NITRIC OXIDE. COEFFICIENT J_{NO}

Altitude (km)	J_{NO} (sec^{-1})	Altitude (km)	J_{NO} (sec^{-1})
100	5.0×10^{-6}	70	2.1×10^{-6}
90	4.9	65	7.5×10^{-7}
85	4.6	60	1.2×10^{-7}
80	4.2	55	4.1×10^{-9}
75	3.3	50	1.1×10^{-11}

the lower mesosphere. However, it is necessary to point out that the complex absorption in the O_2 Schumann–Runge bands [39] influences the NO photodissociation and the values given in Table 12.6 are only a crude approximation. Actually, the photodecomposition of NO is not very well known. As an example, *Fig. 12.4* gives the optical depth at 60 km for the spectral range of the δ bands around 1830 Å. It can be seen that the penetration of the solar radiation [40] in the 10–0 and 9–0 bands of the O_2 Schumann–Runge system is not simple. The structure of the optical depth will, therefore, influence the nitric oxide dissociation.

With the introduction of the photodissociation coefficient J_{NO} the differential equation related to the variation of nitric oxide (12.44) is replaced

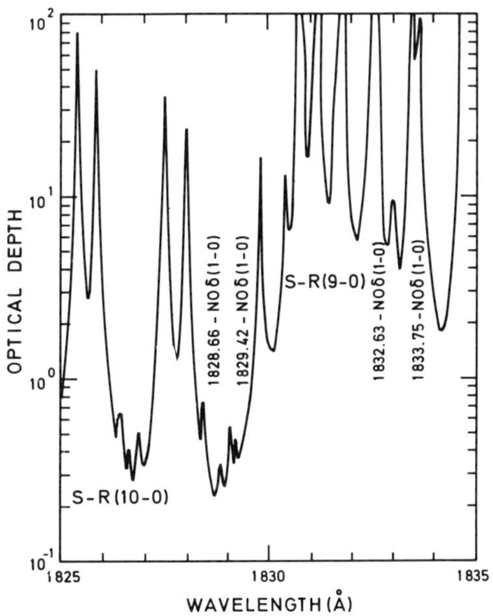

Fig. 12.4 Optical depth at 60 km for an overhead sun. The absorption results from the tail of the 10–0 band and from the head of the 9–0 band of the Schumann–Runge system. The heads of the 1–0 δ band of nitric oxide, leading to photodissociation, are not absorbed at the same altitude.

in the chemosphere by

$$\frac{dn(\mathrm{NO})}{dt} + n(\mathrm{NO})[J_{\mathrm{NO}} + b_6 n(\mathrm{N})]$$
$$= [b_{1b} n(\mathrm{O}) + b_7 n(\mathrm{O_2}) + b_7{}^* n(\mathrm{O_2}{}^*) + b_7^{**} n(\mathrm{O_2^{**}}) + b_8 n(\mathrm{O_3})] n(\mathrm{N}) \qquad (12.61)$$

and the equation which gives the variation of atomic nitrogen, (12.43) becomes

$$\frac{dn(\mathrm{N})}{dt} + n(\mathrm{N})[b_{1b} n(\mathrm{O}) + b_6 n(\mathrm{NO}) + b_7 n(\mathrm{O_2})$$
$$+ b_7{}^* n(\mathrm{O_2}{}^*) + b_7^{**} n(\mathrm{O_2^{**}}) + b_8 n(\mathrm{O_3})] = n(\mathrm{NO}) J_{\mathrm{NO}} + P(\mathrm{N}). \qquad (12.62)$$

By adding (12.61) and (12.62), we obtain the proper expression for changes in the chemosphere:

$$\frac{dn(\mathrm{NO})}{dt} + \frac{dn(\mathrm{N})}{dt} + 2b_6 n(\mathrm{NO}) n(\mathrm{N}) = P(\mathrm{N}) \qquad (12.63)$$

which is identical to (12.46) since the additional processes which were considered involve simultaneously NO and N.

Before discussing (12.61)–(12.63), it is necessary to consider the various rates of nitric oxide formation (or of loss of nitrogen atoms) by (12.14), (12.36), and (12.38)–(12.40).

At low altitudes, in the stratosphere, there is no doubt that the reaction (12.36) of atomic nitrogen with molecular oxygen is certainly an important process. Values for the coefficient are as follows:

Altitude (km):	20	30	40	50
$b_7 n(O_2)$ (sec^{-1}):	3.0×10^{-1}	2.0×10^{-1}	3.0×10^{-1}	1.0×10^{-1}

However, the reaction (12.40) with ozone (see *Fig. 12.5*) may compete with the process involving molecular oxygen.

In the mesosphere, other process may become more important than (12.36) since its value decreases rapidly because of the simultaneous decrease of the temperature and of the O_2 concentration. Selected values are as follows:

Altitude (km):	60	70	80	85
$b_7 n(O_2)$ (sec^{-1}):	1.0×10^{-2}	2.0×10^{-4}	1.6×10^{-6}	7.8×10^{-8}

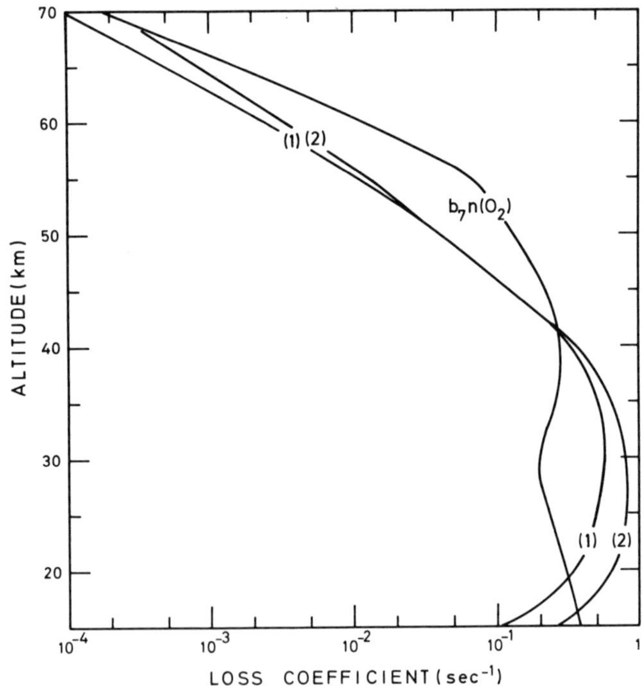

Fig 12..5 Loss coefficient of nitrogen atoms reacting with molecular oxygen and ozone in the stratosphere and lower mesosphere. (1) $b_{8a}n(O_3)$; (2) $b_{8b}n(O_3)$.

The fact that nitric oxide is rapidly produced by a reaction with ozone leads to larger values at the mesopause:

Altitude (km):	60	70	80	85
$b_{8a}n(O_3)$ (sec^{-1}):	2.8×10^{-3}	1.0×10^{-4}	4.3×10^{-6}	1.4×10^{-6}
$b_{8b}n(O_3)$ (sec^{-1}):	3.3×10^{-3}	2.0×10^{-4}	1.7×10^{-5}	8.4×10^{-6}

It is clear that ozone may play a more important role than molecular oxygen in the loss process of nitrogen atoms particularly if the activation energy of the reaction (12.40) is not more than 1 kcal (*see Fig. 12.6*).

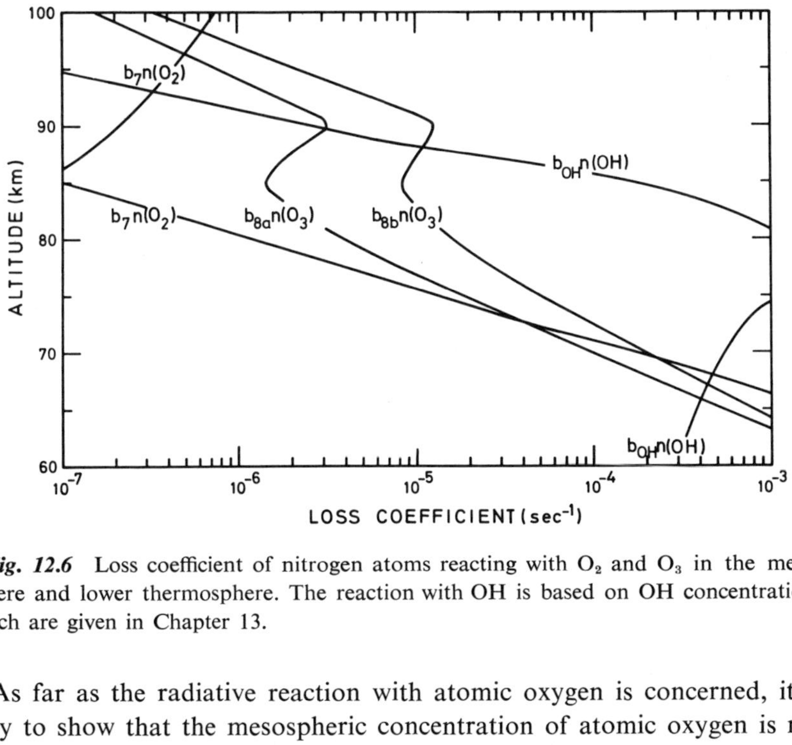

Fig. 12.6 Loss coefficient of nitrogen atoms reacting with O_2 and O_3 in the mesosphere and lower thermosphere. The reaction with OH is based on OH concentrations which are given in Chapter 13.

As far as the radiative reaction with atomic oxygen is concerned, it is easy to show that the mesospheric concentration of atomic oxygen is not sufficient to compete with other processes. With a rate coefficient $b_{1b} = 2 \times 10^{-17}$ cm^3 sec^{-1}, (12.16), a loss coefficient of the order of 2.5×10^{-6} sec^{-1}, requires at least 10^{11} oxygen atoms cm^{-3}, i.e., a value which occurs only in the lower thermosphere.

Finally, in order for (12.38) to introduce an effect in the mesosphere, it is necessary for the rate coefficient b_7^* to reach values greater than 10^{-15} cm^3 sec^{-1}. With $b_7^* = 10^{-15}$ cm^3 sec^{-1} the coefficient $b_7^*n(O_2^*)$ equals 3×10^{-6} sec^{-1} at 85 km which is less than the value of $b_8n(O_3)$. The range of values ($b_7{}^* > 10^{-13}$ cm^3 sec^{-1}) which are considered in the aeronomic study [32] of $O_2(^1\Delta_g)$ cannot be accepted since values of the order of 3×10^{-15} cm^3 sec^{-1} have been found [33, 34]. The alternative mechanism with $O_2(^1\Sigma_g{}^+)$ requires a very high rate coefficient in order to be important. With $n(O_2^{**}) = 10^6$ cm^{-3}, it would be necessary to assume $b_7^{**} \geq 10^{-11}$ cm^3 sec^{-1}, i.e., a kinetic cross section.

Another mechanism for the formation of nitric oxide involves the reaction of nitrogen atoms with the mesospheric radical OH,

$$(b_{OH}): \quad N + OH \rightarrow NO + H + 49 \quad \text{kcal mole}^{-1}. \tag{12.64}$$

This is a very fast reaction with a rate coefficient [41] of the order of 7×10^{-11} cm^3 sec^{-1} at 320°K. If there is no activation energy for such a reaction, the rate coefficient may be written

$$b_{OH} = 4\times10^{-12}T^{1/2} \quad \text{cm}^3\ \text{sec}^{-1}. \tag{12.65}$$

The rapidity of this reaction indicates that it can play the principal role at the mesopause level if the OH concentration reaches even small values of the order of 10^6 cm^{-3} (Fig. 12.6).

In the mesophere, where it is possible to adopt a photochemical equilibrium for atomic nitrogen, (12.63) becomes

$$dn(NO)/dt + 2b_6 n(N)n(NO) = P(N) \tag{12.66}$$

which is identical to (12.47). In order to reach an equilibrium for $n(NO)$ a time of the order $[2b_6n(N)]^{-1} = 2\times10^{10}/n(N)$ sec is required; 5×10^5 nitrogen atoms cm^{-3} lead to 4×10^4 sec or about 1 day.

Thus, accepting $n(N) > 5\times10^5$ cm^{-3} as a working hypothesis and taking photoequilibrium conditions, (12.61) leads to

$$\begin{aligned}\frac{n(NO)}{n(N)} &= \frac{b_{1b}n(O) + b_7 n(O) + b_7{}^*n(O^*) + b_7^{**}n(O^{**}) + b_8 n(O_3) + b_{OH}n(OH)}{b_6 n(N) + J_{NO}} \\ &= \frac{b_7 n_7}{b_6 n(N) + J_{NO}}\end{aligned} \tag{12.67}$$

where $b_7 n_7$ represents all possible loss processes for N. Since $J_{NO} \leq 5\times10^{-6}$ sec^{-1} and $b_6 n(N) \geq 10^{-5}$, an excellent approximation of (12.67) is given by the *chemical equilibrium* expression

$$n(NO) = b_7 n_7/b_6 \equiv n_*(NO). \tag{12.68}$$

In an equilibrium state $dn(\mathrm{NO})/dt = 0$ and $dn(\mathrm{N})/dt = 0$ so that (12.61) and (12.62) lead to

$$2b_6 n(\mathrm{N}) n(\mathrm{NO}) = P(\mathrm{N}) \tag{12.69}$$

and

$$2b_7 n_7 n(\mathrm{N}) = 2n(\mathrm{NO})J_{\mathrm{NO}} + P(\mathrm{N}). \tag{12.70}$$

Using the chemical equilibrium value $n_*(\mathrm{NO})$ given by (12.68), the concentrations of N and NO are obtained from (12.69) and (12.70) in the form

$$n(\mathrm{NO}) = n_*(\mathrm{NO}) \Big/ \left\{ \frac{1}{2} + \left[\frac{n_*(\mathrm{NO})J_{\mathrm{NO}}}{2P(\mathrm{N})} + \frac{1}{4} \right]^{1/2} \right\} \tag{12.71}$$

and

$$n(\mathrm{N}) = \frac{P(\mathrm{N})}{2b_7 n_7} \left\{ \frac{1}{2} + \left[\frac{n_*(\mathrm{NO})J_{\mathrm{NO}}}{2P(\mathrm{N})} + \frac{1}{4} \right]^{1/2} \right\}. \tag{12.72}$$

Two limiting cases can be considered. In the first case, the production of nitrogen atoms from molecular nitrogen is more important than the production resulting from NO photodissociation, i.e., $n_*(\mathrm{NO})J_{\mathrm{NO}} \ll P(\mathrm{N})$. With this condition (12.71) and (12.72) give

$$n(\mathrm{NO}) \equiv n_*(\mathrm{NO}) = b_7 n_7 / b_6 \tag{12.73}$$

and

$$n(\mathrm{N}) \equiv n_*(\mathrm{N}) = P(\mathrm{N})/2b_7 n_7. \tag{12.74}$$

In the second case, the production of nitrogen atoms from molecular nitrogen is assumed to be small and Eqs. (12.71) and (12.72) lead to

$$n(\mathrm{NO}) = n_*(\mathrm{NO})[2P(\mathrm{N})/n_*(\mathrm{NO})J_{\mathrm{NO}}]^{1/2} < n_*(\mathrm{NO}) \tag{12.75}$$

and

$$n(\mathrm{N}) = n_*(\mathrm{N})[n_*(\mathrm{NO})J_{\mathrm{NO}}/2P(\mathrm{N})]^{1/2} > n_*(\mathrm{N}). \tag{12.76}$$

Thus, expression (12.73), which corresponds to a high production of N, leads to a nitric oxide concentration close to the chemical equilibrium value (12.68). When the atomic nitrogen production from N_2 is smaller, (12.75) shows that the concentration $n(\mathrm{NO})$ is smaller than the chemical equilibrium value (12.68). *Figure 12.7* summarizes how the origin of nitric oxide is explained by the dissociation of molecular nitrogen and how the chemical equilibrium value is limited by the reactions with atomic nitrogen.

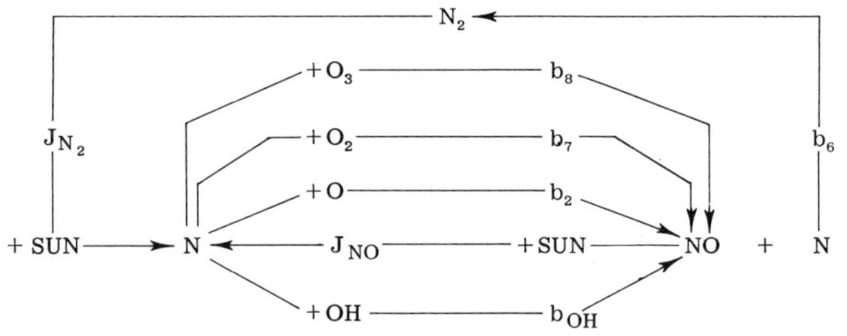

Fig. 12.7 Principal chemical and photochemical reactions involved in the production and loss of nitric oxide and atomic nitrogen in the chemosphere.

12.5 Reactions Involving Atomic Hydrogen

Before studying ionospheric conditions for NO it is necessary to consider chemospheric processes in which hydrogen is involved. In a region such as the mesosphere and the lower thermosphere where H and NO are present, the formation of nitroxyl and its loss are due to a catalytic action of nitric oxide on atomic hydrogen.

The three-body process [15, 16]

$$(b_{16}):\quad H + NO + M \rightarrow HNO + M + 49 \quad \text{kcal mole}^{-1} \tag{12.77}$$

has been the subject of recent laboratory studies since an infrared spectrum has been detected. Absorption and emission bands have been analyzed and detailed investigations have been carried out. With the following rate coefficients:

$$b_{16} = 1 \times 10^{-32} e^{300/T} \quad \text{cm}^6\ \text{sec}^{-1} \tag{12.78}$$

for the total reaction (12.77), and

$$b_{16a} = 7 \times 10^{-19} (273/T)^3 \quad \text{cm}^3\ \text{sec}^{-1} \tag{12.79}$$

for the chemiluminescent reaction ($\lambda\lambda$ 7625–6925–7965–6262 Å), it is clear that the aeronomic effect cannot be important. In the upper mesosphere and lower thermosphere, it would be difficult to reach a production of one photon cm^{-3} sec^{-1} since the condition $n(\text{NO})n(\text{H}) > 10^{18}$ cm^{-6} would be required.

The catalytic action of nitric oxide occurs in the recombination of hydrogen atoms through the process

$$(b_{17}):\quad H + HNO \rightarrow NO + H_2 + 55 \quad \text{kcal mole}^{-1} \tag{12.80}$$

with a rate coefficient

$$b_{17} = 5\times 10^{-13} T^{1/2} e^{-1200/T} \quad \text{cm}^3\ \text{sec}^{-1}. \tag{12.81}$$

The reaction of nitroxyl with a hydroxyl radical

$$(b_{18}):\quad OH + HNO \rightarrow NO + H_2O + 69 \quad \text{kcal mole}^{-1} \tag{12.82}$$

has perhaps a larger rate coefficient such as

$$b_{18} = 5\times 10^{-12} T^{1/2} e^{-1200/T} \quad \text{cm}^3\ \text{sec}^{-1}. \tag{12.83}$$

The reaction of HNO with HO_2

$$(b_{19}):\quad HO_2 + HNO \rightarrow NO + H_2O_2 + 40 \quad \text{kcal mole}^{-1} \tag{12.84}$$

may also occur. The rate coefficient b_{19}, which is not known, cannot be greater than b_{17} and b_{18}. Thus it seems from (12.77), (12.80), (12.82), and (12.84) that the equilibrium ratio

$$\frac{n(HNO)}{n(H)} = \frac{b_{16}n(M)n(NO)}{b_{17}n(H) + b_{18}n(OH) + b_{19}n(HO_2)} \tag{12.85}$$

may have a maximum value in the upper mesosphere

$$n(HNO) \leq [b_{16}n(M)/b_{17}]n(NO). \tag{12.86}$$

The net result of reactions (12.77)–(12.84) is that the nitric oxide concentration is not affected (*see Fig. 12.8*).

Atomic hydrogen is subject to other more important loss processes and its chemiluminescent reaction in the formation of nitroxyl should be difficult

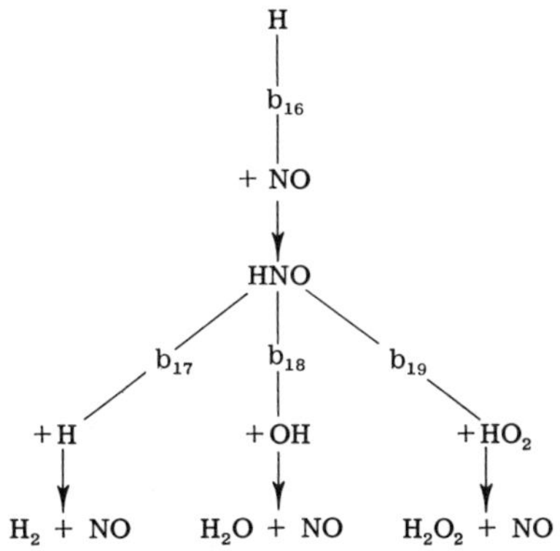

Fig. 12.8 Principal reactions in which nitroxyl is involved in the mesosphere with the catalytic action of nitric oxide.

to detect. In the mesosphere (12.86) leads to

$$n(\mathrm{HNO}) \leq 3\times 10^{-3} n(\mathrm{NO}) \quad \mathrm{cm}^{-3} \tag{12.87}$$

which should be considered as a limit for possible values of the nitroxyl concentration in the mesosphere.

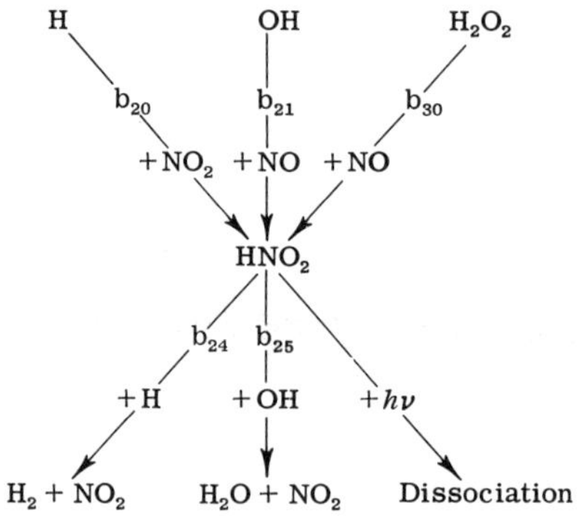

Fig. 12.9 Principal reactions in which nitrous acid is involved in the stratosphere. Reactions with atomic hydrogen which occur only in the mesosphere are negligible. Nitric oxide has a catalytic action in a sunlit atmosphere.

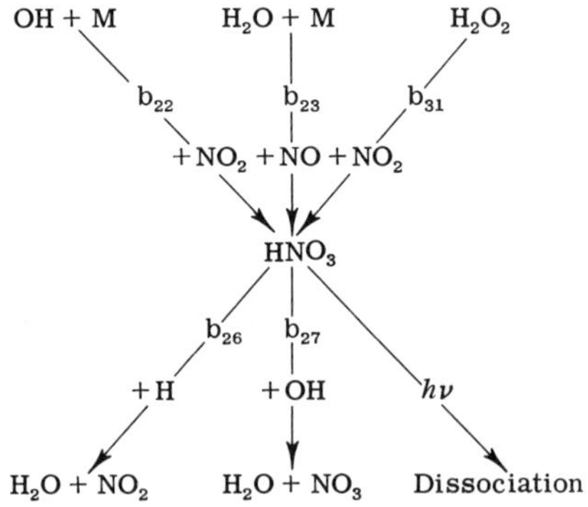

Fig. 12.10 Production and loss processes of nitric acid in the stratosphere. The OH, HO_2, and H_2O_2 concentrations depend on the catalytic action of nitrogen oxides leading to nitric acid.

In addition to (12.77) other possible three-body associations leading to nitrous and nitric acids must be considered (see *Figs. 12.9* and *12.10*). They are

$$(b_{20}):\quad NO_2 + H + M \rightarrow HNO_2 + M + 80 \quad \mathrm{kcal\ mole^{-1}} \tag{12.88}$$

$$(b_{21}):\quad NO + OH + M \rightarrow HNO_2 + M + 60 \quad \mathrm{kcal\ mole^{-1}} \tag{12.89}$$

$$(b_{22}):\quad NO_2 + OH + M \rightarrow HNO_3 + M + 53 \quad \mathrm{kcal\ mole^{-1}} \tag{12.90}$$

$$(b_{23}):\quad NO + HO_2 + M \rightarrow HNO_3 + M + 63 \quad \mathrm{kcal\ mole^{-1}}. \tag{12.91}$$

The rate coefficients for these reactions are not known, but typical values of such three-body rate coefficients are between 10^{-32} and 10^{-31} cm^6 sec^{-1}. Reaction (12.88) leading to nitrous acid is not important since it requires H and NO_2, but it is clear that reactions (12.89)–(12.91) may be important in the stratosphere.

Little is known of the photodissociation of nitrous and nitric acids. Since they absorb [38] in the region of $\lambda > 3000$ Å, the photodissociation coefficients should be included with the loss process due to reaction with H and OH:

$$(b_{24}):\quad H + HNO_2 \rightarrow H_2 + NO_2 + 23 \quad \text{kcal mole}^{-1} \tag{12.92}$$

$$(b_{25}):\quad OH + HNO_2 \rightarrow H_2O + NO_2 + 38 \quad \text{kcal mole}^{-1} \tag{12.93}$$

$$(b_{26}):\quad H + HNO_3 \rightarrow H_2O + NO_2 + 66 \quad \text{kcal mole}^{-1} \tag{12.94}$$

$$(b_{27}):\quad OH + HNO_3 \rightarrow H_2O + NO_3 + 15 \quad \text{kcal mole}^{-1}. \tag{12.95}$$

These reactions are exothermic, but the activation energies are not known.

Finally, other reactions involving hydrogen and nitrogen oxides must be considered. A rapid reaction such as

$$(b_{28}):\quad H + NO_2 \rightarrow NO + OH + 30 \quad \text{kcal mole}^{-1} \tag{12.96}$$

may have a rate coefficient without activation energy

$$b_{28} = 2\times 10^{-12} T^{1/2} \quad \text{cm}^3\,\text{sec}^{-1} \tag{12.97}$$

but has a limited aeronomic role since $O + NO_2$ and $H + O_3$ are more important reactions. However, the effect of nitrogen oxides on the perhydroxyl radical and on hydrogen peroxide should be important. There is no doubt that the reaction

$$(b_{29}):\quad NO + HO_2 \rightarrow OH + NO_2 + 9 \quad \text{kcal mole}^{-1} \tag{12.98}$$

is fast [42] relative to other reactions of HO_2 in similar circumstances. In order to be of major importance with respect to other aeronomic reactions, reaction (12.98) should have a rather low activation energy with a value less than 2.5 kcal, a coefficient such as

$$b_{29a} = 5\times 10^{-12} T^{1/2} e^{-1200/T} \quad \text{cm}^3\,\text{sec}^{-1} \tag{12.99}$$

leads to 1×10^{-12} cm^3 sec^{-1} at 300°K, and with an activation energy of about 1 kcal,

$$b_{29b} = 1.5\times 10^{-11} T^{1/2} e^{-500/T} \quad \text{cm}^3\,\text{sec}^{-1} \tag{12.100}$$

gives 4×10^{-11} cm^3 sec^{-1} at 300°K. Such extreme values indicate that the reaction (12.98) may be compared in the stratosphere with the reaction HO_2–O (see Section 13.1) since they both lead to the production of the hydroxyl radical. In other words, the equilibrium $n(OH)/n(HO_2)$ in the stratosphere may depend on the nitric oxide concentration.

The reactions of hydrogen peroxide with nitrogen oxides have not been investigated, and until quantitative studies have been performed the aeronomic importance of these reactions cannot be assessed. The reactions

$$(b_{30}):\quad H_2O_2 + NO \rightarrow OH + HNO_2 + 11 \quad \text{kcal mole}^{-1} \qquad (12.101)$$

and

$$(b_{31}):\quad H_2O_2 + NO_2 \rightarrow OH + HNO_3 + 4 \quad \text{kcal mole}^{-1} \qquad (12.102)$$

cannot be neglected. Even if the hydrogen peroxide concentration is relatively small in the stratosphere, one would expect that reactions (12.101) and (12.102) play an aeronomic role.

In summary, the essential reactions in which nitrogen oxides are involved with atomic hydrogen, hydroxyl, and perhydroxyl radicals, and also hydrogen peroxide may be fast enough to be of importance in the chemosphere. It seems that the production of HNO_3 is possible by a three-body process in the stratosphere with a rate coefficient between 10^{-14} and 10^{-15} cm^3 sec^{-1} and also by a reaction in which H_2O_2 is involved. Whether or not the rate of reaction (12.102) determines the nitric acid concentration is another matter, since the reaction may be complex and its rate coefficient is still unknown. Therefore it is very important to measure the concentrations of nitrogen dioxide [43] or nitric oxide in the stratosphere simultaneously with that of nitric acid [27, 28] to know the fraction of nitrogen oxides which have been transformed into more complex compounds.

As far as the presence of nitrous acid is concerned, it represents an interesting speculative possibility, since at present its reaction rate coefficients are not known. A rapid photodissociation is not impossible, however.

12.6 Ionospheric Production of N and NO

In the ionosphere atomic nitrogen can be produced indirectly through photoionization of N_2 and O followed by dissociative recombination and ion-neutral reactions [44] with N_2^+ and O^+. The processes involving N_2^+ ions can be written (see Section 12.1):

$$(I_{N_2}):\quad N_2 + h\nu \rightarrow N_2^+ + e \qquad (12.103)$$

$$(\alpha_{N_2}):\quad N_2^+ + e \rightarrow N + N \qquad (12.104)$$

$$(\gamma_3):\quad N_2^+ + O \rightarrow NO^+ + N(^2D) \quad \text{or} \quad N(^4S) \qquad (12.105)$$

$$(\gamma_9):\quad N_2^+ + O_2 \rightarrow N_2 + O_2^+. \qquad (12.106)$$

The fraction of N_2^+ ions leading to excited and normal nitrogen atoms can now be expressed by

$$X = \frac{\alpha_{N_2} n_e + \gamma_3 n(O)}{\alpha_{N_2} n_e + \gamma_3 n(O) + \gamma_9 n(O_2)} \tag{12.107}$$

or with the numerical values (see Chapters 7 and 10) for the coefficients

$$X = \frac{10^3 n_e + n(O)}{10^3 n_e + n(O) + 0.4 n(O_2)}. \tag{12.108}$$

The corresponding equilibrium concentration for N_2^+ is

$$n^+(N_2) = \frac{n(N_2) I_{N_2}}{\alpha_{N_2} n_e + \gamma_3 n(O) + \gamma_9 n(O_2)}. \tag{12.109}$$

The principal processes involving O^+ ions can be written

$$(I_O): \quad O + h\nu \rightarrow O^+ + e \tag{12.110}$$

$$(\gamma_1): \quad O^+ + N_2 \rightarrow NO^+ + N(^4S) \tag{12.111}$$

$$(\gamma_2): \quad O^+ + O_2 \rightarrow O + O_2^+. \tag{12.112}$$

The fraction of O^+ ions leading to atomic nitrogen is then given by

$$Y = \frac{\gamma_1 n(N_2)}{\gamma_1 n(N_2) + \gamma_2 n(O_2)} \tag{12.113}$$

or with the numerical coefficients of Chapter 10 for a constant temperature,

$$Y = \frac{n(N_2)}{n(N_2) + 10 n(O_2)}. \tag{12.114}$$

The corresponding equilibrium concentration for O^+ is

$$n^+(O) = \frac{n(O) I_O}{\gamma_1 n(N_2) + \gamma_2 n(O_2)}. \tag{12.115}$$

In the F1 layer where $n(O) > n(O_2)$, $X \simeq 1$; in the E layer, X is about 0.5 between 100 km and 105 km and reaches only 0.2 at 90 km. Y is about 0.5 at 120 km and increases rapidly upward to its limit of 1.

Molecular oxygen ions are also involved in dissociative recombination

$$(\alpha_{O_2}): \quad O_2^+ + e \rightarrow O + O \tag{12.116}$$

and in the following ion-neutral reactions:

$$(\gamma_5): \quad O_2^+ + NO \rightarrow O_2 + NO^+ \tag{12.117}$$

$$(\gamma_6): \quad O_2^+ + N \rightarrow O + NO^+ \tag{12.118}$$

$$(\gamma_7): \quad O_2^+ + N_2 \rightarrow NO + NO^+. \tag{12.119}$$

Ionization equilibrium conditions then lead to

$$n^+(O_2) = \frac{n(O_2)I_{O_2}}{\alpha_{O_2}n_e + \gamma_5 n(NO) + \gamma_6 n(N)} \tag{12.120}$$

where I_{O_2} is the photoionization coefficient. The reaction (12.119) can be neglected since the rate coefficient is less than 2×10^{-16} cm^3 sec^{-1}. With an electron concentration of 10^5 cm^{-3} and a nitric oxide concentration of 3×10^7 cm^{-3}, Eq. (12.120) shows that $\alpha_{O_2}n_e$ is less than $\gamma_5 n(NO)$. Thus in the E layer, the O_2^+ concentration depends on the reaction with nitric oxide.

With the additional reactions

$$(\gamma_{4a}): \quad O^+ + NO \rightarrow O + NO^+ + 4.35 \quad \text{eV} \tag{12.121}$$

$$(\gamma_{4b}): \quad O^+ + NO \rightarrow N + O_2^+ + 0.16 \quad \text{eV} \tag{12.122}$$

$$(\alpha_{NO}): \quad NO^+ + e \rightarrow O(^3P) + N(^4S) + 2.76 \quad \text{eV} \tag{12.123}$$

$$\rightarrow O(^1D) + N(^4S) + 0.79 \quad \text{eV} \tag{12.124}$$

$$\rightarrow O(^3P) + N(^2D) + 0.38 \quad \text{eV} \tag{12.125}$$

the NO^+ equilibrium condition is given by

$$\begin{aligned}\alpha_{NO}n_e n^+(NO) = {} & n(NO)[I_{NO} + \gamma_{4a}n^+(O) + \gamma_5 n^+(O_2)] \\ & + \gamma_1 n^+(O)n(N_2) + \gamma_3 n^+(N_2)n(O) \\ & + n^+(O_2)[\gamma_6 n(N) + \gamma_7 n(N_2)]\end{aligned} \tag{12.126}$$

where I_{NO} is the photoionization coefficient of nitric oxide.

In the ionosphere, the differential equations (12.61) and (12.62) can now be written with (12.126)

$$\begin{aligned}&\frac{dn(NO)}{dt} + n(NO)[J_{NO} + I_{NO} + b_6 n(N) + (\gamma_{4a} + \gamma_{4b})n^+(O) + \gamma_5 n^+(O_2)] \\ &\quad = b_7 n_7 n(N) + \gamma_7 n^+(O_2)n(N_2)\end{aligned} \tag{12.127}$$

and

$$\frac{dn(\mathrm{N})}{dt} + n(\mathrm{N})[b_6 n(\mathrm{NO}) + b_7 n_7]$$
$$= n(\mathrm{N}_2)[2J_{\mathrm{N}_2} + 2\gamma_1 n^+(\mathrm{O}) + \gamma_7 n^+(\mathrm{O}_2)]$$
$$+ 2\alpha_{\mathrm{N}_2} n^+(\mathrm{N}_2) n_e + n(\mathrm{NO})[J_{\mathrm{NO}} + I_{\mathrm{NO}} + (\gamma_{4a} + \gamma_{4b}) n^+(\mathrm{O})$$
$$+ \gamma_5 n^+(\mathrm{O}_2)] + 2\gamma_3 n(\mathrm{O}) n^+(\mathrm{N}_2). \qquad (12.128)$$

The simultaneous variation of NO and N can be expressed by

$$\frac{dn(\mathrm{NO})}{dt} + \frac{dn(\mathrm{N})}{dt} + 2b_6 n(\mathrm{NO}) n(\mathrm{N})$$
$$= 2n(\mathrm{N}_2)[J_{\mathrm{N}_2} + \gamma_1 n^+(\mathrm{O}) + \gamma_7 n^+(\mathrm{O}_2)]$$
$$+ 2\alpha_{\mathrm{N}_2} n^+(\mathrm{N}_2) n_e + 2\gamma_3 n(\mathrm{O}) n^+(\mathrm{N}_2). \qquad (12.129)$$

With (12.107) and (12.114) the expression (12.129) becomes

$$\frac{dn(\mathrm{NO})}{dt} + \frac{dn(\mathrm{N})}{dt} + 2b_6 n(\mathrm{NO}) n(\mathrm{N})$$
$$= 2n(\mathrm{N}_2)[J_{\mathrm{N}_2} + XI_{\mathrm{N}_2}] + 2YI_{\mathrm{O}} n(\mathrm{O}) + 2n(\mathrm{N}_2)\gamma_7 n^+(\mathrm{O}_2). \qquad (12.130)$$

For equilibrium conditions $[dn(\mathrm{NO})/dt = 0$ and $dn(\mathrm{N})/dt = 0]$ neglecting γ_7

$$b_6 n(\mathrm{NO}) n(\mathrm{N}) = n(\mathrm{N}_2)[J_{\mathrm{N}_2} + XI_{\mathrm{N}_2}] + YI_{\mathrm{O}} n(\mathrm{O}) \qquad (12.131)$$

and

$$b_7 n_7 n(\mathrm{N}) = n(\mathrm{N}_2)[J_{\mathrm{N}_2} + XI_{\mathrm{N}_2}] + YI_{\mathrm{O}} n(\mathrm{O}) + n(\mathrm{NO})[J_{\mathrm{NO}} + I_{\mathrm{NO}}$$
$$+ (\gamma_{4a} + \gamma_{4b}) n^+(\mathrm{O}) + \gamma_5 n^+(\mathrm{O}_2)]. \qquad (12.132)$$

By comparison of these relations with the chemospheric equations (12.69) and (12.70) it is seen that the equivalent atomic nitrogen production resulting from N_2 is

$$P(\mathrm{N}) = 2n(\mathrm{N}_2)[I_{\mathrm{N}_2} + XI_{\mathrm{N}_2}] + 2YI_{\mathrm{O}} n(\mathrm{O}) \qquad (12.133)$$

and that the effective dissociation D_{NO} of nitric oxide is

$$D_{\mathrm{NO}} n(\mathrm{NO}) = n(\mathrm{NO})[J_{\mathrm{NO}} + I_{\mathrm{NO}} + (\gamma_{4a} + \gamma_{4b}) n^+(\mathrm{O}) + \gamma_5 n^+(\mathrm{O}_2)]. \qquad (12.134)$$

The corresponding values of NO and N are given by equations similar to

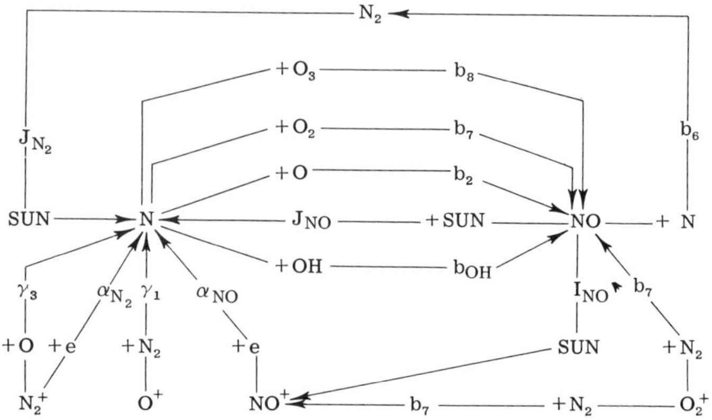

Fig. 12.11 Chemical and ionic reactions involved in the production and loss of nitric oxide and atomic nitrogen in the ionosphere.

(12.71) and (12.72). The ionospheric processes leading to N and NO are summarized in *Fig. 12.11*.

Using data given in Tables 12.2 and 12.3, it is now possible to deduce the equilibrium values of nitric oxide and of atomic nitrogen concentrations. *Figure 12.12* shows the various vertical distributions of the atomic nitrogen

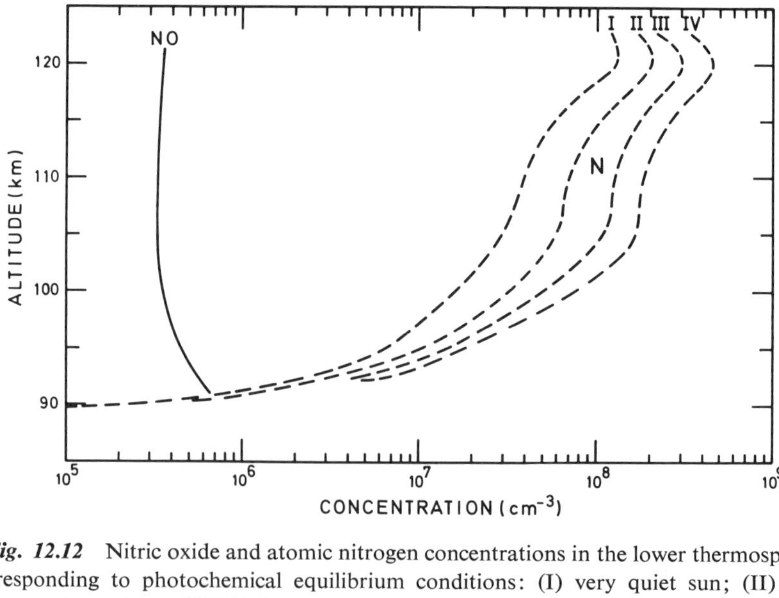

Fig. 12.12 Nitric oxide and atomic nitrogen concentrations in the lower thermosphere corresponding to photochemical equilibrium conditions: (I) very quiet sun; (II) low average solar activity; (III) high average; (IV) maximum. All the dissociative recombination of N_2^+ and NO^+ leads to normal nitrogen atoms.

concentration for various conditions of the solar activity compared to the chemical equilibrium value of the nitric oxide concentration. They correspond to equilibrium conditions for an overhead sun. The nitric oxide concentration is relatively small and that of atomic nitrogen is relatively high. However, equilibrium conditions for atomic nitrogen cannot be easily reached since the time to reach equilibrium conditions in the 100–120 km region is not less than 10^5 sec.

Above the E layer (*Fig. 12.13*) both concentrations depend strongly on the solar activity. For equilibrium conditions, the nitric oxide concentration increases by a factor of about 10 from minimum to maximum sunspot

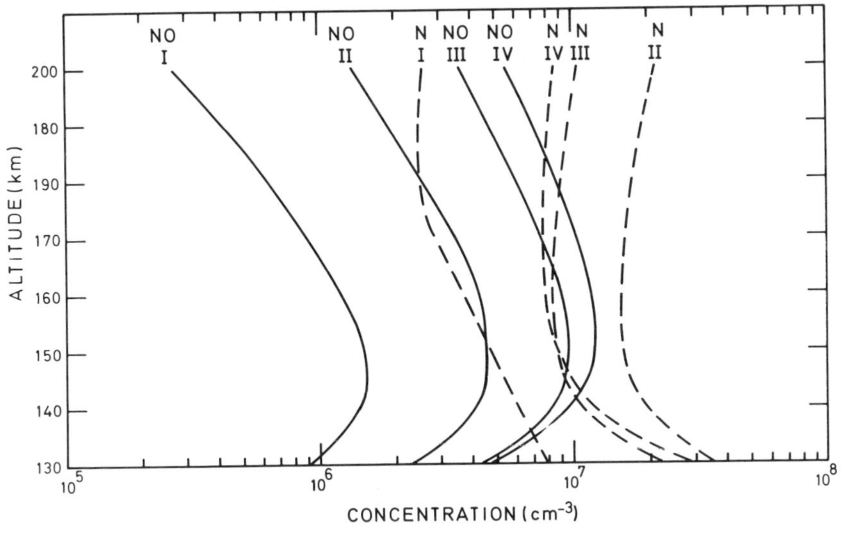

Fig. 12.13 Nitric oxide and atomic nitrogen concentrations above the E layer for solar conditions identical to those of Fig. 12.12.

conditions. Due to the nitric oxide increase related to the temperature increase, the atomic nitrogen concentration is limited; its maximum concentration occurs during a relatively quiet sun.

It is clear that the theoretical results illustrated in Fig. 12.12 and 12.13 cannot represent the observational results [45–49]. The observational concentrations are considerably greater than the equilibrium distributions. The total content of NO above 125 km estimated [45–48] from the intensity of the γ-band dayglow is of the order of 10^{13} cm^{-2} and cannot be less than 10^{14} cm^{-2} above 90 km. The small theoretical concentration of NO between 90 and 120 km (see Fig. 12.12) shows that the calculated content departs drastically from the measurements. The fact that there is a possible agreement above 125 km cannot be accepted; the profiles of NO and N correspond to chemical equilibrium conditions which do not lead to the real aeronomic conditions. At the minimum of the atomic nitrogen concentration, near 150–160 km for an overhead sun, photoequilibrium conditions would be reached after generally less than 1 hr. With a concentration $n(N) = 10^7$ cm^{-3}, the chemical equilibrium conditions would be applied also to nitric oxide. Thus, nitrogen atoms and nitric oxide molecules would react together to form again molecular nitrogen.

Because of the discrepancies between the observed and calculated concentration of nitric oxide in the thermosphere, it must be concluded that all the chemical and ionic processes producing nitrogen atoms cannot explain the aeronomic behavior. A process such as (12.119) leading to a direct production of nitric oxide molecules which has been suggested [8] cannot contribute significantly to an increase of the NO concentration compared with that of atomic nitrogen. Other processes with excited oxygen molecules or even other appropriate molecules such as OH involve normal nitrogen atoms and are not an independent source of nitric oxide. However, they may increase the ratio $n(NO)/n(N)$. Since the product $n(NO) \times n(N)$ remains constant under equilibrium conditions, an increase of $n(NO)$ corresponds to a decrease of $n(N)$ and it is possible to imagine conditions where $n(NO) > n(N)$ with the conclusion that the reaction between N and NO does not contribute to the whole of the loss. A transport of nitric oxide may be involved.

A satisfactory solution must be obtained by a formation of nitric oxide without the direct production of normal nitrogen atoms. This may happen through the production of $N(^2D)$ atoms by the dissociative recombination of NO^+ and N_2^+ and their reaction with molecular oxygen [17–19].

12.7 Effect of Excited Nitrogen Atoms

Possible sources of excited nitrogen atoms in the sunlit hemisphere have been discussed [50] and compared with dayglow observations of 5200 Å emissions. These observations [51–55] indicate that the abundance of the excited $N(^2D)$ depends on dissociative recombination processes and associated ionic reactions. The production of $N(^2D)$ atoms may be due to

1. dissociative recombination of NO^+ [see (12.125)]
2. dissociative recombination of N_2^+ [see (12.3–12.5)]
3. ion–atom interchange with O [see (12.105)].

The deactivation rate coefficient of $N(^2D)$ has been measured in several gases [56]. Molecular oxygen is the major quencher with a rate of 7×10^{-12} $cm^3\ sec^{-1}$ while for N_2 the rate coefficient is only $6\times10^{-15}\ cm^3\ sec^{-1}$. The mechanism in which O_2 is involved leads to a production of NO:

$$(b_7^*):\quad N(^2D) + O_2 \rightarrow NO + O(^3P) \tag{12.135}$$

with a rate coefficient $b_7^* = 5\times10^{-12}\ cm^3\ sec^{-1}$. A simple deactivation such as

$$N(^2D) + O_2 \rightarrow N(^4S) + O_2 \tag{12.136}$$

may be neglected. Accepting the three preceding possibilities for the production of excited atoms $N(^2D)$, we consider that [17]

$$NO^+ + e \rightarrow N(^2D) + O; \quad \text{fraction } F_{NO} \tag{12.137}$$

$$\rightarrow N(^4S) + O; \quad \text{fraction } 1 - F_{NO} \tag{12.138}$$

$$N_2^+ + e \rightarrow N(^2D); \quad \text{fraction } F_{N_2} \tag{12.139}$$

$$\rightarrow N(^4S); \quad \text{fraction } 1 - F_{N_2} \tag{12.140}$$

$$N_2^+ + O \rightarrow NO^+ + N(^2D); \quad \text{fraction } F_O \tag{12.141}$$

$$\rightarrow NO^+ + N(^4S); \quad \text{fraction } 1 - F_O. \tag{12.142}$$

There is no information on the magnitude of F_{NO}, F_{N_2}, and F_O; a maximum value for the nitric oxide production would be $F_{NO} = F_{N_2} = F_O = 1$. Thus $F_{NO} = F_{N_2} = F_O = 0$ leads to Eqs. (12.127) and (12.128) where all dissociative recombinations give only nitrogen atoms without direct production of nitric oxide. It is clear, therefore, that it is impossible to reach any definite conclusions regarding the relative importance of the reactions (12.137)–(12.142) without an exact knowledge of F_{NO}, F_{N_2}, and F_O. Neglecting the term $\gamma_7 n^+(O_2)n(N_2)$, Eqs. (12.127) and (12.128) can be modified by using (12.137) to (12.142). One obtains [17]

$$\begin{aligned}\frac{dn(NO)}{dt} &+ n(NO)[J_{NO} + b_6 n(N) + \gamma_{4b} n^+(O) \\ &+ \{1 - F_{NO}\}\{I_{NO} + \gamma_{4a} n^+(O) + \gamma_5 n^+(O_2)\}] \\ &= b_7 n_7 n(N) + 2XI_{N_2}F_{N_2}n(N_2) + YI_O F_{NO} n(O) \\ &+ \gamma_3 n(O)n^+(N_2)\{F_O + F_{NO} - 2F_{N_2}\} + \gamma_6 n(N)F_{NO}n^+(O_2)\end{aligned} \tag{12.143}$$

and

$$\frac{dn(\mathrm{N})}{dt} + n(\mathrm{N})[b_6 n(\mathrm{NO}) + b_7 n_7 + \gamma_6 F_{\mathrm{NO}} n^+(\mathrm{O_2})]$$
$$= n(\mathrm{NO})[J_{\mathrm{NO}} + \gamma_{4b} n^+(\mathrm{O}) + \{1 - F_{\mathrm{NO}}\}\{I_{\mathrm{NO}} + \gamma_{4a} n^+(\mathrm{O}) + \gamma_5 n^+(\mathrm{O_2})\}]$$
$$+ 2n(\mathrm{N_2})[J_{\mathrm{N_2}} + XI_{\mathrm{N_2}}\{1 - F_{\mathrm{N_2}}\}] + YI_{\mathrm{O}} n(\mathrm{O})\{2 - F_{\mathrm{NO}}\}$$
$$+ \gamma_3 n(\mathrm{O}) n^+(\mathrm{N_2})\{2F_{\mathrm{N_2}} - F_{\mathrm{NO}} - F_{\mathrm{O}}\}. \qquad (12.144)$$

The simultaneous variation of NO and N is identical to (12.129) when γ_7 is neglected. In order to evaluate the possible importance of the $N(^2D) + O_2$ process for the NO and N concentrations it is assumed [17] that $F_{\mathrm{NO}} = 1$, i.e., all the dissociative recombination of NO^+ leads to ^2D atoms. With $F_{\mathrm{N_2}} = 0.5$ an equal amount of ^2D and ^4S atoms results from the N_2^+ dissociative recombination and with $F_{\mathrm{O}} = 0$ it is assumed that $N_2^+ + O$ produces only ^4S atoms. The expressions (12.143) and (12.144) are thus simplified in the form

$$\frac{dn(\mathrm{NO})}{dt} + n(\mathrm{NO})[J_{\mathrm{NO}} + b_6 n(\mathrm{N}) + \gamma_{4b} n^+(\mathrm{O})]$$
$$= b_7 n_7 n(\mathrm{N}) + XI_{\mathrm{N_2}} n(\mathrm{N_2}) + YI_{\mathrm{O}} n(\mathrm{O}) + \gamma_6 n(\mathrm{N}) N^+(\mathrm{O_2}) \qquad (12.145)$$

and

$$\frac{dn(\mathrm{N})}{dt} + n(\mathrm{N})[b_6 n(\mathrm{NO}) + b_7 n_7 + \gamma_6 n^+(\mathrm{O_2})]$$
$$= n(\mathrm{NO})[J_{\mathrm{NO}} + \gamma_{4b} n^+(\mathrm{O})] + n(\mathrm{N_2})[2J_{\mathrm{N_2}} + XI_{\mathrm{N_2}}] + YI_{\mathrm{O}} n(\mathrm{O}). \qquad (12.146)$$

For equilibrium conditions

$$n(\mathrm{N})[b_7 n_7 + \gamma_6 n^+(\mathrm{O_2})] = n(\mathrm{NO})[J_{\mathrm{NO}} + \gamma_{4b} n^+(\mathrm{O})] + n(\mathrm{N_2}) J_{\mathrm{N_2}} \qquad (12.147)$$

and

$$b_6 n(\mathrm{NO}) n(\mathrm{N}) = n(\mathrm{N_2})[J_{\mathrm{N_2}} + XI_{\mathrm{N_2}}] + YI_{\mathrm{O}} n(\mathrm{O}). \qquad (12.148)$$

These equations can be used to compute the nitrogen and nitric oxide distributions shown in *Fig. 12.14* which should be compared with Fig. 12.13. The introduction of $N(^2D)$ leads to an increase of the nitric oxide concentration. However, the concentrations in the lower thermosphere (see *Fig. 12.15*) indicate that photoequilibrium does not represent the observed conditions.

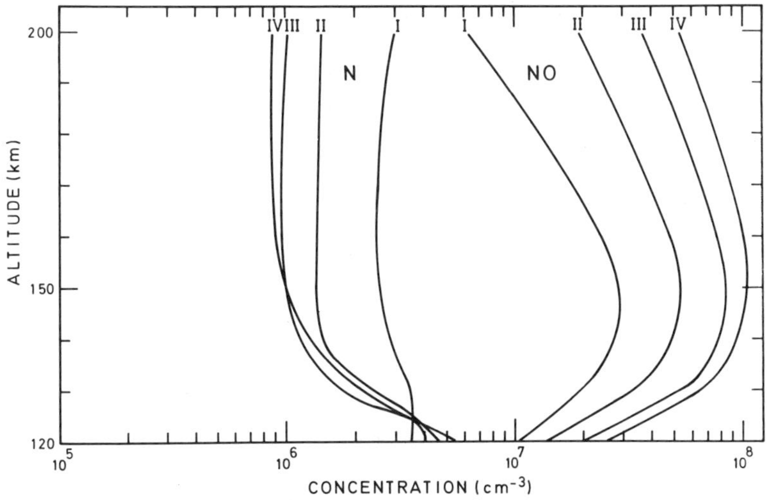

Fig. 12.14 Atomic nitrogen and nitric oxide concentrations for the same solar conditions as in Fig. 12.12, when excited atomic nitrogen is involved.

For precise analysis of the aeronomic behavior a knowledge of the atmospheric conditions (eddy diffusion) is required. At 150–160 km where there is a peak of the ionic production and of the nitric oxide concentration under photochemical equilibrium conditions, it is clear that the removal of NO

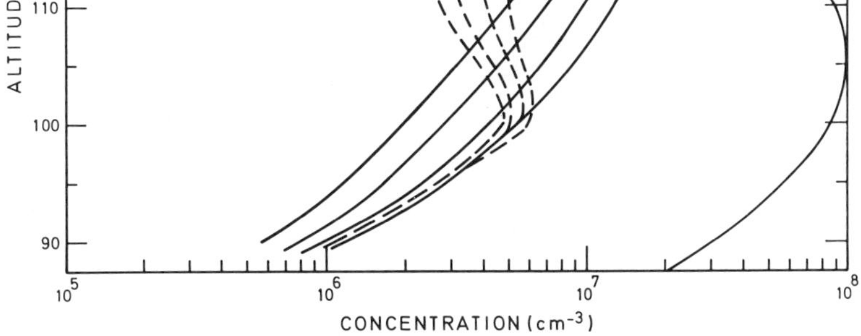

Fig. 12.15 Atomic nitrogen and nitric oxide concentration in the lower thermosphere for various conditions of solar activity. Observational values are represented by ME [49].

molecules is impossible through the normal process N + NO since $n(\mathrm{N}) < n(\mathrm{NO})$. With a dissociation coefficient $J_{\mathrm{NO}} = 5\times 10^{-6}\ \mathrm{sec}^{-1}$, it is necessary to accept the effect of molecular diffusion with an average lifetime less than 2×10^5 sec. However, with a concentration $n(\mathrm{N}) = 1\text{–}2\times 10^6\ \mathrm{cm}^{-3}$, the reciprocal life of NO is $2\text{–}4\times 10^{-5}\ \mathrm{sec}^{-1}$ which requires a molecular diffusion time not less than $(3 \pm 1)\times 10^4$ sec.

With a diffusion coefficient $D(\mathrm{NO})$ between 2×10^8 and $4\times 10^8\ \mathrm{cm}^2\ \mathrm{sec}^{-1}$ corresponding to the region of the NO concentration peak (150–160 km) for an overhead sun, the diffusion time $\tau_{\mathrm{D}} = H^2/D$ is about $(3 \pm 1)\times 10^4$ sec at 150–160 km, i.e., a value which leads to a downward transport. The corresponding diffusion velocity is about 5 km hr^{-1}.

This crude estimate of the transport time by molecular diffusion demonstrates that the nitric oxide concentration in the lower thermosphere does not depend on the chemospheric or ionic processes in which the formation and the removal of nitric oxide are involved in the E layer, but is related to the diffusion transport of the nitric oxide formed in the F layer.

With all significant sources of nitric oxide in the thermosphere above the E layer, the diffusive downward flow on NO molecules is given approximately by (see Chapter 15)

$$F(\mathrm{NO}) = n(\mathrm{NO}) \times \frac{D(\mathrm{NO})}{H}\left[\frac{H}{H(\mathrm{NO})} - \frac{m(\mathrm{NO})}{m}\right] \tag{12.149}$$

in which $m(\mathrm{NO})$ denotes the NO mass, m and H are the atmospheric mass and scale height, and $H(\mathrm{NO})$ and $D(\mathrm{NO})$ are the scale height and the molecular diffusion coefficient of nitric oxide, respectively. If this flux is to replace the nitric oxide destroyed by photodissociation it must be of the order of

$$F(\mathrm{NO}) = \tfrac{1}{2}n(\mathrm{NO})J_{\mathrm{NO}}H(\mathrm{NO}). \tag{12.150}$$

Combining (12.149) and (12.150) we obtain

$$D(\mathrm{NO}) = \frac{J_{\mathrm{NO}}H^2(\mathrm{NO})}{2[1 - m(\mathrm{NO})H(\mathrm{NO})/mH]} \tag{12.151}$$

as the condition of maintaining nitric oxide with a certain scale height. Considering the ratio $H(\mathrm{NO})/H = 0.9$, it is clear from (12.151) that $D(\mathrm{NO}) \geq 2.5\times 10^7\ \mathrm{cm}^2\ \mathrm{sec}^{-1}$ can maintain a vertical distribution of nitric oxide which follows the atmospheric distribution. Such a value of the molecular diffusion coefficient is reached above 120 km, i.e., in the thermosphere (F layer) where NO is produced. Below 120 km, the molecular eddy diffusion coefficient $D(\mathrm{NO})$ is replaced by the eddy diffusion coefficient $K(\mathrm{NO})$ and

$$K(\mathrm{NO}) = \frac{J_{\mathrm{NO}} H^2(\mathrm{NO})}{2[1 - H(\mathrm{NO})/H]}. \tag{12.152}$$

With an atmospheric scale height of the order of 5 km in the lower thermosphere, the eddy diffusion coefficient which is required to maintain a mixing distribution of the nitric oxide concentration is of the order of 5×10^6 cm^2 sec^{-1}.

It seems clear that any attempt to explain the presence of a high nitric oxide content in the lower thermosphere in terms of reaction with atomic nitrogen cannot be successful. Downward transport by molecular diffusion of nitric oxide produced in the F layer by the reaction of the excited nitrogen atom ^{2}D is possible and it will explain NO concentrations greater than 10^7 in the E layer [17–19].

In order to obtain the exact vertical distribution of nitric oxide in the thermosphere the following information should be available:

1. the exact proportion of nitrogen atoms ^{4}S and ^{2}D produced in the dissociative recombination of NO^+ and N_2^+, and in the reaction of O^+ and N_2,
2. precise data on the NO photodissociation,
3. an exact value of the eddy diffusion coefficient in the lower thermosphere.

Since the dayglow emission of atomic nitrogen at λ 5200 Å of the order of 10^8 photons cm^{-2} sec^{-1} requires [50] the production of ^{2}D atoms by dissociative recombination, it seems possible to find after an experimental analysis the production of ^{2}D atoms. In the same way, the nightglow emission [54, 55] of about 10^6 photons cm^{-2} sec^{-1} is also a clear indication of an excitation process by dissociative recombination. A complete analysis of the airglow phenomena is certainly the most direct approach to the problem of evaluating the concentration of N(^{2}D).

REFERENCES

1. D. Rapp and P. Englander-Golden, *J. Chem. Phys.* **43**, 1464 (1965).
2. B. L. Schram, F. J. De Heer, M. J. van der Wiel, and J. Kistemaker, *Physica* **31**, 94 (1965).
3. B. L. Schram, H. R. Moustafa, J. Schutten, and F. J. De Heer, *Physica* **32**, 734 (1966).
4. D. Rapp, P. Englander-Golden, and D. D. Briglia, *J. Chem. Phys.* **42**, 4081 (1965).
5. N. R. Daly and R. E. Powell, *Proc. Phys. Soc.* **89**, 273 (1966).

6. H. F. Winters, *J. Chem. Phys.* **44**, 1472 (1966).
7. B. Van Zijl and G. H. Dunn, *Phys. Rev.* **163**, 43 (1967).
8. M. Nicolet, *Mém. Inst. Roy. Météorol. Belg.* **19**, 162 (1945).
9. D. R. Bates, *Ann. Géophys.* **8**, 194 (1952).
10. M. Nicolet, in *The Solar System* (G. P. Kuiper, ed.), Vol. II, *The Earth as Planet*, pp. 644–712. Univ. of Chicago Press, Chicago, Illinois, 1954.
11. M. Nicolet, *J. Atmos. Terr. Phys.* **7**, 297 (1955).
12. M. Nicolet, in *The Physics and Medicine of the Atmosphere and Space* (O. O. Benson, Jr. and H. Strughold, eds.), pp. 14–47. Wiley, New York, 1960.
13. C. A. Barth, in *Chemical Reactions in the Lower and Upper Atmosphere* (Standford Res. Inst., ed.), pp. 303–326. Wiley (Interscience), New York, 1961.
14. M. Nicolet, *J. Geophys. Res.* **70**, 679 (1965).
15. J. Heicklen and N. Cohen, *Advan. Photochem.* **5**, 157 (1968).
16. H. I. Schiff, *Can. J. Chem.* **47**, 1903 (1969).
17. M. Nicolet, *Planet. Space Sci.* **18**, 1111 (1970).
18. R. B. Norton and C. A. Barth, *J. Geophys. Res.* **75**, 3903 (1970).
19. D. F. Strobel, D. M. Hunten, and M. B. McElroy, *J. Geophys. Res.* **75**, 4307 (1970).
20. C. A. Barth, W. J. Schade, and J. Kaplan, *J. Chem. Phys.* **30**, 347 (1959).
21. R. A. Young and R. L. Sharpless, *J. Chem. Phys.* **39**, 1071 (1963).
22. A. B. Callear and I. W. H. Smith, *Discuss. Faraday Soc.* **37**, 96 (1964).
23. R. W. F. Gross and N. Cohen, *J. Chem. Phys.* **48**, 2582 (1968).
24. N. Jonathan and R. Petty, *Trans. Faraday Soc.* **64**, 1240 (1968).
25. K. H. Becker, W. Groth, and F. Joo, *Ber. Bunsenges. Phys. Chem.* **72**, 157 (1968).
26. M. A. A. Clyne, B. A. Thrush, and R. P. Wayne, *Trans. Faraday Soc.* **60**, 359 (1964).
27. D. G. Murcray, T. G. Kyle, F. H. Murcray, and W. J. Williams, *Nature* (*London*) **218**, 78 (1968).
28. D. G. Murcray, T. G. Kyle, F. H. Murcray, and W. J. Williams, *J. Opt. Soc. Amer.* **59**, 1131 (1969).
29. W. E. Wilson, *J. Chem. Phys.* **46**, 2017 (1967).
30. A. S. Vlastaras and C. A. Winckler, *Can. J. Chem.* **45**, 2837 (1967).
31. K. H. Becker, W. Groth, and D. Kley, *Z. Naturforsch.* **24a**, 1280 (1969).
32. D. M. Hunten and M. B. McElroy, *J. Geophys. Res.* **73**, 2421 (1968).
33. I. D. Clark and R. P. Wayne, *Chem. Phys. Lett.* **3**, 405 (1969).
34. F. D. Findlay, C. J. Fortin, and D. R. Snelling, *Chem. Phys. Lett.* **3**, 204 (1969).
35. L. F. Phillips and H. I. Schiff, *J. Chem. Phys.* **36**, 1509 (1962).
36. M. C. Chen and H. A. Taylor, *J. Chem. Phys.* **34**, 1344 (1961).
37. C. A. Barth and J. Kaplan, in *The Threshold of Space* (M. Zelikoff, ed.), pp. 3–13. Pergamon Press, Oxford, 1957.
38. P. A. Leighton, *Photochemistry of Air Pollution.* Academic Press, New York, 1961.
39. M. Ackerman, F. Biaumé, and G. Kockarts, *Planet. Space Sci.* **18**, 1639 (1970).
40. G. Kockarts, in *Mesospheric Models and Related Experiments* (G. Fiocco, ed.), pp. 160–171. Reidel Publ., Dordrecht, The Netherlands, 1971.
41. I. M. Campbell and B. A. Thrush, *Trans. Faraday Soc.* **64**, 1265 (1968).
42. B. J. Tyler, *Nature* (*London*) **195**, 279 (1962).
43. M. Ackerman and D. Frimout, *Acad. Roy. Belg., Bull. Classe Sci.* **55**, 948 (1969).
44. M. Nicolet, *J. Geophys. Res.* **70**, 691 (1965).
45. C. A. Barth, *J. Geophys. Res.* **69**, 3301 (1964).
46. C. A. Barth, *Ann. Géophys.* **22**, 198 (1966).

47. C. A. Barth, *Planet. Space Sci.* **14**, 623 (1966).
48. J. B. Pearce, *J. Geophys. Res.* **74**, 853 (1969).
49. L. G. Meira, J., *J. Geophys. Res.* **76**, 202 (1971).
50. L. Wallace and M. B. McElroy, *Planet. Space Sci.* **14**, 677 (1966).
51. G. Courtès, *C.R. Acad. Sci. Paris* **231**, 62 (1950).
52. M. Dufay, *C.R. Acad. Sci. Paris* **233**, 419 (1951).
53. D. R. Bates, *Ann. Géophys.* **8**, 194 (1952).
54. G. Weill, *C.R. Acad. Sci. Paris* **264B**, 1685 (1967).
55. G. Hernandez and J. P. Turtle, *Planet. Space Sci.* **17**, 675 (1969).
56. G. Black, T. G. Slanger, G. A. St. John, and R. A. Young, *J. Chem. Phys.* **51**, 116 (1969).

CHAPTER 13

Hydrogen Compounds and Their Effects in the Chemosphere

13.1 Introduction

The role hydrogen compounds play in the stratosphere and mesosphere has been increasingly appreciated in the past decade. Not only are hydrogen atoms important, but also water vapor, methane, molecular hydrogen, hydroperoxyl radicals, and other hydrogen compounds have been found to participate rapidly and effectively in the formation of many minor constituents. In several instances the products of these reactions are of practical significance; for example, the distribution of ozone and atomic oxygen, the formation of the lower ionosphere, and the liberation of atomic hydrogen from the earth's atmosphere.

In this chapter the basic hydrogen compound reactions are discussed with respect to different chemospheric processes.

13.2 Dissociation and Oxidation of Hydrogen Compounds

Hydrogen compounds undergo photolysis when exposed to far ultraviolet radiation. In most cases, however, molecular dissociation in the atmosphere is more rapidly achieved through oxidation reactions involving oxygen atoms. In this section the rates of the photolysis and oxidation reactions are discussed for use in aeronomic analysis.

13.2.1 Dissociation of H_2O

The photodissociation of water vapor occurs slowly when solar radiation of $\lambda > 1750$ Å penetrates into the mesosphere through atmospheric windows of the bands of the Schumann–Runge system. The value of the dissociation rate coefficient in the mesosphere is therefore a function of height. Neglecting for the moment the problem of absorption by O_2, two spectral regions can contribute to photolysis: one with a continuum whose maximum lies at about 1650 Å and the second at wavelengths smaller than 1300 Å associated with superimposed bands. In both continua predissociation is observed. Some of the radiation near 1600 Å is strongly absorbed by the Schumann–Runge continuum of molecular oxygen and cannot penetrate much below 100 km altitude.

The first dissociation limits of H_2O are

$$H_2O + h\nu \rightarrow H_2 + O \qquad \text{for} \quad \lambda < 2565 \text{ Å}, \tag{13.1}$$

$$H_2O + h\nu \rightarrow OH + H \qquad \text{for} \quad \lambda < 2425 \text{ Å}. \tag{13.2}$$

At $\lambda = 1236$ Å, the two processes seem to be present [1] with a ratio of the order of one-third; but this ratio is related to the H_2O pressure [2]. In any case on the long-wavelength side of 1800 Å the absorption cross section is very small and the photodissociation rate of H_2O by (13.2), which is the most important process, is very small in the lower thermosphere and mesosphere due to the effect of molecular oxygen absorption. Taking this absorption into account the photodissociation coefficient at 100 km is only

$$J_{H_2O}(\lambda < 2400 \text{ Å}) = 6.5 \times 10^{-6} \quad \text{sec}^{-1} \tag{13.3}$$

and at the mesopause level (85 km)

$$J_{H_2O}(\lambda < 2400 \text{ Å}) = 4.3 \times 10^{-6} \quad \text{sec}^{-1}. \tag{13.4}$$

Radiation which is not attenuated by O_2 reaches the lower mesosphere, and at 65 km the photodissociation coefficient is [3]

$$J_{H_2O}(\lambda > 1750 \text{ Å}) = 1.2 \times 10^{-7} \quad \text{sec}^{-1}. \tag{13.5}$$

An important part of the photodissociation of H_2O in the upper mesosphere and in the lower thermosphere is due to Lyman α which can reach the 70 km level, as indicated in *Fig. 13.1.* The photodissociation in the Schumann–Runge bands requires a special analysis as discussed in Section 8.4.

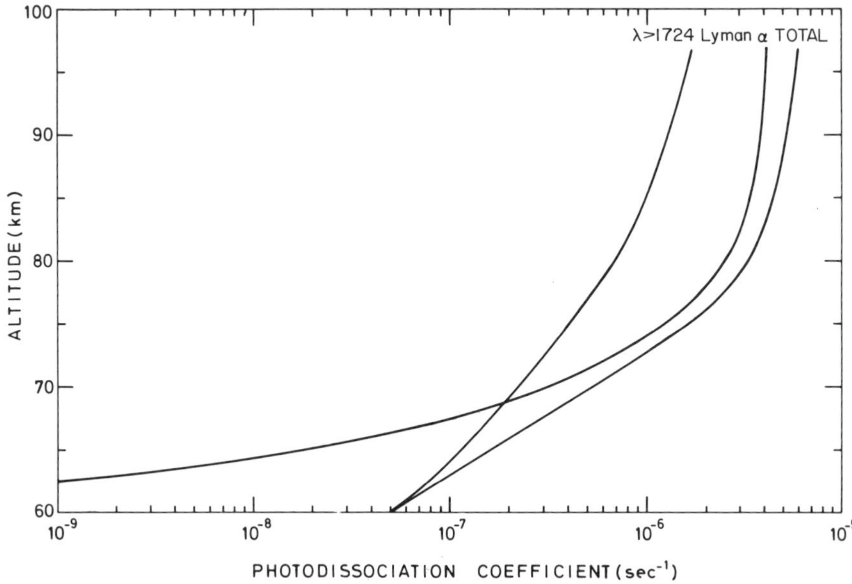

Fig. 13.1 Photodissociation coefficient of water vapor for Lyman α and for $\lambda > 1724$ Å with an overhead sun (Nicolet [3]).

An important result of the photodissociation of H_2O by Lyman α is that the following process is possible:

$$H_2O + h\nu\ (\lambda = 1216\ \text{Å}) \to H(^2S) + OH(^2\Sigma)_{v' \leq 2}. \tag{13.6}$$

Experimental analysis [4] shows that the emission of the OH bands [5] corresponding to fluorescence is between 1 and 10% of the total process. Thus, a fluorescence of OH bands at 3064 Å occurs in the upper atmosphere where Lyman α can penetrate. As shown by Fig. 13.1, the photodissociation by Lyman α represents an important part of the dissociation of water vapor in the mesosphere. For subsequent analysis $J_{H_2O}^{(1)}$ is taken as the rate applied to process (13.1) while $J_{H_2O}^{(2)}$ applies to the sum of dissociations (13.2) and (13.6).

While the photodissociation process of H_2O is important at the mesopause, it is replaced in the lower mesosphere and stratosphere by an oxidation process with atomic oxygen [6–8]. However, the reaction

$$O(^3P) + H_2O \to 2OH - 15 \quad \text{kcal mole}^{-1} \tag{13.7}$$

is endothermic and extremely slow in the stratosphere and mesosphere.

Considering that atomic oxygen in the O(^{1}D) state is present in appreciable quantity in the stratosphere and mesosphere, the reaction [9–14]

$$(a_{13}^*): \quad O(^1D) + H_2O \rightarrow OH + OH_{v\leq 2} + 28.8 \quad \text{kcal mole}^{-1}, \tag{13.8}$$

which has a high rate coefficient of the order of [15]

$$a_{13}^* = (3 \pm 1)\times 10^{-10} \quad \text{cm}^3\ \text{sec}^{-1}, \tag{13.9}$$

must be important. This value has also been obtained for the rate coefficient [16] of O(^{1}D) with O_3:

$$a_{O_3}^* = (3 \pm 1.5)\times 10^{-10} \quad \text{cm}^3\ \text{sec}^{-1} \tag{13.10}$$

for the reaction

$$O(^1D) + O_3 \rightarrow 2O_2 + 139 \quad \text{kcal mole}^{-1}. \tag{13.11}$$

13.2.2 Dissociation of CH_4

Methane can be dissociated [17] by sunlight according to

$$CH_4 + h\nu \rightarrow CH_3 + H \qquad \text{for} \quad \lambda < 2785\ \text{Å} \tag{13.12}$$

$$CH_4 + h\nu \rightarrow CH_2 + H_2 \qquad \text{for} \quad \lambda < 2530\ \text{Å}. \tag{13.13}$$

However, since the absorption cross section is very small at $\lambda > 1500$ Å, the principal photodissociation coefficient is

$$J_{CH_4}\ (\lambda = 1216\ \text{Å}) = 5\times 10^{-6} \quad \text{sec}^{-1}. \tag{13.14}$$

The contribution from Lyman α to the photodissociation of CH_4 is essentially constant in the thermosphere [3] (see *Fig. 13.2*) and is a decreasing function of the optical depth in the mesosphere. It falls off so rapidly below the 70 km level that it is negligible in the lower mesosphere.

In the lower mesosphere and stratosphere oxidation by atomic oxygen is important. The oxidation of methane according

$$O(^3P) + CH_4 \rightarrow OH + CH_3 - 2 \quad \text{kcal mole}^{-1} \tag{13.15}$$

is endothermic by about 2 kcal mole^{-1} and is a slow reaction. Its activation energy is not less than 7 kcal mole^{-1} according to various laboratory measurements [18–21]. With the minimum activation energy, the following rate

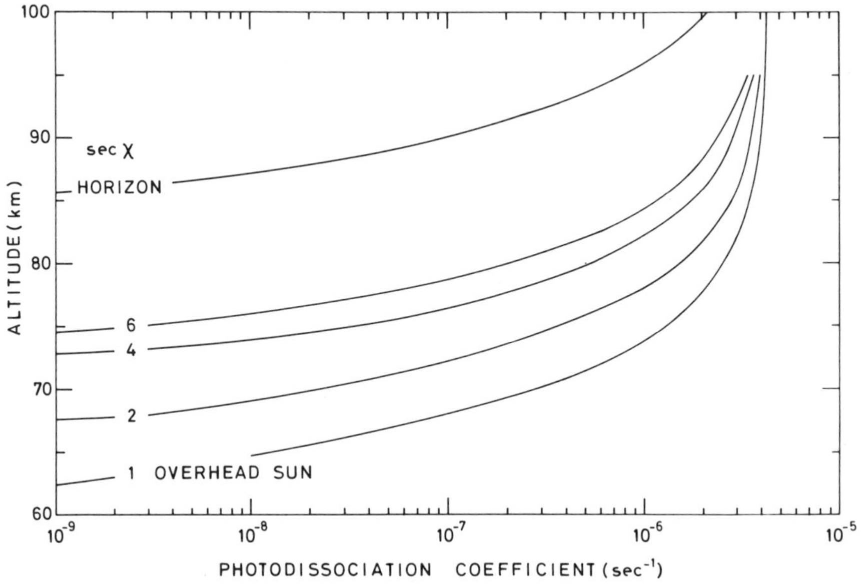

Fig. 13.2 Photodissociation coefficient of methane in the thermosphere and mesosphere for various solar zenith angles (Nicolet [52]).

coefficient can be adopted:

$$a_{\mathrm{O,\,CH_4}} = 3.5 \times 10^{-13} T^{1/2} e^{-3650/T} \quad \mathrm{cm^3\ sec^{-1}} \tag{13.16}$$

for an estimate of the importance of process (13.15). Numerical results are illustrated in *Fig. 13.3.*

Methane can also react with $O(^1D)$ following the process

$$O(^1D) + CH_4 \to CH_3 + OH_{v \leq 4} + 43.5 \quad \mathrm{kcal\ mole^{-1}}\ . \tag{13.17}$$

Experimental evidence [6, 10–14] gives a reaction rate on the order $(3 \pm 1) \times 10^{-10}$ $\mathrm{cm^3\ sec^{-1}}$. From Fig. 13.3, it is clear that the reactions with $O(^1D)$ are of great importance in the whole stratosphere.

13.2.3 Dissociation of H_2

The photodissociation of H_2 is a very slow process and we take $J_{H_2} = 0$.

The reaction of $O(^3P)$ with molecular hydrogen can produce OH molecules according to

$$(a_{24}):\quad O(^3P) + H_2 \to H + OH - 2 \quad \mathrm{kcal\ mole^{-1}} \tag{13.18}$$

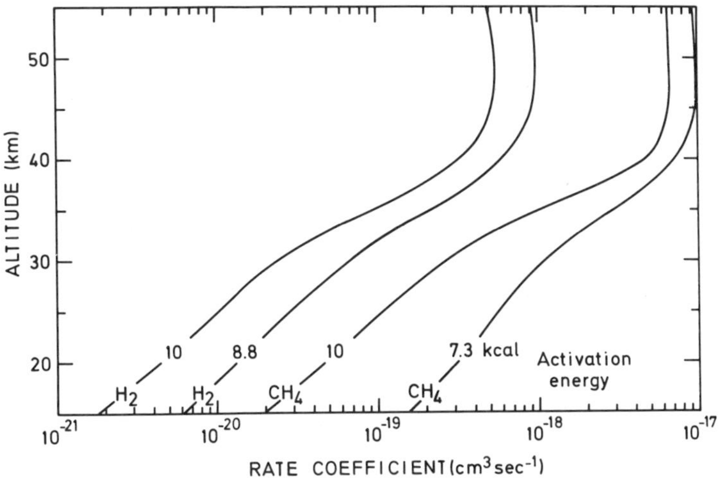

Fig. 13.3 Oxidation rate coefficients of H_2 and CH_4 by atomic oxygen in the stratosphere. Two extreme values for the activation energies have been adopted. $O + H_2 \rightarrow OH + H - 2$ kcal; $O + CH_4 \rightarrow CH_3 + OH - 2$ kcal.

with an activation energy greater than 8.5 kcal mole^{-1}. From experimental values [20, 22, 23] obtained at temperatures higher than 300°K, we consider the possibility of two expressions (see Fig. 13.3) for the rate coefficient:

$$a_{24} = 5 \times 10^{-13} T^{1/2} e^{-4400/T} \quad \text{cm}^3 \text{ sec}^{-1} \tag{13.19}$$

and

$$a_{24} = 2.5 \times 10^{-12} T^{1/2} e^{-5000/T} \quad \text{cm}^3 \text{ sec}^{-1}. \tag{13.20}$$

Thus, molecular hydrogen reacts more slowly with atomic oxygen in its ground state than does methane.

With regard to $O(^1D)$ the reaction

$$(a_{24}^*): \quad O(^1D) + H_2 \rightarrow H + OH_{v \leq 4} + 43.5 \quad \text{kcal mole}^{-1} \tag{13.21}$$

is important in the stratosphere since it has a reaction rate of the order of [11] 10^{-10} cm^3 sec^{-1} at 87°K. A value similar to a_{13}^* can be adopted; i.e.

$$a_{24}^* = (3 \pm 1) \times 10^{-10} \quad \text{cm}^3 \text{ sec}^{-1}. \tag{13.22}$$

13.2.4 Dissociation of H_2O_2

The rate of photodissociation for hydrogen peroxide is known from laboratory measurements [24, 25]. The principal process for $\lambda < 3650$ Å, corresponding to a binding energy of 48.8 kcal mole^{-1} [26], is

$$H_2O_2 + h\nu \rightarrow 2OH(^3\Pi) \tag{13.23}$$

with a photodissociation rate at zero optical depth of [3]

$$J_{H_2O_2} = 1.4 \times 10^{-4} \quad sec^{-1}. \tag{13.24}$$

The alternative process

$$H_2O_2 + h\nu \rightarrow H_2O + O \tag{13.25}$$

does not readily occur and accounts for less than 1% of all photodissociations of H_2O_2 [27, 28].

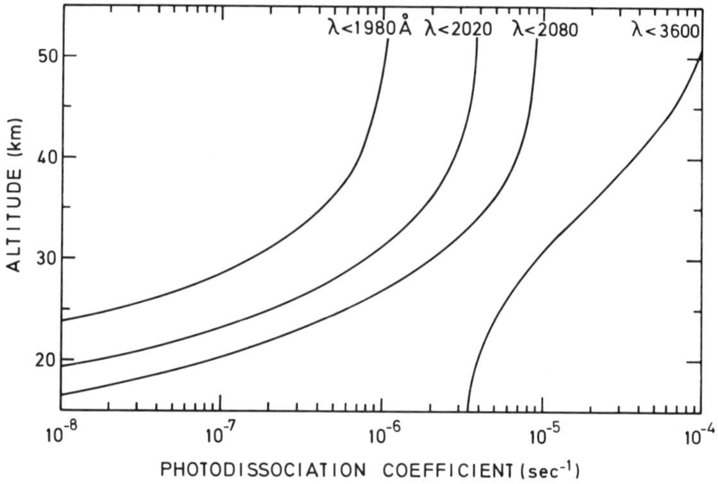

Fig. 13.4 Partial photodissociation coefficients of H_2O_2 under an overhead sun indicate that short ultraviolet radiation cannot play an important role in an OH fluorescence process.

Absorption of far ultraviolet radiation by O_2 introduces a strong altitude dependence in the H_2O_2 photolysis rate, as shown in *Fig. 13.4*. In the stratosphere $J_{H_2O_2}$ becomes smaller than 5×10^{-6} at 20 km.

For $\lambda < 2100$ Å $J_{H_2O_2}$ is less than 10^{-5} sec^{-1} below 55 km altitude. Thus, the process

$$H_2O_2 + h\nu \rightarrow OH(^2\Pi) + OH(^2\Sigma^+) \tag{13.26}$$

operative for $\lambda < 2000$ Å [29] leading to a fluorescence of OH cannot be an important source of airglow emission. However, the H_2O_2 absorption spectrum near Lyman α is poorly known.

13.2.5 Dissociation of OH *and* HO_2

The photodissociation of the hydroxyl molecule

$$OH + h\nu \rightarrow O + H \tag{13.27}$$

is a very slow process and we adopt $J_{OH} = 0$, even though there is a possibility for predissociation of the A $^2\Sigma^+$ state for $v > 0$ [30].

Nothing is known about the photodissociation of perhydroxyl and it is reasonable to assume that a photodissociation according to

$$HO_2 + h\nu\ (\lambda < 4500\ \text{Å}) \rightarrow OH + O \tag{13.28}$$

is possible. Due to the similarity of the structure of HO_2 to that of hydrogen peroxide, for which $J_{H_2O_2} = 1.4 \times 10^{-4}$ sec^{-1}, and also to the large photodissociation coefficient of nitrogen peroxide, $J_{NO_2} = 3.5 \times 10^{-3}$ sec^{-1}, possible values such as

$$10^{-3} \leq J_{HO_2} \leq 10^{-4} \quad \text{sec}^{-1} \tag{13.29}$$

can be suggested.

The photodissociation process

$$HO_2 + h\nu \rightarrow H + O_2, \tag{13.30}$$

where the bonds between H and O atoms are difficult to break, has a very low probability and can be neglected.

For HO_2 the photodissociation process (13.28) is less important above 30 km in the stratosphere and mesosphere than the reaction [31, 32]

$$(a_7){:}\quad O + HO_2 \rightarrow OH_{v \leq 6} + O_2 + 55 \quad \text{kcal mole}^{-1} \tag{13.31}$$

involving atomic oxygen. The rate coefficient

$$a_7 = 3 \times 10^{-12} T^{1/2} \quad \text{cm}^3\ \text{sec}^{-1} \tag{13.32}$$

is adopted in the absence of direct measurements. The value for a_7 has been taken as the same as that for a_5 (described in the next section) to simplify the aeronomic analysis, even though it may be a factor of 2 too large.

13.3 Water Vapor and Methane in the Stratosphere and Mesosphere

The vibrational–rotational bands of the hydroxyl radical OH which appear in the airglow aroused interest twenty years ago in the photochemistry of hydrogen–oxygen compounds [33] and, in particular, of methane [34, 35] and of water vapor [36].

Water vapor, which is quite abundant near ground level with a fractional volume concentration reaching 10^{-2}, has a decreasing mixing ratio with altitude. The water vapor content of the stratosphere is very small and 3×10^{-6} is not an unreasonable value to adopt as an order of magnitude for the fractional volume concentration of water vapor [37–39] in the stratosphere and mesosphere. Methane and molecular hydrogen, which have been found as permanent constituents of the troposphere (see Table 3.1), have their sources at ground level. Their total amounts by volume are, respectively, 1.5×10^{-6} and 0.5×10^{-6} of the atmosphere. Unlike water vapor, however, their fractional concentrations are not affected in the troposphere. Consequently, in the stratosphere the combined H-atom content of CH_4 is equal to that adopted for H_2O. It follows that even if the assumed specific humidity is changed, the total number of available hydrogen atoms cannot be far from the assumed value since almost all hydrogen compounds are eventually dissociated in the high atmosphere.

The basic photolysis and oxidation reactions for H_2O and CH_4 have been introduced in Section 13.2. Additional reactions, however, are of importance for methane and methyl radicals. With OH, the following reactions are possible:

$$OH + CH_4 \rightarrow CH_3 + H_2O + 15 \quad \text{kcal mole}^{-1} \qquad (13.33)$$

$$OH + H_2 \rightarrow H + H_2O + 15 \quad \text{kcal mole}^{-1}. \qquad (13.34)$$

At 300°K an average value [40] of the rate coefficients is 8×10^{-15} cm^3 sec^{-1}, which corresponds to an activation energy greater than 6 kcal mole^{-1} without a steric factor. An activation energy of 8 kcal mole^{-1} leads to a steric factor of about 5×10^{-2}.

The reaction

$$HO_2 + CH_4 \rightarrow CH_3 + H_2O_2 - 15 \quad \text{kcal mole}^{-1} \qquad (13.35)$$

is endothermic by about 15 kcal mole^{-1} and it is difficult to see how perhydroxyl radicals could attack methane.

When a single hydrogen atom is removed from a methane molecule the methyl radical CH_3 is produced. Due to the rapid destruction of CH_3 the net result is a permanent loss of the original methane molecule. The following reaction, yielding formaldehyde [41], can be considered as the most important process for the reaction of O with CH_3 (the rate coefficient at 300°K is greater than 3×10^{-10} cm^3 sec^{-1}):

$$O + CH_3 \rightarrow H + HCHO + 67 \quad \text{kcal mole}^{-1}. \tag{13.36}$$

This may be followed by $HCHO + O \rightarrow CHO + OH$ and can end by $O + HCO \rightarrow OH + CO + 84$ kcal mole^{-1}.

Methyl radicals react also with ozone and molecular oxygen:

$$CH_3 + O_3 \rightarrow CH_3O + O_2 + 66 \quad \text{kcal mole}^{-1} \tag{13.37}$$

and

$$CH_3 + O_2(+M) \rightarrow CH_3O_2(+M) + 26 \quad \text{kcal mole}^{-1}. \tag{13.38}$$

The presence of $O(^1D)$ atoms in the atmosphere leads to the possibility of the production of hydrogen atoms through reactions with CH_4, H_2, and H_2O. With the fractional volume concentrations which have been adopted for H_2 (0.5×10^{-6}), for CH_4 (1.5×10^{-6}), and for H_2O (3×10^{-6}), the total production P(H) of hydrogen atoms by reactions of $O(^1D)$ and

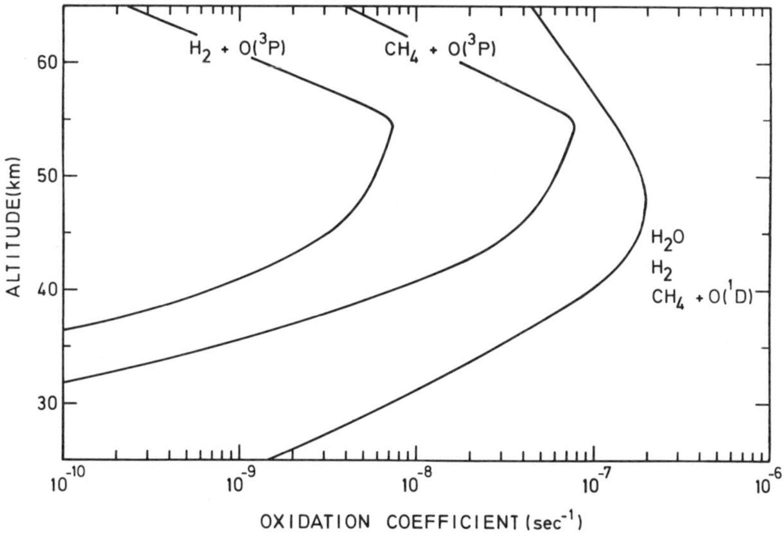

Fig. 13.5 Oxidation coefficients under an overhead sun for H_2, CH_4, and H_2O by $O(^3P)$ and $O(^1D)$ oxygen atoms. A round figure of 3×10^{-10} cm^3 sec^{-1} is adopted for the $O(^1D)$ reactions.

CH_4, H_2, and H_2O is

$$P(\mathrm{H}) = 39\times 10^{-16} n(\mathrm{M}) n[\mathrm{O}(^1\mathrm{D})] \quad \mathrm{cm}^{-3}\,\mathrm{sec}^{-1} \tag{13.39}$$

if $n(\mathrm{M})$ is the total concentration and if the average rate coefficient of 3×10^{-10} cm^3 sec^{-1} is adopted. An increase of a factor of 2 in the rate coefficient of (13.8) and the neglect of reaction (13.17) will also lead to (13.39).

With the $O(^1D)$ concentrations obtained from (11.32) [listed in Table 11.1] it is possible to deduce the oxidation coefficient in the stratosphere and mesosphere. *Figure 13.5* illustrates the variation and differences of reactions with $O(^3P)$ and $O(^1D)$ atoms. It is clear that in the stratosphere all constituents are attacked by $O(^1D)$ and, in addition, at the stratopause the effect of $O(^3P)$ on CH_4 is also important. Reliance cannot, of course, be placed on precise numerical values, but the indicated trend of the results is real.

13.4 Vertical Transport in the Stratosphere and Mesosphere

From the foregoing discussion it is apparent that the mechanism responsible for the production of hydrogen atoms in the stratosphere can be identified with an oxidation process of H_2O, CH_4, and H_2 by atomic oxygen $O(^1D)$ and of CH_4 by $O(^3P)$ atoms near the stratopause level.

At the stratopause level the oxidation of methane by a collision process with atomic oxygen leads to a mean lifetime of 10^7 sec. If the vertical distribution of CH_4 is assumed to be a mixing distribution up to 60 km, the oxidation process leads to a total loss of about 2×10^{10} CH_4 molecules cm^{-2} sec^{-1}. Under mixing conditions, it can be shown [42] that the maximum *diffusion* flow $F_\mathrm{D}(CH_4)$ of methane is of the order of

$$F_\mathrm{D}(\mathrm{CH_4}) = 7\times 10^6 \quad \mathrm{cm}^{-2}\,\mathrm{sec}^{-1} \tag{13.40}$$

which is a negligible fraction of the possible loss by oxidation. Since the tropospheric mixing time is short enough to lead to a uniform vertical distribution of CH_4, the injection rate of CH_4 into the stratosphere is an essential problem which must be considered in detail. Accepting the possible injection of CH_4 from the troposphere into the stratosphere and noting the inadequacy of molecular diffusion, it follows that an eddy diffusion flow of the order of 10^{10} CH_4 molecules cm^{-2} sec^{-1} must be present in order to replace the loss by oxidation. Consequently, the time required to disrupt the observed column content of 2.5×10^{19} CH_4 molecules cm^{-2} is 40 years. Such a time seems to correspond in order of magnitude to the time required for the production [43] of the total CH_4 column content. Thus, it is not immediately clear if the fractional concentration of CH_4 is affected by its transformation into H_2O in the stratosphere.

With regard to water vapor, a ratio $n(H_2O)/n(M) = 3\times 10^{-6}$ for mixing conditions leads to a maximum molecular diffusion flow

$$F_D(H_2O) = 1.1\times 10^7 \quad cm^{-2}\,sec^{-1}. \tag{13.41}$$

The number of hydrogen atoms carried vertically by diffusion of water vapor and CH_4 is about equal.

As far as H_2 is concerned, its maximum diffusion flow for mixing conditions is

$$F_D(H_2) = 1.2\times 10^7 \quad cm^{-2}\,sec^{-1} \tag{13.42}$$

which is of the same magnitude as for water vapor transport. Thus, the upward flux $F_D(H)$ of hydrogen atoms in hydrogen compounds due to molecular diffusion [44] is

$$3\times 10^7 \leq F_D \leq 8\times 10^7 \quad \text{atoms cm}^{-2}\,sec^{-1} \tag{13.43}$$

for mixing conditions.

The amounts of hydrogen present in the stratosphere as H_2O, CH_4, and H_2 are controlled mainly by phenomena at the earth's surface and by conditions near the tropopause. Chemical reactions in the stratosphere determine the hydrogen content of the mesosphere and lower thermosphere through the presence of vertical eddy diffusion.

It will be shown in Chapter 15 that in the presence of eddy diffusion, the maximum vertical speed w of constituent of mass m_1 can be approximated by

$$w = \frac{D}{H(M)}\left[\left(\frac{H}{H_1} - \frac{m_1}{m}\right) + \Lambda\left(\frac{H}{H_1} - 1\right)\right] \tag{13.44}$$

in which Λ is the ratio of the eddy diffusion coefficient K to the molecular diffusion coefficient D, H is the atmospheric scale height, H_1 is the scale height of the diffusing minor constituent, and m is the mean molecular mass.

The maximum upward current of CH_4 molecules, due to eddy and molecular diffusion in the stratosphere, is approximately given by

$$F_{CH_4} = n(CH_4)w = \frac{1\times 10^{13} n(CH_4)}{n(M)}\left[\left(\frac{H(M)}{H(CH_4)} - 0.55\right) + \Lambda\left(\frac{H(M)}{H(CH_4)} - 1\right)\right] \quad cm^{-2}\,sec^{-1} \tag{13.45}$$

and must correspond to the integrated CH_4 loss rate of

$$F_{CH_4} = \tfrac{1}{2} a_{O-CH_4} n(O) n(CH_4) H(CH_4) \tag{13.46}$$

in order to replace the methane which is destroyed chemically in a sunlit atmosphere by reaction with oxygen atoms (^3P or ^1D). Using numerical values in (13.45) and (13.46), it is found that eddy diffusion is necessary to maintain CH_4 molecules in the upper stratosphere. If the ratio of the local scale heights $H(CH_4)/H(M) \doteq \frac{1}{2}$ it is found that an eddy diffusion coefficient of 4×10^4 cm^2 sec^{-1} is sufficient near the stratopause [3] (*Fig. 13.6*).

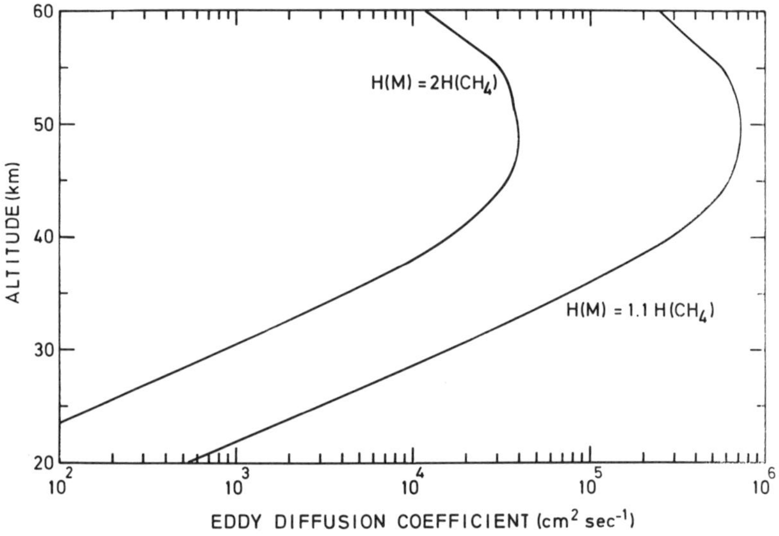

Fig. 13.6 Eddy diffusion coefficient necessary to maintain methane with scale heights $H(CH_4) = 0.5H(M)$ and $H(CH_4) = 0.9H(M)$ in the stratosphere (Nicolet [3]).

However, a mixing distribution of CH_4, i.e., a scale height, $H(M) = 1.1$ $H(CH_4)$, implies values of Λ reaching 3×10^3 near the stratopause. Thus, a permanent eddy diffusion coefficient greater than 5×10^5 cm^2 sec^{-1} (Fig. 13.6) is required in order to maintain a vertical distribution of CH_4 following the atmospheric distribution.

Above the stratopause, a mixing distribution of CH_4 could be maintained through (13.46) by an eddy diffusion coefficient less than 10^5 cm^2 sec^{-1}. However, the additional process of photodissociation of CH_4 in the mesosphere by Lyman α (see 13.14 and Fig. 13.2) must result in a flux given by

$$F_{CH_4} = \tfrac{1}{2} n(CH_4) J_{CH_4} H(CH_4). \tag{13.47}$$

Combining (13.45) and (13.47) it is found in the upper mesosphere ($\geq$75 km) and lower thermosphere that the eddy diffusion coefficient cannot be less than 10^5 cm^2 sec^{-1} to maintain a scale height $H(CH_4) = H(M)/2$. A mixing distribution of CH_4 requires not less than 2×10^6 to 5×10^6 cm^2 sec^{-1} for the eddy diffusion coefficient.

By an argument similar to that used in connection with the distribution of methane, it may be found that H_2 undergoes similar processes. Hydrogen molecules are destroyed chemically through reaction with excited oxygen atoms O* with a rate coefficient $a^* \simeq 3\times10^{-10}$ cm^3 sec^{-1}. With expressions equivalent to (13.45) and (13.46) the height integrated loss of H_2 can be matched to the maximum vertical diffusive flux to give

$$2\times10^{-14}a_{24}^* n(O^*)n(M)H(M)$$
$$= \frac{H(M)}{H(H_2)}\left[\left(\frac{H(M)}{H(H_2)} - 0.034\right) + \Lambda\left(\frac{H(M)}{H(H_2)} - 1\right)\right]. \qquad (13.48)$$

Figure 13.7 shows that condition (13.48) is met for $H(M) = 2H(H_2)$ when the eddy diffusion reaches 2×10^4 cm^2 sec^{-1} at the stratopause level. A fractional volume ratio $n(H_2)/n(M)$ which does not decrease with altitude is only satisfied for an eddy diffusion coefficient reaching more than 10^5 cm^2 sec^{-1} in the neighborhood of the stratopause.

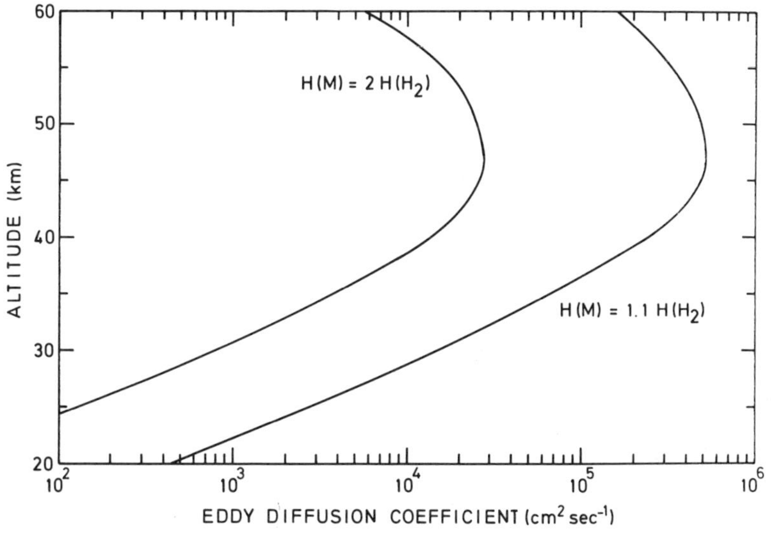

Fig. 13.7 Eddy diffusion coefficient necessary to maintain molecular hydrogen with scale heights $H(H_2) = 0.5H(M)$ and $H(H_2) = 0.9H(M)$ in the stratosphere.

Since H_2O is also subject to a major dissociative process through oxidation by excited oxygen atoms, the vertical distribution of H_2O, when considered as an equilibrium between dissociation and atmospheric mixing effects, can be expressed as

$$5\times 10^{-14} a_{13}^{*} n(\mathrm{O}^{*}) n(\mathrm{M}) H(\mathrm{M}) = \frac{H(\mathrm{M})}{H(\mathrm{H_2O})}\left[\left(\frac{H(\mathrm{M})}{H(\mathrm{H_2O})} - 0.6\right) + \Lambda\left(\frac{H(\mathrm{M})}{H(\mathrm{H_2O})} - 1\right)\right]. \quad (13.49)$$

Numerical results corresponding to $H(\mathrm{M}) = 2\,H(\mathrm{H_2O})$ and $H(\mathrm{M}) = 1.1\,H(\mathrm{H_2O})$ are illustrated in *Fig. 13.8* where eddy diffusion coefficients at the

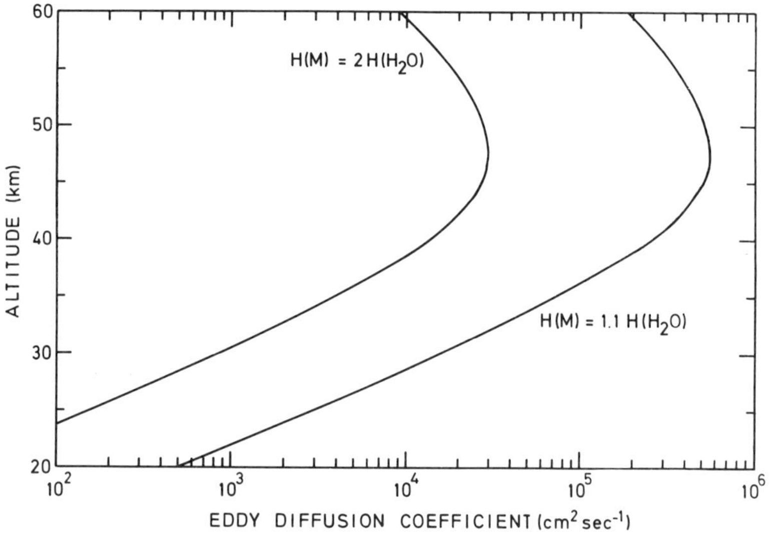

Fig. 13.8 Eddy diffusion coefficient necessary to maintain water vapor with scale heights $H(\mathrm{H_2O}) = 0.5H(\mathrm{M})$ and $H(\mathrm{H_2O}) = 0.9H(\mathrm{M})$ (Nicolet [52]).

stratopause are not less than 2×10^4 and 5×10^5 cm^2 sec^{-1}, respectively. In the mesosphere, where the photodissociation of water vapor occurs, the eddy diffusion coefficient must rise to 5×10^6 cm^2 sec^{-1} to maintain a mixing distribution at the mesopause.

The photodissociation of water vapor in the mesosphere is not a rapid process (see Fig. 13.1) since the mean lifetime of H_2O at the mesopause level for an overhead sun is about 5×10^5 sec. The upward diffusion flow is important and by an argument similar to that used in connection with the distribution of methane, it may be found that an eddy diffusion coefficient of about 2×10^5 cm^2 sec^{-1} can maintain a scale height $H(H_2O) = \frac{1}{2} H(M)$. If $H(H_2O) = 0.9\ H(M)$, then the eddy diffusion coefficient must be greater than 5×10^6 cm^2 sec^{-1} above the mesopause level.

Eddy diffusion coefficients in the atmosphere are not known with precision, but it seems that values greater than 10^6 cm^2 sec^{-1} are typical of the lower thermosphere.

From the foregoing discussion, it follows that the destruction rate for CH_4 and H_2 and H_2O through a reaction with oxygen atoms in the ^1D state is maximum in the upper stratosphere. If the concentration of CH_4, H_2, or H_2O at such altitudes is not determined by a local equilibrium between dissociation and recombination, but by dissociation and atmospheric mixing effects, an eddy diffusion coefficient (or any equivalent large scale motion in the atmosphere) of not less than 10^5 cm^2 sec^{-1} is required to maintain a constant volume ratio near the stratopause. It seems, therefore, that $K \geq 10^5$ cm^2 sec^{-1} should be accepted to simulate a complete atmospheric mixing of CH_4 and H_2. In any case, in the upper stratosphere the mean lifetime of methane and hydrogen molecules is less than a year. Near 50 km, the following round figure

$$\tau(H_2) = 1\times10^7 \quad \text{sec} \tag{13.50}$$

must be considered as a characteristic time to be associated with atmospheric motions which would maintain those molecules in mixing distributions.

Near 20 km, turbulent effects leading to an eddy diffusion coefficient of the order of 5×10^5 cm^2 sec^{-1} have been observed but it must be pointed out that turbulence as strong as this is a rare event in the lower stratosphere. Nevertheless, wind shears are probably quite strong in the upper stratosphere, particularly in winter, so that eddy diffusion coefficients of more than 10^5 cm^2 sec^{-1} are likely. However, they occur over only a small portion of the earth's surface.

It seems, therefore, that it is extremely difficult to maintain a constant mixing ratio of methane and molecular hydrogen in the stratosphere. The stratospheric scale height of these two constituents depends on atmospheric motions with small or large scales.

As far as water vapor is concerned, it is possible to maintain mixing distribution without significant vertical transport since H_2O destruction is offset by reactions in which H, OH, HO_2, and H_2O_2 lead to H_2O.

13.5 Reactions Involving a Single Hydrogen Atom

A number of reactions can occur when hydrogen compounds are present in an oxygen atmosphere. We consider here those reactions involving free hydrogen atoms, then proceed to reactions where a single combined hydrogen atom is present.

The hydroxyl radical can be formed directly by the reaction [36, 45]

$$(a_0):\quad H + O + M \rightarrow OH + M + 101.3 \quad \text{kcal mole}^{-1} \tag{13.51}$$

for which a conventional rate coefficient may be taken as $a_0 = 2\times 10^{-32}$ cm^6 sec^{-1}.

There is spectroscopic evidence that such a reaction leads to the emissions of the $OH(^2\Sigma^+) \rightarrow OH(^2\Pi)$ system since the dissociation energy of 101.3 kcal mole^{-1} is sufficient to excite $^2\Sigma$ states. Consideration must also be given to the possibility of an airglow emission at 3064 Å which is blended with Herzberg bands [46] of O_2.

In the atmosphere, a more important reaction involving molecular oxygen is

$$(a_1):\quad H + O_2 + M \rightarrow HO_2 + M + 46 \quad \text{kcal mole}^{-1} \tag{13.52}$$

with a rate coefficient [47]

$$a_1 = 3.3\times 10^{-33} e^{800/T} \quad \text{cm}^6\ \text{sec}^{-1}. \tag{13.53}$$

Since chemiluminescence may occur during such a reaction, an analysis should be made in order to see if an airglow emission could actually be present.

Another equally important loss process for atomic hydrogen is

$$(a_2):\quad H + O_3 \rightarrow O_2 + OH_{v\leq 9} + 77 \quad \text{kcal mole}^{-1} \tag{13.54}$$

with a rate coefficient [31, 48, 49]

$$a_2 = 1.5\times 10^{-12} T^{1/2} \quad \text{cm}^3\ \text{sec}^{-1}. \tag{13.55}$$

This reaction has been known [36] since 1950 as the reaction which leads to the OH vibration–rotation bands observed in the airglow from 5000 Å to 5 μm with a total intensity of about 5×10^{12} photons cm^{-2} sec^{-1}. The parallel reaction leading to HO_2

$$(a_3):\quad H + O_3 \rightarrow HO_2 + O + 22 \quad \text{kcal mole}^{-1} \tag{13.56}$$

is less important than (13.54) since it has an activation energy. The following rate coefficient, which has not yet been measured, is adopted:

$$a_3 = 1.5\times10^{-11}T^{1/2}e^{-2000/T} \quad \text{cm}^3\ \text{sec}^{-1}\,. \tag{13.57}$$

For aeronomic conditions it is clear that the reactions (13.52) and (13.54) with rate coefficients a_1 and a_2 are more important than (13.51) and (13.56) with a_0 and a_3.

The production of hydroxyl radicals also can be followed by the three-body association

$$(a_4):\quad \text{OH} + \text{O} + \text{M} \rightarrow \text{HO}_2 + \text{M} + 63 \quad \text{kcal mole}^{-1} \tag{13.58}$$

with the conventional value of the rate coefficient

$$a_4 = 2\times10^{-32} \quad \text{cm}^6\ \text{sec}^{-1}. \tag{13.59}$$

However, the bimolecular process

$$(a_5):\quad \text{OH} + \text{O} \rightarrow \text{H} + \text{O}_2 + 16.6 \quad \text{kcal mole}^{-1} \tag{13.60}$$

is more important since its rate coefficient [31, 49, 50]

$$a_5 = 3\times10^{-12}T^{1/2} \quad \text{cm}^3\ \text{sec}^{-1} \tag{13.61}$$

leads to a value of 5×10^{-11} cm^3 sec^{-1} at 300°K.

Reactions (13.54) and (13.60) show that hydrogen acts as a catalyst for the loss of odd oxygen atoms and will affect the atomic oxygen and ozone distributions.

Two other reactions involve hydroxyl radicals and ozone. The first is

$$(a_6):\quad \text{OH} + \text{O}_3 \rightarrow \text{HO}_2 + \text{O}_2 + 39 \quad \text{kcal mole}^{-1} \tag{13.62}$$

which, according to recent measurements [52], has a very small rate coefficient

$$a_6 < 10^{-16} \quad \text{cm}^3\ \text{sec}^{-1}. \tag{13.63}$$

The following second reaction between ozone and perhydroxyl radicals could be involved for the OH catalyzed chain decomposition of ozone:

$$(a_{6a}):\quad \text{HO}_2 + \text{O}_3 \rightarrow \text{OH} + 2\text{O}_2 + 31 \quad \text{kcal mole}^{-1}. \tag{13.64}$$

There is no experimental evidence for such a reaction [32]; in any case, its rate coefficient must be very small [3] since such a reaction requires the simultaneous breakage of the two bonds (OH—O and O_2—O). The fact that OH radicals are involved in the catalyzed chain decomposition of O_3 in laboratory investigation can be explained by the presence of vibrationally excited OH molecules, which are a major reaction product. When OH is in any vibrational level above $v = 0$, the reaction

$$OH^* \ (\geq 10.2 \text{ kcal mole}^{-1}) + O_3 \rightarrow H + 2O_2 + (>1 \text{ kcal mole}^{-1}) \tag{13.65}$$

becomes exothermic and is followed by the rapid reaction (13.54) leading to $OH_{v \leq 9}$. In the atmosphere, such OH* excited molecules cannot be significant and the role of HO_2 as a chain carrier is doubtful. Therefore, reactions (13.62) and (13.64) will be neglected in the following analysis.

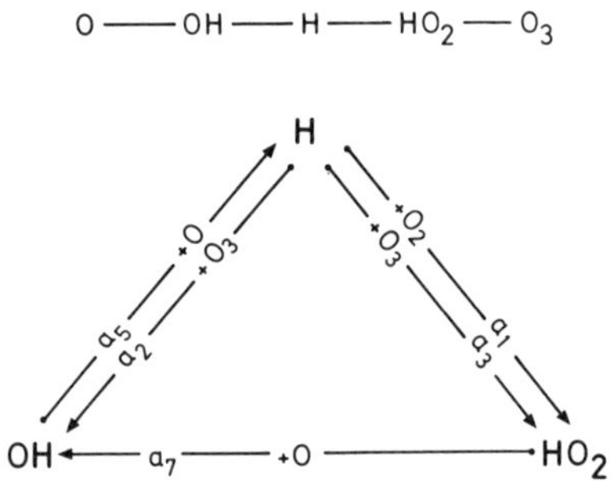

Fig. 13.9 Principal reactions involving H, OH, and HO_2 in an atmosphere with O_2, O, and O_3.

The principal reactions between atomic hydrogen, hydroxyl, and perhydroxyl radicals in an ozone–atomic oxygen atmosphere are summarized in *Fig. 13.9.* It is clear that the atomic hydrogen concentration must increase with the atomic oxygen concentration and that the perhydroxyl radical concentration increases with the ozone concentration.

13.6 Effect of Hydrogen on the Distributions of O and O_3

Expressions for the equilibrium ratio of $n(OH)/n(H)$ and $n(HO_2)/n(H)$ can now be easily obtained if it is assumed that only reactions (13.28), (13.31), and (13.51)–(13.65) are involved. The production and loss of atomic hydrogen lead to

$$\frac{n(OH)}{n(H)} = \frac{a_0 n(O)n(M) + a_1 n(O_2)n(M) + a_2 n(O_3) + a_3 n(O_3)}{a_5 n(O)}. \tag{13.66}$$

Numerical values show that $a_0 n(O)n(M)$ can be neglected. With the adopted value (13.57) for a_3, $a_3 n(O_3)$ is much smaller than $a_1 n(O_2) n(M)$ and the simplest expression for the ratio $n(OH)/n(H)$ is given by

$$\frac{n(OH)}{n(H)} = \frac{a_1 n(O_2) n(M) + a_2 n(O_3)}{a_5 n(O)}. \tag{13.67}$$

The production and loss of HO_2 lead to the equilibrium ratio

$$\frac{n(HO_2)}{n(H)} = \frac{a_1 n(O_2) n(M) + a_3 n(O_3)}{J_{HO_2} + a_7 n(O)} + \frac{n(OH) a_4 n(O) n(M)}{n(H)[J_{HO_2} + a_7 n(O)]}. \tag{13.68}$$

Neglecting $a_3 n(O_3)$ and $a_4 n(O) n(M)$, the equilibrium ratio $n(HO_2)/n(H)$ becomes

$$\frac{n(HO_2)}{n(H)} = \frac{a_1 n(O_2) n(M)}{a_7 n(O)} \tag{13.69}$$

in the atmospheric regions where $a_7 n(O) > 10^{-3} > J_{HO_2}$.

Assuming that $a_5 = a_7$, expressions (13.67) and (13.69) lead to

$$\frac{n(HO_2)}{n(OH)} = \frac{a_1 n(M) n(O_2)}{a_1 n(M) n(O_2) + a_2 n(O_3)}. \tag{13.70}$$

Numerical values [52] show that the atomic hydrogen concentration is greater than the concentrations of OH and HO_2 in the upper mesosphere. The ratio $n(HO_2)/n(OH)$, which is of the order of 0.8, indicates that both radicals have about the same concentration in the mesosphere. *Figure 13.10* shows the distribution of the most important terms involved in the ratios $n(OH)/n(H)$ and $n(HO_2)/n(H)$.

When hydrogen compounds are considered, the differential equation related to atomic oxygen becomes [see Eq. (11.7)]

$$\begin{aligned} \frac{dn(O)}{dt} &+ 2k_1 n(M) n^2(O) \\ &+ [k_2 n(M) n(O_2) + k_3 n(O_3) + a_5 n(OH) + a_7 n(HO_2)] n(O) \\ &= 2n(O_2) J_2 + n(O_3) J_3, \end{aligned} \tag{13.71}$$

in which the reactions with coefficient labeled a are the additional processes in which atomic oxygen is involved in a hydrogen–oxygen atmosphere (Fig. 13.9). From this equation, one sees that oxygen atoms may disappear by reactions with OH and HO_2. Since $a_5 \simeq a_7 \simeq 4 \times 10^{-11}$ cm^3 sec^{-1}, a direct effect is not important at low altitudes where the term $k_2 n(M) n(O_2)$

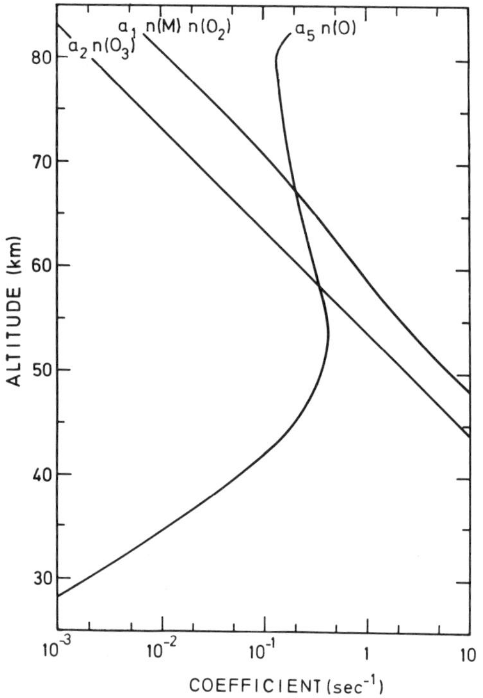

Fig. 13.10 Principal coefficients related to H, OH, and HO_2 reactions with ozone molecules and oxygen atoms in the stratosphere and mesosphere. Values are for daytime equilibrium conditions.

(see Table 11.1) reaches its highest values. But, in the mesosphere and particularly the upper mesosphere, possible effects can arise. At 75 km, $k_2n(M)n(O_2) \leq 10^{-3}$ sec^{-1} and $a_5n(OH)$, with 10^7 OH molecules cm^{-3}, is of the same order of magnitude. At 85 km (the mesopause), 10^6 OH molecules cm^{-3} are present and are important to the loss of oxygen atoms.

Since values of the order of 10^7 or 10^6 for the concentrations of OH (and HO_2) are possible at 75 km and 85 km respectively, oxygen concentrations such as those represented in Fig. 11.14 may be considered as maximum values. Nevertheless, since at the mesopause level atomic oxygen disappears slowly, the actual concentration $n(O)$ depends also on the vertical transport from the thermosphere, and not only on reaction mechanisms.

With regard to the ozone to atomic oxygen ratio in the thermosphere, Eq. (11.10) can be rewritten as

$$\frac{n(O_3)}{n(O)} = \frac{k_2n(M)n(O_2)}{J_3 + k_3n(O) + [a_2 + a_3]n(H)} \triangleq \frac{k_2n(M)n(O_2)}{J_{3a}} \tag{13.72}$$

since ozone is always in equilibrium with atomic oxygen. By introducing

(13.72) into (13.71) the variation of atomic oxygen is given by

$$\frac{dn(\mathrm{O})}{dt} + 2n^2(\mathrm{O})\left[k_1 n(\mathrm{M}) + \frac{k_3 k_2 n(\mathrm{M}) n(\mathrm{O}_2)}{J_{3\mathrm{a}}}\right] + n(\mathrm{O})\left[a_5 n(\mathrm{OH}) + a_7 n(\mathrm{HO}_2) + \frac{k_2 n(\mathrm{M}) n(\mathrm{O}_2)}{J_{3\mathrm{a}}}[a_2 + a_3] n(\mathrm{H})\right] = 2n(\mathrm{O}_2) J_2. \tag{13.73}$$

Such an equation corresponds to photochemical conditions in the lower thermosphere and in the mesosphere. Numerical calculations at 80 km show (see Table 11.1) that

$$k_1 n(\mathrm{M}) = 1 \times 10^{-18} \quad \mathrm{cm^3\ sec^{-1}} \simeq k_2 k_3 n(\mathrm{M}) n(\mathrm{O}_2)/J_3, \tag{13.74}$$

i.e., that the direct recombination rate of oxygen atoms is of the same order of magnitude as their association rate to form ozone. But, such rate coefficients are too small to lead to a photochemical equilibrium. As far as the term involving H, OH, and HO_2 is concerned, the effect leads to a decrease of the ozone and atomic oxygen concentrations. Nevertheless, chemical equilibrium cannot be reached for atomic oxygen in the thermosphere, and its vertical distribution is subject to the transport conditions in the lower thermosphere.

Since the effect of atomic hydrogen on the vertical distribution of ozone at the mesopause level is not negligible, its inclusion is based on the hypothesis [52] that H in the lower thermosphere is subject to a uniform mixing. Eddy diffusion conditions are assumed to be sufficiently important compared with loss reactions. Thus, the following conditions are adopted:

$$n(\mathrm{H}) = n(\mathrm{N}_2)/2 \times 10^5 \quad \mathrm{cm^{-3}}. \tag{13.75}$$

In other words, $n(\mathrm{H}) = 3.5 \times 10^7\ \mathrm{cm^{-3}}$ at 100 km leads to about 7×10^8 atoms at the mesopause. With such values of $n(\mathrm{H})$, the term $a_2 n(\mathrm{H})$, which is given in *Table 13.1*, is not negligible and is of the same order as $J_3 = 10^{-2}$ $\mathrm{sec^{-1}}$ near the mesopause. The atomic hydrogen concentration at 100 km is an approximate value which falls in the range of the values necessary to explain the Lyman α observations about 100 km (see Chapter 15).

Considering that a possible approximation at the mesopause level for Eq. (13.73) is

$$\frac{dn(\mathrm{O})}{dt} + n(\mathrm{O})\left[a_2 \frac{k_2 n(\mathrm{M}) n(\mathrm{O}_2)}{J_{3\mathrm{a}}} n(\mathrm{H}) + a_5 n(\mathrm{OH}) + a_7 n(\mathrm{HO}_2)\right] = 2n(\mathrm{O}_2) J_2 \tag{13.76}$$

TABLE 13.1

ATOMIC HYDROGEN CONCENTRATION AND COEFFICIENT $a_2n(\mathrm{H})$ IN THE LOWER THERMOSPHERE

Altitude (km)	$n(\mathrm{H})$ (cm^{-3})	$a_2n(\mathrm{H})$ sec^{-1}	Altitude (km)	$n(\mathrm{H})$ (cm^{-3})	$a_2n(\mathrm{H})$ sec^{-1}
100	3.5×10^7	0.07×10^{-2}	92	1.56×10^8	0.31×10^{-2}
99	4.2	0.085	91	1.90	0.38
98	5.0	0.10	90	2.34	0.47
97	6.0	0.12	89	2.88	0.56
96	7.2	0.14	88	3.56	0.71
95	8.7	0.17	87	4.42	0.88
94	1.5×10^8	0.21	86	5.52	1.10
93	1.28	0.26	85	6.91	1.38

and even

$$\frac{dn(\mathrm{O})}{dt} + a_2n(\mathrm{H}) \frac{k_2n(\mathrm{M})n(\mathrm{O}_2)}{J_{3\mathrm{a}}} n(\mathrm{O}) = 2n(\mathrm{O}_2)J_2 \tag{13.77}$$

since atomic hydrogen is sufficiently important, it is possible to define a time

$$\tau_0 = \frac{J_3 + a_2n(\mathrm{H})}{2a_2n(\mathrm{H})k_2n(\mathrm{M})n(\mathrm{O}_2)} \tag{13.78}$$

necessary to reach an important fraction of the equilibrium value. With numerical values, τ_0 equals 8.4×10^3 sec at 85 km. Thus, when atomic hydrogen plays a role at the mesopause level, an equilibrium value of $n(\mathrm{O})$ can be reached in about 2 hr. With all numerical parameters used here and for an overhead sun, $n(\mathrm{O})_{85\ \mathrm{km}} = 3 \times 10^{10}\ \mathrm{cm}^{-3}$ and

$$n(\mathrm{O}_3) = \frac{k_2n(\mathrm{M})n(\mathrm{O}_2)n(\mathrm{O})}{J_3 + a_2n(\mathrm{H})} = 10^8 \quad \mathrm{cm}^{-3}. \tag{13.79}$$

In the lower thermosphere there is a transition region where the atomic oxygen concentration must increase rapidly, since at 90 km the time to reach a photoequilibrium is not less than 10^5 sec.

After sunset, in the lower thermosphere,

$$\frac{dn(\mathrm{O}_3)}{dt} + n(\mathrm{O}_3)a_2n(\mathrm{H}) = k_2n(\mathrm{M})n(\mathrm{O}_2)n(\mathrm{O}). \tag{13.80}$$

Provided $n(O)$ does not vary appreciably with time, the relevant solution to the differential equation (13.80) is simply

$$n(O_3) = n(O_3)_{t=0}e^{-a_2 n(H)t} + \frac{k_2 n(M) n(O_2) n(O)}{a_2 n(H)} [1 - e^{-a_2 n(H)t}]. \quad (13.81)$$

Since $a_2 n(H)t = 1$ after less than 100 sec at 100 km, the ozone concentration in the nighttime thermosphere can be conveniently written as

$$n(O_3) = \frac{k_2 n(M) n(O_2) n(O)}{a_2 n(H)}. \quad (13.82)$$

Thus, the thermospheric ozone is directly related to the values adopted for the concentrations of atomic oxygen and atomic hydrogen.

Since the daytime ratio $n(O_3)/n(O)$ is

$$\frac{n(O_3)}{n(O)} = \frac{k_2 n(M) n(O_2)}{J_3 + a_2 n(H)} \quad (13.83)$$

an increase of the ozone concentration in the thermosphere is possible from daytime conditions (13.83) to nighttime conditions (13.82).

When hydrogen reactions are involved, Eq. (11.8) governing the rate of change of the ozone concentration becomes

$$\frac{dn(O_3)}{dt} + [J_3 + k_3 n(O) + (a_2 + a_3) n(H)] n(O_3) = k_2 n(M) n(O_2) n(O). \quad (13.84)$$

If it is assumed that atomic oxygen is in photochemical equilibrium in the stratosphere and also in the lower mesosphere, Eq. (13.71) leads to the approximation

$$n(O) = \frac{2n(O_2)J_2 + n(O_3)J_3}{k_2 n(M) n(O_2) + k_3 n(O_3) + a_5 n(OH) + a_7 n(HO_2)} \quad (13.85)$$

since the term proportional to $k_1 n(M)$ can be neglected. Introducing (13.85) in (13.84), the differential equation for the ozone concentration becomes

$$\begin{aligned} \frac{dn(O_3)}{dt} &+ \frac{n^2(O_3)}{D} k_3 \{2J_3 + [a_2 + a_3] n(H)\} \\ &+ \frac{n(O_3)}{D} \{J_3 [a_5 n(OH) + a_7 n(HO_2)] \\ &+ [a_2 + a_3] n(H) [k_2 n(M) n(O_2) + a_5 n(OH) + a_7 n(HO_2)]\} \\ &= \frac{2n(O_2)J_2}{D} [k_2 n(M) n(O_2) - k_3 n(O_3)] \end{aligned} \quad (13.86)$$

where

$$D = k_2n(M)n(O_2) + k_3n(O_3) + a_5n(OH) + a_7n(HO_2). \tag{13.87}$$

At sufficiently low altitudes, i.e., in the stratosphere and in the major part of the mesosphere,

$$J_3 > [a_2 + a_3]n(H) \tag{13.88}$$

and

$$k_2n(M)n(O_2) > k_3n(O_3) + a_5n(OH) + a_7n(HO_2). \tag{13.89}$$

With the conditions (13.88) and (13.89) the rate of change of the concentration $n(O_3)$ can be conveniently rewritten as:

$$\begin{aligned} \frac{dn(O_3)}{dt} &+ 2n^2(O_3)\frac{J_3k_3}{k_2n(M)n(O_2)} \\ &+ n(O_3)\left\{[a_2 + a_3]n(H) + \frac{a_5J_3n(OH)}{k_2n(M)n(O_2)} + \frac{a_7J_3n(HO_2)}{k_2n(M)n(O_2)}\right\} \\ &= 2n(O_2)J_2. \end{aligned} \tag{13.90}$$

When the effect of hydrogen compounds is neglected in (13.90), one obtains the conventional equilibrium relation

$$n^2(O_3) = \frac{k_2n(M)n^2(O_2)J_2}{k_3J_3}. \tag{13.91}$$

With the hydrogen compounds, (13.90) leads to the equilibrium relation

$$n^2(O_3) = \frac{k_2n(M)n^2(O_2)J_2}{k_3J_3\left[1 + \frac{[a_2 + a_3]k_2n(M)n(O_2)}{2k_3J_3n(O_3)} + \frac{a_5n(OH)}{2k_3n(O_3)} + \frac{a_7n(HO_2)}{2k_3n(O_3)}\right]}. \tag{13.92}$$

If (13.92) is compared with (13.91), it is clear that the effect of hydrogen compounds is equivalent to an increase [3, 52] of the ozone photodissociation coefficient. The equivalent O_3 photodissociation coefficient J_{3A} is written

$$J_{3A} = J_3(1 + A) \tag{13.93}$$

where A is given by

$$A = [a_2 + a_3]\frac{n(H)k_2n(M)n(O_3)}{2J_3k_3n(O_3)} + \frac{a_5n(OH)}{2k_3n(O_3)} + \frac{a_7n(HO_2)}{2k_3n(O_3)} \tag{13.94}$$

The conventional equation for equilibrium (13.91) is now replaced by

$$n^2(O_3) = \frac{k_2 n(M) n^2(O_2) J_2}{k_3 J_{3A}}. \tag{13.95}$$

With the adopted vertical distribution of ozone (see Fig. 11.2) the correction term $1 + A$ is given in the stratosphere by

$$(1 + A)^{1/2} \simeq \left\{1 + \frac{a_7[n(OH) + n(HO_2)]}{2k_3 n(O_3)}\right\}^{1/2} \simeq 1.5 \tag{13.96}$$

where $a_7 = a_5$.

An illustration of such an effect is given [3] by *Fig. 13.11* which shows that below 35 km the vertical distribution of the ozone concentration corresponds to a departure from photochemical conditions and transport processes determine the vertical distribution.

Above the stratopause level, it is necessary to consider all the terms

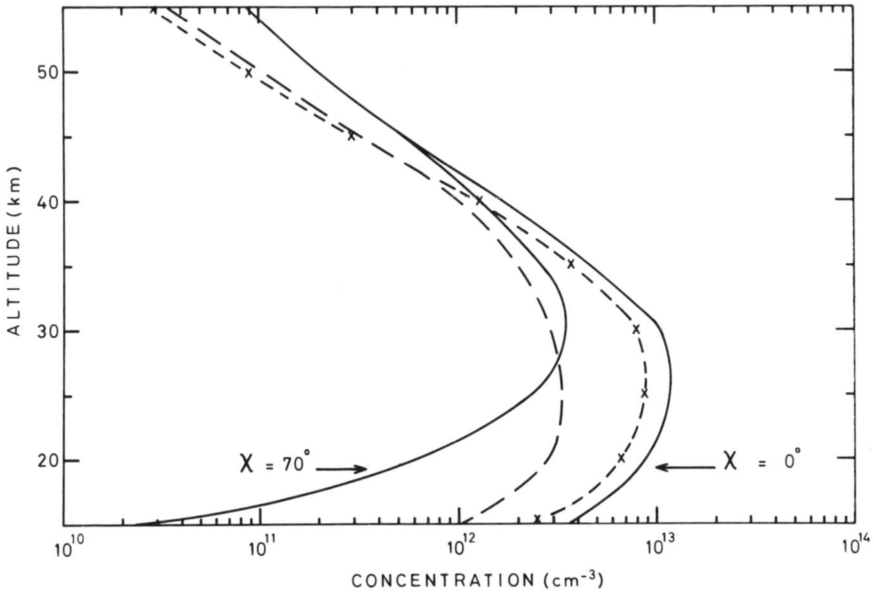

Fig. 13.11 Ozone distribution in the stratosphere for equilibrium conditions without the effect of hydrogen compounds ($\chi = 0°$ and $70°$). Effect of OH and HO_2 reactions when $n(H_2O)/n(M) = 3 \times 10^{-6}$ and $n(CH_4)/n(M) = 1.5 \times 10^{-6}$. The O_3 distribution adopted in Fig. 11.2 is shown for comparison. (——) ozone; (- - × - -) O_3 equilibrium with $H_2O/M = 3 \times 10^{-6}$, $CH_4/M = 1.5 \times 10^{-6}$; (- - -) O_3 adopted model (Nicolet [3]).

of (13.94) so that

$$[1 + A]^{1/2} = \left[1 + \frac{a_2 n(\mathrm{H}) k_2 n(\mathrm{M}) n(\mathrm{O}_2)/J_3 + a_5 n(\mathrm{OH}) + a_7 n(\mathrm{HO}_2)}{2k_3 n(\mathrm{O}_3)}\right]^{1/2}. \tag{13.97}$$

The correction term $[1 + A]^{1/2}$ increases from 1.5 at the stratopause to about 100 at the mesopause. *Figure 13.12* illustrates [3] the difference between an oxygen atmosphere and a hydrogen–oxygen atmosphere. There is a considerable decrease of the atomic oxygen concentration in the mesosphere and consequently a parallel decrease of the ozone concentration.

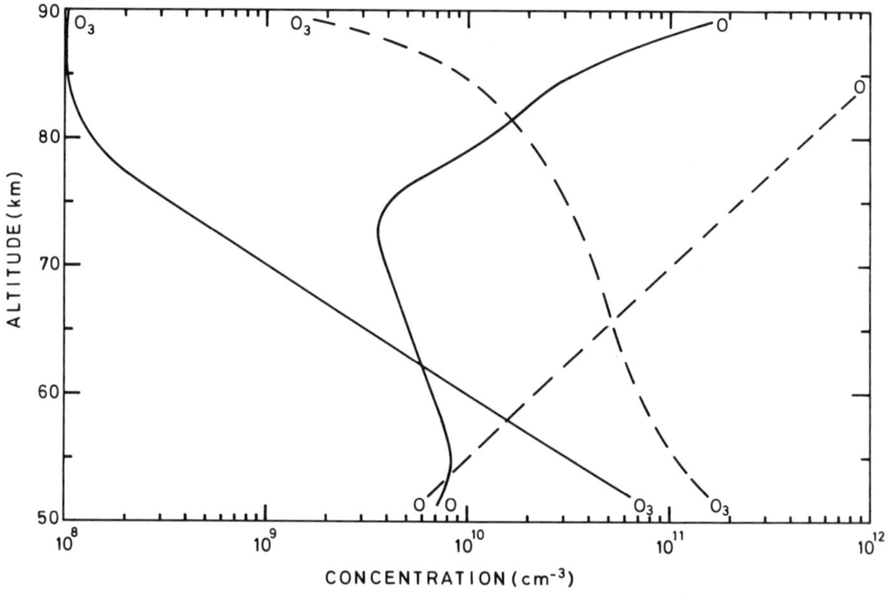

Fig. 13.12 Comparison of the ozone and atomic oxygen concentration for photoequilibrium conditions with and without the effect of the hydrogen compounds. Solid line, O_3 and O with H effect; dashed line, equilibrium for O_3 and O without H (Nicolet [3]).

13.7 Rates of Change for the Hydrogen Compounds

Before deducing general equations for the rates of change of H, H_2, OH, H_2O, H_2O_2, and HO_2 it is necessary to discuss the importance of additional reactions of hydrogen compounds.

The reformation of water vapor through a three-body reaction like

$$(a_{14}):\quad H + OH + M \rightarrow H_2O + M + 118 \quad \text{kcal mole}^{-1} \tag{13.98}$$

is not important compared to other aeronomic processes, even if its rate coefficient may be high [53];

$$a_{14} = 2 \times 10^{-31} \quad \text{cm}^6 \text{ sec}^{-1}. \tag{13.99}$$

The reaction of hydrogen atoms and perhydroxyl radicals leads to two hydroxyl radicals

$$(a_{15}):\quad H + HO_2 \rightarrow OH + OH_{v\leq 4} + 39 \quad \text{kcal mole}^{-1} \tag{13.100}$$

for which the following rate coefficient can be adopted:

$$a_{15} = 1.5 \times 10^{-12} T^{1/2} \quad \text{cm}^3 \text{ sec}^{-1} \tag{13.101}$$

with practically no activation energy. Such a reaction was first detected [54, 55] by its vibrational emission from OH ($v \leq 4$).

The reaction between two hydroxyl radicals leads to the reformation of water vapor

$$(a_{16}):\quad OH + OH \rightarrow H_2O + O + 17 \quad \text{kcal mole}^{-1} \tag{13.102}$$

with an activation energy not greater than 2 kcal. Thus, a minimum rate coefficient in the stratosphere and mesosphere is given by

$$a_{16} = 5 \times 10^{-12} T^{1/2} e^{-1000/T} \quad \text{cm}^3 \text{ sec}^{-1}. \tag{13.103}$$

In the same way, the reaction between hydroxyl and perhydroxyl radicals leads to H_2O:

$$(a_{17}):\quad OH + HO_2 \rightarrow H_2O + O_2 + 72 \quad \text{kcal mole}^{-1} \tag{13.104}$$

with a small activation energy ($\sim$1 kcal mole^{-1}). The value

$$a_{17} = 3 \times 10^{-12} T^{1/2} e^{-500/T} \quad \text{cm}^3 \text{ sec}^{-1} \tag{13.105}$$

may be considered as a maximum value for aeronomic applications.

Additional reactions lead to formation of H_2O, but they are not important in the mesosphere and stratosphere compared with the reactions already given. The three-body association

$$(a_{18}):\quad H_2 + O + M \rightarrow H_2O + M + 116 \quad \text{kcal mole}^{-1} \tag{13.106}$$

should have a small rate coefficient such as

$$a_{18} < 10^{-32} \quad \text{cm}^6 \text{ sec}^{-1}. \tag{13.107}$$

The exothermic reaction between H_2 and OH

$$(a_{19}):\quad H_2 + OH \rightarrow H_2O + H + 15 \quad \text{kcal mole}^{-1} \tag{13.108}$$

requires a high activation energy and its maximum value is

$$a_{19} = 1.5 \times 10^{-12} T^{1/2} e^{-2500/T} \quad \text{cm}^3 \text{ sec}^{-1}. \tag{13.109}$$

Finally, in addition to (13.100), the reaction between hydrogen atoms and perhydroxyl radicals may lead to the reformation of H_2O:

$$(a_{20}):\quad H + HO_2 \rightarrow H_2O + O + 55 \quad \text{kcal mole}^{-1} \tag{13.110}$$

with a rate coefficient smaller than a_{15},

$$a_{20} = 1.5 \times 10^{-12} T^{1/2} e^{-2000/T} \quad \text{cm}^3 \text{ sec}^{-1}. \tag{13.111}$$

In addition to the various processes of the type just described which are related to the reformation of H_2O, there are a great variety of reactions involving molecular hydrogen. The three-body reaction

$$(a_{21}):\quad H + H + M \rightarrow H_2 + M + 103.2 \quad \text{kcal mole}^{-1} \tag{13.112}$$

is known with a rate coefficient of the order

$$a_{21} = 1 \times 10^{-32} \quad \text{cm}^6 \text{ sec}^{-1}. \tag{13.113}$$

The bimolecular reaction

$$(a_{22}):\quad H + OH \rightarrow H_2 + O + 1.9 \quad \text{kcal mole}^{-1} \tag{13.114}$$

requires a high activation energy of 7–8 kcal mole^{-1}. At low temperatures the rate coefficient is between

$$a_{22} = 2 \times 10^{-13} T^{1/2} e^{-3400/T} \quad \text{cm}^3 \text{ sec}^{-1} \tag{13.115}$$

and

$$a_{22} = 1 \times 10^{-12} T^{1/2} e^{-4000/T} \quad \text{cm}^3 \text{ sec}^{-1}. \tag{13.116}$$

It appears that a more important mode of molecular hydrogen formation occurs through the reaction of hydrogen atoms with perhydroxyl radicals:

$$(a_{23}):\quad H + HO_2 \rightarrow H_2 + O_2 + 57 \quad \text{kcal mole}^{-1} \tag{13.117}$$

since it is possible to assume an activation energy not greater than 2 kcal mole^{-1} and a_{23} can be written as

$$a_{23} = 5\times 10^{-12} T^{1/2} e^{-1000/T} \quad \text{cm}^3\,\text{sec}^{-1}. \tag{13.118}$$

After considering all the reactions in which H_2O and H_2 are involved, the equations governing the rates of change of the concentration $n(H_2O)$ and $n(H_2)$ are

$$\frac{dn(H_2O)}{dt} + n(H_2O)[J^{(1)}_{H_2O} + J^{(2)}_{H_2O} + a^*_{13}n(O^*)]$$
$$= n(H)[a_{14}n(M)n(OH) + a_{20}n(HO_2)] + n(OH)[a_{16}n(OH) + a_{17}n(HO_2)]$$
$$+ n(H_2)[a_{19}n(OH) + a_{18}n(M)n(O)] \tag{13.119}$$

and, as an acceptable approximation [52] (see *Fig. 13.13*)

$$\frac{dn(H_2O)}{dt} + n(H_2O)[J^{(2)}_{H_2O} + a^*_{13}n(O^*)]$$
$$= a_{16}n^2(OH) + n(HO_2)[a_{17}n(OH) + a_{20}n(H)] \tag{13.120}$$

where $n(O^*)$ is the concentration of $O(^1D)$ and $J^{(2)}_{H_2O}$ corresponds to (13.2) and (13.6).

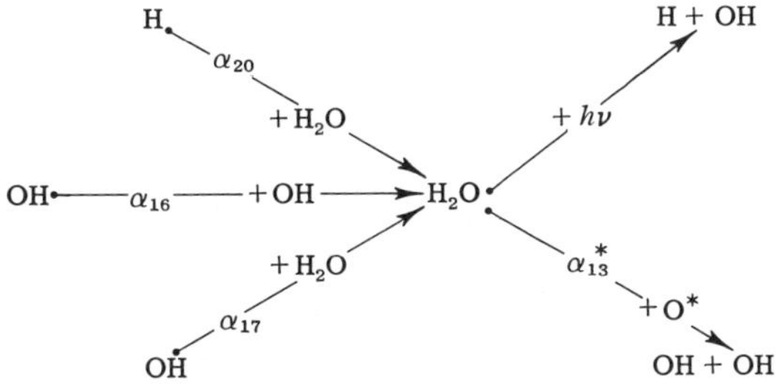

Fig. 13.13 Principal reactions in which H_2O is involved in the stratosphere, mesosphere, and lower thermosphere; photodissociation is the important loss process above the stratopause and a^*_{13} dominates below the stratopause.

As far as H_2 is concerned, its differential equation is

$$\frac{dn(H_2)}{dt} + n(H_2)\{a_{24}^* n(O^*) + [a_{18} n(M) + a_{24}] n(O) + a_{19} n(OH)\}$$
$$= n(H_2O) J_{H_2O}^{(1)} + n(H)[a_{21} n(M) n(H) + a_{22} n(OH) + a_{23} n(HO_2)] \tag{13.121}$$

and with an excellent approximation (see *Fig. 13.14*)

$$\frac{dn(H_2)}{dt} + [a_{24} n(O) + a_{24}^* n(O^*)] n(H_2) = n(H_2O) J_{H_2O}^{(1)} + a_{23} n(H) n(HO_2). \tag{13.122}$$

The reactions involving the dissociation of H_2O and H_2 lead to H atoms and OH and HO_2 radicals which are easily in equilibrium. In contrast, H_2O and H_2 are subject to more complex conditions.

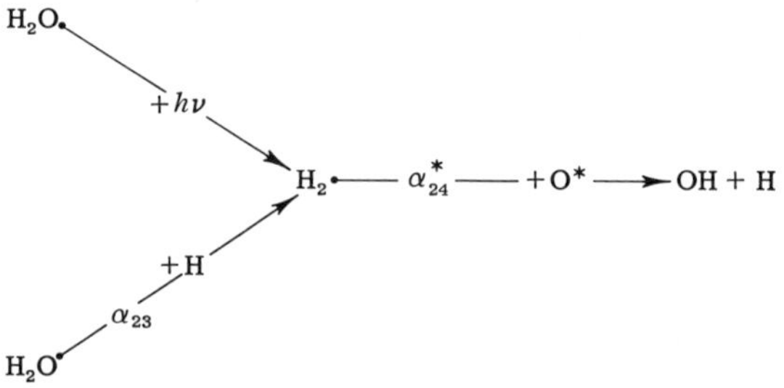

Fig. 13.14 Principal reactions in which H_2 is involved in the stratosphere, mesosphere, and lower thermosphere; a_{23} is an important production process below the mesopause and a_{24}^* is an important loss process below the stratopause.

Water vapor cannot be in chemical equilibrium in the stratosphere and mesosphere since its dissociation time is too long. Nevertheless, its behavior is completely different from that of H_2 and CH_4 for which there is no possibility of immediate reformation after an attack by the oxidation process. All products of the oxidation and dissociation of H_2, CH_4, and H_2O lead to the formation of water vapor through H, OH, and HO_2. When a uniform mixing is adopted for H_2O in the stratosphere, the hypothesis is made that H_2O is involved in sufficiently rapid catalytic action so that an anomalously high eddy diffusion coefficient is not required. At the mesopause and in the lower thermosphere, where the H_2O reformation is slow, eddy diffusion leads to a vertical transport of H_2O molecules.

Finally, since it is extremely difficult to assume that permanent eddy diffusion coefficients are more than 10^5 cm^2 sec^{-1} in the upper stratosphere, the possibility of a decrease of the ratios $n(H_2)/n(M)$ and $n(CH_4)/n(M)$ must be considered in the stratosphere. In Figs. 13.6 and 13.7 the curves at the left display the situation for a minor constituent scale height which is only 50% of the atmospheric scale height. In that case, the required eddy diffusion coefficient is still greater than 10^4 cm^2 sec^{-1} near the stratopause. It seems, therefore, that one may assume that in the region above 35 km H_2 and CH_4 are oxidized and lead to H_2O, introducing the possibility of an increase of the mixing ratio $n(H_2O)/n(M)$.

The processes described above are not complete since it is known that nitrogen oxides may play a role [56, 57]. For example, the reaction

$$(a_{25}):\quad H + NO_2 \rightarrow OH + NO + 30 \quad \text{kcal mole}^{-1} \tag{13.123}$$

is rapid with a rate coefficient

$$a_{25} = 2\times 10^{-12} T^{1/2} \quad \text{cm}^3\ \text{sec}^{-1}. \tag{13.124}$$

In the lower stratosphere, it is possible that the ratio $n(HO_2)/n(OH)$ depends on a fast reaction like

$$(a_{26}):\quad HO_2 + NO \rightarrow OH + NO_2 + 9 \quad \text{kcal mole}^{-1} \tag{13.125}$$

with an activation energy of about 2.5 kcal mole^{-1} leading to a rate coefficient of about 10^{-11} cm^3 sec^{-1} at 500°K. If the hypothesis is made that the activation energy for (13.125) is small, a rate coefficient such as

$$a_{26} = 3\times 10^{-12} T^{1/2} \quad \text{cm}^3\ \text{sec}^{-1} \tag{13.126}$$

can be adopted in order to reach not less than 5×10^{-11} cm^3 sec^{-1} at 273°K. Thus, a leading role could be played by (13.125) if $n(NO) > 10^8$ cm^{-3}. If this is the case, Eq. (13.70) should be replaced in the lower stratosphere by (see *Fig. 13.15*)

$$\frac{n(HO_2)}{n(OH)} = \frac{\dfrac{a_1 n(M) n(O_2) a_5 n(O)}{a_1 n(M) n(O_2) + a_2 n(O_3)}}{a_7 n(O) + a_{26} n(NO)}. \tag{13.127}$$

The description of the reaction chains which contribute to the formation and destruction of H, OH, and HO_2 is inadequate when hydrogen peroxide is neglected *a priori*. The only low activation energy two-body process yielding hydrogen peroxide is

$$(a_{27}):\quad HO_2 + HO_2 \rightarrow H_2O_2 + O_2 + 42 \quad \text{kcal mole}^{-1} \tag{13.128}$$

for which a rate coefficient similar to a_{16} is adopted, i.e.,

$$a_{27} = 5 \times 10^{-12} T^{1/2} e^{-1000/T} \quad \text{cm}^3 \text{ sec}^{-1}. \tag{13.129}$$

Such a process is a loss for HO_2 and Eq. (13.127) is not the complete $n(HO_2)/n(H)$ expression for the stratosphere.

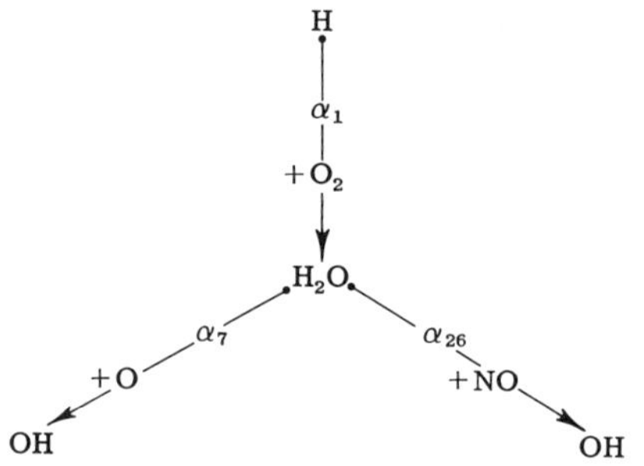

Fig. 13.15 Principal reactions determining the ratio of hydroxyl and perhydroxyl radicals in the mesosphere and stratosphere. Nitric oxide should play a role in the stratosphere by reaction a_{26}.

In addition to the photodissociation loss discussed in Section 13.2.4, the following reactions can also be considered:

$$(a_{29a}): \quad H + H_2O_2 \rightarrow H_2O + OH + 69 \quad \text{kcal mole}^{-1} \tag{13.130}$$

$$(a_{29b}): \quad H + H_2O_2 \rightarrow H_2 + HO_2 + 14 \quad \text{kcal mole}^{-1}. \tag{13.131}$$

Experimental evidence [58] seems to indicate that the activation energies of (13.130) and (13.131) are important and the rates

$$a_{29a} = 5 \times 10^{-12} T^{1/2} e^{-3000/T} \quad \text{cm}^3 \text{ sec}^{-1} \tag{13.132}$$

$$a_{29b} = 5 \times 10^{-12} T^{1/2} e^{-4500/T} \quad \text{cm}^3 \text{ sec}^{-1} \tag{13.133}$$

show that the reactions between H and H_2O_2 are not sufficiently rapid to be of importance.

The reaction between a hydroxyl radical and hydrogen peroxide

$$(a_{30}): \quad OH + H_2O_2 \rightarrow H_2O + HO_2 + 30 \quad \text{kcal mole}^{-1} \tag{13.134}$$

is important since [59]

$$a_{30} = 4 \times 10^{-13} T^{1/2} e^{-600/T} \quad \text{cm}^3 \text{ sec}^{-1}. \tag{13.135}$$

The principal production and loss processes of HO_2 are now illustrated in *Fig. 13.16* and the complete equation for the HO_2/OH ratio is

$$\frac{n(HO_2)}{n(OH)} = \frac{\dfrac{a_1 n(M) n(O_2) a_5 n(O)}{a_1 n(O_2) n(M) + a_2 n(O_3)} + a_{30} n(H_2O_2)}{J_{HO_2} + a_7 n(O) + a_{17} n(OH) + a_{26} n(NO) + 2a_{27} n(HO_2)}. \tag{13.136}$$

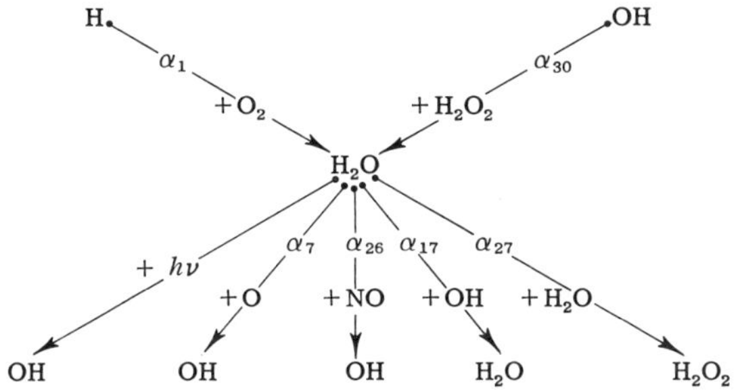

Fig. 13.16 Reactions in which HO_2 is involved in the stratosphere and mesosphere.

If it is assumed that only (13.128)–(13.136) are involved in the production and destruction of H_2O_2, the differential equation for H_2O_2 is [52]

$$\frac{dn(H_2O_2)}{dt} + n(H_2O_2)\{J_{H_2O_2} + [a_{29a} + a_{29b}]n(H) + a_{30} n(OH)\} = a_{27} n^2(HO_2) \tag{13.137}$$

and as a first approximation in a sunlit atmosphere,

$$\frac{dn(H_2O_2)}{dt} + n(H_2O_2) J_{H_2O_2} = a_{27} n^2(HO_2). \tag{13.138}$$

Experimental evidence does not rule out $O(^1D)$–HO_2 reactions such as

$$(a_{31}^*): \quad O(^1D) + HO_2 \rightarrow OH + O_2 + 60 \quad \text{kcal mole}^{-1} \tag{13.139}$$

which could have zero activation energy and occur at a rate similar to the reactions of $O(^1D)$ with O_3 and H_2. If the rate

$$a_{31}^* = 10^{-10} \quad \text{cm}^3\,\text{sec}^{-1} \tag{13.140}$$

is adopted, however, the coefficient $a_{31}^* n(O\,^1D) < 10^{-7}$ sec^{-1} is less important than $J_{H_2O_2} > 5\times 10^{-6}$ sec^{-1} (see *Fig. 13.17*).

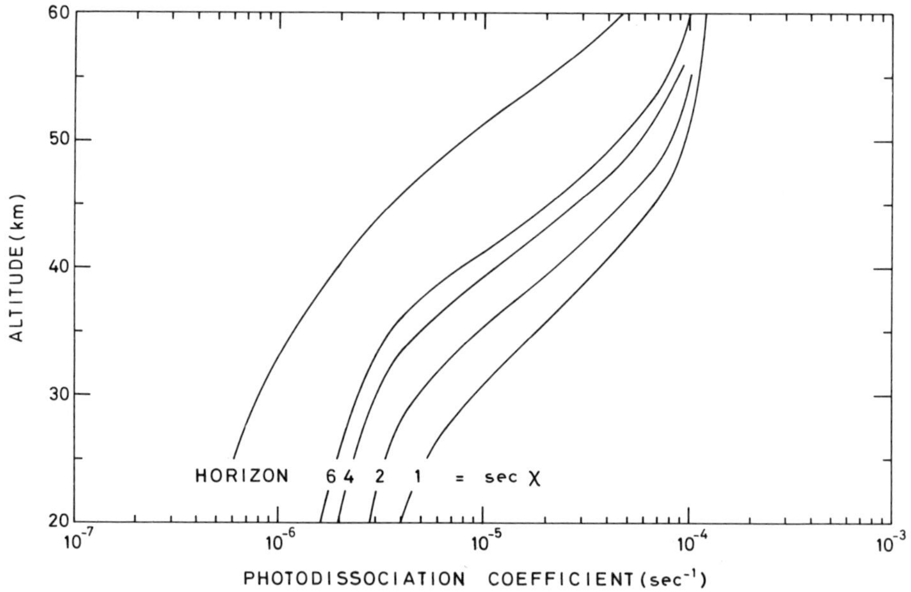

Fig. 13.17 Photodissociation coefficient of hydrogen peroxide in the mesosphere and stratosphere for various solar zenith angles (Nicolet [3]).

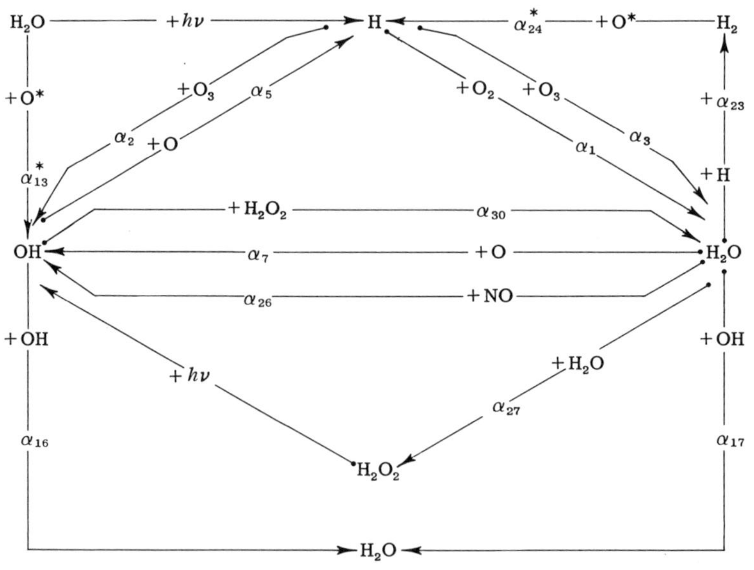

Fig. 13.18 Fundamental H_2O reaction scheme.

In place of the simple scheme presented in Fig. 13.9 it is necessary to consider the whole system (*Fig. 13.18*) not only with the dissociation of H_2O into OH and H and the oxidation of H_2O leading to two OH radicals, but also with respect to the reformation of water vapor by reactions between OH and HO_2.

The equations governing the rates of change of the concentrations $n(H)$, $n(HO_2)$, and $n(OH)$ are [52]

$$\begin{aligned}\frac{dn(\mathrm{H})}{dt} &+ n(\mathrm{H})\{a_0 n(\mathrm{M})n(\mathrm{O}) + a_1 n(\mathrm{M})n(\mathrm{O_2}) + [a_2 + a_3]n(\mathrm{O_3}) \\ &+ [a_{14}n(\mathrm{M}) + a_{22}]n(\mathrm{OH}) + [a_{15} + a_{20} + a_{23}]n(\mathrm{HO_2}) \\ &+ 2a_{21}n(\mathrm{M})n(\mathrm{H}) + [a_{29a} + a_{29b}]n(\mathrm{H_2O_2})\} \\ &= n(\mathrm{OH})[a_5 n(\mathrm{O}) + a_{19}n(\mathrm{H_2})] + a_{24}n(\mathrm{H_2})n(\mathrm{O}) \\ &+ a_{24}^* n(\mathrm{H_2})n(\mathrm{O^1D}) + n(\mathrm{H_2O})J_{\mathrm{H_2O}}^{(2)},\end{aligned} \tag{13.141}$$

$$\begin{aligned}\frac{dn(\mathrm{HO_2})}{dt} &+ n(\mathrm{HO_2})\{J_{\mathrm{HO_2}} + a_{6a}n(\mathrm{O_3}) + a_7 n(\mathrm{O}) + [a_{15} + a_{20} + a_{23}]n(\mathrm{H}) \\ &+ a_{17}n(\mathrm{OH}) + 2a_{27}n(\mathrm{HO_2})\} \\ &= [a_1 n(\mathrm{M})n(\mathrm{O_2}) + a_3 n(\mathrm{O_3})]n(\mathrm{H}) \\ &+ [a_4 n(\mathrm{M})n(\mathrm{O}) + a_6 n(\mathrm{O_3}) + a_{30}n(\mathrm{H_2O_2})]n(\mathrm{OH}),\end{aligned} \tag{13.142}$$

$$\begin{aligned}\frac{dn(\mathrm{OH})}{dt} &+ n(\mathrm{OH})\{[a_5 + a_4 n(\mathrm{M})]n(\mathrm{O}) + [a_{22} + a_{24}n(\mathrm{M})]n(\mathrm{H}) + 2a_{16}n(\mathrm{OH}) \\ &+ a_6 n(\mathrm{O_3}) + a_{14}n(\mathrm{M})n(\mathrm{H}) + a_{17}n(\mathrm{HO_2}) + a_{30}n(\mathrm{H_2O_2})\} \\ &= n(\mathrm{H})[a_0 n(\mathrm{M}) + a_2 n(\mathrm{O_2}) + a_{15}n(\mathrm{HO_2})] \\ &+ n(\mathrm{HO_2})[J_{\mathrm{HO_2}} + a_{6a}n(\mathrm{O_3}) + a_7 n(\mathrm{O}) + a_{15}n(\mathrm{H})] \\ &+ n(\mathrm{H_2})a_{24}n(\mathrm{O}) + n(\mathrm{H_2O})J_{\mathrm{H_2O}}^{(2)} \\ &+ a^*(^1\mathrm{D})[n(\mathrm{CH_4}) + n(\mathrm{H_2O}) + n(\mathrm{H_2})]n(\mathrm{O}\ ^1\mathrm{D}).\end{aligned} \tag{13.143}$$

For daytime conditions, the principal reactions involve atomic oxygen and ozone. The principal terms shown in Fig. 13.9 indicate that Eq. (13.141) can be approximated as

$$\frac{dn(\mathrm{H})}{dt} + n(\mathrm{H})\{a_1 n(\mathrm{M})n(\mathrm{O_2}) + [(a_2 + a_3)n(\mathrm{O_3})]\} = n(\mathrm{OH})a_5 n(\mathrm{O}). \tag{13.144}$$

Since the coefficient of $n(H)$ is more than 10^{-3} sec^{-1} at the mesopause, equilibrium conditions can be accepted in the mesosphere and stratosphere.

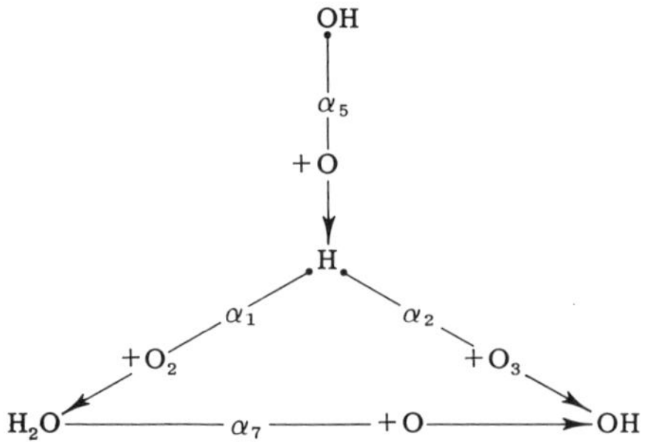

Fig. 13.19 Reaction scheme leading to a loss of atomic oxygen by atomic hydrogen and hydroxyl radicals.

Thus

$$\frac{n(\mathrm{OH})}{n(\mathrm{H})} = \frac{a_1 n(\mathrm{M}) n(\mathrm{O_2}) + a_2 n(\mathrm{O_3})}{a_5 n(\mathrm{O})} \tag{13.145}$$

which is Eq. (13.67) deduced directly from simple equilibrium conditions (see *Fig. 13.19*). It is clear from the numerical values of the ratio $n(\mathrm{OH})/n(\mathrm{H})$ given in *Table 13.2* that atomic hydrogen becomes negligible compared with

TABLE 13.2

RATIO OF THE ATOMIC HYDROGEN AND HYDROXYL RADICAL CONCENTRATIONS IN THE STRATOSPHERE AND MESOSPHERE FOR DAYTIME CONDITIONS

Altitude (km)	$n(\mathrm{O_3})/n(\mathrm{O})$	$n(\mathrm{OH})/n(\mathrm{H})$
15	2.0×10^{7}	1.9×10^{11}
20	3.1×10^{6}	1.9×10^{9}
25	4.8×10^{5}	4.5×10^{7}
30	7.4×10^{4}	1.5×10^{6}
35	8.5×10^{3}	4.5×10^{4}
40	8.3×10^{2}	1.9×10^{3}
45	8.6×10^{1}	1.7×10^{2}
50	1.5×10^{1}	2.9×10^{1}
55	3.8×10^{0}	6.7×10^{0}
60	1.5×10^{0}	3.5×10^{0}
65	6.3×10^{-1}	1.8×10^{0}
70	2.5×10^{-1}	8.3×10^{-1}
75	8.4×10^{-2}	2.9×10^{-1}
80	1.0×10^{-2}	2.6×10^{-2}
85	3.3×10^{-3}	4.5×10^{-3}

the hydroxyl radical in the stratosphere. The mesosphere is a transition region where the ratio decreases by about a factor of 10^4.

Equation (13.142) governing the rate of change of the concentration $n(HO_2)$ can conveniently be simplified to

$$\frac{dn(HO_2)}{dt} + n(HO_2)a_7 n(O) = a_1 n(M)n(O_2)n(H) \tag{13.146}$$

since (see Fig. 13.10) $a_7 n(O) = a_5 n(O)$ and both are not less than 10^{-1} sec^{-1} in the whole mesosphere. Thus, for equilibrium conditions in the mesosphere

$$\frac{n(HO_2)}{n(H)} = \frac{a_1 n(M)n(O_2)}{a_7 n(O)} \tag{13.147}$$

which is identical to Eq. (13.69). This ratio represents the normal ratio $n(OH_2)/n(H)$ in the mesosphere. But, if nitric oxide is considered to react with HO_2 via (13.125), the ratio $n(HO_2)/n(OH)$ does not increase in the same way in the stratosphere. It is then given by (13.127), where $a_{26}n(NO)$ becomes greater than $a_7 n(O)$ in the lower stratosphere. Even if reaction (13.125) between HO_2 and NO is significant, it is not yet possible to give a correct order of magnitude to its rate coefficient. Thus, the following simple formula may be used in the mesosphere:

$$\frac{n(HO_2)}{n(OH)} = \frac{a_1 n(M)n(O_2)}{a_1 n(M)n(O_2) + a_2 n(O_3)}. \tag{13.148}$$

Numerical results for (13.148) are given in *Table 13.3*. In the stratosphere, the effect of nitric oxide [57] must be involved below 35 km.

Considering the total rate of change of the concentrations $n(H)$, $n(OH)$, and $n(HO_2)$ to be given by the sum of (13.141)–(13.143), the following equation is obtained:

$$\begin{aligned}\frac{1}{2}\left[\frac{dn(H)}{dt} + \frac{dn(OH)}{dt} + \frac{dn(HO_2)}{dt}\right]& \\ + a_{16}n^2(OH) + a_{17}n(OH)n(HO_2) + a_{27}n^2(HO_2) + n(H)\{a_{21}n(M)n(H)& \\ + [a_{14}n(M) + a_{22}]n(OH) + [a_{20} + a_{23}]n(HO_2)\}& \\ = n(H_2O)J^{(2)}_{H_2O} + n(H_2O_2)J_{H_2O_2} + a_{24}n(O)n(H_2)& \\ + a^*(^1D)[n(CH_4) + n(H_2O) + n(H_2)]n(O^1D).&\end{aligned} \tag{13.149}$$

At altitudes where atomic hydrogen is the most important hydrogen constituent, the ratios $n(OH)/n(H) = x$ and $n(HO_2)/n(H) = y$ [which have

TABLE 13.3

RATIOS $n(HO_2)/n(H)$ AND $n(HO_2)/n(OH)$ IN THE MESOSPHERE FOR DAYTIME CONDITIONS

Altitude (km)	$n(HO_2)/n(H)$	$n(HO_2)/n(OH)$
50	2.1×10^{1}	7.3×10^{-1}
55	4.8×10^{0}	7.2×10^{-1}
60	2.7×10^{0}	7.8×10^{-1}
65	1.4×10^{0}	8.2×10^{-1}
70	7.1×10^{-1}	8.5×10^{-1}
75	2.5×10^{-1}	8.5×10^{-1}
80	2.1×10^{-2}	8.1×10^{-1}
85	2.8×10^{-3}	6.3×10^{-1}

been determined in Tables 13.2 and 13.3] may be introduced in (13.149), which becomes

$$\frac{dn(\mathrm{H})}{dt} + 2n^2(\mathrm{H})a_{16}[y + (x + y)^2] = 2P \tag{13.150}$$

where it has been assumed for simplicity that $a_{16} = a_{23} = a_{27} = a_{17}/2$ and where P is the direct and indirect production of hydrogen atoms. At the mesopause level, atomic hydrogen is produced by the photodissociation of H_2O by Lyman α. At lower altitudes, the reaction of $O(^1D)$ atoms with hydrogen compounds is the most important source.

If we adopt, at 85 km, $2P = 3.6 \times 10^3$ cm^{-3} sec^{-1} with $2a_{16} = 7.2 \times 10^{-13}$ cm^3 sec^{-1}, and $[y + (x + y)^2] = 7.75 \times 10^{-4}$, the time which is required to reach 50% of the equilibrium value at the mesopause is

$$\tau_{50\%}(\mathrm{H}) = 0.275\{Pa_{16}[y + (x + y)^2]\}^{-1/2} = 3.8 \times 10^5 \quad \mathrm{sec} \tag{13.151}$$

corresponding to at least 4 days. Thus, there is a certain justification to adopt a mixing distribution for atomic hydrogen above the mesopause as the first approximation for its vertical distribution.

Within the mesosphere, the time necessary to reach equilibrium conditions for atomic hydrogen is decreased; at 80 km, it is about 7×10^4 sec and only 8×10^3 sec near 75 km. Again, in the neighborhood of the mesopause there is a transition region where there is a departure from photochemical equilibrium conditions. Thus, it should be pointed out that the effect of atmospheric motions must be simultaneously taken into account for atomic hydrogen and oxygen in this transition region. Detailed computations are scarcely justified, however, since there is no theoretical way to decrease the uncertainty in the knowledge of the rate coefficients and of the atmospheric motions.

Within the mesosphere and in the whole stratosphere, the general Eq. (13.149) becomes

$$\frac{dn(\mathrm{OH})}{dt} + \frac{dn(\mathrm{HO_2})}{dt} + 2a_{16}n^2(\mathrm{OH}) + 2a_{17}n(\mathrm{OH})n(\mathrm{HO_2}) + 2a_{27}n^2(\mathrm{HO_2}) = 2P \qquad (13.152)$$

since $n(\mathrm{H}) < n(\mathrm{OH}) + n(\mathrm{HO_2})$ and, in order to simplify for an approximate calculation,

$$\frac{dn(\mathrm{OH})}{dt} + \frac{dn(\mathrm{HO_2})}{dt} + 2a_{16}[n(\mathrm{OH}) + n(\mathrm{HO_2})]^2 = 2P. \qquad (13.153)$$

The calculation is made as follows: In the lower stratosphere it is not convenient to use $n(\mathrm{H_2O})$ as a basic parameter, but rather $n(\mathrm{CH_4}) + n(\mathrm{H_2}) + n(\mathrm{H_2O})$, the total concentrations of minor constituents involving hydrogen. Thus, we can write

$$\begin{aligned} &n(\mathrm{O}\ {}^1\mathrm{D})\times[4n(\mathrm{CH_4}) + 2n(\mathrm{H_2O}) + 2n(\mathrm{H_2})]\times 3\times 10^{-10} \\ &= n(\mathrm{O}\ {}^1\mathrm{D})\times[4\times 1.5\times 10^{-6} + 2\times 3\times 10^{-6} + 2\times 5\times 10^{-7}]n(\mathrm{M})\times 3\times 10^{-10} \\ &= n(\mathrm{O}\ {}^1\mathrm{D})\times 13\times 10^{-6}n(\mathrm{M})\times 3\times 10^{-10} \quad \mathrm{cm^{-3}\,sec^{-1}} \end{aligned} \qquad (13.154)$$

where $n(\mathrm{M})$ is the total concentration and 3×10^{-10} cm^3 sec^{-1} is the mean reaction coefficient with H_2O, CH_4, and H_2.

Such a hydrogen atom production as given by (13.154) will be maintained up to the mesopause even if there is a decrease of CH_4 (or H_2) mixing ratio at the stratopause. This crude estimation is employed here since an implicit assumption is made concerning the molecular diffusion of hydrogen atoms which is needed to support an escape rate of 10^8 H atoms cm^{-2} sec^{-1}. This assumption corresponds to a mixing ratio in the upper stratosphere $n(\mathrm{H_2O})/n(\mathrm{M}) = 6.5\times 10^{-6}$, i.e., about twice the conventional value adopted in the lower stratosphere.

The effect of the photodissociation of H_2O by Lyman α must be added to the oxidation process above 70 km. Up to 70 km, the principal dissociation process of H_2O is due to the oxidation by $O(^1D)$; at 75 km the direct photodissociation plays a role although the variation with the solar zenith angle is still important. At 80 km and above, the problem is complicated owing to the fact that there must be a transport of H_2O by eddy diffusion from below. At 70 km and below the oxidation by $O(^1D)$ must be compared with the photodissociation by $\lambda > 1750$ Å.

A complete knowledge (see Fig. 13.8) of the eddy diffusion coefficient is needed in order to determine the exact vertical distribution of H_2O. If the eddy diffusion coefficient is greater than 10^5 cm^{-2} sec^{-1}, it is certain that the scale height of water vapor $H(H_2O)$ is greater than $\frac{1}{2}H(M)$, the atmospheric scale height. Eddy diffusion coefficients between 10^6 and 10^7 cm^2 sec^{-1} indicate that $n(H_2O)/n(M)$ will not fall off sharply through the lower thermosphere.

Considering only the effect of the oxidation by $O(^1D)$ atoms given by (13.154), and applying Eq. (13.153) from 15 km to 70 km, it is possible to deduce the photochemical conditions of OH and HO_2 in the stratosphere and lower mesosphere. The essential result (see *Table 13.4*) shows the total concentration $n(OH) + n(HO_2)$ is not less than 10^7 cm^{-3} and reaches 10^8 cm^{-3} in the upper stratosphere. Numerical values which are given in Table 13.4 correspond to an overhead sun condition; different conditions

TABLE 13.4

PRODUCTION OF OH AND HO_2 (P), TOTAL CONCENTRATION $[n(OH) + n(HO_2)]$ FOR AN OVERHEAD SUN, AND THE TIME TO REACH 50% OF THE EQUILIBRIUM VALUE

Altitude (km)	P (cm^{-3} sec^{-1})	$n(OH) + n(HO_2)$ (cm^{-3})	Equilibrium time (sec)
15	1.4×10^3	4.7×10^7	9.3×10^3
20	5.9×10^3	8.8×10^7	4.1×10^3
25	1.5×10^4	1.3×10^8	2.4×10^3
30	3.4×10^4	1.8×10^8	1.4×10^3
35	6.5×10^4	2.1×10^8	8.8×10^2
40	1.0×10^5	2.3×10^8	6.1×10^2
45	1.0×10^5	2.2×10^8	5.9×10^2
50	5.5×10^4	1.6×10^8	8.0×10^2
55	2.1×10^4	1.0×10^8	1.3×10^3
60	7.3×10^3	6.9×10^7	2.6×10^3
65	2.3×10^3	4.8×10^7	5.6×10^3
70	7.6×10^2	3.4×10^7	1.3×10^4

are illustrated [3] in *Fig. 13.20* for various solar zenith angles. Variations in the stratosphere are particularly important in its lower part, since they depend strongly on the O(^{1}D) concentration. Above 35 km the total concentrations $n(OH) + n(HO_2)$ does not change very much except when the sun is near the horizon.

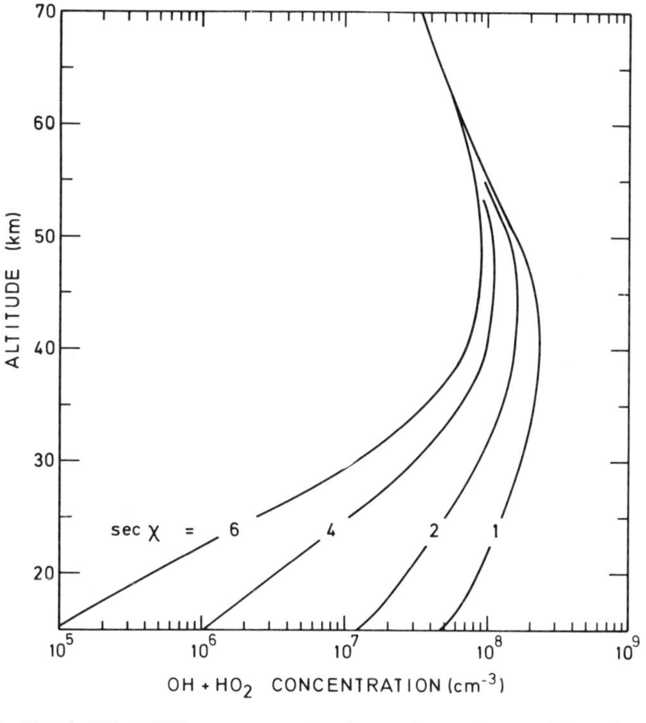

Fig. 13.20 Total OH + HO_2 concentration for various solar zenith angles and equilibrium conditions (Nicolet [3]).

After such an analysis, it is possible to explain the physical situation in the thermosphere even if the detailed structure cannot yet be regarded as determined with precision. It may easily be seen from *Table 13.5*, where the concentrations of atomic oxygen and atomic hydrogen are fixed by mixing conditions, that ozone is still important above the mesopause while the hydroxyl and perhydroxyl radical concentrations are very small. At and below the mesopause, 85–80 km, the aeronomic conditions are such that they correspond to a transition region, and all concentrations of Table 13.5 in this altitude range are intermediate values between those resulting from mixing and from photochemical equilibrium. Such an aspect is clearly

TABLE 13.5

COMPARISON OF THE CONCENTRATION OF O_3, OH, AND HO_2 FOR FIXED CONCENTRATIONS OF ATOMIC OXYGEN AND HYDROGEN IN THE LOWER THERMOSPHERE AND IN THE UPPER MESOSPHERE

Altitude (km)	$n(O)$ (cm^{-3})	$n(O_3)$ (cm^{-3})	$n(H)$ (cm^{-3})	$n(OH)$ (cm^{-3})	$n(HO_2)$ (cm^{-3})
100	3.2×10^{11}	1.6×10^{6}	3.5×10^{7}	1.7×10^{2}	6.5×10^{0}
95	3.3×10^{11}	1.3×10^{7}	8.7×10^{7}	3.2×10^{3}	1.4×10^{2}
90	3.0×10^{11}	1.1×10^{8}	2.3×10^{8}	7.0×10^{4}	4.5×10^{3}
85	3.0×10^{10}	1.0×10^{8}	6.9×10^{8}	2.0×10^{6}	5.3×10^{5}
80	1.4×10^{10}	1.4×10^{8}	8.6×10^{8}	2.2×10^{7}	1.7×10^{7}
75	3.8×10^{9}	3.2×10^{8}	5.1×10^{7}	1.8×10^{7}	1.3×10^{7}
70	4.1×10^{9}	1.0×10^{7}	1.0×10^{7}	9.5×10^{6}	7.0×10^{6}

seen [3] in *Fig. 13.21* where there is discontinuity in the curves around 80 km. Such a discontinuity as it is shown in Figure 13.21 must be considered only as an illustration of the extreme variability of physical conditions below the mesopause level. In other words, it must be stressed that any theoretical

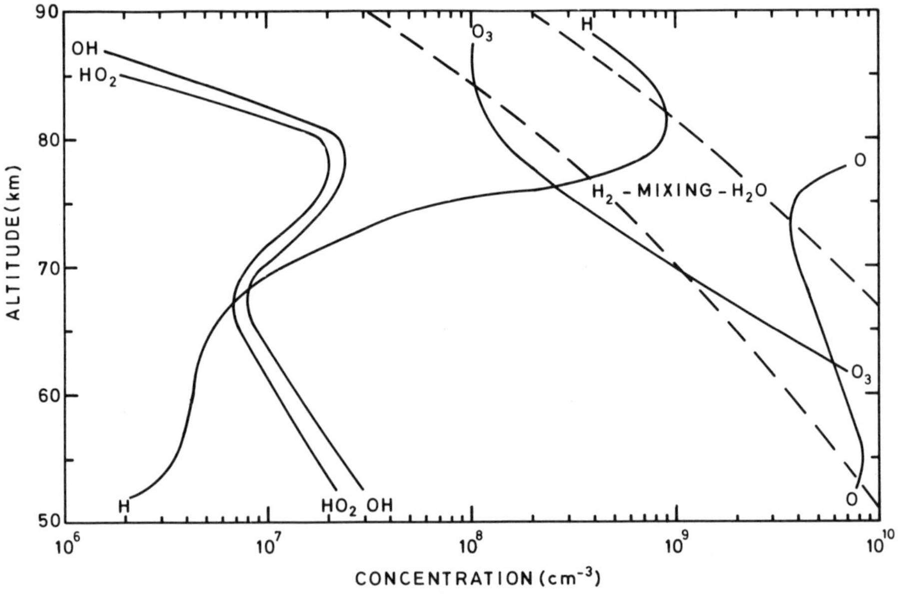

Fig. 13.21 Example of H, OH, and HO_2 vertical distribution in the mesosphere for daytime equilibrium conditions and resulting O_3 and O distributions (Nicolet [3]).

analysis of this part of the higher mesosphere and lower thermosphere is idealized and differs from the true atmosphere in that a discontinuity occurs which varies between day and night and with latitude and season. Nevertheless, near 70 km, in the middle of the mesosphere where $n(H) \simeq n(HO_2) \simeq n(OH)$, photochemical conditions seem to be well established. The results of *Table 13.6* represent HO_2, OH, and H concentrations [3] for photo-

TABLE 13.6

HO_2, OH, AND H CONCENTRATIONS FOR PHOTOEQUILIBRIUM CONDITIONS OF AN OVERHEAD SUN

Altitude (km)	$n(HO_2)$ (cm^{-3})	$n(OH)$ (cm^{-3})	$n(H)$ (cm^{-3})
15	$\leq 3.4 \times 10^7$	$>4.4 \times 10^3$	—
20	$\leq 5.8 \times 10^7$	$>3.5 \times 10^4$	—
25	$\leq 7.4 \times 10^7$	$>2.1 \times 10^5$	—
30	$\leq 9.1 \times 10^7$	$>1.2 \times 10^6$	8.0×10^{-1}
35	$\leq 9.4 \times 10^7$	$\geq 5.9 \times 10^6$	1.2×10^2
40	$\leq 7.8 \times 10^7$	2.6×10^7	1.1×10^4
45	4.5×10^7	5.1×10^7	2.4×10^5
50	2.8×10^7	4.4×10^7	1.2×10^6
55	1.6×10^7	2.8×10^7	3.3×10^6
60	1.2×10^7	1.9×10^7	4.4×10^6
65	8.7×10^6	1.3×10^7	6.0×10^6
70	7.0×10^6	9.5×10^6	1.0×10^7

equilibrium conditions corresponding to an overhead sun. Below 40 km, however, it is difficult to determine the exact value of the ratio $n(HO_2)/n(OH)$ since the effect of nitric oxide given in Eq. (13.127) cannot be determined. *Figure 13.22* illustrates [3] the possible extreme equilibrium conditions of OH and HO_2. One effect of nitric oxide is to increase the minimum value of $n(OH)$ which is represented in Fig. 13.22 and to decrease $n(HO_2)$, the total remaining constant (Fig. 13.20).

Attention may be drawn to the fact that the photodissociation of hydrogen peroxide has not been included in the production function P in (13.153). This means that it is assumed that $n(H_2O_2)$ is too small to lead to a production of 10^3 OH molecules cm^{-3} sec^{-1}. Without the effect of nitrogen oxides, the HO_2 concentration would reach more than 10^9 cm^{-3} in the lower stratosphere (see Fig. 13.22). Without a knowledge of the rate coefficient, however, it seems that it is better to start from the total concentration $n(OH)$

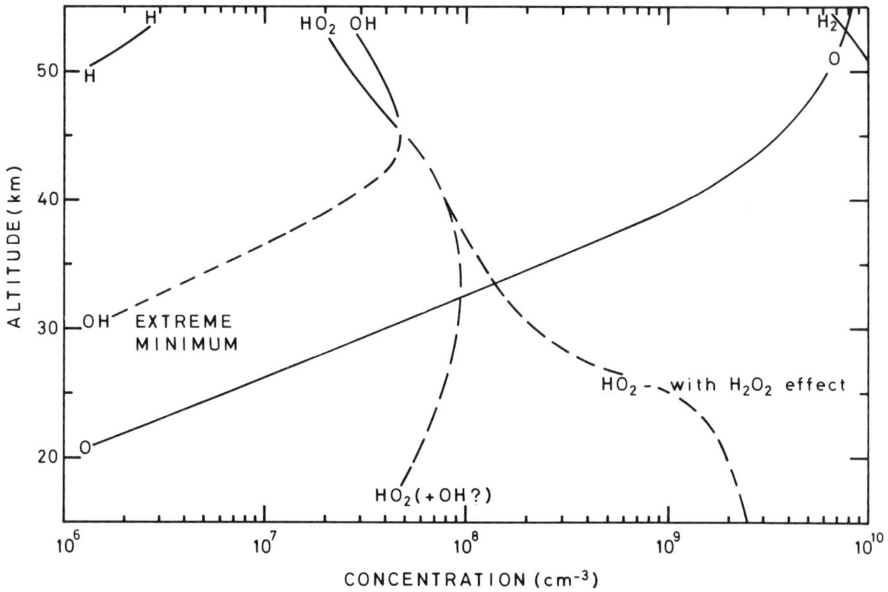

Fig. 13.22 Example of OH and HO_2 concentrations in the stratosphere for conditions of overhead sun and equilibrium (Nicolet [3]).

$+ n(HO_2)$ as given in Fig. 13.20 and to consider extreme values as given by the OH and HO_2 curves in Fig. 13.22 as unlikely values.

It is clear that more consistent observational data are needed and that an improvement of the knowledge of reaction rates is still required. Eddy diffusion coefficients should also be determined in an accurate way since they influence the vertical distribution of the minor constituents [60–62]. Finally, time dependent solutions of the chemospheric components always require certain assumptions for the boundary values and it is not always easy to fix those values unambiguously. The chemosphere is and remains an open field for important studies.

REFERENCES

1. J. R. McNesby, I. Tanaka, and H. Okabe, *J. Chem. Phys.* **36**, 605 (1962).
2. M. Cottin, J. Masanet, and C. Vermeil, *J. Chim. Phys. Paris* **63**, 959 (1966).
3. M. Nicolet, *Ann. Géophys.* **26**, 531 (1970).
4. T. Carrington, *J. Chem. Phys.* **41**, 2012 (1964).
5. N. Neuimin and A. Terenin, *Acta Physicochem. USSR* **5**, 465 (1936).
6. R. Engleman, Jr., *J. Amer. Chem. Soc.* **87**, 4193 (1965).
7. W. D. McGrath and R. G. W. Norrish, *Proc. Roy. Soc.* **A254**, 317 (1960).

8. N. Basco and R. G. W. Norrish, *Proc. Roy. Soc.* **A260**, 293 (1961).
9. R. D. Cadle, *Disc. Faraday Soc.* **37**, 66 (1964).
10. W. B. De More and O. F. Raper, *J. Chem. Phys.* **46**, 2500 (1967).
11. W. B. De More, *J. Chem. Phys.* **47**, 2777 (1967).
12. W. B. De More, *J. Phys. Chem.* **73**, 391 (1969).
13. W. B. De More, *J. Chem. Phys.* **52**, 4309 (1970).
14. D. Biedenkapp, L. G. Hartshorn, and E. J. Bair, *Chem. Phys. Lett.* **5**, 379 (1970).
15. G. Paraskevopoulos and R. J. Cvetanovic, *Chem. Phys. Lett.* **9**, 603 (1971).
16. R. D. Snelling and E. J. Bair, *J. Chem. Phys.* **47**, 228 (1967).
17. R. W. Ditchburn, *Proc. Roy. Soc.* **A229**, 44 (1955).
18. R. D. Cadle and E. R. Allen, *J. Phys. Chem.* **69**, 1611 (1965).
19. J. M. Brown and B. A. Thrush, *Trans. Faraday Soc.* **63**, 630 (1967).
20. A. A. Westenberg and N. de Haas, *J. Chem. Phys.* **46**, 490 (1967).
21. E. L. Wong and A. E. Potter, Jr., *Can. J. Chem.* **45**, 367 (1967).
22. E. L. Wong and A. E. Potter, Jr., *J. Chem. Phys.* **43**, 3371 (1965).
23. I. M. Campbell and B. A. Thrush, *Trans. Faraday Soc.* **64**, 1265 (1968).
24. H. C. Urey, L. H. Dawsey, and F. O. Rice, *J. Amer. Chem. Soc.* **51**, 1371 (1929).
25. R. K. Holt, C. R. McLane, and O. Oldenberg, *J. Chem. Phys.* **16**, 225, 638 (1948).
26. P. A. Giguère, *J. Chem. Phys.* **30**, 322 (1959).
27. N. R. Greiner, *J. Chem. Phys.* **45**, 99 (1966); **46**, 2450 (1967).
28. L. J. Stief and V. J. De Carlo, *J. Chem. Phys.* **50**, 1234 (1969).
29. K. H. Becker, W. Groth, and D. Kley, *Z. Naturforsch.* **20a**, 748 (1965).
30. W. H. Smith, *J. Chem. Phys.* **53**, 792 (1970).
31. F. Kaufman, *Ann. Géophys.* **20**, 106 (1964).
32. W. B. DeMore, *J. Chem. Phys.* **46**, 813 (1967).
33. D. R. Bates and M. Nicolet, *C.R. Acad. Sci. Paris* **230**, 1943 (1950).
34. D. R. Bates and M. Nicolet, *Publ. Astron. Soc. Pacific* **62**, 106 (1950).
35. D. R. Bates and A. E. Witherspoon, *Mon. Not. Roy. Astron. Soc.* **112**, 101 (1952).
36. D. R. Bates and M. Nicolet, *J. Geophys. Res.* **55**, 301 (1950).
37. E. J. Williamson and J. Houghton, *Quart. J. Roy. Met. Soc.* **91**, 330 (1965).
38. H. J. Mastenbrook, *J. Atmos. Sci.* **25**, 299 (1968).
39. N. Sissenwine, D. D. Grantham, and H. A. Salmela, *J. Atmos. Sci.* **25**, 1129 (1968).
40. N. R. Greiner, *J. Chem. Phys.* **48**, 1413 (1968).
41. H. Niki, E. E. Daby, and B. Weinstock, *J. Chem. Phys.* **48**, 5729 (1968).
42. M. Nicolet, *Discuss. Faraday Soc.* **37**, 7 (1964).
43. T. Koyama, *J. Geophys. Res.* **68**, 3971 (1963).
44. D. R. Bates and M. Nicolet, *Planet. Space Sci.* **13**, 905 (1965).
45. M. Nicolet, *Problèmes météorologiques de la stratosphère et de la mésosphère* (M. Nicolet, ed.), pp. 441–471. Presses Univ. de France, Paris, 1965.
46. F. W. P. Götz and M. Nicolet, *J. Geophys. Res.* **56**, 577 (1951).
47. M. A. A. Clyne and B. A. Thrush, *Proc. Roy. Soc.* **A275**, 559 (1963).
48. L. F. Phillips and H. I. Schiff, *J. Chem. Phys.* **37**, 1233 (1962).
49. F. Kaufman, *Can. J. Chem.* **47**, 1917 (1969).
50. M. A. A. Clyne and B. A. Thrush, *Proc. Roy. Soc.* **A275**, 544 (1963).
51. K. F. Langley and W. D. McGrath, *Planet. Space Sci.* **19**, 413 (1971).
52. M. Nicolet, in *Mesospheric Models and Related Experiments* (G. Fiocco, ed.), pp. 1–55. Reidel Publ., Dordrecht, The Netherlands, 1971.
53. J. L. J. Rosenfeld and T. M. Sugden, *Combust. Flame* **8**, 44 (1964).

54. J. K. Cashion and J. C. Polanyi, *J. Chem. Phys.* **30**, 316 (1959).
55. P. E. Charters and J. C. Polanyi, *Can. J. Chem.* **38**, 1742 (1960).
56. M. Nicolet, *J. Geophys. Res.* **70**, 679 (1965).
57. M. Nicolet, *Bull. Acad. Roy. Belg., Cl. Sc.* **57**, 935 (1971).
58. N. Hata and P. A. Giguère, *Can. J. Chem.* **44**, 869 (1966).
59. N. R. Greiner, *J. Phys. Chem.* **72**, 406 (1968).
60. B. G. Hunt, *J. Geophys. Res.* **71**, 1385 (1966).
61. E. Hesstvedt, *Geophys. Publ. Norv.* **27**, n. 4 (1968).
62. T. Shimazaki and A. R. Laird, *J. Geophys. Res.* **75**, 3221 (1970).

APPENDIX A

A Guide to Reactions and Reaction Coefficients

Reaction coefficient	Reaction	Equation number
γ_{34}	$Ar^+(^2P) + CO(^1\Sigma) \rightarrow Ar(^1S) + CO^+(^2\Sigma)$	10.201
γ_{33}	$Ar^+(^2P) + CO_2(^1\Sigma) \rightarrow Ar(^1S) + CO_2^+(^2\Pi)$	10.199
γ_{36}	$Ar^+ + H_2 \rightarrow H + ArH^+$	10.205
γ_{31}	$Ar^+(^2P) + N_2(^1\Sigma) \rightarrow Ar(^1S) + N_2^+(^2\Pi)$	10.195
γ_{35}	$Ar^+(^2P) + NO(^2\Pi) \rightarrow Ar(^1S) + NO^+(^1\Sigma)$	10.203
γ_{32}	$Ar^+(^2P) + O_2(^3\Sigma) \rightarrow Ar(^1S) + O_2^+(^2\Pi)$	10.197
γ_{37}	$ArH^+ + H_2 \rightarrow Ar + H_3^+$	10.206
γ_{30}	$C^+(^2P) + CO_2(^1\Sigma) \rightarrow CO(^1\Sigma) + CO^+(^2\Sigma)$	10.193
γ_{29}	$C^+(^2P) + O_2(^3\Sigma) \rightarrow O(^3P) + CO^+(^2\Sigma)$	10.191
	$CH_3 + O \rightarrow H + HCHO$	13.36
	$CH_3 + O_2(+M) \rightarrow CH_3O_2(+M)$	13.38
	$CH_3 + O_3 \rightarrow CH_3O + O_2$	13.37
	$CH_4 + O(^3P) \rightarrow OH + CH_3$	13.15
	$CH_4 + O(^1D) \rightarrow CH_3 + OH$	13.17
	$CH_4 + OH \rightarrow CH_3 + H_2O$	13.33
γ_{28}	$CO^+ + CO_2 \rightarrow CO + CO_2^+$	10.181
γ_{39}	$CO^+ + H_2 \rightarrow COH^+ + H$	10.183
γ_{53}	$CO^+ + N \rightarrow NO^+ + C$	10.185
γ_{55}	$CO^+(^2\Sigma) + NO(^2\Pi) \rightarrow CO(^1\Sigma) + NO^+(^1\Sigma)$	10.189
γ_{54}	$CO^+(^2\Sigma) + O(^3P) \rightarrow CO(^1\Sigma) + O^+(^4S)$	10.187
γ_{27}	$CO^+ + O_2 \rightarrow CO + O_2^+$	10.179

Reaction coefficient	Reaction	Equation number
γ_{51a}	$CO_2^+ + H \rightarrow CO_2 + H^+$	10.174
γ_{51b}	$CO_2^+ + H \rightarrow COH^+ + O$	10.175
γ_{40}	$CO_2^+ + H_2 \rightarrow CO_2H^+ + H$	10.172
γ_{26a}	$CO_2^+ + O \rightarrow O_2^+ + CO$	10.169
γ_{26b}	$CO_2^+ + O \rightarrow O^+ + CO_2$	10.170
γ_{26}	$CO_2^+ + O_2 \rightarrow CO_2 + O_2^+$	10.167
δ_{27}	$CO_3^- + NO \rightarrow NO_2^- + CO_2$	10.349
δ_{28}	$CO_3^- + NO_2 \rightarrow NO_3^- + CO_2$	10.351
δ_{26}	$CO_3^- + O \rightarrow O_2^- + CO_2$	10.347
δ_{30}	$CO_4^- + NO \rightarrow NO_3^- + CO_2$	10.356
δ_{29}	$CO_4^- + O \rightarrow CO_3^- + O_2$	10.353
δ_{29}	$CO_4^- + O \rightarrow O_3^- + CO_2$	10.354
ϕ_2	$e + O \rightarrow O^- + h\nu$	10.272
ϕ_1	$e + O_2 + O_2 \rightarrow O_2^- + O_2$	10.270
ϕ_3	$e + O_3 \rightarrow O^- + O_2$	10.274
a_{34}	$H + H + Ar \rightarrow H_2 + Ar$	10.18
a_{21}	$H + H + M \rightarrow H_2 + M$	13.112
b_{17}	$H + HNO \rightarrow NO + H_2$	12.80
b_{24}	$H + HNO_2 \rightarrow H_2 + NO_2$	12.92
b_{26}	$H + HNO_3 \rightarrow H_2O + NO_2$	12.94
a_{23}	$H + HO_2 \rightarrow H_2 + O_2$	13.117
a_{20}	$H + HO_2 \rightarrow H_2O + O$	13.110
a_{15}	$H + HO_2 \rightarrow OH + OH_{v\leq 4}$	13.100
a_{29b}	$H + H_2O_2 \rightarrow H_2 + HO_2$	13.131
a_{29a}	$H + H_2O_2 \rightarrow H_2O + OH$	13.130
b_{16}	$H + NO + M \rightarrow HNO + M$	12.77
a_{25}	$H + NO_2 \rightarrow OH + NO$	13.123
b_{28}	$H + NO_2 \rightarrow NO + OH$	12.96
a_0	$H + O + M \rightarrow OH + M$	13.51
a_{22}	$H + OH \rightarrow H_2 + O$	13.114
a_{14}	$H + OH + M \rightarrow H_2O + M$	13.98
a_1	$H + O_2 + M \rightarrow HO_2 + M$	13.52
		10.49
a_3	$H + O_3 \rightarrow HO_2 + O$	13.56
a_2	$H + O_3 \rightarrow O_2 + OH_{v\leq 9}$	13.54
		10.60
γ_{52}	$H^+ + CO_2 \rightarrow COH^+ + O$	10.177

Reaction coefficient	Reaction	Equation number
	$H^+(^1S) + O(^3P) \rightarrow OH^+(^3\Pi) \rightarrow OH^+(^3\Sigma) + h\nu$	10.6
	$HO_2 + CH_4 \rightarrow CH_3 + H_2O_2$	13.35
b_{19}	$HO_2 + HNO \rightarrow NO + H_2O_2$	12.84
a_{27}	$HO_2 + HO_2 \rightarrow H_2O_2 + O_2$	13.128
a_{26}	$HO_2 + NO \rightarrow OH + NO_2$	13.125
a_{31}^*	$HO_2 + O(^1D) \rightarrow OH + O_2$	13.139
a_{6a}	$HO_2 + O_3 \rightarrow OH + 2O_2$	13.64
a_{24}	$H_2 + O(^3P) \rightarrow H + OH$	13.18
a_{24}^*	$H_2 + O(^1D) \rightarrow H + OH^*_{v\leq 4}$	13.21
a_{18}	$H_2 + O + M \rightarrow H_2O + M$	13.106
a_{19}	$H_2 + OH \rightarrow H_2O + H$	13.108
γ_{42}	$H_2O^+ + H_2 \rightarrow H + H_3O^+$	10.212
γ_{45}	$H_2O^+ + H_2O \rightarrow H_3O^+ + OH$	10.218
γ_{44}	$H_2O^+ + O_2 \rightarrow H_2O + O_2^+$	10.216
b_{30}	$H_2O_2 + NO \rightarrow OH + HNO_2$	12.101
b_{31}	$H_2O_2 + NO_2 \rightarrow OH + HNO_3$	12.102
γ_{38}	$H_3^+ + Ar + Ar \rightarrow Ar + ArH_3^+$	10.207
α_D	$H_3O^+ + e \rightarrow H + OH^*$	10.100
$\varkappa_{19}$	$H_3O^+ + H_2O + M \rightarrow H_3O^+ \cdot H_2O + M$	10.260
$\varkappa_{21}$	$H_3O^+ \cdot H_2O + H_2O + M \rightleftharpoons H_3O^+ \cdot (H_2O)_2 + M$	10.262
$\varkappa_{22}$	$H_3O^+ \cdot (H_2O)_2 + H_2O + M \rightleftharpoons H_3O^+ \cdot (H_2O)_3 + M$	10.263
$\varkappa_{20}$	$H_3O^+ \cdot OH + H_2O \rightarrow H_3O^+ \cdot H_2O + OH$	10.261
γ_{50}	$He^+ + CO \rightarrow He + O + C^+$	10.165
γ_{49}	$He^+(^2S) + CO_2(^1\Sigma) \rightarrow He(^1S) + CO_2^+(^2\Pi)$	10.163
	$\rightarrow He + CO + O^+$	
	$\rightarrow He + O + CO^+$	
γ_{19}	$He^+(^2S) + NO(^2\Pi) \rightarrow He(^1S) + O(^3P) + N^+(^3P)$	10.155
γ_{18}	$He^+(^2S) + N_2 \rightarrow He(^1S) + N(^4S) + N^+(^3P)$	10.153
γ_{18}^*	$He^+(^2S) + N_2 \rightarrow He(^1S) + N_2^+(C\,^2\Sigma)_{v=3,4}$	10.154
γ_{17}	$He^+(^2S) + O_2(^3\Sigma) \rightarrow He(^1S) + O(^3P) + O^+(^4S)$	10.151
	$He^+(^2S) + O_2(^3\Sigma) \rightarrow He(^1S) + O_2^+(c\,^4\Sigma)_{v=0}$	10.156
γ_{17}^*	$He^+(^2S) + O_2(^3\Sigma) \rightarrow He(^1S) + O_2^+(^2\Pi)$	10.152
$k(N, O)$	$N(^4S) + O(^3P) \rightleftharpoons NO(a\,^4\Pi) \rightarrow NO(C\,^2\Pi)$	10.13
	$\rightarrow NO(X\,^2\Pi) + h\nu$	
$k(N_2, {}^5\Sigma_g^+)$	$N(^4S) + N(^4S) + M \rightleftharpoons N_2(^5\Sigma_g^+) + M$	10.40
$k(N_2, N_2)$	$N + N + N_2 \rightarrow N_2 + N_2$	10.38
$k(N_2, O^1S)$	$N + N + O \rightarrow N_2 + O(^1S)$	10.31

Reaction coefficient	Reaction	Equation number
$b_{1\delta}$	$N(^4S) + O(^3P) \rightarrow NO(^4\Pi) \rightarrow NO(C\,^2\Pi) \rightarrow NO(X\,^2\Pi)$	12.8
		10.8
b_{1a}	$N + O + M \rightarrow NO + M$	12.6
$b_{1\beta}$	$N + O + M \rightarrow NO^* + M \rightarrow NO(B\,^2\Pi) + M$	12.14
$k(NO, N_2)$	$N + O + N_2 \rightarrow NO + N_2$	10.35
$k(NO, O^1S)$	$N + O + O \rightarrow NO + O(^1S)$	10.33
		11.5
$k(N, O_2)$	$N + O_2 \rightarrow O + NO^*$	10.72
	$N(^2D) + O_2 \rightarrow N(^4S) + O_2$	12.136
$b_7{}^*$	$N(^2D) + O_2 \rightarrow NO + O(^3P)$	12.135
b_6	$N + NO \rightarrow N_2(v'' \leq 12) + O$	12.34
		10.75
b_5	$N + NO_2 \rightarrow O + N_2O$	12.31
	$\rightarrow NO + NO$	12.32
	$\rightarrow N_2 + O_2$	12.33
b_{OH}	$N + OH \rightarrow NO + H$	12.64
b_7	$N + O_2 \rightarrow NO^*(v \leq 6) + O$	12.36
$b_7{}^*$	$N + O_2(^1\Delta_g) \rightarrow NO^*(v'' \leq 11) + O$	12.38
b_7^{**}	$N + O_2(^1\Sigma_g{}^+) \rightarrow NO(v'' \leq 14) + O$	12.39
b_8	$N + O_3 \rightarrow NO + O_2$	12.40
γ_{25}	$N^+(^3P) + CO(^1\Sigma) \rightarrow N(^4S) + CO^+(^2\Sigma)$	10.148
γ_{24}	$N^+(^3P) + CO_2(^1\Sigma) \rightarrow N(^4S) + CO_2{}^+(^2\Pi)$	10.147
	$N^+(^3P) + O(^3P) \rightarrow NO^+(^1\Pi) \rightarrow NO^+(^1\Sigma) + h\nu$	10.9
	$N^+(^3P) + N(^4S) \rightarrow N_2{}^+(^2\Pi) \rightarrow N_2{}^+(^2\Sigma) + h\nu$	10.10
γ_{13}	$N^+(^3P) + NO(^2\Pi) \rightarrow N(^4S) + NO^+(^1\Sigma)$	10.144
γ_{11}	$N^+(^3P) + NO(^2\Pi) \rightarrow O(^3P) + N_2{}^+(^2\Sigma)$	10.143
γ_{15}	$N^+(^3P) + O(^3P) \rightarrow N(^4S) + O^+(^4S)$	10.146
γ_{12}	$N^+(^3P) + O_2(^3\Sigma) \rightarrow N(^4S) + O_2{}^+(^2\Pi)$	10.138
γ_{12}^*	$N^+(^3P) + O_2(^3\Sigma) \rightarrow O(^3P) + NO^+(^1\Sigma)$	10.139
b_{23}	$NO + HO_2 + M \rightarrow HNO_3 + M$	12.91
b_{29}	$NO + HO_2 \rightarrow OH + NO_2$	12.98
b_{21}	$NO + OH + M \rightarrow HNO_2 + M$	12.89
b_{11}	$NO + NO_3 \rightarrow 2NO_2$	12.51
b_4	$NO + O_3 \rightarrow O_2 + NO_2{}^*$	10.62
$\varkappa_{13}$	$NO^+ + CO_2 + Ar \rightarrow NO^+ \cdot CO_2 + Ar$	10.243
$\varkappa_{11}$	$NO^+ + CO_2 + He \rightarrow NO^+ \cdot CO_2 + He$	10.241
$\varkappa_{12}$	$NO^+ + CO_2 + N_2 \rightarrow NO^+ \cdot CO_2 + N_2$	10.242

Reaction coefficient	Reaction	Equation number
$\alpha_D(NO^+)$	$NO^+ + e \rightarrow N + O$	10.80
	$NO^+(^1\Sigma^+) + e \rightarrow N(^4S) + O(^3P)$	10.81
		12.123
	$NO^+(^1\Sigma^+) + e \rightarrow N(^4S) + O(^1D)$	10.82
		12.124
	$NO^+(^1\Sigma^+) + e \rightarrow N(^2D) + O(^3P)$	10.83
		12.125
$\varkappa_{15}$	$NO^+ + H_2O + N_2 \rightarrow NO^+ \cdot H_2O + N_2$	10.249
$\varkappa_9$	$NO^+ + N_2 + He \rightarrow NO^+ \cdot N_2 + He$	10.237
$\varkappa_{10}$	$NO^+ + O_2 + He \rightarrow NO^+ \cdot O_2 + He$	10.238
$\varkappa_{14}$	$NO^+ \cdot CO_2 + H_2O \rightarrow NO^+ \cdot H_2O + CO_2$	10.247
	$NO^+ \cdot H_2O + H \rightarrow NO + H_3O^+$	10.259
$\varkappa_{16}$	$NO^+ \cdot H_2O + H_2O + M \rightleftharpoons NO^+ \cdot (H_2O)_2 + M$	10.251
$\varkappa_{17}$	$NO^+ \cdot (H_2O)_2 + H_2O + M \rightleftharpoons NO^+ \cdot (H_2O)_3 + M$	10.252
$\varkappa_{18}$	$NO^+ \cdot (H_2O)_3 + H_2O \rightarrow H_3O^+ \cdot (H_2O)_2 + HNO_2$	10.253
δ_{40}	$NO^- + O_2 \rightarrow O_2^- + NO$	10.378
b_{20}	$NO_2 + H + M \rightarrow HNO_2 + M$	12.88
b_{15}	$NO_2 + NO_2 + M \rightarrow N_2O_4 + M$	12.59
b_{22}	$NO_2 + OH + M \rightarrow HNO_3 + M$	12.90
δ_{33}	$NO_2^- + H \rightarrow OH^- + NO$	10.362
δ_{34}	$NO_2^- + H_2O + NO \rightarrow NO_2^- \cdot H_2O + NO$	10.365
δ_{32}	$NO_2^- + NO_2 \rightarrow NO_3^- + NO$	10.360
δ_{31}	$NO_2^- + O_3 \rightarrow NO_3^- + O_2$	10.358
b_{12}	$NO_3 + NO_2 \rightarrow N_2O_5$	12.53
b_{13}	$NO_3 + NO_2 \rightarrow NO + NO_2 + O_2$	12.55
b_{14}	$NO_3 + NO_3 \rightarrow 2NO_2 + O_2$	12.57
δ_{35}	$NO_3^- + H_2O + M \rightarrow NO_3^- \cdot H_2O + M$	10.367
δ_{36}	$NO_3^- + O_2 + M \rightarrow NO_3^- \cdot O_2 + M$	10.368
δ_{37}	$NO_3^- \cdot O_2 + H_2O \rightarrow NO_3^- \cdot H_2O + O_2$	10.369
	$N_2 + e \rightarrow N_2^+ + 2e$	9.60
	$\rightarrow N_2^{2+} + 3e$	9.61
	$\rightarrow N + N^+ + 2e$	9.62
γ_{23}	$N_2^+(^2\Sigma) + CO(^1\Sigma) \rightarrow N_2(^1\Sigma) + CO^+(^2\Sigma)$	10.136
γ_{22}	$N_2^+(^2\Sigma) + CO_2(^1\Sigma) \rightarrow N_2(^1\Sigma) + CO_2^+(^2\Pi)$	10.134
γ_3	$N_2^+(^2\Sigma) + O(^3P) \rightarrow N(^4S) + NO^+(^1\Sigma)$	10.125
		12.105
γ_8	$N_2^+(^2\Sigma) + O(^3P) \rightarrow N_2(^1\Sigma) + O^+(^4S)$	10.127

Reaction coefficient	Reaction	Equation number
γ_{10}	$N_2^+(^2\Sigma) + N(^4S) \rightarrow N_2(^1\Sigma) + N^+(^3P)$	10.130
γ_{16}	$N_2^+(^2\Sigma) + NO(^2\Pi) \rightarrow N_2(^1\Sigma) + NO^+$	10.132
γ_9	$N_2^+(^2\Sigma) + O_2(^3\Sigma) \rightarrow N_2(^1\Sigma) + O_2^+(^2\Pi)$	10.128
		12.106
γ_{14}	$N_2^+(^2\Sigma) + O_2(^3\Sigma) \rightarrow NO(^2\Pi) + NO^+(^1\Sigma)$	10.131
$\alpha_D(N_2^+)$	$N_2^+ + e \rightarrow N^* + N^{**}$	10.86
		12.104
	$N_2^+(^2\Sigma_g^+) + e \rightarrow N(^4S) + N(^4S)$	10.87
		12.2
	$N_2^+(^2\Sigma_g^+) + e \rightarrow N(^4S) + N(^2D)$	10.88
		12.3
	$N_2^+(^2\Sigma_g^+) + e \rightarrow N(^4S) + N(^2P)$	10.89
		12.4
	$N_2^+(^2\Sigma_g^+) + e \rightarrow N(^2D) + N(^2D)$	10.90
		12.5
	$O(^3P) + H(^2S) \rightarrow OH(^2\Sigma) \rightarrow OH(^2\Pi) + h\nu$	10.7
a_7	$O + HO_2 \rightarrow OH_{v\leq 6} + O_2$	13.31
	$O(^3P) + H_2O \rightarrow 2OH$	13.7
	$O(^3P) + OH(X\,^2\Pi) \rightarrow H + O_2^*(^3\Sigma_g^-)$	10.71
$k(O, NO)$	$O + NO \rightarrow NO_2 + h\nu$	10.15
b_2	$O + NO + M \rightarrow NO_2^* + M$	12.17
$k(O, NO)$	$O + NO + N_2 \rightarrow NO_2 + N_2$	10.46
b_3	$O + NO_2 \rightarrow NO + O_2^*$	12.23
b_{10}	$O + NO_2 + M \rightarrow NO_3 + M$	12.49
k_4	$O + O \rightarrow O_2 + h\nu$	4.3
k_1	$O + O + M \rightarrow O_2 + M$	4.2
$k(O_2, b\,^1\Sigma_g^+)$	$O(^3P) + O(^3P) + N_2 \rightarrow O_2(b\,^1\Sigma_g^+) + N_2$	10.24
		11.6
$k(O_2, A\,^3\Sigma_u^+)$	$O(^3P) + O(^3P) + N_2 \rightarrow O_2(A\,^3\Sigma_u^+) + N_2$	10.22
		11.6
$k(O_2, O\,^1S)$	$O + O + O \rightarrow O_2 + O(^1S)$	10.29
		11.4
$k(O_2, N_2)$	$O + O + N_2 \rightarrow O_2 + N_2$	10.20
$k(O, O_2)$	$O + O_2 + N_2 \rightarrow O_3 + N_2$	10.43
k_3	$O(^3P) + O_3 \rightarrow 2O_2$	10.65
		11.17
	$O(^3P) + O_3 \rightarrow O_2^*(^3\Sigma_g^-) + O_2(^1\Delta_g)$	10.67

Reaction coefficient	Reaction	Equation number
a_{13}^{*}	$O(^1D) + H_2O \rightarrow OH + OH_{v\leq 2}$	13.8
	$O(^1D) + N_2(X\,^1\Sigma_g{}^+) \rightarrow N_2(X\,^1\Sigma_g{}^+) + O(^3P)$	11.26
	$O(^1D) + O_2(X\,^3\Sigma_g{}^-) \rightarrow O_2(b\,^1\Sigma_g{}^-) + O(^3P)$	11.27
	$O(^1D) + O_2(X\,^3\Sigma_g{}^-) \rightarrow O_2(a\,^1\Delta_g) + O(^3P)$	11.28
	$O(^1D) + O_2(X\,^3\Sigma_g{}^-) \rightarrow O_2(X\,^3\Sigma_g{}^-) + O(^3P)$	11.29
	$O(^1D) + O_3 \rightarrow 2O_2$	13.11
γ_{21}	$O^+(^4S) + CO_2(^1\Sigma) \rightarrow CO(^1\Sigma) + O_2{}^+(^2\Pi)$	10.114
	$O^+(^4S) + H(^2S) \rightleftharpoons O(^3P_J) + H^+(1p)$	9.81
$\gamma_4{}^*$	$O^+(^4S) + NO(^2\Pi) \rightarrow N(^4S) + O_2{}^+(^2\Pi)$	10.112
		12.122
γ_4	$O^+(^4S) + NO(^2\Pi) \rightarrow O(^3P) + NO^+(^1\Sigma)$	10.111
		12.121
γ_{47}	$O^+ + NO_2 \rightarrow NO_2{}^+ + O$	10.116
γ_1	$O^+(^4S) + N_2(^1\Sigma) \rightarrow N(^4S) + NO^+(^1\Sigma)$	10.104
		12.111
$\gamma_1{}^*$	$O^+(^2D) + N_2(X\,^1\Sigma_g, v' = 0) \rightarrow N_2{}^+(A\,^2\Pi_u, v' = 1) + O(^3P)$	10.108
γ_{48}	$O^+ + N_2O \rightarrow N_2O^+ + O$	10.117
	$O^+(^4S) + O(^3P) \rightarrow O_2{}^+(^2\Pi) \rightarrow O_2{}^+(^2\Pi) + h\nu$	10.11
γ_2	$O^+(^4S) + O_2(^3\Sigma) \rightarrow O(^3P) + O_2{}^+(^2\Pi)$	10.105
		12.112
λ_{10}	$O^- + CO \rightarrow CO_2 + e$	10.306
δ_{10}	$O^- + CO_2 + CO_2 \rightarrow CO_3{}^- + CO_2$	10.316
δ_{11}	$O^- + CO_2 + He \rightarrow CO_3{}^- + He$	10.318
λ_8	$O^- + H_2 \rightarrow H_2O + e$	10.302
λ_{11}	$O^- + N \rightarrow NO + e$	10.308
λ_6	$O^- + O \rightarrow O_2 + e$	10.298
λ_9	$O^- + NO \rightarrow NO_2 + e$	10.304
δ_8	$O^- + NO_2 \rightarrow NO_2{}^- + O$	10.312
λ_7	$O^- + O_2(^1\Delta_g) \rightarrow O_3 + e$	10.300
δ_9	$O^- + O_2 + O_2 \rightarrow O_3{}^- + O_2$	10.314
δ_7	$O^- + O_3 \rightarrow O_3{}^- + O$	10.310
k_2	$O_2 + O + M \rightarrow O_3 + M$	4.4
		11.16
	$O_2{}^*(^3\Sigma_g{}^-) + O_3 \rightarrow 2O_2(^3\Sigma_g{}^-) + O(^1D)$	10.68
$\alpha_D(O_2{}^+)$	$O_2{}^+ + e \rightarrow O^* + O^*$	10.93
		12.116
	$O_2{}^+(^2\Pi_g) + e \rightarrow O(^3P) + O(^3P)$	10.94

Reaction coefficient	Reaction	Equation number
$\alpha_D(O_2^+)$	$O_2^+(^2\Pi_g) + e \rightarrow O(^3P) + O(^1D)$	10.95
	$O_2^+(^2\Pi_g) + e \rightarrow O(^3P) + O(^1S)$	10.96
	$O_2^+(^2\Pi_g) + e \rightarrow O(^1D) + O(^1D)$	10.97
	$O_2^+(^2\Pi_g) + e \rightarrow O(^1D) + O(^1S)$	10.98
$\varkappa_1$	$O_2^+ + O_2 + M \rightarrow O_4^+ + M$	10.222
$\varkappa_4$	$O_2^+ + H_2O + N_2 \rightarrow O_2^+ \cdot H_2O + N_2$	10.227
$\varkappa_5$	$O_2^+ + H_2O + O_2 \rightarrow O_2^+ \cdot H_2O + O_2$	10.228
γ_6	$O_2^+(^2\Pi) + N(^4S) \rightarrow O(^3P) + NO^+(^1\Sigma)$	10.122 12.118
γ_5	$O_2^+(^2\Pi) + NO(^2\Pi) \rightarrow O_2(^3\Sigma) + NO^+(^1\Sigma)$	10.120 12.117
γ_7	$O_2^+(^2\Pi) + N_2(^1\Sigma) \rightarrow NO(^2\Pi) + NO^+(^1\Sigma)$	10.124 12.119
$\varkappa_6$	$O_2^+ \cdot H_2O + H_2O \rightarrow H_3O^+ + OH + O_2$	10.231
$\varkappa_7$	$O_2^+ \cdot H_2O + H_2O \rightarrow H_3O^+ \cdot OH + O_2$	10.232
$\varkappa_8$	$O_2^+ \cdot H_2O + H_2O + O_2 \rightarrow O_2^+ \cdot (H_2O)_2 + O_2$	10.235
δ_4	$O_2^- + CO_2 + CO_2 \rightarrow CO_4^- + CO_2$	10.290
δ_6	$O_2^- + H_2O + O_2 \rightarrow O_2^- \cdot H_2O + O_2$	10.294
λ_2	$O_2^- + N \rightarrow NO_2 + e$	10.278
λ_1	$O_2^- + O \rightarrow O_3 + e$	10.276
δ_2	$O_2^- + NO_2 \rightarrow NO_2^- + O_2$	10.286
λ_3	$O_2^- + O_2(^1\Delta_g) \rightarrow O_2 + O_2 + e$	10.280
δ_5	$O_2^- + O_2 + CO_2 \rightarrow CO_4^- + O_2$	10.292
δ_3	$O_2^- + O_2 + O_2 \rightleftharpoons O_4^- + O_2$	10.288
δ_1	$O_2^- + O_3 \rightarrow O_3^- + O_2$	10.284
δ_{41}	$O_2^- \cdot H_2O + CO \rightarrow CO_4^- + H_2O$	10.380
δ_{42}	$O_2^- \cdot H_2O + NO \rightarrow NO_3^- + H_2O$	10.382
b_4	$O_3 + NO \rightarrow NO_2 + O_2$	12.25
b_9	$O_3 + NO_2 \rightarrow NO_3 + O_2$	12.47
k_3	$O_3 + O \rightarrow O_2 + O$	4.6
δ_{13}	$O_3^- + CO_2 \rightarrow CO_3^- + O_2$	10.321
δ_{18}	$O_3^- + H_2O + O_2 \rightarrow O_3^- \cdot H_2O + O_2$	10.331
δ_{14}	$O_3^- + NO \rightarrow NO_3^- + O$	10.323
δ_{15}	$O_3^- + NO_2 \rightarrow$ products (NO_2^- and/or NO_3^-)	10.325
δ_{12}	$O_3^- + O \rightarrow O_2^- + O_2$	10.320
δ_{16}	$O_3^- + SO_2 \rightarrow SO_3^- + O_2$	10.327
δ_{17}	$O_3^- + SO_2 \rightarrow SO_4^- + O$	10.329

Reaction coefficient	Reaction	Equation number
$\varkappa_3$	$O_4^+ + H_2O \rightarrow O_2^+ \cdot H_2O + O_2$	10.224
$\varkappa_2$	$O_4^+ + O \rightarrow O_2^+ + O_3$	10.223
δ_{24}	$O_4^- + CO \rightarrow CO_3^- + O_2$	10.343
δ_{19}	$O_4^- + CO_2 \rightarrow CO_4^- + O_2$	10.333
δ_{22}	$O_4^- + H_2O \rightarrow O_2^- \cdot H_2O + O_2$	10.340
δ_{20}	$O_4^- + NO \rightarrow NO_3^- + O_2$	10.335
δ_{21}	$O_4^- + O \rightarrow O^- + O_2 + O_2$	10.338
	$\rightarrow O_3^- + O_2$	10.337
δ_{23}	$O_4^- + N_2 \rightarrow O_2^- \cdot N_2 + O_2$	10.341
δ_{25}	$O_4^- + N_2O \rightarrow O_2^- \cdot N_2O + O_2$	10.344
b_{18}	$OH + HNO \rightarrow NO + H_2O$	12.82
b_{25}	$OH + HNO_2 \rightarrow H_2O + NO_2$	12.93
b_{27}	$OH + HNO_3 \rightarrow H_2O + NO_3$	12.95
a_{17}	$OH + HO_2 \rightarrow H_2O + O_2$	13.104
	$OH + H_2 \rightarrow H + H_2O$	13.34
a_{30}	$OH + H_2O_2 \rightarrow H_2O + HO_2$	13.134
a_5	$OH + O \rightarrow H + O_2$	13.60
a_4	$OH + O + M \rightarrow HO_2 + M$	13.58
a_6	$OH + O_3 \rightarrow HO_2 + O_2$	13.62
	$OH^*(\geq 10.2 \text{ kcal mole}^{-1}) + O_3 \rightarrow H + 2O_2$	13.65
a_{16}	$OH + OH \rightarrow H_2O + O$	13.102
γ_{41}	$OH^+ + H_2 \rightarrow H + H_2O^+$	10.210
γ_{46}	$OH^+ + H_2O \rightarrow H_3O^+ + O$	10.220
γ_{43}	$OH^+ + O_2 \rightarrow OH + O_2^+$	10.214
δ_{39}	$OH^- + CO_2 + M \rightarrow HCO_3^- + M$	10.372
λ_{13}	$OH^- + H \rightarrow H_2O + e$	10.376
δ_{38}	$OH^- + NO_2 \rightarrow NO_2^- + OH$	10.370
λ_{12}	$OH^- + O \rightarrow HO_2 + e$	10.374

Author Index for Part A

Numbers in parentheses are reference numbers and indicate that an author's work is referred to although his name is not cited in the text. Numbers in italics show the page on which the complete reference is listed.

C

I

J

N

Subject Index for Parts A and B

The page numbers referring to Part A (Chapters 1 through Appendix A) are roman. The page numbers referring to Part B (Chapters 14 through Appendix B) are italic.

F

I

J

K

L

M

N

O

V

W

X